THINKING LIKE AN ENGINEER

AN ACTIVE LEARNING APPROACH

THINKING LIKE AN ENGINEER

AN ACTIVE LEARNING APPROACH

Elizabeth A. Stephan
Clemson University

David R. Bowman
Clemson University

William J. Park
Clemson University

Benjamin L. Sill
Clemson University

Matthew W. Ohland
Purdue University

PEARSON

Boston Columbus Indianapolis New York San Francisco Upper Saddle River
Amsterdam Cape Town Dubai London Madrid Milan Munich Paris Montréal Toronto
Delhi Mexico City São Paulo Sydney Hong Kong Seoul Singapore Taipei Tokyo

Vice President and Editorial Director, ECS: *Marcia J. Horton*
Senior Editor: *Holly Stark*
Editorial Assistant: *Keri Rand*
Vice President, Production: *Vince O'Brien*
Marketing Manager: *Tim Galligan*
Marketing Assistant: *Mack Patterson*
Senior Managing Editor: *Scott Disanno*
Production Project Manager: *Clare Romeo*
Senior Operations Specialist: *Alan Fischer*
Operations Specialist: *Lisa McDowell*
Art Director: *Kenny Beck*

Interior Designer: *LCI Design*
Cover Designer: *Kenny Beck*
Cover Illustration/Photo(s): *Red Guitar/Evgeny Guityaev/*
   Shutterstock; X-ray of Guitar/Gustoimages/Science Photo Library
Manager, Visual Research: *Beth Brenzel*
Image Permission Coordinator: *Debbie Latronica*
Manager, Cover Visual Research & Permissions: *Karen Sanatar*
Composition: *MPS Limited, A Macmillan Company*
Overseas Project Management: *Anoop Chaturvedi*
Printer/Binder: *Edwards Brothers, Inc.*
Typeface: 10/12 Times Ten Roman

Many of the designations by manufacturers and seller to distinguish their products are claimed as trademarks. Where those designations appear in this book, and the publisher was aware of a trademark claim, the designations have been printed in initial caps or all caps.

The author and publisher of this book have used their best efforts in preparing this book. These efforts include the development, research, and testing of theories and programs to determine their effectiveness. The author and publisher make no warranty of any kind, expressed or implied, with regard to these programs or the documentation contained in this book. The author and publisher shall not be liable in any event for incidental or consequential damages with, or arising out of, the furnishing, performance, or use of these programs.

Pearson Education Ltd., *London*
Pearson Education Singapore, Pte. Ltd.
Pearson Education Canada, Inc.
Pearson Education—Japan
Pearson Education Australia PTY, Limited
Pearson Education North Asia, Ltd., *Hong Kong*
Pearson Educación de Mexico, S.A. de C.V.
Pearson Education Malaysia, Pte. Ltd.
Pearson Education, Inc., *Upper Saddle River, New Jersey*

Contact the Library of Congress for CIP data.

Prentice Hall
is an imprint of

www.pearsonhighered.com

10 9 8 7 6 5 4 3 2
ISBN-13: 978-0-13-606442-8
ISBN-10: 0-13-606442-6

CONTENTS

PREFACE

At our university, all students who wish to major in engineering begin in the General Engineering Program, and after completing a core set of classes, they can declare a specific engineering major. Within this core set of classes, students are required to take math, physics, chemistry and a two-semester engineering sequence. Over the past 10 years, our courses have evolved to address not only the changing qualities of our students, but also the changing needs of our customers. The material taught in our courses is the foundation upon which the upper level courses depend for the skills necessary to master more advanced material. It was for these freshman courses that this text was created.

We didn't set out to write a textbook: we simply set out to find a better way to teach our students. Our philosophy was to help students move from a mode of learning, where everything was neatly presented as lecture and handouts where the instructor was looking for the "right" answer, to a mode of learning driven by self-guided inquiry. We wanted students to advance beyond "plug-and-chug" and memorization of problem-solving methods—to ask themselves if their approaches and answers make sense in the physical world. We couldn't settle on any textbooks we liked without patching materials together—one chapter from this text, four chapters from this one—so we wrote our own notes. Through them, we tried to convey that engineering isn't always about having the answer—sometimes it's about asking the right questions, and we want students to learn how to ask those sorts of questions. Real-world problems rarely come with all of the information required for their solutions. Problems presented to engineers typically can't be solved by looking at how someone else solved the exact same problem. Part of the fun of engineering is that every problem presents a unique challenge and requires a unique solution. Engineering is also about arriving at an answer and being able to justify the "why" behind your choice, and equally important, the "why not" of the other choices.

We realized quickly, however, that some students are not able to learn without sufficient scaffolding. Structure and flexibility must be managed carefully. Too much structure results in rigidity and unnecessary uniformity of solutions. On the other hand, too much flexibility provides insufficient guidance, and students flounder down many blind alleys, thus making it more difficult to acquire new knowledge. The tension between these two must be managed constantly. We are a large public institution, and our student body is very diverse. Our hope is to provide each student with the amount of scaffolding they need to be successful. Some students will require more background work than others. Some students will need to work five problems, and others may need to work 50. We talk a great deal to our students about how each learner is unique. Some students need to listen to a lecture; some need to read the text over three times, and others just need to try a skill and make mistakes to discover what they still don't understand. We have tried to provide enough variety for each type of learner throughout the text.

> Tell me and I forget. Teach me and I may remember. Involve me and I learn.
>
> *Benjamin Franklin*

> Education is what survives when what has been learnt has been forgotten.
>
> *B. F. Skinner*

> The larger the island of knowledge, the longer the shoreline of wonder.
>
> *Ralph W. Sockman*

Over the years, we have made difficult decisions on exactly what topics, and how much of each topic, to teach. We have refined our current text to focus on mastering five areas, each of which is introduced below.

Part 1: Engineering Essentials

You could be an engineering major if: when your professor asks you where your homework is, you claim to have accidentally determined its momentum so precisely, that according to Heisenberg, it could be anywhere in the universe.

Original source unknown

The best way to have a good idea is to have a lot of ideas.

Linus Pauling

A scientist describes what is, and an engineer creates what never was.

Theodore von Karman

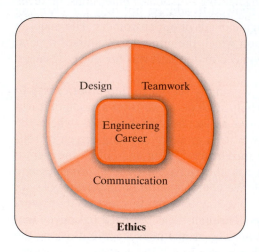

There are two threads that bind the first four chapters in Engineering Essentials together. The first is expressed in the section title: all are essential for a successful career in engineering. The other is communications, as will be discussed further below.

First, as an aspiring engineer, it is important that students attempt to verify that engineering is not only a career that suits their abilities, but also one in which they will find personal reward and satisfaction.

Second, practicing engineers often make decisions that will affect not only the lives of people but also the viability of the planetary ecosystem that affects all life on Earth. Without a firm grounding in making decisions based on ethical principles, there is an increased probability that undesirable or even disastrous consequences may occur.

Third, most engineering projects are too large for one individual to accomplish alone; thus, practicing engineers must learn to function effectively as a team, putting aside their personal differences and combining their unique talents, perspectives, and ideas to achieve the goal.

Finally, communications bind it all together. Communication, whether written, graphical, or spoken, is essential to success in engineering.

Part 2: Problem Paradigms

Engineering problems are under-defined; there are many solutions, good, bad and indifferent. The art is to arrive at a good solution. This is a creative activity, involving imagination, intuition and deliberate choice.

Ove Arup

When the only tool you own is a hammer, every problem begins to resemble a nail.

Abraham Maslow

It is better to know some of the questions than all of the answers.

James Thurber

Part 2 of the book, Problem Paradigms, starts off where all good problem solving should—with estimation. It's always best to have a good guess at any problem before trying to solve it more precisely. SOLVEM provides a framework for solving problems that encourages creative observation as well as methodological rigor. Graphing Guidelines is included because graphs are needed both to understand a system and to communicate it to others. A chapter on interpolation supports the interpretation of graphs and tabular data. Univariate statistics and statistical process control wraps up this part of the book by providing a way for engineering students to describe both distributions and trends.

Part 3: Ubiquitous Units

The wise man knows all things, as far as possible, although he has not knowledge of each of them in detail.

Aristotle

A mind stretched to a new idea never goes back to its original dimension.

Oliver Wendell Holmes

The world can be described using relatively few dimensions. We need to know what these are and how to use them to analyze engineering situations. Dimensions, however, are worthless in allowing engineers to find the numeric solution to a problem. Understanding units is essential to determine the correct numeric answers to problems. Different disciplines use different units to describe phenomena (particularly with respect to the properties of materials such as viscosity, thermal conductivity, density and so on). Engineers must know how to convert from one unit system to another. Knowledge of dimensions allows engineers to improve their problem-solving abilities by revealing the interplay of various parameters.

Part 4: Scrupulous Spreadsheets

A theory may be so rich in descriptive possibilities that it can be made to fit any data.
Philip Johnson-Laird

What is commonly overlooked in using the computer is the fact that the central goal of design is still to obviate failure, and thus it is critical to identify exactly how a structure may fail. The computer cannot do this by itself.

Henri Petroski

When choosing an analysis tool to teach students, our first pick is Excel™. Students enter college with varying levels of experience with Excel. To allow students who are novice users to learn the basics without hindering more advanced users, we have placed the basics of Excel in the Appendix material, which is available online. To help students determine if they need to review the Appendix material, an activity has been included in the introductions to Chapter 13 (Worksheets), Chapter 14 (Graphing) and Chapter 15 (Trendlines) to direct students to Appendices B, C, and D, respectively.

Once students have mastered the basics, each chapter in this section provides a deeper usage of Excel in each category. Some of this material extends beyond a simple introduction to Excel, and often, we teach the material in this unit by jumping around, covering half of each chapter in the first semester, and the rest of the material in the second semester course.

Chapter 15 introduces students to the idea of similarities among the disciplines, and how understanding a theory in one application can often aid in understanding a similar theory in a different application. We also emphasize the understanding of models (trendlines) as possessing physical meaning. Chapter 16 discusses a process for determining a mathematical model when presented with experimental data, and some advanced material on dealing with limitations of Excel.

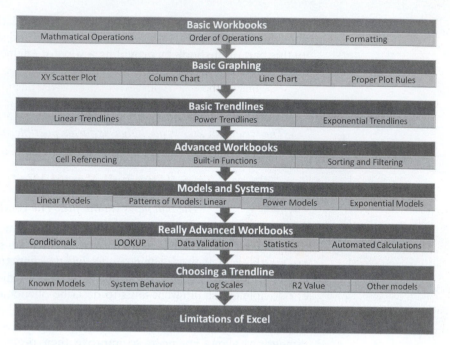

Part 5: Punctilious Programming

If debugging is the process of removing bugs, then programming must be the process of putting them in.

Edsger W. Dijkstra

Measuring programming progress by lines of code is like measuring aircraft building progress by weight.

Bill Gates

Part 5 (Punctilious Programming) covers a variety of topics common to any introductory programming textbook. In contrast to a traditional programming textbook, this

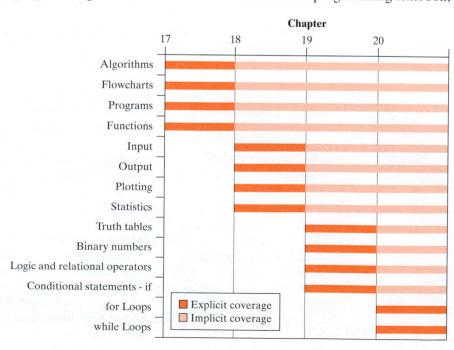

section approaches each topic from the perspective of how each can be used in unison with the others as a powerful engineering problem-solving tool. The topics presented in Part 5 are introduced as if the student has no prior programming ability and are continually reiterated throughout the remaining chapters.

For this textbook we chose MATLAB™ as the programming language because it is commonly used in many engineering curricula. The topics covered provide a solid foundation of how computers can be used as a tool for problem solving and provide enough scaffolding for transfer of programming knowledge into other languages commonly used by engineers (such as C/C++/Java.)

The "Other" Stuff We've Included...

The bad news is time flies. The good news is you're the pilot.

Michael Altshuler

All who have accomplished great things have had a great aim, have fixed their gaze on a goal which was high, one which sometimes seemed impossible.

Orison Swett Marden

Do or do not. There is no try.

Yoda

Throughout the book, we have included sections on surviving engineering, time management and goal setting, personality types and study skills. We did not group them into a single chapter, but have scattered them throughout the section introductions to assist students on a topic when they are most likely to need it. For example, we find students are much more open to discuss time management in the middle of the semester rather than the beginning.

In addition, we have called upon many practicing and aspiring engineers to help us explain the "why" and "what" behind engineering. They offer their "Wise Words" throughout this text. We have included our own set of "Wise Words" as the introduction to each topic here as a glimpse of what inspired us to include certain topics.

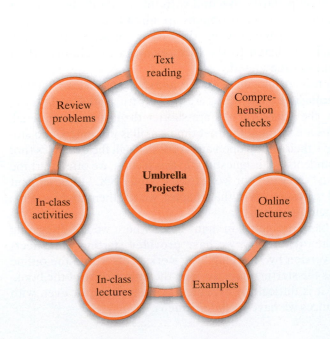

How to Use this Text

As we have alluded to previously, this text contains many different types of instruction to address different types of learners. There are two main components to this text: hard copy and online.

In the hardcopy, the text is presented topically rather than sequentially, but hopefully with enough autonomy for each piece to stand alone. For example, we routinely discuss only part of the Excel material in our first semester course, and leave the rest to the second semester. There are some topics (such as interpolation) that we simply tell our students that we expect them to know and do not spend any class time discussing. We hope this will give you the flexibility to choose how deeply into any given topic you wish to dive, depending on the time you have, the starting abilities of your students, and the outcomes of your course. More information about topic sequence options can be found in the instructor's manual.

Within the text, there are several checkpoints for students to see if they understand the material. Within the reading are **Comprehension Checks**, with the answers are provided in the back of the book. Our motivation for including Comprehension Checks within the text rather than include them as end of section questions is to maintain the active spirit of the classroom within the reading, allowing the students to self-evaluate their understanding of the material in preparation for class—to enable students to be self-directed learners, we must encourage them to self-evaluate regularly. At the end of each chapter, **In-Class Activities** are given to reinforce the material in each chapter. In-Class Activities exist to stimulate active conversation within pairs and groups of students working through the material. We generally keep the focus on student effort, and ask them to keep working the problem until they arrive at the right answer. This provides them with a set of worked out problems, using their own logic, before they are asked to tackle more difficult problems. The **Review** sections at the end of each part provide additional questions, often combining skills within the part to help students climb to the next level of understanding. By providing these three types of practice, students are encouraged to reflect on their understanding in *preparing for class, during class*, and *at the end of each chapter and part* as they prepare to transfer their knowledge to other areas. Finally in the instructor's manual, we have provided a series of **Umbrella Projects** to allow students to apply skills that they have mastered to larger-scope problems. We have found the use of these problems extremely helpful in providing context for the skills that they learn throughout the unit.

Understanding that every student learns differently, we have included several media components in addition to traditional text. Each section within each chapter has an accompanying set of **lecture slides**. Within these slides, the examples presented are unique from those in the text to provide another set of sample solutions. The slides are presented with **voiceover**, which has allowed us to move away from traditional in-class lecture. We expect the students to listen to the slides outside of class, and then in class we typically spend time working problems, reviewing assigned problems, and providing **"wrap-up" lectures**, which are mini-versions of the full lectures to summarize what they should have gotten from the assignment. We expect the students to come to class with questions from the reading and lecture that we can then help clarify. We find with this method, the students pay more attention, as the terms and problems are already familiar to them, and they are more able to verbalize what they don't know. Furthermore, they can always go back and listen to the lectures again to reinforce their knowledge as many times as they need.

Some sections of this text are difficult to lecture, and students will learn this material best by **working through examples**. This is especially true with Excel, so you will notice that many of the Excel lectures are shorter than previous material. With Excel, the examples are scripted the first time a skill is presented, and students are expected to have their laptop open and work through the examples (not just read them). When students ask us questions in this section, we often start the answer by asking them to "show us your work from Chapter XX." If the student has not actually worked the examples in that chapter, we tell them to do so first; often, this will answer their questions.

After the first few basic problems, in many cases where we are discussing more advanced skills than data entry, we have **provided starting worksheets and code** in the online version by "hanging" the worksheets within the online text. Students can access the starting data through the online copy of the book. In some cases, though, it is difficult to explain a skill on paper, or even with slides, so for these instances we have included **videos**.

The following text markers indicate corresponding online materials are available.

 Excel file

 Word file

 PowerPoint file

 Video file

Finally, for the communication section, we have provided **templates** for several types of reports and presentations. These can also be accessed in the Pearson eText version.

Additional Resources for Instructors

Instructor's Manual—Available to all adopters, this provides a complete set of solutions for all activities and homework exercises. For the in-class activities, suggested guided-inquiry questions along with time frame guidelines are included. Suggested content sequencing and descriptions of how to couple assignments to the umbrella projects are also provided.

PowerPoints—A complete set of lecture PowerPoint slides, available with voiceover or as standard slides, make course planning as easy as possible.

Sample Exams—Available to all adopters, these will assist in creating tests and quizzes for student assessment.

All requests for instructor resources are verified against our customer database and/or through contacting the requestor's institution. Contact your local Pearson/Prentice Hall representative for additional information.

What Does Thinking Like an Engineer Mean?

We are often asked about the title of the book. We thought we'd take a minute and explain what this means, to each of us. Our responses are included in alphabetical order.

Author team: Ohland, Park, Stephan, Bowman, Sill

For me, thinking like an engineer is about creatively finding a solution to some problem. In my pre-college days, I was very excited about music. I began my musical pursuits by learning the fundamentals of music theory by playing in middle school band and eventually worked my way into different bands in high school (orchestra, marching band, jazz band) and branching off into teaching myself how to play guitar. I love playing and listening to music because it gives me an outlet to create and discover art. I pursued engineering for the same reason; as an engineer, you work in a field that creates or improves designs or processes. For me, thinking like an engineer is exactly like thinking like a musician—through my fundamentals, I'm able to be creative, yet methodical, in my solutions to problems.

D. Bowman, Computer Engineer

Thinking like an engineer is about solving problems with whatever resources are most available—or fixing something that has broken with materials that are just lying around. Sometimes, it's about thinking ahead and realizing what's going to happen before something breaks or someone gets hurt—particularly in thinking about what it means to fail safe—to design how something will fail when it fails. Thinking like an engineer is figuring out how to communicate technical issues in a way that anyone can understand. It's about developing an instinct to protect the public trust—an integrity that emerges automatically.

M. Ohland, Civil Engineer

To me, understanding the way things work is the foundation on which all engineering is based. Although most engineers focus on technical topics related to their specific discipline, this understanding is not restricted to any specific field, but applies to everything! One never knows when some seemingly random bit of knowledge, or some pattern discerned in a completely disparate field of inquiry, may prove critical in solving an engineering problem. Whether the field of investigation is Fourier analysis, orbital mechanics, Hebert boxes, personality types, the Chinese language, the life cycle of mycetozoans, or the evolution of the music of Western civilization, the more you understand about things, the more effective an engineer you can be. Thus, for me, thinking like an engineer is intimately, inextricably, and inexorably inter-twined with the Quest for Knowledge. Besides, the world is a truly fascinating place if one bothers to take the time to investigate it.

W. Park, Electrical Engineer

Engineering is a bit like the game of golf. No two shots are ever exactly the same. In engineering, no two problems or designs are ever exactly the same. To be successful, engineers need a bag of clubs (math, chemistry, physics, English, social studies) and then need to have the training to be able to select the right combination of clubs to move from the tee to the green and make a par (or if we are lucky, a birdie). In short, engineers need to be taught to THINK.

B. Sill, Aerospace Engineer

I like to refer to engineering as the color grey. Many students enter engineering because they are "good at math and science." I like to refer to these disciplines as black and white—there is one way to integrate an equation and one way to balance a chemical re-action. Engineering is grey, a blend of math and science that does not necessarily have one clear answer. The answer can change depending on the criteria of the problem. Thinking like an engineer is about training your mind to conduct the methodical process of problem solving. It is examining a problem from many different angles, considering the good, the bad and the ugly in every process or product. It is thinking creatively to discover ways of solving problems, or preventing issues from becoming problems. It's about finding a solution in the grey and presenting it in black and white.

E. Stephan, Chemical Engineer

Lead author note: When writing this preface, I asked each of my co-authors to answer this question. As usual, I got a wide variety of interpretations and answers. This is typical of the way we approach everything we do, except that I usually try and mesh the responses into one voice. In this instance, I let each response remain unique. As you progress throughout this text, you will (hopefully) see glimpses of each of us inter-woven with the one voice. We hope that through our uniqueness, we can each reach a different group of students and present a balanced approach to problem solving, and, hopefully, every student can identify with at least one of us.   —Beth Stephan

Clemson University
Clemson, SC
Fall, 2009

ACKNOWLEDGEMENTS

When we set out to formalize our instructional work, we wanted to portray engineering as a reality, not the typical flashy fantasy portrayed by most media forums. We called on many of our professional and personal relationships to help us present engineering in everyday terms. During a lecture to our freshman, Dr. Ed Sutt [PopSci's 2006 Inventor of the Year for the HurriQuake Nail] gave the following advice. *A good engineer can reach an answer in two calls: the first, to find out who the expert is; the second, to talk to the expert.* Realizing we are not experts, we have called on many folks to contribute articles. To our experts who contributed articles for this text, we thank: Dr. Lisa Benson, Dr. Neil Burton, Jan Comfort, Solange Dao, Troy Nunmaker, and Jessica Pelfry.

To Dr. Lisa Benson, thank you for allowing us to use "Science as Art" for the basis of many photos that we have chosen for this text. To explain "Science as Art": *Sometimes, science and art meet in the middle. For example, when a visual representation of science or technology has an unexpected aesthetic appeal, it becomes a connection for scientists, artists and the general public. In celebration of this connection, Clemson University faculty and students are challenged to share powerful and inspiring visual images produced in laboratories and workspaces for the "Science as Art" exhibit.* For more information, please visit www.scienceasart.org. To the creators of the art, thank you for letting us showcase your work in this text: Martin Beagley, Dr. Caye Drapcho, Eric Fenimore, Dr. Scott Husson, Dr. Jaishankar Kutty, Dr. Kathleen Richardson, and Dr. Ken Webb. A special thanks to Kautex Machines and Chuck Flammer, and to Russ Werneth for getting us the great Hubble teamwork photo.

To the Rutland Institute for Ethics at Clemson University: The four-step procedure outlined in Chapter 2 on Ethics is based on the toolbox approach presented in the Ethics Across the Curriculum Seminar. Our thanks to Dr. Daniel Wueste, Director, and the other Rutlanders (Kelly Smith, Stephen Satris and Charlie Starkey) for their input into this chapter.

To Jonathan Feinberg and all the contributors to the Wordle (http://www.wordle.net) project, thank you for the tools to create for the Wordle images in the introduction sections. We hope our readers enjoy this unique way of presenting information, and are inspired to create their own Wordle!

To our friends and former students who contributed their Wise Words: Tyler Andrews, Corey Balon, Ed Basta, Brittany Brubaker, Tim Burns, Clark Chewing, Ashley Childers, Jeremy Comardelle, Matt Cuica, Jeff Dabling, Ed D'Avignon, Brian Dieringer, Michael Dukes, Lauren Edwards, Andrew Flowerday, Stacey Forkner, Lisa Gascoigne, Victor Gallas Cervo, Khadijah Glast, Colleen Hill, Tom Hill, Becky Holcomb, Beth Holloway, Selden Houghton, Allison Hu, Jason Huggins, Ryan Izard, Lindy Johnson, Darryl Jones, Rob Kriener, Jim Kronberg, Maria Koon, Rachel Lanoie, Danielle Lanigan, Mai Lauer, Marie Long, Stacy Luell, Jack Meena, Matt Morgan, Alan Passman, Candace Pringle, Mike Peterson, Derek Rollend, Jake

Sadie, Janna Sandel, Kaycie Smith, Ellen Styles, Adam Thompson, Wendy Valtadoros, Devin Walford, Russ Werneth and Aynsley Zollinger.

To our fellow faculty members, for providing inspiration, ideas, and helping us find countless mistakes: Dr. Lisa Benson, Dr. Steve Brandon, John Minor, and Dr. Julie Trenor. You guys are the other half of this team that makes this the best place on earth to work! We could not have done this without you.

To the staff of the GE program, we thank you for your support of us and our students: Kelli Blankenship, Lib Crockett, Linda Law, Chris Porter, Janeen Putman, and Susan Smith. To Chuck Heck, Linda Nilson, Barbara Weaver, and the rest of the Teaching with Technology group for helping us discover better ways to teach, spark our innovation, and create such wonderful classroom environments. To the administration at Clemson, we thank you for your continued support of our program: Associate Dean Dr. Randy Collins, Interim Chair Dr. Melanie Cooper, Dean Dr. Esin Gulari, Provost Dr. Dori Helms, Former Dean Dr. Tom Keinath, and Former Dean Dr. Steve Melshimer. Special thanks to President Jim Barker for his inspirational leadership of staying the course and giving meaning to "One Clemson."

To the thousands of students who used this text in various forms over the years—thanks for your patience, your suggestions, and your criticism. You have each contributed not only to the book, but to our personal inspirations to keep doing what we do.

To the great folks at Prentice Hall: Scott Disanno, Greg Dulles, Mary Lou Nohr and Dan Sandin—this project would not be a reality without all your hard work. To Eric Hakanson, without that chance meeting this project would not have begun! Thank you to Meg Tiedemann and Bill Clements for their work on early custom editions of these materials. Thank you to all the reviewers who provided such valuable feedback to help us improve. Thanks to Holly Stark for her belief in this project and in us! A special thanks to Clare Romeo for editing the text about a zillion times and putting up with all of our seemingly endless questions! We would not have made it through this without all of the PH team efforts and encouragement!

Finally, on a Personal Note

DRB: Thanks to my parents and sister for supporting my creative endeavors with nothing but encouragement and enthusiasm. To my grandparents, who value science, engineering, and education to be the most important fields to study. To my co-authors, who have reaffirmed my interest in pursuing engineering education as a career. To everyone I've given the excuse of "book stuff": I owe you all a good beer.

MWO: My wife Emily has my love, admiration, and gratitude for all she does, including holding the family together. For my children, who share me with my students—Charlotte, whose "old soul" touches all who take the time to know her; Carson, who is quietly inspiring; and Anders, whose love of life and people endears him to all. I acknowledge my father Theodor, who inspired me to be an educator; my mother Nancy, who helped me understand people; my sister Karen, who lit a pathway in engineering; my brother Erik, who showed me that one doesn't need to be loud to be a leader; and my mother-in-law Nancy Winfrey, who shared the wisdom of a long career. I recognize those who helped me create an engineering education career path: Fred Orthlieb, Civil and Coastal Engineering at the University of Florida, Marc Hoit, Duane Ellifritt, Cliff Hays, Mary Grace Kantowski, and John Lybas, the NSF's SUCCEED Coalition, Tim Anderson, Clemson's College of Engineering and Science and General Engineering, Steve Melsheimer, Ben Sill, and Purdue's School of Engineering Education.

WJP: Choosing only a few folks to include in an acknowledgment is a seriously difficult task, but I have managed to reduce it to five. First, Beth Stephan has been the guiding force behind this project, without whom it would never have come to fruition. In addition, she has shown amazing patience in putting up with my shenanigans and my weird perspectives. Next, although we exist in totally different realities, my parents have always supported me, particularly when I was a newly married, destitute graduate student fresh off the farm. Third, my son Isaac, who has the admirable quality of being willing to confront me with the truth when I am behaving badly, and for this I am grateful. Finally, and certainly most importantly, to Lila, my partner of more than one-third century, I owe a debt beyond anything I could put into words. Although life with her has seldom been easy, her influence has made me a dramatically better person.

BLS: To my amazing family, who always picked up the slack when I was off doing "creative" things, goes all my gratitude. To Anna and Allison, you are wonderful daughters who both endured and "experienced" the development of many "in class, hands on" activities—know that I love you and thank you. To Lois who has always been there with her support and without whining for over 40 years, all my love. Finally, to my co-authors who have tolerated my eccentricities and occasional tardiness with only minimum grumbling, you make a great team.

EAS: To my co-authors, for tolerating all my strange demands, my sleep-deprived ravings and the occasional "I need this now" hysteria. You guys are the best! To all those folks who cared for my family while this project consumed my every waking minute, I am eternally indebted. Special thanks to Glenn and Sara—you did so much for us, and still call and ask even though the answer is "The book thing . . ." To my mom, Kay and Denny—thanks for your love and support. To Khadijah, you can do anything you put your mind to. To Brock and Katie, I love you both a bushel and a peck. You are the best kids in the world. Finally, to Sean . . . for all the times you held it all together, kept things running, and brought me a Diet Coke—I love you more than I can say. "Show a little faith, there's magic in the night . . ."

Thinking Like an Engineer Reviewers

John W. Nicklow	Southern Illinois University
Peter Orono	Indiana University Purdue
Christopher J. Rowe	Vanderbilt University
Elliot P. Douglas	University of Florida
Brian D. Koehler	North Carolina State University
Sherry Snyder	University of Colorado at Boulder
Patricia Brackin	Rose-Hulman Institute of Technology
Kelly Crittenden	Louisiana Tech University
Monica Cardella	Purdue University
John R. Biddle	Cal Poly Pomona University
Irene Ferrara	Penn State Altoona College
Stephen Krause	Arizona State University
Jessica Townsend	Olin College
Martha Selby	Iowa State University
John T. Demel	The Ohio State University
Paul Kominsky	University of Michigan
Howard M. Fulmer	Villanova University
Andrew Lau	Penn State University
Susan Freeman	Northeastern University

PART 1

ENGINEERING ESSENTIALS

Success is not the key to happiness. Happiness is the key to success. If you love what you are doing, you will be successful.

ALBERT SCHWEITZER

As the reader of this textbook, you're no doubt in a situation where you have an idea you want to be an engineer. Someone or something put into your head this crazy notion—that you might have a happy and successful life working in the engineering profession. Maybe you're good at math or science, or you want a job where creativity is as important as technical skill. Maybe someone you admire works as an engineer. Maybe you are looking for a career that will challenge you intellectually, or maybe you like to solve problems.

You may recognize yourself in one of these statements from practicing engineers on why they chose to pursue an engineering degree.

I chose to pursue engineering because I enjoyed math and science in school, and always had a love for tinkering with electronic and mechanical gadgets since I was old enough to hold a screwdriver.
S. Houghton, Computer Engineer

I chose to pursue engineering because I always excelled in science and math and I really enjoy problem solving. I like doing hands-on activities and working on "tangible" projects.
M. Koon, Mechanical Engineer

I wanted to pursue engineering to make some kind of positive and (hopefully) enduring mark on the world.
J. Kronberg, Electrical Engineer

I excelled in math and science, my dad was an engineer, and it was a field with a lot of opportunity for women.
D. Lanigan, Industrial Engineer

My parents instilled a responsibility to our community in us kids. As an engineer, I can serve my community through efficient and responsible construction while still satisfying my need to solve challenging problems.
J. Meena, Civil Engineer

I chose to pursue engineering because I knew it would be challenging, and like to push myself. I wanted an opportunity to do great and meaningful things. Ever since I can remember I have had a curiosity of the world around me, wanting to know how things work.

M. Morgan, Mechanical Engineer

I am a first-generation college student and I wanted to have a strong foundation when I graduated from college.

C. Pringle, Industrial Engineer

Engineering is a highly regarded and often highly compensated profession that many skilled high school students choose to enter for the challenge, engagement, and ultimately the reward of joining the ranks of the esteemed engineers of the world. But what, exactly, does an engineer do? This is one of the most difficult questions to answer because of the breadth and depth of the engineering field. So, how do the experts define engineering?

The National Academy of Engineering defines engineering as follows:

Engineering has been defined in many ways. It is often referred to as the "application of science" because engineers take abstract ideas and build tangible products from them. Another definition is "design under constraint," because to "engineer" a product means to construct it in such a way that it will do exactly what you want it to, without any unexpected consequences.

According to the Merriam-Webster online dictionary:

Engineering is the application of science and mathematics by which the properties of matter and the sources of energy in nature are made useful to people.

More or less, engineering is a broad, hard-to-define field requiring knowledge of science and mathematics and other fields to turn ideas into reality. The ideas and problems posed to engineers often don't require a mastery-level knowledge of any particular scientific field, but instead require the ability to put together all of the pieces learned in those fields. N. W. Dougherty, a former professor of civil engineering at the University of Tennessee said:

The ideal engineer is a composite . . . He is not a scientist, he is not a mathematician, he is not a sociologist or a writer; but he may use the knowledge and techniques of any or all of these disciplines in solving engineering problems.

Because engineers solve real-life problems, their ultimate motivation is to work toward making life better for everyone. In "The Heroic Engineer" (*Journal of Engineering Education*, January 1997) by Taft H. Broome (Howard University), and Jeff Peirce (Duke University), those authors claimed:

Engineers who would deem it their professional responsibility to transcend self-interests to help non-experts advance their own interests may well prove indispensable to free societies in the twenty-first century.

Broome and Peirce go on to explain that the traits and behaviors of engineers can be compared to those of a hero. The motivation of any hero is to save someone's life; engineers create products, devices, and methods to help save lives. Heroes intervene to protect from danger; engineers devise procedures, create machines, and improve processes to protect people and the planet from danger. While learning an engineering discipline can be challenging, the everyday engineer doesn't see it as an obstacle: it's merely an opportunity to be a hero.

Scattered throughout this text, you will find quotes from practicing engineers often referred to as "Wise Words" inside highlighted text boxes. As a good engineering team would, we recognize we (the authors) are not experts at all things, and request input and advice when needed. We asked engineers we know working at "everyday engineering" jobs to reflect on the choices they made in school and during their careers. We hope you benefit from their collective knowledge. When asked for advice to give to an incoming freshman, one gave the following reply, summing up this section better than we ever could have imagined.

> *[A career in engineering] is rewarding both financially and personally. It's nice to go to work and see some new piece of technology—to be on the cutting edge. It's also a great feeling to know that you are helping improve the lives of other people. Wherever there has been a great discovery, an engineer is to thank. That engineer can be you.*

A. Thompson, Electrical Engineer

ENGINEERING IS AN . . . ITCH!

Contributed by: Dr. Lisa Benson, Assistant Professor of Engineering and Science Education, Clemson University

There are a lot of reasons why you are majoring in engineering. Maybe your goal is to impress someone, like your parents, or to defy all those who said you'd never make it, or simply to prove to yourself that you have it in you. Maybe your goal is to work with your hands as well as your mind. Maybe you have no idea why you're here, but you know you like cars. There are about as many goals as there are students, and they serve to motivate students to learn. Some goals are better motivators than others.

Lots of experts have studied goals and how they affect what students do in school. Not surprisingly, there are as many ideas and theories about goals as there are experts. But most experts agree on the idea that there are mastery goals and performance goals.

> *Students who are **mastery oriented** try to do things well because they want to do their best. They are not driven by external factors like grades or praise, but instead they seek to learn things because they want to really understand them and not just get the correct answer.*

> *Students who are **performance oriented** seek to earn good grades to reflect how hard they've worked. They study because they know it will get them something—a scholarship, an above-average grade, or praise from their parents. Since grades are tied to their sense of achievement, students who are performance oriented tend to feel discouraged and anxious when they earn low grades. They tend to want to memorize and pattern-match to solve problems, rather than learn the underlying concepts and methods.*

Most students have been performance oriented throughout high school. In college, you will be more successful if you start thinking in terms of mastery. If you seek to really understand what you are learning in your classes, performance (i.e., good grades) will follow. But performance (a grade, an award, or praise) isn't everything, and it will not be enough to keep you motivated when projects and coursework are challenging.

There is nothing like the feeling when you finally understand something that you didn't get before. Sometimes it's an "aha!" moment, and sometimes it's a gradual dawning. The feeling is like an itch you can't scratch—you'll want to keep at it once you get it. *When you are motivated to understand and master something, you're taking pride in your achievement of conquering the material, not just getting a good grade on an exam.* And you're going to keep scratching that itch. Keep "scratching" at the material—working, practicing, drilling skills if you need to, whatever it takes—to master that . . . itch!

LEARNING OBJECTIVES

The overall learning objectives for this unit include the following:

- Conduct research on ethical issues related to engineering; formulate and justify positions on these issues;
- Demonstrate an ability to design a system, component, or process to meet desired needs;
- Demonstrate an ability to function on multidisciplinary teams; and
- Communicate technical information effectively by composing clear and concise oral presentations and written descriptions of experiments and projects.

CHAPTER 1
EVERYDAY ENGINEERING

Most students who start off in a technical major know very little about their chosen field. This is particularly true in engineering, which is not generally present in the high school curriculum. Students commonly pick engineering and science majors because someone suggested them. In this section we help you ask the right questions about your interests, skills, and abilities; we then show you how to combine the answers with what you learn about engineering and science in order to make the right career decision.

1.1   CHOOSING A CAREER

In today's society, the careers available to you upon your college graduation are numerous and diverse. It is often difficult as a young adult to determine exactly what occupation you want to work at for the rest of your life because you have so many options. As you move through the process, there are questions that are appropriate to ask. You can't make a good decision without accurate information. No one can (or should) make the decision for you: not your relatives, professors, advisors or friends. Only you know what feels right and what doesn't. You may not know all the answers to your questions right away. That means you'll have to get them by gathering more information from outside resources and through your personal experience. Keep in mind that choosing your major and ultimately your career is a process. You constantly evaluate and reevaluate what you learn and experience. A key component is whether you feel challenged or overwhelmed. True success in a profession is not measured in monetary terms; it is measured in job satisfaction . . . enjoying what you do, doing what you enjoy. As you find the answers, you can choose a major that leads you into a career path that you enjoy and be successful.

Before you decide, answer the following questions about your tentative major choice. Start thinking about the questions you can't answer and look for ways or resources to get the information you need. It may take a long time before you know, and that's okay!

- What do I already know about this major?
- What courses will I take to earn a degree in this major?
- Do I have the appropriate academic preparation to complete this major? If not, what will I have to do to acquire it?
- Am I enjoying my courses . . . do I feel challenged or stressed?
- What time demands are involved? Am I willing to spend the time it takes to complete this major?
- What kinds of jobs will this major prepare me for? Which sound most interesting?
- What kinds of skills will I need to do the job I want? Where can I get them?

This process will take time. Once you have the information, you can make a choice. Keep in mind, nothing is set in stone—you can change your mind!

1.2 CHOOSING ENGINEERING AS A CAREER

In the Engineering Essentials introduction, we gave several examples of why practicing engineers wanted to pursue a career in engineering. Here are a few more:

- I was always in to tinkering with things and I enjoyed working with computers from a young age. Math, science, and physics came very natural to me in high school. For me it was an easy choice.

  J. Comardelle, Computer Engineer

- My initial instinct for a career path was to become an engineer. I was the son of a mechanical engineer, performed well in science and mathematics during primary education, and was always "tinkering" with mechanical assemblies.

  M. Cuica, Mechanical Engineer

- I chose engineering for a lot of the same reasons that the "typical" entering freshman does—I was good at math and science. I definitely did not know that there were so many types of engineering and to be honest, was a little overwhelmed by the decision I needed to make of what type of engineering was for me.

  L. Edwards, Civil Engineer

- I wasn't really sure what I wanted to do. My parents were not college graduates so there was not a lot of guidance from them, so my high school teachers influenced me a lot. I was taking advanced math and science classes and doing well in them. They suggested that I look into engineering, and I did.

  S. Forkner, Chemical Engineer

- I enjoyed science and problem solving. I thought an engineering degree would be great preparation for the job market and give me a lot of options when it came time to find a job.

  W. Valtadoros, Mechanical Engineer

Table 1-1 describes the authors' perspective on how various engineering and science disciplines might contribute to different industries or innovations. This table is only an interpretation by a few engineers and doesn't handle every single possibility of how an engineer might contribute toward innovation. For example, an industrial engineer might be called in to work on an energy project to share a different perspective on energy efficiency. The broad goal of any engineering discipline is to solve problems, so there's often a need for a different perspective to possibly shed new light toward an innovative solution.

The kind of problems engineers will be solving in the 21st century are summarized in the Grand Challenges for Engineering, which is a document published by the National Academy of Engineering. These 14 Grand Challenges are just an idea on the kind of problems all of the different engineering disciplines might be working to solve. For example, one grand challenge is to advance health informatics—since there isn't a "health informatics engineering" discipline, this goal could only be met through collaboration of a variety of engineering disciplines. For further information on the Grand Challenges for Engineering, search for "grand challenges" and "NAE" in your favorite search engine.

Table 1-1 Sample career paths and possible majors. Shaded boxes indicate a good starting point for further exploration.

CAREERS	ENGINEERING										SCIENCE				
	Aerospace	Bioengineering	BioSystems	Civil	Chemical	Materials	Electric/computer	Environmental	Industrial	Mechanical	Chemistry	Computer Science	Geology	Mathematics	Physics
GENERAL															
Energy Industry					X			X		X	X		X		X
Machines		X					X			X					
Manufacturing				X	X	X			X	X					
Materials	X	X			X	X				X	X				
Structures	X			X		X				X					
Technical Sales				X	X		X				X	X			X
SPECIFIC															
Rocket/airplane	X					X				X					X
Biomedical		X	X				X				X				
Coastal Engineering			X	X				X					X		
Computing							X					X		X	
Cryptography							X					X		X	
Defense	X						X			X		X		X	X
Environment			X	X	X			X			X		X		X
Fiber optics						X	X						X		
Forensics					X		X				X				
Groundwater				X				X					X		
Human Factors	X						X		X		X				
Industrial Sensors	X						X		X	X					
Intelligent Systems							X		X			X		X	
Management	X	X							X	X					
Operations Research									X			X		X	
Outdoor Work			X	X				X					X		
Pharmaceutical		X	X		X						X				
Plastics					X	X				X					
Robotics	X						X			X		X			
Semiconductors						X	X								X
Telecommunications							X					X		X	
Transportation	X			X					X	X					
Waste Management			X	X	X			X							

> **MORE WISE WORDS: HOW DID YOU CHOOSE A MAJOR IN COLLEGE?**
>
> Since I knew I wanted to design computers, I had a choice between electrical and computer engineering. I chose computer engineering so I could learn about both the hardware and software. It was my interests in computers and my high school teachers that were the biggest influence in my decision.
>
> *E. D'Avignon, CpE*
>
> My first choice in majors was Mechanical Engineering. I changed majors after taking a drafting class in which I did well enough to get a job teaching the lab portion, but I did not enjoy the work. After changing to Electrical and Computer Engineering, I took a Statics and Dynamics course as part of my required coursework and that further confirmed my move as I struggled with that material.
>
> *A. Flowerday, EE*
>
> I was interested in aerospace engineering, but chose to pursue mechanical engineering, a broader field, that would keep more job options available to me upon graduation. I knew that I could place an emphasis on aerospace within my undergraduate degree, or pursue it further with a graduate degree. I chose ME because I thought it would give me a solid engineering background with a lot of flexibility for change in my future career.
>
> *M. Long, ME*
>
> Coming into college, I knew I wanted to pursue a career in medicine after graduation. I also knew that I did not want to major in chemistry, biology, etc. Therefore, bioengineering was a perfect fit. It provides a challenging curriculum while preparing me for medical school at the same time. In addition, if pursuing a career in medicine does not go according to plan, I know that I will also enjoy a career as a bioengineer.
>
> *D. Walford, BioE*

1.3   CHOOSING A SPECIFIC ENGINEERING FIELD

The following paragraphs briefly introduce several different types of engineering majors. By no means is this list completely inclusive.

Bioengineering (BioE) or Biomedical Engineering (BME)

BioE and BME apply engineering principles to the understanding and solution of medical problems. Bioengineers are involved in research and development in all areas of medicine, from investigating the physiological behavior of single cells to designing implants for the replacement of diseased or traumatized body tissues. Bioengineers design new instruments, devices, and software, assemble knowledge from many scientific sources to develop new procedures and conduct research to solve medical problems.

Typical bioengineers work in such areas as artificial organs, automated patient monitoring, blood chemistry sensors, advanced therapeutic and surgical devices, clinical laboratory design, medical imaging systems, biomaterials, and sports medicine.

Use multidisciplinary experience to help develop and commercialize new bioelectronic technologies and products; Invent new devices and technologies; Prepare presentation graphics and information; Coordinate with the FDA; Help implement Good Manufacturing Practices, compiling engineering files for products prior to sales launch.

J. Kronberg, EE

Bioengineers are employed in universities, industry, hospitals, research facilities, and government. In industry they may be part of a team serving as a liaison between engineers and clinicians. In hospitals, they select appropriate equipment and supervise equipment performance, testing, and maintenance. In government agencies, they are involved in safety standards and testing.

I work with scientists and engineers to protect their innovations by writing patent applications describing their inventions and presenting the applications before the United States Patent & Trademark Office. I also assist clients in determining whether another party is infringing their patents and help my clients to avoid infringing other's patents.

M. Lauer,
Environmental
Engineer

Biosystems Engineering (BE)

BE is the field of engineering most closely allied with advances in biology that intersect with engineering, including the biological production of pharmaceutical drugs and bio-fuels (hydrogen, ethanol) and design of biological machines, biosensors, and even biological batteries. Scientific emphasis is shifting toward the biosciences. Biosystems engineers participate in the application of these biotech advances to man's daily life in areas ranging from environment to medical sciences to agriculture.

Here are some activities of Biosystems Engineers:

- Design bioprocesses and systems for biopharmaceutical, bioplastics, biofuels, and food processing industries
- Develop sites and designs for storm water and sedimentation abatement
- Protect environmental quality and natural resources through environmental remediation/engineering
- Provide engineering expertise for agriculture, food production, and manufacturing
- Provide engineering support for biotechnology and biomedical activities
- Pursue medical or veterinary school or graduate school in wide range of disciplines

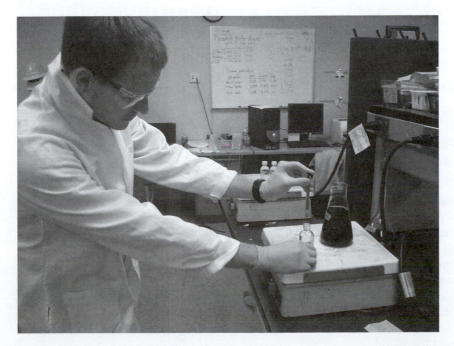

My research is part of a Water Research Foundation project which is investigating the formation of emerging disinfection byproducts in drinking water treatment.

Disinfection byproducts (DBPs) are undesirable, toxic compounds that are formed when water is chlorinated. I am investigating the effects of pH, bromide and iodide concentrations, and preoxidants on the formation of a specific family of DBPs.

D. Jones, BE

I am a project manager for new product development.

I oversee and coordinate the various activities that need to be completed in order to get a new product approved and manufactured, and ultimately in the hands of our consumers.

S. Forkner, ChE

Chemical Engineering (ChE)

ChE incorporates strong emphasis on three sciences: chemistry, physics, and mathematics. Chemical engineers are involved in the research and development, manufacture, sales, and use of chemicals, pharmaceuticals, electronic components, food and consumer goods, petroleum products, synthetic fibers and films, pulp and paper, and many other products. They work on environmental remediation and pollution prevention, as well as in medical and health-related fields. Chemical Engineers:

- Conduct research and develop new products; develop and design new manufacturing processes
- Earn additional degrees to practice medicine or patent, environmental, or corporate law
- Sell and provide technical support for sophisticated chemical products to customers
- Solve environmental problems; work in biotechnology
- Troubleshoot and solve problems in chemical manufacturing facilities

My team is responsible for implementing the engineered design in the field.

We install, tune, test, and accept into operations all of the electronics that allow customers to use our state of the art fiber optic network to run voice, video, and data for their residential needs.

L. Gascoigne, CE

My current responsibilities include

– Analysis of traffic signal operations and safety for municipal and private clients;

– Preparation of traffic impact studies;

– Review of plans and traffic studies for municipalities and counties;

– Design of traffic signal installations and traffic signing projects.

C. Hill, CE

Civil Engineering (CE)

CE involves the planning, design, construction, maintenance and operation of facilities and systems to control and improve the environment for modern civilizations. This includes projects of major importance such as bridges, transportation systems, buildings, ports, and water distribution systems.

Here are just a few of many opportunities available for CEs:

- Design and analyze structures ranging from small buildings to skyscrapers to offshore oil platforms
- Design dams and building foundations; develop new materials for pavements, buildings, and bridges
- Design improved transportation systems
- Design water distribution and removal systems
- Develop new methods to improve safety, reduce cost, speed construction, and reduce environmental impact
- Provide construction and project management services for large engineered projects throughout the world

I am responsible for assisting in the management of commercial and healthcare projects for Brasfield & Gorrie. Working closely with the owner and architect, I maintain open lines of communication and aim to provide exceptional service to the entire project team from the preconstruction phase of the project through construction. I assist in establishing and monitoring procedures for controlling the cost, schedule, and quality of the work in accordance with the construction contract.

L. Edwards, CE

I develop, manage, and support all software systems. I also deal with system scalability, customer satisfaction, and data management.

J. Comardelle, CpE

Computer Engineering (CpE)

CpE spans the fields of computer science and engineering, giving a balanced view of hardware, software, hardware-software tradeoffs, and basic modeling techniques that represent the computing process involving the following technologies:

- Communication system design
- Computer interface design
- Computer networking
- Digital signal processing applications
- Digital system design
- Imbedded computer design
- Process instrumentation and control
- Software design

Digital designer working on the read channel for hard disk drives. The read channel is the portion of the controller SOC (system on a chip) that decodes the analog signal read from the hard disk and converts it to digital data.

I am responsible for writing Verilog RTL code, verification, synthesis into gates, and meeting timing requirements of my blocks.

E. D'Avignon, CpE

As a Radiation Effects Engineer I test the performance of electronic components in a specific application exposed to different types of radiation. Responsibilities include interfacing with design and system engineers, creating test plans, performing testing and data analysis, and authoring test reports.

A. Passman, EE

I manage global programs that help develop leadership capabilities and skills of our current and future leaders. I am a consultant, a coach, a mentor, and a guide. If leaders are interested in improving how they lead and the impact they have on their employees and on company results, we work with them to identify the best ways for them to continue their development.

A. Hu, EE

Electrical Engineering (EE)

EE ranges from the generation and delivery of electrical power to the use of electricity in integrated circuits. The rapid development of technology, based on integrated circuit devices, has enabled the pervasive use of computers in command, control, communication, and computer-aided design. Some systems Electrical Engineers work on include:

- Communication system design
- Control systems—from aircraft to automotive
- Electrical power generation and distribution
- Electromagnetic waves
- Integrated circuit design
- Process instrumentation and control
- Robotic systems design
- Telecommunications

My group supports [a major automotive manufacturer] decisions pertaining to where to put new plants around the world, what products to build in them, and at what volumes.

In particular, my work involves understanding what the other auto manufacturers are planning for the future (footprint, capacity, technology, processes, etc.), so that information can be used to affect decisions about how to compete around the globe.

M. Peterson, EE

Industrial Engineering (IE)

IE deals with the design and improvement of systems, rather than with the objects and artifacts that other engineers design. A second aspect of industrial engineering is the involvement of people in these systems—from the people involved in the design and production to the people who are ultimate end users. Industrial engineers use mathematical, physical, social sciences and engineering, combined with the analytical and design methods to design, install, and improve complex systems that provide goods and services to our society. Industrial Engineers are involved in:

- Analyze and model complex work processes to find system improvements
- Analyze how combinations of people and machines work together
- Analyze how the surroundings affect the worker, and design to reduce the negative effects of this environment
- Develop mathematical and computer models of how systems operate and interact
- Improve production and service processes from the perspectives of quality, productivity, and cost
- Work on teams with other professionals in manufacturing, service industries, government agencies

My primary job responsibilities include maintaining, up-grading, and designing all the computer systems and IT infrastructure for the Vermont Railroad. I handle all the servers and take care of network equipment. When needed, I also program customized applications and websites for customers or our own internal use. I also serve as a spare conductor and locomotive engineer when business needs demand.

S. Houghton, CpE

As the Business Leader for Central Florida at a major power company, I develop and man-age a $42 million budget. I ensure that our engineering project schedule and budget match, and report on variances monthly. I also conduct internal audits and coach employees on Sarbanes-Oxley compliance requirements.

R. Holcomb, IE

Currently I am working on a project to determine patient priorities for evacuations from healthcare facilities during emergencies. The assumption of an evacuation is that there will be enough time to transfer all of the patients, but in the event of limited resources, there may not be enough time to move all of the patients to safety. Further—and depending on the emergency type—it may be an increased risk to transport some patient types. Based on certain objectives, we are developing guidelines to most ethically determine a schedule for choosing patients for emergency evacuations.

A. Childers, IE

Material Science Engineering (MSE)

MSE focuses on the manufacture and study of materials. Nature supplied only 92 naturally occurring elements to serve as building blocks to construct all modern conveniences. A Materials Engineer works to unlock the relationship between atomic, molecular, and larger-scale structure and the resultant properties. This category includes majors such as Ceramic Engineering or Metallurgical Engineering.

Here is a partial list of products designed and manufactured by MSE:

- Brick, tile, and whitewares research and manufacturing for the home and workplace
- Ceramic spark plugs, oxygen sensors, and catalytic converters that optimize engine performance
- Insulating tiles that keep the extreme heat of launch and reentry away from the space shuttle astronauts
- Metal and ceramic materials that enable biomedical implants and prosthetics
- Microwave ceramics that stabilize and filter cellular phone reception
- Superconducting ceramics that provide zero-loss transfer of electricity over great distances
- Ultra-pure glass optical fibers that carry telephone conversations and Internet communications

Consulting Metallurgical Engineer, duties include

- Consulting firm management/administration
- Failure analysis;
- Subcontracted metals testing services;
- Metallurgical quality systems design/auditing;
- Metallurgical expert in litigation cases; and
- Materials selection and design consultant, in-process and final inspection and testing services.

E. Basta,
Metallurgical Engr

Implement technology to protect national assets against adversaries.

J. Dabling, ME

Mechanical Engineering (ME)

ME involves areas related to machine design, manufacturing, energy production and control, materials, and transportation. Areas supported by Mechanical Engineers include:

- Construction
- Energy production and control
- Environmental systems
- Food production
- Management
- Materials processing
- Medicine
- Military service
- Propulsion and transportation systems
- Technical sales

In my job as a management consultant, I address CEO-level management decisions as part of a project team by helping clients identify, analyze and solve business-related problems. My responsibilities include generating hypotheses, gathering and analyzing data, conducting benchmarking and best practices assessments, recommending actions, and working with clients to develop implementation plans.

M. Cuica, ME

As Plant Engineering Manager I report directly to the Plant Manager. My primary responsibilities are managing all the capital investments; providing technical support and expertise to the plant leadership team; and mentoring and developing the plant's engineering staff and technical resources.

J. Huggins, ME

I am a salesman, so at the end of the day I'm looking to grow my market share while trying to protect the market share I already have. I help companies maintain a safe, reliable, and efficient steam and condensate system by utilizing the many products and services that we have to offer. This is mostly done by designing and installing upgrades and improving my customer's existing steam systems.

T. Burns, ME

I am an aerospace engineering manager responsible for developing unique astronaut tools and spacewalk procedures and for testing and training for NASA's Hubble Space Telescope servicing missions. My job ranges from tool and procedure design and development to underwater scuba testing to real-time, on-console support of Space Shuttle missions.

R. Werneth, ME

WISE WORDS: WHAT DID YOU DO YESTERDAY MORNING AT WORK?

I worked on completing a failure analysis report for an industrial client.

E. Basta, CME

I reviewed the results of the overnight simulation runs. There were several failures, so I analyzed the failures and devised fixes for the problems.

E. D'Avignon, CpE

On any given day, my morning might be spent this way: in meetings, at my computer (e-mail, drafting documents/reports), making phone calls, talking to other project members, running a test on the manufacturing lines. Not glamorous, but necessary to solve problems and keep the project moving forward.

S. Forkner, ChE

I continued to design a 150′ x 120′ warehouse/office building on a nuclear expansion project.

T. Hill, CE

Yesterday morning I started the day by checking and responding to e-mail. I then designed a spreadsheet to assist in more precisely forecasting monthly expenditures.

R. Holcomb, IE

I attended the Plant Morning Meeting and the Boardmill Leadership Team Meeting, following by the Plant Budget Meeting. In between meetings I returned e-mails and project-related phone calls. Typically I spend about 50% of my time in meetings. I use the information I receive at these meetings to direct and focus the efforts of the engineering staff.

J. Huggins, ME

Testing some failed prototype biostimulators returned by a trial user, to determine why and how they failed and how to prevent it from happening in coming production versions.

J. Kronberg, EE

Yesterday, I worked on a patent infringement opinion involving agricultural seeding implements, a Chinese patent office response for a component placement and inspection machine used in circuit board manufacturing, and a U.S. patent office response for database navigation software.

M. Lauer, Enviro E

In my current position, I spend much of my time reading technical manuals and interface control documents. I attended a meeting detailing lightning protection for the Ares rocket.

E. Styles, EE

1.4 | GATHERING INFORMATION

You will need to gather a lot of information in order to answer your questions about engineering or any other major. Many resources are available on your campus and through the Internet.

The Career Center

Most universities have a centralized campus Career Center. The staff specializes in helping students explore various occupations and make decisions. They offer testing and up-to-date information on many career fields. Professional counselors are available by appointment to assist students with job and major selection decisions.

Career Websites

To learn more about engineering and the various engineering fields, you can find a wealth of information from engineering professional societies. Each engineering field has a professional society dedicated to promoting and disseminating knowledge about that particular discipline. Table 1-2 provides a list of most major engineering fields and the professional society in the United States with which it is associated. In some cases, more than one society is connected with different subdisciplines. Other regions of the world may have their own professional societies.

Perusing the various societies' websites can provide you with information invaluable in helping you decide on a future career. We have not given URLs for the societies, since these sometimes change. To find the current address, simply use an online search engine with the name of the society.

In addition, a few engineering societies are not specific to a discipline, but to their membership.

- National Society of Black Engineers (NSBE)
- Society of Hispanic Professional Engineers (SHPE)
- Society of Women Engineers (SWE)
- Tau Beta Pi, The Engineering Honor Society (TBP)

Most engineering schools have student chapters of the relevant organizations on campus. These organizations provide an excellent opportunity for you to learn more about your chosen discipline or the ones you are considering, and they also help you meet other students with similar interests. Student membership fees are usually nominal, and the benefits of membership far outweigh the small cost.

Active participation in these societies while in school not only gives you valuable information and experience but also helps you begin networking with professionals in your field and enhances your résumé.

Table 1-2 Website research starting points

Society	Abbreviation
American Ceramic Society	ACerS
Association for Computing Machinery	ACM
Audio Engineering Society	AES
American Institute of Aeronautics and Astronautics	AIAA
American Institute of Chemical Engineers	AIChE
American Nuclear Society	ANS
American Society of Agricultural and Biological Engineers	ASABE
American Society of Civil Engineers	ASCE
American Society for Engineering Education	ASEE
American Society of Metals International	ASM Int'l.
American Society of Mechanical Engineers	ASME
Biomedical Engineering Society	BMES
Engineers Without Borders	EWB
Institute of Biological Engineering	IBE
Institute of Electrical and Electronics Engineers	IEEE
Institute of Industrial Engineers	IIE
Institute of Transportation Engineers	ITE
Materials Research Society	MRS
National Academy of Engineers	NAE
National Society of Black Engineers	NSBE
National Society of Professional Engineers	NSPE
Society of Automotive Engineers International	SAE Int'l.
Society of Hispanic Professional Engineers	SHPE
Society of Petroleum Engineers	SPE International
Society of Plastics Engineers	SPE
Society of Women Engineers	SWE
Tau Beta Pi, The Engineering Honor Society	TBP

WISE WORDS: ADVICE ABOUT SOCIETY PARTICIPATION

Get involved!!!! It is so much fun! Plus, you're going to meet a ton of cool people doing it!

T. Andrews, CE

While academics should remain your first priority throughout your college career, getting involved with a student organization is very beneficial. I joined the American Society of Civil Engineers as a sophomore. This group afforded me the opportunity to learn about current industry practices. It also created opportunities for me to network with other engineers throughout the field. This networking opportunity has proven valuable in searching for jobs.

M. Dukes, CE

My advice to students willing to participate on student activities is for them to not be shy when going to a student organization for the first couple of times. It takes time to get well known and feel comfortable around new people, but don't let that prevent you from being part of a student organization that can bring many benefits to you. Always have a positive attitude, be humble, and learn to listen to others; these last advices are traits which you will use in your professional life.

V. Gallas Cervo, ME

My advice to first-year students is to not get involved with too many organizations all at once. It is easy to get distracted from your class work with all the activities on campus. Focus on a couple and be a dedicated officer in one of them. This way you have something to talk about when employers see it on your resume. I would recommend that you are involved with one organization that you enjoy as a hobby and one organization that is a professional organization.

D. Jones, BE

The most important thing to do when joining any group is to make sure you like the people in it. This is probably even more important than anything the group even does. Also make sure that if the group you're joining has a lot of events they expect you to be at, you have the time to be at those events.

R. Kriener, EE

Clubs might have been dorky in high school, but they are different in college. They will advance you socially, academically, and professionally.

D. Lanigan, IE

Make the time to participate in student activities. If possible, try to get a leadership position in one of the activities because it will be useful in interviews to talk about your involvement. While employers and grad schools may not be impressed with how you attended meetings occasionally as a general member, they will be interested to hear about the projects that you worked on and the challenges that you faced in a leadership position.

K. Smith, ChE

Get involved! College is about more than just academics. Participating in student activities is a lot of fun and makes your college experience more memorable. I've made so many friends not just at my school, but all over the country by getting involved. It's also a great way to develop leadership and interpersonal skills that will become beneficial in any career.

A. Zollinger, CE

1.5 PURSUING STUDENT OPPORTUNITIES

In addition to the traditional educational experience, many students seek experience outside of the classroom. Many engineering colleges and universities have special departments that help place students in programs to gain real engineering work experience or provide them with a culturally rich study environment. Ask a professor or advisor if your university provides experiences similar to those described in this section.

Cooperative Education

Contributed by: Dr. Neil Burton, Director of
Cooperative Education Program, Clemson University

People learn things in a lot of different ways. Some people learn best by reading; others by hearing someone else speak; and still others by participating in a group discussion. One very effective form of learning is called **experiential learning**, also referred to as engaged learning in some places. As the name suggests, experiential learning means learning through experience, and there's a very good chance you used this method to learn how to ride a bike, bake a cake, change a flat tire, or perform any other complex process that took some practice to perfect. The basic assumption behind experiential learning is that you learn more by doing than by simply listening or watching.

Becoming a good engineer is a pretty challenging process, so it seems only natural that experiential learning would be especially useful to an engineering student. In 1906 Herman Schneider, the Dean of Engineering at the University of Cincinnati, developed an experiential learning program for engineering students because he felt students would understand the material in their engineering classes much better if they had a chance to put that classroom knowledge into practice in the workplace. Schneider called this program **Cooperative Education**, and more than 100 years later, colleges and universities all over the world offer cooperative education (co-op) assignments to students in just about every major, although engineering students remain the primary focus of most co-op programs.

There are many different kinds of co-op programs, but all of them offer engineering students the chance to tackle real-world projects with the help and guidance of experienced engineers. One common model of cooperative education allows students to alternate semesters of co-op with semesters of school. In this model, students who accept co-op assignments spend a semester working full-time with a company, return to school for the following academic term, go back out for a second co-op **rotation**, return to school the following term, and continue this pattern until they have spent enough time on assignment to complete the co-op program.

On my co-op, I worked on the hazard analysis for the new Ares I launch vehicle that NASA is designing to replace the Space Shuttles when they retire in 2010.

Basically, we looked at the design and asked: what happens if this part breaks, how likely is it that this part will break, and how can we either make it less likely for the part to break or give it fault tolerance so the system can withstand a failure.

J. Sandel, on co-op

> The value of the cooperative education was to apply the classroom material to real-world applications, develop an understanding of the expectations post-grad, and provide the opportunity for a trial run for a future career path in a low-risk environment.
>
> *M. Cucia, ME*

> I wanted some practical experience, and I wasn't exactly sure what career I wanted pursue when I graduated. My experience at a co-op set me on a completely different career path than I had been on previously.
>
> *K. Smith, ChE*

Students learn a lot about engineering during their co-op assignments, but there are many other benefits as well. Engineering is a tough discipline, and a co-op assignment can often help a student determine if he or she is in the right major. It is a lot better to figure out that you don't want to be an engineer before you have to take Thermodynamics or Heat Transfer! Students who participate in co-op also have a chance to develop some great professional contacts, and these contacts are very handy when it comes time to find a permanent job at graduation. The experience students receive while on a co-op assignment is also highly valued by employers who want to know if a student can handle the challenges and responsibilities of a certain position. In fact, many students receive full-time job offers from their co-op employers upon graduation.

Perhaps the most important benefit a co-op assignment can provide is improved performance in the classroom. By putting into practice the theories you learn about in class, you gain a much better understanding of those theories. You may also see something on your co-op assignment that you will cover in class the following semester, putting you a step ahead of everyone else in the class. You will also develop time management skills while on a co-op assignment, and these skills should help you complete your school assignments more efficiently and effectively when you return from your assignment.

Companies who employ engineers often have cooperative education programs because co-op provides a number of benefits to employers as well as students. While the money companies pay co-op students may be double or even triple what those students would earn from a typical summer job, it is still much less than companies would pay full-time engineers to perform similar work. Many employers also view cooperative education as a recruiting tool—what better way to identify really good employees than to bring aboard promising students and see how they perform on co-op assignments!

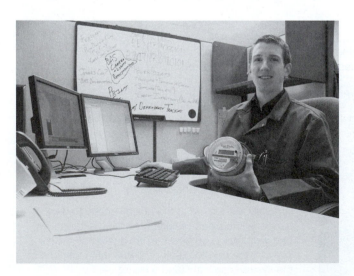

I did my research, and it really made sense to pursue a co-op—you get to apply the skills you learn in class, which allows you to retain the information much better, as well as gain an increased understanding of the material.

As for choosing a co-op over an internship, working for a single company for an extended period of time allows students to learn the ropes and then progress to more intellectually challenging projects later in the co-op. And, if you really put forth your best effort for the duration of your co-op, you could very well end up with a job offer before you graduate!

R. Izard, on co-op

WISE WORDS: WHAT DID YOU GAIN FROM YOUR CO-OP OR INTERNSHIP EXPERIENCE?

I was able to learn how to practically apply the knowledge I was gaining from college. Also, the pay allowed me to fund my schooling.

B. Dieringer, ME

The best part about that experience was how well it meshed with my courses at the time. My ability to apply what I was learning in school every day as well as to take skills and techniques I was learning from experienced engineers and use them toward the projects I was working on in school was invaluable.

A. Flowerday, CpE

My internships were a great introduction to the professional work place—the skills and responsibilities that are expected; the relationships and networks that are needed.

S. Forkner, ChE

I decided to pursue a co-op because I had trouble adapting to the school environment in my first years and taking some time off to work at a company seemed a good way to rethink and reorganize myself I made a good decision taking some time off, since it allowed me to learn a lot more about myself, how I work, how I learn and how I operate. I learned that the biggest challenges where only in my mind and believing in me was, and still is, the hardest thing.

V. Gallas Cervo, ME

My internship taught me that I could be an engineer—and a good one too. I had a lot of self-doubt before that experience, and I learned that I was better than I thought I was.

B. Holloway, ME

Being able to immediately apply the things I learned in school to real-world applications helped reinforce a lot of the concepts and theories. It also resulted in two job offers after graduation, one of which I accepted.

J. Huggins, ME

This was an amazing opportunity to get a real taste for what I was going to be doing once I graduated. I began to realize all the different types of jobs I could have when I graduated, all working in the same field. In addition, my work experience made my resume look 100 times more appealing to potential employers. The experience proved that I could be a team player and that I could hit the ground running without excessive training.

L. Johnson, CE

It gave me a chance to see how people work together in the "real world" so that I could learn how to interact with other people with confidence, and also so that I could learn what kind of worker or manager I wanted to be when I "grew up."

M. Peterson, ME

Without a doubt, the best professional decision of my life. After my co-op rotation finished, I approached school as more like a job. Furthermore, co-oping makes school easier! Imagine approaching something in class that you have already seen at work!

A. Thompson, EE

Internship

*Contributed by: Mr. Troy Nunamaker, Director of Internships
and Educational Career Services, Clemson University*

Internships offer the unique opportunity to gain career-related experience in a variety of settings. Now, more than ever, employers look to hire college graduates with internship experience in their field.

Why Choose an Internship?

- Bridge classroom applications to the professional world
- Build a better resume
- Possibly receive higher full-time salary offers upon graduation
- Gain experience and exposure to an occupation or industry
- Network and increase marketability
- Potentially fulfill academic requirements and earn money

Employers indicate that good grades and participation in student activities are not enough to help students land a good, full-time job. In today's competitive job market, the students with career-related work experience are the students getting the best interviews and job offers. As an added bonus, many companies report that over 70 percent of full-time hires come directly from their internship program.

Searching for an Internship

Although a number of students will engage in an internship experience during their freshman and sophomore years, most students pursue an internship during their junior and senior years. Some students might participate in more than one internship during their college career. Allow plenty of time for the search process to take place and be sure to keep good records of all your applications and correspondences.

- **Figure out what you are looking for.** You should not start looking for an internship before you have answered the following questions:
  - What are my interests, abilities, and values?
  - What type of organization or work environment am I looking for?
  - Are there any geographical constraints, or am I willing to travel anywhere?

- **Start researching internship opportunities.** Start looking one to two semesters before your desired start date. You should utilize as many resources as possible in order to have the broadest range of options.
  - Visit your campus's career center office to do the following: meet with a career counselor; attend a workshop on internships; find out what positions and

I wanted to be able to apply what I had learned in classes to solve real engineering problems. It is also good when searching for a permanent job after graduation. It lets potential employers know you've gotten a taste of what it's like to hold a real engineering job. I didn't want to take longer than four years to graduate, so I interned over the summer.

I performed an outfall inventory for Anderson County, South Carolina, stormwater infrastructure by locating and attributing outfall points, using a Mobile Matrix GPS device.

I also did some work using AutoCAD to create drafts for retention pond outlet designs and stream restoration projects.

S. Luell, on internship

resources are available; and look for internships postings through the career center's recruiting system and website.

- Attend a career fair on your campus or in your area. Also, if there are specific companies where you would like to work, contact them directly and find out if they offer internships.
- Network. Network. Network. Only about a quarter of internship opportunities are actually posted. Talk to friends, family, and professors and let them know that you are interested in an internship.

■ **Narrow down the results and apply for internships.** Look for resources on your campus to help with developing a résumé and cover letter that are applicable to internships. *Each résumé and cover letter should then be tailored for specific applications.* As part of the application process, do not be surprised if a company requests additional documents such as references, transcripts, writing samples, and formal application packets.

■ **Wait for responses.** It may take up to a month to receive any responses to your applications. One to two weeks after you have submitted your application, call the organization to make sure they have all the required documents.

■ **Interview for positions.** Once you have your interviews scheduled, stop by the career center to see what resources they have available to help you prepare for the interview. Do not forget to send a thank-you note within 24 hours of the interview, restating your interest in the position.

Accepting an Internship

Once you have secured an internship, look to see if academic internship coursework is available on your campus so that the experience shows up on your transcripts. If you were rejected from any organizations, take it as a learning experience and try to determine what might have made your application stronger.

POINTS TO PONDER

Am I eligible for an internship?
Most companies look to hire rising juniors and seniors, but a rising sophomore or even a freshman with relevant experience and good grades is also a strong candidate.

Will I be paid for my internship?
The pay rate will depend on your experience, position, and the individual company. However, most engineering interns receive competitive compensation; averages are $14–$20 per hour.

When should I complete an internship?
Contrary to some popular myths, an internship can be completed not only during summers, but also during fall and spring semesters. Be sure to check with your campus on how to maintain your student enrollment status while interning.

Will I be provided housing for an internship?
Do not let the location of a company deter you. Some employers will provide housing, while others will help connect you with resources and fellow interns to find an apartment in the area.

I think that first year-students in engineering should do a co-op or internship. It was extremely valuable to my education. Now that I am back in the classroom, I know what to focus on and why what I am learning is important.

Before the experience I did not know what I wanted to do with my major, and I didn't fully understand word problems that were presented in a manner that applied to manufacturing or real life. Being in industry and working in a number of different departments, I figured out that I liked one area more than any other area, and that is where I am focusing my emphasis area studies during my senior year.

K. Glast, IE

WISE WORDS: DESCRIBE A PROJECT YOU WORKED ON DURING YOUR CO-OP OR INTERNSHIP

The large project I worked on was an upgrade of an insulin production facility. A small project I worked on for two weeks was the design of a pressure relief valve for a heat exchanger in the plant.

D. Jones, BE

I have been working on a series of projects, all designed to make the production of electric power meters more efficient. I am rewriting all the machine vision programs to make the process more efficient and to provide a more sophisticated graphical user interface for the operators. These projects have challenged me by requiring that I master a new "machine vision" programming language, as well as think in terms of efficiency rather than simply getting the job done.

R. Izard, ME

I designed and prototyped removable filter covers as well as filter medium for a pressure switch installed on the rear differential of most "work trucks." The design was very successful and was being finalized for production as I left the company.

M. Morgan, ME

I worked on several LEAN Manufacturing projects. Lean Manufacturing is a systematic approach to identify and eliminate waste (extra time, material waste) through continual improvements. It seeks to produce higher-quality products with less work. We worked on these projects for several different manufacturing processes. I even was certified as a Lean Practitioner by the company.

D. Lanigan, IE

The site I worked at designs and manufactures radar systems (among others). During my internship I wrote C code that tests the computer systems in a certain radar model. The code will eventually be run by an operator on the production floor before the new radars are sent out to customers.

D. Rollend, EE

One of my last projects I worked on during my first term was building a new encoder generator box used to test the generator encoder on a wind turbine. The goal was to make a sturdier box that was organized inside so that if something had broken, somebody who has no electrical skills could fix it. I enjoyed this project because it allowed me to use my skills I have learned both from school and my last internship.

C. Balon, EE

Study Abroad

Contributed by: Ms. Jessica R. Pelfrey, Program Coordinator for
International Programs, College of Engineering and Science, Clemson University

In today's global economy, it is important for engineering students to recognize the importance of studying abroad. A few reasons to study abroad . . .

- Taking undergraduate courses abroad is an exciting way to set your résumé apart from those of your peers. Prospective employers will generally inquire about your international experiences during the interview process, giving you the chance to make a lasting impression that could be beneficial.
- Studying abroad will give you a deeper, more meaningful understanding of a different culture. These types of learning experiences are not created in traditional classrooms in the United States, and cannot be duplicated by traveling abroad on vacation.
- Students who study abroad generally experience milestones in their personal development, as a result of stepping outside of their comfort zone.
- There is no better time to study abroad than now! Students often think that they will have the opportunity to spend significant amounts of time traveling the world after graduation. In reality, entering the workforce typically becomes top priority.
- Large engineering companies tend to operate on a global scale. For instance, a company's headquarters may be in the United States, but that company may also control factories in Sweden, and have parts shipped to them from Taiwan. Having an international experience under your belt will give you a competitive edge in your career because you will have global knowledge that your co-workers may not possess.

While programs differ between universities, many offer a variety of choices:

- **Exchange Programs:** Several institutions are part of the Global Engineering Education Exchange, or GE[3]. This consortium connects students from top engineering schools in the United States with foreign institutions in any of 18 countries. In an ideal situation, the number of international students on exchange in the United States would be equal to the number of United States students studying abroad during the same time frame.

  Generally, students participating in these types of exchange programs will continue to pay tuition at their home institution; however, you should check your university's website for more information. Also, by entering "Global E3" into a web browser, you will be able to access the website to determine if your institution currently participates, which foreign institutions offer courses taught in English, and which schools offer courses applicable to your particular major. If your institution is not a member of GE[3], consult with someone in the international, or study abroad, office at your institution to find out about other exchange opportunities.

- **Faculty-Led Programs:** It is not unusual for faculty members to connect with institutions or other professionals abroad to establish discipline-specific, study-abroad programs. These programs typically allow students to enroll in classes during the summer at their home institution, then travel abroad to complete the coursework. Faculty-led programs offer organized travel, lodging, administration, and excursions, making the overall experience hassle-free. Consult with professors in your department to find out if they are aware of any programs are already in place at your institution.

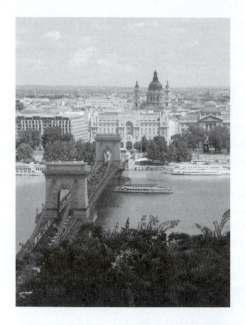

I studied abroad twice. The first time I spent a semester in The Netherlands, experiencing a full immersion in the Dutch culture and exploring my own heritage, and the second time I spent a summer in Austria taking one of my core chemical engineering classes. Both countries are beautiful and unique places and they will always hold a special place in my heart.

Through studying abroad, I was able to expand my own comfort zone by encountering novel situations and become a more confident individual. Although the experiences were amazing and the memories are truly priceless, the biggest thing I gained from studying abroad was that I was able to abandon many perceptions about other cultures and embrace new perspectives.

R. Lanoie, studying abroad in Netherlands and Austria

- **Third-Party Providers:** Many universities screen and recommend providers of programs for students. If there's a place you want to go and your university does not have an established program at that particular location, you will probably find a study program of interest to you by discussing these options with the study-abroad office at your university.

- **Direct Enrollment:** If you are interested in a particular overseas institution with which your home university does not have an established program, there is always the option of direct enrollment. This process is basically the same as applying to a university in the United States. The school will likely require an admissions application, a purpose statement, transcripts, and a letter of reference. Be cautious if you choose this route for study abroad, as it can be difficult to get credits transferred back to your home institution. It is a good idea to get your international courses

I think study abroad is a wonderful learning experience everyone should have. In going abroad you get to meet people you would never meet otherwise, and experience things you never thought you would see.

I learned about independently navigating a new country, as well as adjusting to foreign ways of doing things. It is interesting to take classes that you would have taken at home and experience them a completely different way.

Studying abroad teaches you how to easily adjust to new situations and allows you to learn about cultures you had no idea of before.

B. Brubaker, studying abroad in Scotland

preapproved prior to your departure. On a positive note, it can be cheaper to directly enroll in a foreign institution than it is to attend your home university. This option is best suited for students that want to go abroad for a semester or full year.

Typically, students go abroad during their junior year, though recently there has been an increase in the number of second-year student participation. The timing of your experience should be agreed upon by you and your academic advisor. Each program offers different international incentives, and some are geared for better opportunities later in your academic career, whereas others may be better toward the beginning.

If you would like to go abroad for a semester or year but do not feel entirely comfortable with the idea, why not get your feet wet by enrolling in a summer program first? This will give you a better idea of what it is that you are looking for without overwhelming you. Then, you can plan for a semester or year abroad later in your academic career. In fact, more students are electing to spend a semester or year of study abroad, and increasingly more opt for an academic semester plus a semester internship combination.

There are three basic principles to follow when deciding on a location:

- **Personal Preference:** Some people are more interested in Asia than Europe, or maybe Australia instead of Latin America. There are excellent opportunities worldwide, regardless of the location.

- **Program Opportunities:** Certain countries may be stronger or have more options in certain fields. For example, Germany is well known for innovations in mechanical engineering, while the Japanese tend to be more widely recognized for their efforts in computer engineering. Listen to your professors and weigh your options.

- **Language:** You may not feel like you are ready to study in a foreign country, speaking and reading in a foreign language. There are numerous institutions that offer courses taught in English. However, if you have taken at least two years of the same foreign language, you should be knowledgeable enough to succeed in courses taught in that language. Don't let your fears restrict you!

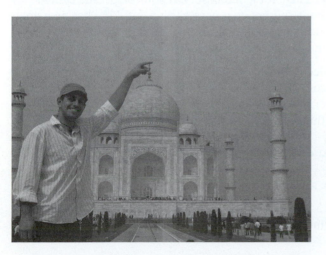

India is quite an eye-opening country. All of your senses work on overload just so you can take everything in at once. My time studying in India was simply phenomenal. I had an excellent opportunity placed before me to travel to India to earn credit for Electrical Engineering courses and I would have been a fool to pass up that chance. I regret nothing, I would do it again, and I would urge everyone with the slightest inkling of studying abroad to put their worries aside and have the experience of a lifetime.

J. Sadie, studying abroad in India

CHAPTER 2
ETHICS

Every day, we make numerous ethical decisions, although most are so minor that we do not even view them as such.

- When you drive your car, do you knowingly violate the posted speed limit?
- When you unload the supermarket cart at your car, do you leave it in the middle of the parking lot, or spend the extra time to return it to the cart corral?
- You know that another student has plagiarized an assignment; do you rat him or her out?
- A person with a mental disability tries to converse with you while waiting in a public queue. Do you treat him with respect or pretend he doesn't exist?
- In the grocery, a teenager's mother tells her to put back the package of ice cream she brought to the cart. The teenager walks around the corner and places the ice cream on the shelf with the soft drinks and returns to the buggy. Do you ignore this, approach the teenager and politely explain that leaving a package of ice cream in that location will cause it to melt thus increasing the cost of groceries for everyone else, or do you replace it in the freezer yourself?
- When going through a public door, do you make a habit of looking back to see if releasing the door will cause it to slam in someone's face?
- You notice a highway patrolman lying in wait for speeders. Do you flash your lights at other cars to warn them?
- A cashier gives you too much change for a purchase. Do you correct the cashier?
- You are on the lake in your boat and notice a person on a JetSki® chasing a great blue heron across the lake. The skier stops at a nearby pier. Do you pilot your craft over to the dock and reprimand him for harassing the wildlife?

On a grand scale, none of these decisions is particularly important, although some might lead to undesirable consequences. However, as an aspiring engineer, you may face numerous decisions in your career that could affect the lives and well-being of thousands of people. Just like almost everything else, practice makes perfect, or at least better. The more you practice analyzing day-to-day decisions from an ethical standpoint, the easier it will be for you to make good decisions when the results of a poor choice may be catastrophic.

In very general terms, there are two reasons people try to make ethical decisions.

- They wish to make the world a better place for everyone—in a single word, altruism.
- They wish to avoid unpleasant consequences, such as fines, incarceration, or loss of job.

In an ideal society, the second reason would not exist. However, history is replete with examples of people, and even nations, who do not base their decisions solely on whether or not they are acting ethically. Because of the common occurrence of unethical behavior

> Good people do not need laws to tell them to act responsibly, while bad people will find a way around the laws.
>
> *Plato*

and the negative impact it has on others, almost all societies have developed rules, codes, and laws to specify what is and is not acceptable behavior, and the punishments that will be meted out when violations occur.

The major religions all have fairly brief codes summarizing how one should conduct their life. Some examples are given below; other examples exist as well.

- Judaism, Christianity, and derivatives thereof (e.g., Latter Day Saints) have the Decalogue, or Ten Commandments.
- Islam has the Five Pillars in addition to a slightly modified and reorganized form of the Decalogue.
- Buddhism has the Noble Eightfold Path.
- Bahá'í has twelve Social Principles.
- In Hinduism, Grihastha dharma has four Goals.

Secular codes of conduct go back more than four millennia to the Code of Ur-Nammu. Although by today's standards, some of the punishments in the earliest codes seem harsh or even barbaric, it was one of the earliest known attempts to codify crimes and corresponding punishments.

Admittedly, although not specifically religious in nature, these codes are usually firmly rooted in the prevailing religious thought of the time and location. Through the centuries, such codes and laws have been expanded, modified, and refined so that most forms of serious antisocial behavior are addressed and consequences for violations specified. These codes exist from a local to a global level. Several examples are given below.

- Most countries purport to abide by the Geneva Conventions, which govern certain types of conduct on an international scale.
- Most countries have national laws concerning murder, rape, theft, etc.
- In the United States, it is illegal to purchase alcohol unless you are 21 years of age. In England, the legal age is 18.
- In North and South Dakota, you can obtain a driver's license at age 14½. In most other states, the legal age is 16.
- It is illegal to say "Oh boy!" in Jonesboro, Georgia.
- Some cities, such as Santa Fe, New Mexico, have ordinances prohibiting use of cell phones while driving.

EXCERPTS FROM THE CODE OF UR-NAMMU

- If a man commits a murder, he must be killed.
- If a man commits a robbery, he will be killed.
- If a man commits a kidnapping, he is to be imprisoned and pay 15 shekels of silver.
- If a man knocks out the eye of another man, he shall weigh out ½ a mina of silver.
- If a man has cut off another man's foot, he is to pay ten shekels.
- If someone severed the nose of another man with a copper knife, he must pay two-thirds of a mina of silver.
- If a man knocks out a tooth of another man, he shall pay two shekels of silver.

2.1   ETHICAL DECISION-MAKING

Some ethical decisions are clear-cut. For example, essentially everyone (excluding psychopaths) would agree that it is unethical to kill someone because you do not like her hat. Unfortunately, many real-world decisions that we must make are far from "black and white" issues, instead having many subtle nuances that must be considered to arrive at what one believes is the "best" decision.

There is no proven algorithm or set of rules that one can follow to guarantee that the most ethical decision possible is being made in any particular situation. However, numerous people have developed procedures that can guide us in considering questions with ethical ramifications. A four-step procedure is discussed here, although there are various other approaches.

<table>
<tr><td>

Step 1: Determine *What* the issues are and *Who* might be affected by the various alternative courses of action that might be implemented.

</td></tr>
</table>

Step 1: Determine *What* the issues are and *Who* might be affected by the various alternative courses of action that might be implemented.

We will refer to the *Who* as **stakeholders**. Note that at this point we are not trying to determine how the stakeholders will be affected by any particular plan of action.

- The issues (What) can refer to a wide variety of things, including, for example, personal freedom, national security, quality of life, economic issues, fairness, and equality.
- The term stakeholders (Who) does not necessarily refer to people, but might be an individual, a group of people, an institution, or a natural system, among other things.

EXAMPLE 2-1

Consider the question of whether to allow further drilling for oil in the Alaska National Wildlife Refuge (ANWR). List several issues and stakeholders.

Issues:

- *Oil independence*
- *The price of gasoline*
- *Possible impacts on the ecosystem*

Stakeholders:

- *Oil companies*
- *The general population of the United States*
- *Other countries from whom we purchase oil*
- *The flora and fauna in ANWR*
- *The native people in Alaska*

<table>
<tr><td>

Step 2: Consider the effects of alternative courses of action from different perspectives:

- Consequences
- Intent
- Character

</td></tr>
</table>

Step 2: Consider the effects of alternative courses of action from different perspectives.

Here, we look at three perspectives.

Perspective 1: Consequences

When considering this perspective, ask how the various stakeholders will be affected by each alternative plan being contemplated. In addition, attempt to assign a relative level of importance (weight) to each effect on each stakeholder. For instance, an action that might affect millions of people adversely is almost always more important than an action that would cause an equivalent level of harm to a dozen people.

EXAMPLE 2-2

Should all United States children be fingerprinted when entering kindergarten and again each third year of grade school (3, 6, 9, 12)? Identify the stakeholders and consequences.

Stakeholders:

- *All United States children*
- *All United States citizens*
- *Law enforcement*
- *The judicial system*
- *The United States constitution*

Consequences:

- *Provides a record to help identify or trace missing children (not common, but possibly very important in some cases).*
- *Affords an opportunity for malicious use of the fingerprint records for false accusation of crime or for identity theft (probability unknown, but potentially devastating to affected individuals).*
- *Could help identify perpetrators of crimes, thus improving the safety of law-abiding citizens (importance varies with type of crime).*
- *Raises serious questions concerning personal freedoms, possibly unconstitutional (importance, as well as constitutionality, largely dependent on the philosophy of the person doing the analysis).*

This list could easily be continued.

Fingerprint technology has advanced in recent years with the implementation of computer recognition for identification. Originally in the United States, the Henry Classification System was used to manually match fingerprints based on three main patterns: arches, loops, and whorls (shown below from left to right).

Today, the Automated Fingerprint Identification System (AFIS) uses algorithmic matching to compare images. Future work of AFIS systems is in the adoption and creation of secure multitouch devices like mobile computers and tablets that can identify different security levels for the operator of the device. For example, a multitouch computer owner might be able to issue permissions to an administrator that might not be available to a five-year-old, all without providing a single password!

Perspective 2: Intent

The intentions of the person doing the acting or deciding are considered in this perspective, sometimes called the "rights" perspective. Since actions based on good intentions can sometimes yield bad results, and vice versa, the intent perspective avoids this possible pitfall by not considering the outcome at all, only the intentions.

It may be helpful when considering this perspective to recall Immanuel Kant's Categorical Imperative: "Act only according to that maxim whereby you can at the same time will that it should become a universal law." To pull this out of the eighteenth century, ask yourself these questions:

1. Is the action I am taking something that I believe everyone should do?
2. Do I believe that this sort of behavior should be codified in law?
3. Would I like to be on the receiving end (the victim) of this action?

EXAMPLE 2-3

Should you download music illegally over the Internet?

Rephrasing this question using the suggestions above yields:

(a) Should everyone illegally download the music they want if it is there for the taking?

(b) Should the laws be changed so that anyone who obtains a song by any means can post it on the web for everyone to get for free?

(c) If you were a struggling musician trying to pay the bills, would you like your revenue stream to dry up because everyone who wanted your music got it for free?

Character: the inherent complex of attributes that determines a persons moral and ethical actions and reactions.

Perspective 3: Character

This perspective considers the character of a person who takes the action under consideration. There are different ways of thinking about this. One is to simply ask: Would a person of good character do this? Another is to ask: IF I do this, does it enhance or degrade my character? Yet another way is to ask yourself if a person you revere as a person of unimpeachable character (whoever that might be) would take this action.

EXAMPLE 2-4

Your friends are deriding another student behind her back because she comes from a poor family and does not have nice clothes.

Do you:

(a) join in the criticism?

(b) ignore it, pretend it is not happening, or simply walk away?

(c) tell your friends that they are behaving badly and insist that they desist?

- *Which of these actions would a person of good character take?*
- *Which of these actions would enhance your character and which would damage it?*
- *What would the founder of your religion do? (Moses or Jesus or Buddha or Mohammed or Bahá'u'lláh or Vishnu or whoever.) If you are not religious, what would the person who, in your opinion, has the highest moral character do?*

Step 3: Correlate perspectives.

Step 3: Correlate perspectives

Now look back at the results of considering the issue from the three perspectives. In many cases, all three perspectives will lead to the same or a similar conclusion. When this occurs, you have a high level of confidence that the indicated action is the best choice from an ethical standpoint.

If the three perspectives do not agree, you may wish to reconsider the question. It may be helpful to discuss the issue with people whom you have not previously consulted in this matter. Did you omit any factors? For complicated issues, it is difficult to make sure you have included all possible stakeholders and consequences. Did you properly assign weights to the various aspects? Upon reconsideration, all three perspectives may converge.

If you cannot obtain convergence of all three perspectives, no matter how hard you try to make sure you left nothing out, then go with two out of three.

Step 4: Act—Do *I* have the courage to do what I know is right?

Step 4: Act

This is often the hardest step of all to take, since ethical action often requires courage. The whistle-blower who risks losing her job, the young man confronting the tanks in Tiananmen Square, the elected official standing up for what she knows to be right when it will probably cost her the next election, or even something as mundane as risking the ridicule of your friends because you refuse to go along with whatever questionable activities they are engaging in for "fun." Ask yourself the question: "Do *I* have the courage to do what I know is right?"

EXAMPLE 2-5

Your company has been granted a contract to develop the next generation of electronic cigarette, also known as a "nicotine delivery system," and you have been assigned to the design team. Can you in good conscience contribute your expertise to this project?

Step 1: *Identify the issues (What) and the stakeholders (Who).*

Issues:

- *Nicotine is poisonous and addictive.*
- *These devices eliminate many of the harmful components of tobacco smoke.*
- *Laws concerning these devices range from completely legal, to classification as a medical device, to banned, depending on country.*
- *There are claims that such devices can help wean tobacco addicts off nicotine.*
- *The World Health Organization does not consider this an effective means to stop smoking.*
- *Whether an individual chooses to use nicotine should be a personal decision, since its use does not generally degrade a person's function in society.*
- *The carrier of the nicotine (80%–90% of the total inhaled product) is propylene glycol, which is relatively safe, but can cause skin and eye irritation, as well as other adverse effects in doses much larger than would be obtained from this device.*
- *A profit can be made from nicotine products or anti-smoking devices.*

Stakeholders:

- *You (your job and promotions)*
- *Your company and stockholders (profit)*
- *Cigarette manufacturers and their employees and stockholders (lost revenue)*
- *Tobacco farmers (less demand)*
- *The public (less second-hand smoke)*
- *The user (various health effects, possibly positive or negative)*

Step 2: *Analyze alternative courses of action from different perspectives.*

1. *Consequences*

   - *You may lose your job or promotion if you refuse.*
   - *If you convince management to abandon the project, the company may lose money.*
   - *If you succeed brilliantly, your company may make money hand over fist, and you receive a promotion.*
   - *If the project goes ahead, the possibility of future lawsuits exists.*
   - *Users' health may be damaged.*
   - *Users' dependence on nicotine may either increase or decrease.*

NOTE

In the interest of brevity, this is not an exhaustive analysis but shows the general procedure.

2. *Intent*

- *Should everyone use electronic cigarettes, or at least condone their use?*
- *Should use of electronic cigarettes be unrestricted by law?*
- *Would I like to risk nicotine addiction because of using these devices?*
- *Would I be able to kick my tobacco habit by using these devices?*

3. *Character*

- *Would a person of good character develop this device, use it, or condone its use?*
- *Would work on this project (thus implicitly condoning its use) or use of the device itself enhance or degrade my character?*
- *Would my personal spiritual leader, or other person I revere, condone development or use of this product?*

Step 3: *Correlate perspectives*

Here we enter the realm of subjective judgment. The individual author responsible for this example has a definite personal answer, but it is in the nature of ethical decision-making that different people will often arrive at different results in good conscience. You would have to weigh the various factors (including any that have been overlooked or knowingly omitted) to arrive at your own conclusion. We refuse to dictate a decision to you.

Step 4: *Act on your decision*

If your decision was that working on this project poses no threat to your soul (if you happen to believe in such), probably little courage is required to follow through, since your career may blossom, or at least not be curtailed.

On the other hand, if you believe that the project is unethical, you need to have the intestinal fortitude to either attempt to change the minds of management or refuse to work on the project, both of which may put your career at risk.

2.2 ENGINEERING CREED

Ethical decisions in engineering have, in general, a narrow focus specific to the problems that arise when designing and producing products or services of a technical nature. Engineers and scientists have, by the very nature of their profession, a body of specialized knowledge that is understood only vaguely, if at all, by most of the population. This knowledge can be used for tremendous good in society, but can also cause untold mischief when used by unscrupulous practitioners. Various engineering organizations have thus developed codes of conduct specific to the profession. Perhaps the most well known is the Code of Ethics for Engineers developed by the National Society of Professional Engineers (NSPE). The entire NSPE Code of Ethics is rather long, so we list only the Engineer's Creed and the Fundamental Canons of the Code here.

Engineer's Creed

As a Professional Engineer, I dedicate my professional knowledge and skill to the advancement and betterment of human welfare. I pledge:

- To give the utmost of performance;
- To participate in none but honest enterprise;
- To live and work according to the laws of man and the highest standards of professional conduct;
- To place service before profit, the honor and standing of the profession before personal advantage, and the public welfare above all other considerations.

    In humility and with need for Divine Guidance, I make this pledge.

Fundamental Canons

Engineers, in the fulfillment of their professional duties, shall:

- Hold paramount the safety, health, and welfare of the public.
- Perform services only in areas of their competence.
- Issue public statements only in an objective and truthful manner.
- Act for each employer or client as faithful agents or trustees.
- Avoid deceptive acts.
- Conduct themselves honorably, responsibly, ethically, and lawfully so as to enhance the honor, reputation, and usefulness of the profession.

    The complete code can easily be found online at a variety of sites. When this book went to press, the URL for the Code of Ethics on the NSPE site was http://www.nspe.org/Ethics/CodeofEthics/index.html

In-Class Activities

ICA 2-1

For each of the following situations, indicate whether you think the action is ethical or unethical or you are unsure. Do not read ahead; do not go back and change your answers.

Situation	Ethical	Unethical	Unsure
1. Not leaving a tip after a meal because your steak was not cooked to your liking			
2. Speeding 5 miles per hour over limit			
3. Killing a roach			
4. Speeding 15 miles per hour over limit			
5. Having plastic surgery after an accident			
6. Killing a mouse			
7. Driving 90 miles per hour			
8. Using Botox			
9. Not leaving a tip after a meal because the waiter was inattentive			
10. Killing a healthy cat			
11. Driving 90 miles per hour taking an injured child to the hospital			
12. Killing a healthy horse			
13. Dyeing your hair			
14. Killing a person			
15. Having liposuction			

ICA 2-2

For each of the following situations, indicate whether you think the action is ethical or unethical or you are unsure. Do not read ahead; do not go back and change your answers.

Situation	Ethical	Unethical	Unsure
1. Using time at work to IM your roommate			
2. Accepting a pen and pad of paper from a company trying to sell a new computer system to your company			
3. Obtaining a fake I.D. to purchase alcohol			
4. Using time at work to plan your friend's surprise party			
5. Accepting a wedge of cheese from a company trying to sell a new computer system to your company			
6. Taking a company pen home from work			
7. Taking extra time at lunch once a month to run a personal errand			
8. Accepting a set of golf clubs from a company trying to sell a new computer system to your company			
9. Drinking a beer while underage at a party in your dorm			
10. Using the company copier to copy your tax return			
11. Drinking a beer when underage at a party, knowing you will need to drive yourself and your roommate home			
12. Taking extra time at lunch once a week to run a personal errand			
13. Borrowing company tools			
14. Going to an NC-17 rated movie when underage			
15. Accepting a Hawaiian vacation from a company trying to sell a new computer system to your company			

ICA 2-3

For each of the following situations, indicate whether you think the action is ethical or unethical or you are unsure. Do not read ahead; do not go back and change your answers.

Situation	Ethical	Unethical	Unsure
1. Acting happy to see an acquaintance who is spreading rumors about you			
2. Letting a friend who has been sick copy your homework			
3. Shortcutting by walking across the grass on campus			
4. "Mooning" your friends as you drive by their apartment			
5. Registering as a Democrat even though you are a Republican			
6. Cheating on a test			
7. Shortcutting by walking across grass behind a house			
8. Saying that you lunched with a co-worker, rather than your high school sweetheart, when your spouse asks who you ate lunch with			
9. Helping people with their homework			
10. Shortcutting by walking through a building on campus			
11. Not telling your professor that you accidentally saw several of the final exam problems when you visited his office			
12. Suppressing derogatory comments about the college because the dean has asked you not to say anything negative when he invited you to meet with an external board evaluating the college			
13. Letting somebody copy your homework			
14. Shortcutting by walking through a house			
15. Not telling your professor that your score on a test was incorrectly totaled as 78 instead of the correct 58			

ICA 2-4

For each of the following situations, indicate how great you feel the need is in the world to solve the problem listed. Do not read ahead; do not go back and change your answers.

Situation	Urgent	Great	Somewhat	Little	None
1. Teaching those who cannot read or write					
2. Helping starving children in poor nations					
3. Helping people locked in prisons					
4. Helping to slow population growth					
5. Helping to reduce dependence on foreign oil					
6. Helping to reduce greenhouse gas emissions					
7. Helping people persecuted for sexual orientation					
8. Helping to reduce gun ownership					
9. Helping those who are mentally disabled					
10. Helping to supply laptops to poor children					
11. Helping prevent prosecution of "victimless" crimes					
12. Helping to end bigotry					
13. Helping to prevent development of WMD (weapons of mass destruction)					
14. Helping prosecute "hate" crimes					
15. Helping to eliminate violence in movies					
16. Helping homeless people in your community					
17. Helping people with AIDS					
18. Helping people in warring countries					
19. Helping endangered species					

ICA 2-5

Discuss the possible actions, if any, that you would take in each of the following situations. In each case, use the four-step analysis procedure presented in Section 2.1 to help determine an appropriate answer.

(a) Your roommate purchased a theme over the Internet and submitted it as his own work in English class.

(b) Your project team has been trying to get your design to work reliably for two weeks, but it still fails about 20 percent of the time. Your teammate notices another team's design that is much simpler, that is easy to build, and that works almost every time. Your teammate wants your group to build a replica of the other team's project at the last minute.

(c) You notice that your professor forgot to log off the computer in lab. You are the only person left in the room.

(d) The best student in the class, who consistently wrecks the "curve" by making 15 to 20 points higher than anyone else on every test, accidentally left her notes for the course in the classroom.

(e) You have already accepted and signed the paperwork for a position as an intern at ENGR-R-US. You then get an invitation to interview for an intern position (all expenses paid) at another company in a city you have always wanted to visit. What would you do? Would you behave differently if the agreement was verbal, but the papers had not been signed?

(f) One of your professors has posted a political cartoon with which both you and your friend vehemently disagree. The friend removes the cartoon from the bulletin board and tears it up.

ICA 2-6

Discuss the possible actions, if any, that you would take in each of the following situations. In each case, use the four-step analysis procedure presented in Section 2.1 to help determine an appropriate answer.

(a) You witness several students eating lunch on a bench on campus. When they finish, they leave their trash on the ground.

(b) You see a student carving his initials in one of the largest beech trees on campus.

(c) You see a student writing graffiti on a trash dumpster.

(d) There is a squirrel in the road ahead of a car you are driving. You know that a squirrel's instinct is to dart back and forth rather than run in a straight line away from a predator (in this case a vehicle) making it quite likely it will dart back into the road at the last instant.

(e) You find a wallet containing twenty-three $100 bills. The owner's contact information is quite clear. Does your answer change if the wallet contained three $1 bills?

ICA 2-7

Engineers often face workplace situations in which the ethical aspects of the job should be considered.

Table 2-1 lists a variety of types of organizations that hire engineers, and one or more possibly ethical issues that might arise.

Pick several of the organizations from the table that interest you (or those that are assigned by your professor) and answer the following:

(a) Can you think of other ethical problems that might arise at each of these organizations?

(b) Apply the four-step ethical decision-making procedure to gain insight into the nature of the decision to be made. In some cases, you may decide that an ethical issue is not really involved, but you should be able to justify why it is not.

(c) List 10 other types of organizations at which engineers would confront ethical problems, and explain the nature of the ethical decisions to be made.

(d) How does one find a balance between profit and environmental concerns?

(e) Under what circumstances should an engineer be held liable for personal injury or property damage caused by the products of his or her labor?

(f) Under what situations would you blow the whistle on your superior or your company?

(g) Should attorneys specializing in personal injury and property damage litigation be allowed to advertise, and if so, in what venues?

Table 2-1 Industries and issues for ICA 2-7

Organization/Occupation	Possible Issues
Alternative Energy Providers	Use of heavy metals in photovoltaic systems Effect of wind generators on bird populations Aesthetic considerations (e.g., NIMBY) Environmental concerns (e.g., Three Gorges project)
Environmental Projects	Fertile floodplains innundated by dams/lakes Safety compromised for cost (e.g., New Orleans levees) Habitat destruction by projects Habitat renovation versus cost (e.g., Everglades)
Chemical Engineers	Toxic effluents from manufacturing process Pesticide effect on ecosystem (e.g., artificial estrogens) Insufficient longitudinal studies of pharmaceuticals Non-biodegradable products (e.g., plastics)
Civil Engineers	Runoff/erosion at large projects Disruption of migration routes (freeways) Quality of urban environments Failure modes of structures
Computer Engineers	Vulnerability of software to malware Intellectual property rights (e.g., illegal downloads) Safety issues (e.g., programmed medical devices, computer-controlled transportation)
Electrical Engineers	Toxic materials in batteries Cell phone safety concerns Power grid safety and quick restoration in crises Possible use to break the law (e.g., radar detectors)
Electric Power Industry	Shipment of high sulfur coal to China Disposal of nuclear waste Environmental issues (e.g., spraying power-line corridors)
Food Processing Industry	Health possibly compromised by high fat/sugar/salt products Use of genetically engineered organisms Sanitation (e.g., *Escherichia coli*, *Salmonella*) Use of artificial preservatives
Manufacturing Companies	Manufacturing in countries with poor labor practices Lax safety standards in some countries Domestic jobs lost Environmental pollution due to shipping distances Trade imbalance
Industrial Engineers	Quality/safety compromised by cost Efficiency versus quality of working environment Management of dangerous tools and materials
Mechanical Engineers	Automotive safety versus cost Environmental impact of fossil fuels Robot failsafe mechanisms

CHAPTER 3
DESIGN AND TEAMWORK

Regardless of your selected engineering discipline or career path, communication is critical for survival as an engineer. Due to the complexity of many analysis and design projects, it is necessary for all engineers to operate effectively on a team. For a project to be completed successfully, there needs to be a synergy of all entities engaged, including design and teamwork. This chapter introduces the two together to emphasize the importance of teamwork in the design process.

3.1 THE DESIGN PROCESS

As quoted from their website:

In 1932 ABET was established as the Engineers' Council for Professional Development (ECPD). ECPD was formed to fill the apparent need for a "joint program for up-building engineering as a profession." Currently, ABET accredits some 2,700 programs at more than 550 colleges and universities nationwide. Accreditation is a non-governmental, peer-review process. ABET accreditation is assurance that a college or university program meets the quality standards established by the profession for which it prepares its students.

Design is a creative process that requires problem definition, idea searches, solutions development, and solutions sharing. Design is inherently multiobjective, so one expects that any problem addressed will have multiple solutions. While a particular solution might address some objectives very well, other objectives might not be well suited at all. The goal is to identify the design that meets the most important objectives. To evaluate ideas and communicate them to others, engineers commonly sketch possible solutions and even build models of their work.

There are many different versions of "the design process," but they have a lot of similarity. Recognizing that design is a creative process, it is more important to think about "a design process." The process we will use is the one used to design engineering education itself. ABET (pronounced with a long "A") was once the Accreditation Board for Engineering and Technology but is known now by just the acronym because it accredits computing and applied science programs as well. The engineering profession is self-governing; the Engineering Accreditation Commission of ABET is made up of member engineering societies, and the Commission operates under a philosophy of outcomes assessment. This approach focuses on determining the outcomes we desire in engineering graduates and allowing the engineering degree program the flexibility of determining how to achieve those outcomes and demonstrating how they have been achieved.

The design process for engineering programs is made up of two iterative processes, shown in Figure 3-1. The iterative loop on the left comes first because a new program would begin there; it includes getting input from constituencies, determining educational objectives, and evaluating or assessing how well those objectives are being achieved. In other engineering processes, these steps might be called something like "problem definition." "Constituencies" may also be called users, clients, stakeholders, or other terms. The process is iterative because it is important to confirm that the constituencies are pleased with the results, to adapt to changing needs, and to achieve the continuous improvement expected by the engineering profession.

Figure 3-1 The ABET Design Approach.

With the problem identified (knowing our educational objectives), the iterative process on the right side of Figure 3-1 begins. This process occurs primarily in the designer's workspace, whatever that is. In the case of designing engineering curriculum, the process takes place within the walls of the university or college. Knowing the objectives, the design team determines the outcomes that will accomplish those objectives, how those outcomes will be achieved, how they will be assessed, and what indicators will demonstrate success before any students are actually taught. Once students have had learning experiences (including extracurricular experiences), evaluation and assessment guide both processes into another cycle.

The step "determine outcomes required to achieve objectives" is called "problem definition" or "specification" in many other design processes. This step is critical because it shapes all the others. "Determine how the outcomes will be achieved" is a particularly creative step in the process and is commonly referred to as "generating ideas," "innovating," "developing possible solutions," or something similar, and might include "research." "Determining how the outcomes will be assessed" is a step that might not be mentioned if most agree on how to measure the success of a design, but in designs with more complex objectives, designers must think carefully about what will be measured and how, finishing up with determining indicators of success. These steps are commonly called "analysis" in a more general design process, breaking down the design to examine its assumptions, benefits, and risks. The remaining part of the process is one of "prototyping," "implementation," and "testing."

What follows in this chapter describes the tools engineers use in this process.

3.2 BRAINSTORMING IN THE DESIGN PROCESS

Brainstorming Rules:

- More is better
- Variety is better
- Do not criticize

Most high school graduates have heard the term **brainstorming**, and many have participated in a process called by that name. This term refers to one particular process by which ideas are generated. Idea generation follows three common rules: encourage a lot of ideas, encourage a wide variety of ideas, and do not criticize. The third rule is important if the first two are to be achieved. It has been said: "The best way to get a good idea is to get lots of ideas!"

Where is quality in this process? We have often been encouraged on "quality, *not* quantity." While some idea-generation methods facilitate the identification of a small number of quality solutions, in brainstorming, "quality *is* quantity." The nature of generating and evaluating ideas is that the greater the number of ideas generated, the more likely that several high-quality ideas are found.

The following story is taken from Alex Osborn's *Applied Imagination* (out of print), illustrating the importance of generating ideas before evaluating them.

> *In November 1952, in Washington State, the local telephone company had to clear the frost from 700 miles of telephone lines to restore long distance service. The company believed strongly enough in the importance of variety in the process that ALL the company's employees were asked to participate in a brainstorming session, executives and secretaries, engineers and linemen.*
>
> *After some time of idea generation, it was clear that the participants needed a break. One of the sessions overheard one lineman say to another at the water fountain, "Ever had a bear chase you up one of the poles? That would sure shake the ice off the lines!" The facilitator encouraged the lineman to repeat himself when the session reconvened.*
>
> *The facilitator hoped that the unusual suggestion would encourage new ideas, and the lineman sheepishly offered his suggestion. "How should we encourage the bears to climb all the poles?" asked the facilitator. "We could put honey pots at the tops of all the poles!" shouted someone else in the room. "How should we put honey pots on the tops of all the poles?" asked the facilitator. "Use the company helicopters that the executives use!" piped in another participant. "Hmm," said one of the company secretaries calmly. "When I served in the Korean War, I was impressed by the force of the downdraft off helicopter blades. I wonder if that would be enough to shake the ice off the power lines."*

Another great example of brainstorming is baking soda. Originally marketed for culinary use, today it is found in hundreds of products. How to you sell more product? Find more ways to use it!

The idea was so intriguing that it was tested immediately—it provided a successful and economically viable solution. This story clearly illustrates the benefit of variety and the value of avoiding criticism in a brainstorming session. The rest of the story illustrates how essential quantity is to the process.

A problem-solving group composed of five veteran air force helicopter pilots was convened to address this same problem. Each was unfamiliar with the solution that had already been implemented. It was hoped that because of their background, they would eventually arrive at the same solution. In fact, they did, but it was the 36th idea on their list. If they had stopped after generating 35 ideas, they may very well have had an acceptable solution, but it might not have been as elegant or as successful as the proven solution.

3.3   EXPERIMENTAL DESIGN: PERIOD ANALYSIS

Experiments enable engineers to come up with a creative solution to a problem and test the validity of the proposed idea. An experiment is a test of a proposed explanation of a problem. A good design of an experiment is a critical part of the scientific method.

What Constitutes the Scientific Method?

1. Observation: Observe the problem and note items of interest.
2. Hypothesis: Search for a known explanation of the phenomenon or attempt to formulate a new explanation.
3. Prediction: Create a model or prediction of behavior based on that hypothesis.
4. Experiment: Test your predictions. If necessary, modify your hypothesis and re-test.

Why Is Experimental Design Important?

As you move through your college career, you will be inundated with many equations and theories. These are useful in solving a wide variety of problems. However, as you will see, often the equations are really only useful in solving the most basic type of problems.

As an example, suppose you are interested in the speed of a ball as it rolls across the floor after rolling down a ramp. In physics, you will learn the equations of motion for bodies moving under the influence of gravity. If you are good, you can use these to examine rolling balls. What you will quickly find, however, is that numerous complicating factors make it difficult to apply the basic equations to obtain an adequate answer. Let's suppose you are interested in smooth balls (such as racquetballs), rough balls (tennis balls), heavy balls (bowling balls), and lightweight balls (ping pong balls). The simplified equations of motion predict that all these will behave in essentially the same way. You will discover, however, that the drag of the air affects the ping pong ball, the fuzz affects the tennis ball, and the flexible nature of the racquetball will allow it to bounce at steep ramp angles. It is difficult to predict the behavior analytically. Often, one of the quickest ways to learn about the performance of such complex situations is to conduct experiments.

What Are Experimental Measurements?

Most scientific experiments involve measuring the effect of variability of an attribute of an object. In an experiment, the **independent variable** is the variable that is controlled. The **dependent variable** is a variable that reacts to a change in the independent variable. A **control variable** is part of the experiment that can vary but is held constant to let the experimenter observe the influence of the independent variable on the dependent variable. Keeping control variables constant throughout an experiment eliminates any confounding effects resulting from excess variability.

Any measurement acquired in an experiment contains two important pieces of information. First, the measurement contains the actual value measured from the instrument. In general, a measurement is some physical dimension that's acquired with some man-made data-collection instrument. As with any man-made device, there may be some imperfection that can cause adverse effects during data collection. Thus, the second piece of information that goes along with any measurement is the level of uncertainty.

Any uncertainty in measurement isn't strictly by instrumentation error. Systematic error is any error resulting from human or instrumentation malfunction. Random

Do you know what these unique measurement instruments do?

- Durometer
- Dynamometer
- Euidometer
- Galvanometer
- Gyroscope
- Goniometer
- Manometer
- Opisometer
- Pycnometer
- Sphygmo-manometer
- Tachymeter
- Thiele tube
- Transit telescope

error is caused by the limits of the precision of the data-collection device. It is possible to minimize the systematic error in an experiment, but random error cannot be completely eliminated.

What Measurements Do You Need to Make?

You need to develop a coherent experimental program. You should make enough measurements to answer any anticipated questions, but you don't usually have the time or money to test every possible condition. Points to consider:

- What are the parameters of interest?
- What is the range of these parameters—minimum values, maximum values?
- What increments are reasonable for testing (every 10 degrees, every 30 seconds, etc.)?
- What order is best to vary the parameters? Which should be tested first, next, etc.?

Here is an acronym (PERIOD) that can help you remember these important steps. As an example, it is applied to the problem of the ramp and rolling balls, described above.

P   *Parameters of interest determined*
- Parameter 1 is the ramp angle.
- Parameter 2 is the distance up the ramp that we release the ball.
- Parameter 3 is the type of ball.

E   *The range of the parameters established*
- Ramp angle can vary between 0 and 90 degrees in theory, but in reality can only vary between 10 degrees (if too shallow, ball won't move) and 45 degrees (if too steep, ball will bounce).
- The distance we release the ball up the ramp can vary between 0 and 3 feet in theory, assuming that the ramp is 3 feet long. We can't release the ball too close to the bottom of the ramp or it won't move. In reality, we can only vary between 0.5 feet and 3 feet.
- We will test as many types of balls as we have interest in.

R   *Repetition of each test specified*
- The ramp angle will be set according to the height of the ramp from the floor, so there is not much room for error in this measurement; only one measurement is needed for such geometry.
- Each placement of the ball before release will vary slightly and may cause the ball to roll slightly differently down the ramp; this is probably the most important factor in determining the speed, so three measurements at each location are needed.
- We will assume that every ball is the same, and the actual ball used will not change the outcome of the experiment; only one ball of each type is needed.

I   *Increments of each of these parameters to test specified*
- We will test every 10 degrees of ramp angle, starting at 10 degrees and ending at 40 degrees.
- We will release the balls at a height of ½, 1, 1½, 2, 2½ and 3 feet up the ramp.
- We will test five types of balls: racquetball, baseball, tennis ball, ping pong ball, and bowling ball.

O   *Order to vary the parameters determined*
- We will set the ramp angle and then test one ball type by releasing it at each of the four different distances up the ramp.

- We will repeat this process three times for each ball.
- We will then repeat this process for each type of ball.
- We will then changed the ramp angle by 10 degrees and repeat the process.
- This process is repeated until all conditions have been tested.

D   *Do the tests by measuring the speed for each condition*

How Many Measurements Are Too Many?

It is always important to determine before you start how many measurements that you need to make. Sometimes you can be too ambitious and end up developing an experimental program that will take too much effort or cost too much money. If this is the case, then you need to decide which increments can be relaxed, to reduce the number of overall measurements.

The number of measurements (N) you will need to make can be easily calculated by the following equation for a total of n parameters:

N = (# increments parameter 1 * number of repetitions for parameter 1) *

(# increments parameter 2 * number of repetitions for parameter 2) * . . .

(# increments parameter n * number of repetitions for parameter n) * . . .

Continuing the examples given above, the number of actual measurements that we need to make is calculated as

N = (4 angles)*(6 distances * 3 repetitions)*(5 types of balls) = 360 measurements

In this example, 360 measurements may be extreme. If we examine our plan, we can probably make the following changes without losing experimental information:

- We decide to test every 10 degrees of ramp angle, starting at 20 degrees and ending at 40 degrees. This will lower the angle testing from 4 angles to 3 angles.
- We will release the balls at a height of 1, 2, 2½ and 3 feet up the ramp. This will lower the distances from 6 to 4.
- We will test three types of balls: racquetball, ping pong ball, and bowling ball. This will lower the type of balls from 5 to 3.

The number of actual measurements that we now need to make is calculated as

N = (3 angles)*(4 distances * 3 repetitions)*(3 types of balls) = 108 measurements

This result seems much more manageable to complete than 360!

3.4   PROJECT TIMELINE

To complete a project successfully, on schedule and satisfying all constraints, careful planning is required. The following steps should help your team plan the completion of project work.

Step 1: Create a Project Timeline

The first consideration is the project's due date. All team members need to note this on a calendar. Examine the due date within the context of other assignments and classes. For example, is there a calculus test in the fourth week? When is the first English paper due?

Next, look at the project itself and break into individual tasks and subtasks. Create a list, making it as specific, detailed and thorough as possible. Your list should include

- All tasks needed to complete the project
- Decisions that need to be made at various times
- Any supplies/equipment that will need to be obtained

Carefully consider the order in which the tasks should be completed. Does one task depend on the results of another? *Then, working backwards from the project due date, assign each task, decision, or purchase its own due date on the calendar.*

Finally, your team should consider meeting at least once a week at a consistent time and location for the duration of the semester. More meetings will be necessary, but there should be at least one time per week when the entire team can get together and review the project status. A standing meeting time will prevent issues of "I didn't know we were going to meet" or "I didn't get the message."

> Choose a weekly team meeting time and STICK TO IT!

> Divide the workload and require updates at your weekly meeting.
>
> If a team member is not completing assigned tasks as required, speak with your professor. Do not wait until the end of the project to raise concerns about a teammate who is AWOL.

Step 2: Create a Responsibility Matrix

List the project's tasks and subtasks one by one down the left side of the paper. Then, create columns beneath each team member's name, written side by side across the top. Put a check mark in the column beneath the name of the member who agrees to perform each task. It then becomes the responsibility of that team member to successfully perform the task by the due date that was agreed upon in Step 1. An alternate grid is shown in Table 3-1.

In assigning the tasks, consider the complexity and time required for the job. One team member may have five small tasks while another may have one major task, with the goal being an equal distribution of effort. A second team member should be assigned to each task to assist or check the work completed by the first team member. Be sure all team members are comfortable with the assignments.

Table 3-1 Sample responsibility matrix

Task	Completed by . . .	Checked by . . .
Purchase supplies	Pat & Chris by 9/15 team mtg	
Write initial proposal	Terry–Email to Robin by 9/22	Robin by 9/25 team mtg
Conduct preliminary calculations on height	Robin–Email to Pat by 9/22	Pat by 9/25 team mtg
Build prototype in lab	All–Lab: 7–9 pm, 9/28	

Step 3: Consider Team Dynamics

Communication: The success of any project depends to a great extent on how well the team members communicate. Don't hesitate to share ideas and suggestions with the group and consider each member's input carefully. Do not be afraid to admit that you are having difficulties with a task or that the task is taking longer than expected. Be ready and willing to help one another.

Trust and respect: Remember the team is working toward a shared objective. Therefore, you must choose to trust and respect one another. Treat fellow team members with simple courtesy and consideration. Follow through with promises of completed tasks, remembering the team is counting on your individual contributions. Try to deal honestly and openly with disagreements. However, do not hesitate to ask for help from faculty if problems begin to escalate.

| Review your matrix each week and adjust as needed. |

Nothing is carved in stone: It is important to plan the project as carefully as possible; however, unforeseen problems can still occur. Treat both the Project Timeline and Responsibility Matrix as working documents. Realize they were created to serve as guides, not as inflexible standards. Watch the project progress relative to the timeline, and don't hesitate to redesign, reallocate, or reschedule should the need arise.

Finally, don't forget to have fun!

3.5 CRITERIA AND EVALUATION

We can generate and evaluate problem definitions, implementation strategies, and even the methods we use to evaluate our implementations. Brainstorming problem definitions stresses how the way we define a problem affects the solutions we consider. Is the problem that there is ice on the electrical lines, that there is water on the lines that can freeze, or is it that the lines and surrounding air are too cold? Looking at the problem these three different ways suggests different solution approaches. The second might suggest covering the lines to keep them dry or coating them with a material with special hydrophobic properties. The third might suggest heating the lines or moving them to somewhere warmer (maybe underground).

There are countless examples of how redefining a problem has been commercially profitable. When Dum Dum Pops® are made, the process is continuous, which causes the manufacture of some lollipops that combine two of the flavors (http://en.wikipedia.org/wiki/Dum_Dum_Pop). That might have been considered a problem, and some companies would have designed their manufacturing process so as to stop and clean the machinery between flavor runs. Instead, the Dum Dum company wraps these "combination" flavor lollipops and labels them "Mystery" flavor. By their defining the problem a different way, a creative solution emerged, saving money and providing a market attraction. In all designs, the way in which the problem is defined will affect the set of solutions explored. Once a particular solution approach is selected, it might be appropriate to generate and evaluate ideas for how to implement that solution.

Criteria: Deciding What is Important

First, the problem definition must be clear so that we can generate an appropriate set of criteria. Then, when we stop generating solutions and begin evaluating the various solutions, we use those **criteria** to narrow our choices objectively. We might be particularly concerned about cost, so certain approaches become less attractive. Other proposed solutions might be hazardous, thus further narrowing our options. The criteria for evaluating potential solutions must be well defined before evaluation begins so that all ideas can be considered fairly—this helps avoid arguments based on hidden criteria.

Defining criteria before brainstorming solutions seems as if it might threaten the "don't criticize" objective—proposed solutions could be steered to match the expectations established by the criteria. Unfortunately, while defining criteria first might constrain idea generation, generating solutions first might cause participants to develop an opinion of a "best solution" and then force their criteria choices to fit their preferred idea. In this way, the process of generating solutions and the process of generating criteria to evaluate those solutions are linked. The second interaction is more damaging to the design process, so we will focus on identifying constraints and criteria first.

Comprehension Check 3-1

We frequently express criteria in terms that are not clear and measurable. Write a clear criterion to replace each of these vague criteria for the products.

Product	Computer	Automobile	Bookshelf
Inexpensive	Less than $300		
Small			
Easy to assemble			Requires only a screwdriver
Aesthetically pleasing			
Lightweight			
Safe			
Durable			
Environmentally friendly		Has an estimated MPG of at least 50	

In general, using fewer criteria keeps things simple. A rule of thumb is that if you can think of 10 criteria that are meaningful, you should consider only the two most important criteria to compare solutions. In this way, you can be sure that the important criteria maintain their importance in your decision-making. Seek consensus on what the most important criteria are. Experts, specialists, managers, customers, research, etc., can help you focus on the most important criteria. In certain situations, many criteria are used—for example, when a magazine rates consumer products, extra criteria should be included to be sure that all the criteria important to the magazine's readers have been considered.

Another situation in which a large number of criteria are included occurs when very complex decisions are being evaluated. The U.S. Green Building Council has long lists of criteria used for certifying various types of projects according to the Leadership in Energy and Environmental Design (LEED) Green Building Rating System. A subset of the LEED criteria for homes is shown below.

LEED is an internationally recognized green building certification system, providing third-party verification that a building or community was designed and built using strategies aimed at improving performance across all the metrics that matter most: energy savings, water efficiency, CO2 emissions reduction, improved indoor environmental quality, and stewardship of resources and sensitivity to their impacts.

From the U.S. Green Building Council website: www.usgbc.org.

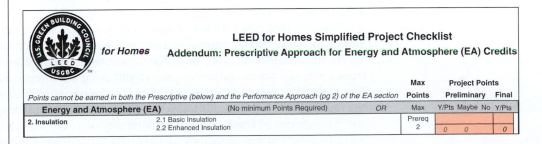

These checklist items are neither clear nor measurable, however, as shown above. The guide to the rating system provides the additional detail needed to make the insulation criteria both clear and measurable. Note that these criteria build from existing standards, meaning that even greater levels of detail can be found elsewhere (e.g., Chapter 4 of the 2004 International Energy Conservation Code and The National Home Energy Rating Standards).

Prerequisites

2.1 Basic Insulation. Meet all the following requirements:

a) Install insulation that meets or exceeds the R-value requirements listed in Chapter 4 of the 2004 International Energy Conservation Code. Alternative wall and insulation systems, such as structural insulated panels (SIPs) and insulated concrete forms (ICFs), must demonstrate a comparable R-value, but thermal mass or infiltration effects cannot be included in the R-value calculation.

b) Install insulation to meet the Grade II specifications set by the National Home Energy Rating Standards (**Table 16**). Installation must be verified by an energy rater or Green Rater conducting a predrywall thermal bypass inspection, as summarized in **Figure 3**.

Note: For any portion of the home constructed with SIPs or ICFs, the rater must conduct a modified visual inspection using the ENERGY STAR Structural Insulated Panel Visual Inspection Form.

Credits

2.2 Enhanced Insulation (2 points). Meet the following requirements:

a) Install insulation that exceeds the R-value requirements listed in Chapter 4 of the 2004 International Energy Conservation Code by at least 5%. Alternative wall and insulation systems, such as structural insulated panels (SIPs) and insulated concrete forms (ICFs), must demonstrate a comparable R-value, but thermal mass or infiltration effects cannot be included in the R-value calculation.

b) Install insulation to meet the Grade I specifications set by the National Home Energy Rating Standards (**Table 16**). Installation must be verified by an energy rater or Green Rater conducting a predrywall thermal bypass inspection as summarized in **Figure 3**.

Note that the LEED criteria specify only those characteristics that are important to the overall goal of energy and the environment. This leaves the choice of the type of insulation (fiberglass, cellulose) up to the designer and the consumer. Because homes, outdoor facilities, and factories face different challenges, different LEED criteria apply. It is important to recognize which choices are "options" rather than criteria— even the most energy efficient factory cannot substitute for a house. Further, the LEED criteria allow different thresholds for different size houses, so as not to penalize multifamily households—even though multifamily households will require more material to build and more energy to operate, the marginal cost is not as high as building another home.

When designing a product or a process, you must first listen to what a client wants and needs. The "client" might be a department such as manufacturing, marketing, or accounting; a customer; your boss; the government, etc. Subsequently, you will need to translate those wants and needs into engineering specifications. Pay particular attention to separating the wants (criteria) from the needs (requirements). Requirements can be imposed legally (e.g., Corporate Average Fuel Economy regulations, building codes), they can be from within your company (e.g., the finished product must not cost more than $50 to manufacture), or they may come from your client (e.g., the product must be made from at least 90% recycled material).

Comprehension Check 3-2

You have been asked to improve the fuel efficiency of an automobile by 20 percent. Convert this request into engineering criteria. What changes might be made to the automobile to achieve this objective?

Evaluation: Deciding on Options

Simply identifying criteria isn't enough information for a decision—each proposed solution must be evaluated against the criteria. The first step will be to eliminate any solutions that do not meet the minimum requirements—those solutions are out of bounds and need not be considered further.

Once the minimum requirements have been satisfied, there are many ways of applying the remaining criteria to select a solution. One approach is to make **pairwise comparisons**. In this approach, you use a table for each criterion to summarize how each of the solutions compares to all the others. An example is shown in Table 3-2 for the criterion "safety." Among the table entries, 0 indicates that that option in that column is worse than the option in that row; 1 indicates that both options rank equally; and 2 indicates that that option in that column is better than the option in that row.

Since a high level of safety is preferred, "better" means "safer." In the example, the first column indicates that option 1 is less safe than option 2, but the same as option 3. The resulting totals indicate that option 2 ranks best in terms of safety. To complete the process, generate similar tables for each of the criteria being considered and then sum the totals to identify the best solution.

Table 3-2 Comparing options

Safety	Option 1	Option 2	Option 3
Option 1		0	1
Option 2	0		0
Option 3	1	2	
Total	1	2	1

A disadvantage of the pairwise comparisons approach is that all criteria have equal weight, whereas some criteria are likely to be more important than others. An alternative approach is to use **weighted benefit analysis**, shown in Table 3-3. In this approach, each option is scored against each of the criteria.

Table 3-3 Options with a weighted benefit analysis

	Weights	Option 1	Option 2	Option 3
Cost	2	2	6	10
Safety	8	10	4	6
Weight	10	7	7	2
Wow	5	2	4	6
Totals		4+80+70+10=164	12+32+70+20=134	20+48+20+30=118

Table 3-4 Sample scoring rubric

Score	Meaning
0	Not satisfactory
1	Barely applicable
2	Fairly good
3	Good
4	Very good; ideal

In the table below, the weights are in the first column and each option has been assigned a score from 0 to 10, indicating how well that option meets each criterion. This approach may be inconsistent in that a "7" for one rater may be different from a "7" for another rater, so it can help to better define the scale. For example, the options may be scored on the scale shown in Table 3-4 as to how well the option fits each criterion.

Reality: Checking It Out

When a solution has been identified as the best fit to the criteria, it is best to stop for a reality check before moving forward. After the evaluation process, some ideas will be left on the cutting room floor. Do any of these merit further consideration? Are there

---

WISE WORDS: IN YOUR JOB, DO YOU WORK ALONE OR ON A TEAM?

I often work alone on projects and analyses, but I do need to interact with client company teams, sometimes leading the team as an outside expert, but still a "member" of the company I am trying to help. Some of the teams I will work with have also have mechanical, chemical, industrial, and process engineers involved.

E. Basta, MSE

I work on a team made up of all electrical and computer engineers.

E. D'Avignon, CpE

Team: chemist, package engineer, process development engineer (chemical engineer), industrial designer, line engineer (mechanical engineer or chemical engineer), process engineer (chemical engineer), planner, quality engineer, marketing, and market research.

S. Forkner, ChE

I work mostly alone on my assignments. I am given tasks and I had to find solutions for my problems on my own. Only when needed I consult someone for questions and guidance on my task.

V. Gallas Cervo, ME

I worked the first 5 years of my career as a "sole contributor" in an engineering role. Since then, I have worked in a team setting managing technical employees.

L. Gascoigne, CE

I work with a chemical engineer, an instrumentation engineer, a mechanical engineer, and a piping engineer.

D. Jones, BE

Every project involves a team. My typical team includes surveyors, structural engineers, environmental scientists (wetlands, endangered species), permit specialists, electrical engineers, land appraisers, archaeologists, and architects.

J. Meena, CE

I work alone, but seek advice from my management team.

C. Pringle, IE

I work on a team including mechanical, electrical, aeronautical and systems hardware designers, software designers, integrators, power engineers, and other liaisons like myself.

E. Styles, EE

The Hubble Space Telescope servicing mission project is made up of a team of mechanical design engineers, human factors engineers, environmental test engineers, thermal engineers, underwater test engineers, safety engineers, mechanical technicians, documentation specialists, space scientists, and systems engineers.

R. Werneth, ME

important elements of those ideas that can be incorporated into the chosen design? If the reality check reveals that an idea really should not have been eliminated, then a change in the selection criteria may be appropriate.

After deciding, you will implement your chosen solution. Undoubtedly, both carrying out the design and using the design once it is complete will provide new information about how the design might be improved. In this way, design tends to be iterative—design, build, test, redesign, and so on. Even when a design performs as expected, it may be important to test a model extensively or even build multiple models to be sure that the design is reliable.

3.6 WORKING IN TEAMS

> The ability to work on a team is one of the most critical traits an engineer needs. Even if you're the greatest engineer in the world, you will not know all of the answers or have all of the right ideas. We can always learn something from our peers.
>
> *A. Thompson, EE*

You have been formed into "teams" not "groups."

Group: A number of people who come together at the same place, at the same time.
Team: *Individuals cooperating to accomplish a common goal.*

As a student and in the workplace, you will complete some assignments individually and complete some as part of a team. When you work independently, you are mostly free to choose when and how you will work. When you work as part of a team, make sure the team has ground rules for how it will operate. Any time several people are asked to work closely together, there is a potential for much good from a diversity of ideas and skills but there is also a potential for conflict. Because conflict can be both productive and unproductive, you need to manage it.

Team Behavior

The most critical task for a team, particularly a new team, is to establish its purpose, process (its way of doing things), and a means of measuring team progress. Here are several topics regarding team behavior that you may wish to consider.

> We trained hard but it seemed that every time we were beginning to form teams we would be reorganized. I was to learn later in life that we tend to meet every situation by reorganizing, and a wonderful method it can be for creating the illusion of progress while producing confusion, inefficiency, and demoralization.
>
> *Petronius*

- **Ground rules:** Each team needs to come to a consensus about acceptable and unacceptable individual as well as team behavior.
- **Decision making:** Teams by necessity make decisions. Each team needs to decide how these decisions will be made. For example, will they be done through consensus, majority vote (either secret or show of hands), or by other methods?
- **Communication:** This is often one of the hardest parts of working effectively as a team. Team members need to recognize the value of real listening and constructive feedback. During the course of team meetings, *every* team member needs to participate *and* listen.
- **Roles:** You may adopt various roles on your team. In a long-term project, roles should rotate so that everyone has a chance to learn each role.
- **Participation:** Decide as a team how work will be distributed. Your team should also consider how to handle shifts in workload when a team member is sick or otherwise unavailable.
- **Values:** The team as a whole needs to acknowledge and accept the unique insights that each team member can contribute to their work.
- **Outcomes:** Discuss and agree on what types of measures will be used to determine that the team has reached its final goal.

Teammate Evaluation: Practicing Accountability

As previously discussed, Engineering is a self-governing profession. ABET is charged with accrediting engineering programs in the United States through membership from the profession. Among a set of required outcomes of engineering graduates, ABET requires that engineering students graduate with an ability to function on multidisciplinary teams. To ensure that each student achieves this outcome, individual accountability is needed. Realizing that much of the activity of a team happens when the team is meeting privately (without a professor), an effective and increasingly common way of addressing this tenet is to have team members rate one another's performance.

It is important to learn how to be an effective team member now, because most engineering work is done in teams, and studies show that most engineering graduates will have supervisory responsibility (at least project management) within five years of graduation. You have worked in teams before, so you have probably noticed that some team members are more effective than others. Consider these three snapshots of interactions with engineering students. All three are true stories.

- Three team members approach the professor, concerned that they haven't seen the fourth team member yet. The professor speaks with the student, who quickly becomes despondent, alerting the professor to a number of serious burdens the student is bearing. The professor alerts an advisor and the student gets needed help.
- A student, acting as the team spokesperson, tells the professor that one team member never comes to meetings. The professor speaks to the nonparticipating student, who expects to be contacted about meetings by cell phone, not email. The professor explains that the student's expectations are unrealistic.
- A student is insecure about being able to contribute during team activities. After the team's ratings of that student are in, the professor talks to the class about the importance of participating and the different ways students can contribute to a team. In the next evaluation, the student receives the highest rating on the team.

Peer evaluations are a useful way for team members to communicate to one another and to their professor about how the members of a team are performing. Reviewing and evaluating job performance is a marketable skill and is as useful to the employee seeking a job or a job advancement as it is to the supervisor.

Comprehension Check 3-3

In the following scenario, a team of four students (Pat, Chris, Terry, and Robin) is getting ready to complete peer evaluations at the end of the semester.

Pat   Struggling to maintain a C in the course. Often asked for explanations during the team sessions, offered ideas when able, always showed up for meetings, and prepared beforehand to the best of ability. Is friendly and willing to work.

Chris   Missed a number of meetings with no excuse, and when showing up, had almost always not looked at the assignment. Contributed very little during the meetings. Was late to one meeting because he claimed no one sent the meeting time. Missed a meeting because the alarm clock didn't go off. Didn't bring that week's work because he figured Terry would want to redo it anyway.

Terry   Very bright, usually first to figure out solutions to difficult problems, and not particularly cooperative. Tended to just work out the solutions and discourage attempts of others to contribute. Reluctant to spend time explaining things to the others, and particularly impatient with Pat's questions. Doesn't like to explain "obvious" things.

Robin   Not as bright as Terry but super-responsible, spent a lot of time giving one-on-one help to Pat and (early in the semester before finally getting fed up) to Chris. Always on time for meetings, and often volunteers to call and re-mind everyone (especially Chris) when the next meeting is.

As you consider the scenario, individually write your assessment of how each character helps or hinders the functioning of the team.

Underwater Training for a Hubble Space Telescope Servicing Mission. NASA engineers (on SCUBA) and astronauts (in modified space suits) take advantage of the effects of neutral buoyancy to practice replacing a Wide Field Camera in preparation for a Hubble Space Telescope (HST) servicing mission. Extensive teamwork is required in a 6.2-million gallon pool in Houston, TX to develop, refine, and practice the procedures to be used on spacewalks in orbit. Engineers from Goddard Space Flight Center in Greenbelt, MD and Johnson Space Center in Houston work together to perfect the procedures and hardware, including specialized astronaut tools. The team uses models representing the flight items for conducting end-to-end tasks in the neutral buoyancy facility. This unique example of engineering teamwork has resulted in five very successful HST servicing missions involving complex astronaut spacewalks. The challenge to the team is to develop the nominal and contingency procedures and tools on the ground (and in the water!) to be used for mission success with HST in orbit 300 miles up.

R. Werneth

In-Class Activities

ICA 3-1

With your team, compose a plan to build the longest bridge possible using the K'Nex™ pieces provided by your instructor. The longest part of your bridge will be defined as the longest stretch of K'Nex pieces that are not touching another surface (table, floor, chair, etc.).

During the planning phase (15 minutes), your team will only be allowed to connect as many pieces as your team has hands. For example, a team of 4 will only be allowed to have 8 K'Nex pieces connected together at any one moment. In addition, every team member must have their hand on a K'Nex piece if it is connected together during planning. As soon as your team has finished the planning phase, disconnect all K'Nex pieces and place them in the provided container. The container will be shaken before you begin, so don't bother attempting to order the pieces in any way.

During the building phase (60 seconds), the restriction on the number of connected pieces goes away, but your team will not be allowed to talk. Your instructor will say go, and then stop after 60 seconds has elapsed. At the end of the building phase, your team must step away from the bridge and remove all hands and other body parts from the K'Nex bridge. If any pieces falls after time is called, you are not allowed to stabilize the structure in any way.

ICA 3-2

With your team, come up with a plan to build the tallest tower possible using the K'Nex™ pieces provided by your instructor. The tallest portion of your tower will be defined as the longest stretch of K'Nex pieces that are not touching another surface (table, floor, chair, etc.).

During the planning phase (15 minutes), your team will only be allowed to connect as many pieces as your team has hands. For example, a team of 4 will only be allowed to have 8 K'Nex pieces connected together at any one moment. In addition, every team member must have their hand on a K'Nex piece if it is connected together during planning. As soon as your team has finished the planning phase, disconnect all K'Nex pieces and place them in the provided container. The container will be shaken before you begin, so don't bother attempting to order the pieces in any way.

During the building phase (60 seconds), the restriction on the number of connected pieces goes away, but your team will not be allowed to talk. Your instructor will say go, and then stop after 60 seconds has elapsed. At the end of the building phase, your team must step away from the bridge and remove all hands and other body parts from the K'Nex tower. If any pieces falls after time is called, you are not allowed to stabilize the structure in any way.

Mini Design Projects

This section provides a wide range of design projects, varying in difficulty and time commitment. Your instructor may assign projects required for your specific course and provide more details.

Category I: Demonstrate a Physical Law or Measure a Material Property

1. Prove the law of the lever.
2. Demonstrate conservation of energy (PE + KE = constant).
3. Determine the coefficient of static and sliding friction for a piece of wood.
4. Prove that the angle of incidence is equal to the angle of reflection.
5. Demonstrate momentum conservation (F = ma).
6. Demonstrate the ideal gas law.
7. Obtain a series of data points from an experiment you conduct that, when plotted, exhibit a normal distribution.
8. Show that forces can be resolved into horizontal and vertical components.
9. Find the center of gravity of an irregular piece of plywood.
10. Show that for circular motion, $F = mv^2/r$.
11. Show that for circular motion, $v = \omega r$.
12. Measure the effective porosity of a sand sample.
13. Prove the law of the pendulum.
14. Prove Hooke's law for a spring.
15. Prove Hooke's law for a metal rod (in deflection).
16. Measure the coefficient of thermal expansion for a solid rod or bar.
17. Estimate the heat capacity for several objects; compare with published results.
18. Prove Archimedes law of buoyancy.
19. Determine the value of pi experimentally.
20. Prove the hydrostatic pressure distribution.
21. Relate the magnetic strength to the radius.
22. Determine the density and specific gravity of a rock.

Category II: Solve a Problem

1. What is the volume of a straight pin?
2. Determine the thickness of a specified coin or a piece of paper.
3. How many pennies are needed to sink a paper cup in water?
4. Determine the specific gravity of your body.
5. What is the volumetric flow rate from your shower?
6. Use a coat hanger to make a direct reading scale for weight.

Category III: Design a Solution

1. THE GREAT EGG DROP

   You have no doubt seen the "Odyssey of the Mind" type of assignment in which you are to design protection for an egg that is to be dropped from some height without being broken. This assignment is to have the same end product (i.e., an unbroken egg) but in a different way.

You cannot protect the egg in any way but are allowed to design "something" for it to land on. You will be allowed 3 drops per team, and will be assigned at random the heights from which you will drop the egg. The egg must free-fall after release. If the egg breaks open, the height will be taken as zero. If the egg shell cracks, the height from which you dropped it will be divided by 2. If you drop the egg and miss the catching apparatus, that is your tough luck (and a zero height will be used)—suggesting that you need to devise a way to always hit the "target."

Your grade will be determined by

Ranking = (height from which egg is dropped)/(weight of catching apparatus)

After demonstrations, the average ranking number for each team will be calculated, and the value truncated to an integer. The teams will then be ranked from highest value to lowest. The heights and weights will be measured in class. The actual grade corresponding to your class ranking will be determined by your instructor.

2. TREE HEIGHT

We have been contacted by a power company to conduct a study of tree height and interference with high voltage wires. Among other requirements, the company is looking for a quick, easy, and inexpensive method to measure the height of a tree.

Your project is to develop different methods of measuring the height of a tree. As a test case, use a tree designated by your instructor. You may *not* climb the tree as one of the methods! In addition, you may not harm the trees or leave any trace of your project behind.

To sell your methods to the customer, you must create a poster. The poster should showcase your measurement methods, including instructions on how to conduct the experiment, any important calculations, graphs or photos, and your resulting measurements. The poster will be graded on neatness, organization, spelling and grammar and mechanics, formatting, and strength of conclusions. You may use any piece of poster board commercially available, or a "science fair" board. The poster can be handwritten, or typed and attached, or . . . here is a chance to use your creativity! If you present more than one method, you must indicate and justify your "best" choice.

Your grade will be determined as follows:

- *Method #1:*   30 points
- *Method #2:*   20 points
- *Method #3:*   10 points
- *Presentation Board:*   40 points

3. FIRE EXTINGUISHER

Make a fire extinguisher for a candle. The candle will be lit, and the extinguisher will put out the flame at a predetermined time after the candle is lit (say, 20 seconds). The only thing that the participant can do to start the time is to light the candle. The candle can be mounted anywhere you like. The results will be scored as follows:

You will be allowed three trials. You will be allowed to use your best trial for grading.

You must extinguish the candle between 19 and 21 seconds; for every second or fraction thereof outside this range you will lose 5 points.

4. CLEPSYDRA

Construct a clepsydra (water clock). When you bring it to class to demonstrate its performance, the following test will be used:

You will have three times to measure: a short time, a medium time, and an extended period. The actual times to be measured will be given to you at the time of demonstration, and you will have 2 minutes to set up your apparatus.

- Short time: Between 10 and 30 seconds (in 2-second increments)
- Medium: Between 1 and 4 minutes (in 30-second increments)
- Extended: Between 5 and 10 minutes (in 1-minute increments)

You will start the clock and tell the timekeeper to begin. You will then call out the times for each of the three intervals and the timekeeper will record the actual times. Your clock must "run" for the total time.

Your grade will be determined by the average percent error of your timings.

For example, if the specified "medium" time was 2 minutes and 30 seconds and you said "mark" at an actual time of 2 minutes and 50 seconds, the absolute value of the percent error would be $(20 \text{ s})/(150 \text{ s}) = 13\%$. The absolute values of the three errors will be summed, divided by 3, and subtracted from 100% to get a numerical grade.

5. ON TARGET

Each team will design, build, and test a device that will allow you to successfully hit a target with a table tennis ball. The target will be a flat sheet of poster paper placed on the floor with a bull's-eye and two other rings around it for scores of 100, 90, and 80 with a score of 60 for hitting the paper. The target will be placed at a location of 15 feet from the point at which you release the ball. Once the ball is released, you cannot touch it again, and it must be airborne before it hits the target. When demonstrating your device, you will not be allowed any trial run.

Your grade will be determined as follows:

Average numerical score + bonus (10, 8, or 6 points for creativity and simplicity)

The class (each person) will be given a slip of paper on which they will rank their top three teams with respect to creativity and also with respect to simplicity. These will be tallied and the top 3 teams in each category will receive 10, 8, or 6 bonus points.

6. KEEPING TIME

Each team is to build a "clock." When you say "go," a stopwatch will be started, and you are to tell the timekeeper when 10 seconds, 30 seconds, and 60 seconds have elapsed. Differences between the actual times and the predicted times at the three checkpoints will be noted and the percent error calculated. Average the absolute values of these three errors and subtract from 100 to obtain your final score. You may not use any store-bought device that is designed to measure time. No electronic devices may be used.

Your final grade will be determined as follows:

Average numerical score + bonus points (10, 8, or 6 for creativity and simplicity)

The class (each person) will be given a slip of paper on which they will rank their top three teams with respect to creativity and simplicity. These will be tallied and the top three teams in each category will receive 10, 8, or 6 bonus points.

Category IV: Additional Projects

1. Develop a device that can be placed into a container of water and used to measure the pressure as a function of depth. Take measurements and plot them against theory for a hydrostatic pressure distribution.

2. A 2-liter soft-drink container is sitting on the floor, nearly full of water and open to the atmosphere. Where could you locate an orifice in the side of the bottle so that the jet of water which squirts out will have the maximum range? Keep the bottle filled by continuously pouring water in the container as the tests are conducted. Justify your answer with theory by discussion rather than equations.

3. Design and build a device that will allow a ping pong ball to hit a target (small circle) between 5 and 15 feet away from the point at which you release the ball with the device on the floor. Points will be given for accuracy (distance from the target center). From a hat you will draw two slips of paper: a short distance (3 to 7 feet) and a long distance (8 to 15 feet). The slips of paper will have values in 1-foot increments on them (3, 4, 5, 6, or 7 feet for the short distances, and similar for the long distances). When you set up your device, you will be given 2 minutes to set the device for the first test and 2 minutes to set the device for the second (long distance) test.

4. Build a vehicle that will travel over a flat surface (hallway in the building). Points will be awarded for the distance travelled divided by the total (initial) weight of the vehicle. The vehicle must move under its own internal "engine"—the team can only release the stationary vehicle when the test begins. You will have two attempts, and the best value will be recorded. Your instructor may impose an allowable maximum weight. No batteries or electricity can be used.

5. Build a thermometer. You are to be able to measure the temperature of cold water in a bucket, room temperature, and the temperature of hot water in a container (degrees Celsius). The total percent errors will be summed (absolute values), averaged, and subtracted from 100.

6. Without moving more than 10 feet from your initial location, position a person a distance of 50 feet (or 100 feet) away from your initial location; bonus points for doing it several ways; the person must initially start beside you. You cannot be connected to the other person in any way (e.g., string, rope). Points are given for accuracy.

7. Roll an object of your design down an inclined plane (provided and the same for all participants). The object is to knock over a small piece of wood placed at a distance of 5 feet from the base of the incline and 5 feet to one side.

8. We are interested in rolling plastic drink bottles across the floor. Rather than hold a race to see which bottle will roll the fastest, we want to determine which will roll the farthest. It is important that each bottle be given a fair chance, so the starting conditions must be the same for each. Each team will use the same ramp (18 inches wide by 24 inches long) and supported by the 4″ dimension of a 2″ × 4″ board. The bottles to be used will be clear-plastic soft-drink bottles. The 2-liter size is probably the best, but smaller bottles could also be used. The objective is to determine the answer to questions. You must develop a defendable test program, carry out the tests, present your results in an easy-to-understand manner, and defend your conclusions.

    (a) How much water should the bottle contain in order to roll the farthest (until it stops)?
    (b) How much water should the bottle contain in order to roll the shortest distance (until it stops)?
    (c) As a part of your test program, you will release two bottles simultaneously on the ramp (with differing amounts of water in them, including one empty). Don't let them roll all the way until they stop, but catch them about 1 foot after they leave the bottom of the ramp. Which moves the fastest, which the slowest, and why?

CHAPTER 4
ENGINEERING COMMUNICATION

It is a common joke that most engineers cannot construct a grammatically correct sentence, and there is all too much truth in this anecdote. In reality, the most successful engineers have developed good communication skills, not only oral and written, but also those involving multimedia formats. You might have the best idea in the known universe, but at some point you are going to have to convince someone to supply the $200,000,000 to develop it. You *must* be able to communicate effectively not only that you have this great idea for a practical antigravity device, but also that it will actually work and that you are the person to lead the team developing it.

Our intent here is not to make you expert communicators, but to at least make you aware of the importance of good communications in engineering, as well as to give you a bit of guidance and practice developing these skills.

WISE WORDS: HOW IMPORTANT ARE COMMUNICATION SKILLS AT YOUR JOB?

When working with clients, we deliver our approach, analysis structure, status, findings, and final deliverable by presentation.

M. Ciuca, ME

Absolutely essential. While I thankfully have a job that allows me to dig into the math and analysis, I still deal with a lot of people. Being able to communicate effectively, where two people (or more) really understand what each is saying is very important on a complex project, and is often more difficult than one would think; or at least it involves more active participation than many are inclined to put into it.

J. Dabling, ME

I have to write design documentation, edit customer specs, produce design review presentations and sometimes present to our customers.

E. D'Avignon, CpE

Being able to connect with customers, and internal team members, is the means to develop relationships and win new business.

B. Dieringer, ME

Absolutely critical. I can think of people that I have worked with who do not have these skills, and they are not easy or fun to work with.

S. Forkner, ChE

Social skills are crucial to success when working at a company. Many times the person who is most successful is not necessarily the one who has the best or brightest ideas, but rather the individual who has solid ideas and is able to communicate in a manner that allows others to easily understand the vision and path to goal achievement.

L. Gascoigne, CE

Good social skills are very important because they improve productivity, team work, and goodwill. I have also found socials skills can be a deciding factor in job advancement.

R. Holcomb, IE

4.1   BASIC PRESENTATION SKILLS

Since most students consider giving an oral presentation a more daunting task than submitting written documents, we focus on live presentations first, although many of the suggestions apply to all forms of engineering communication.

Many years ago, one of us was responsible for a program to recruit high school students into engineering. Each engineering department made a short presentation to the visiting students, extolling the glories of its particular discipline. One department sent its most personable and able communicator about half the time, and Professor X came the other half. Both used the same set of PowerPoint™ slides, but when Professor X showed up, every single student seemed to be completely brain-dead within three minutes. It was *awful*! The other professor maintained their rapt attention for the entire 15 minutes, with supposedly the same presentation.

With this in mind, you need to focus on several factors when planning a presentation. Note that the first item in our list is *who* the audience is, although the other factors mentioned are equally important.

Preplanning

5 W's and 1 H
– Who
– What
– Where
– When
– Why
– How

- **Who is my audience?** Know the age group, demographics, prior knowledge about the topic, and what positions or opinions they may hold.
- **What is my purpose?** What do I hope to accomplish? What response do I expect? What will the audience get out of my speech?
- **Where is all the equipment I need?** Where will the talk be held?
- **When am I on the program agenda?** Will I be the first presenter (when audience is most alert) or the last one before lunch (when they're becoming restless) or after lunch (when they're sleepy)? What will I need to do to keep my listeners attentive?
- **Why am I giving this talk?** Why is the audience here?
- **How long should I talk?** Remember that few people can focus for more than 20 minutes. Trim your talk so that people will ask for more information rather than thinking "When will he sit down?"

Preparing the Verbal Elements

The preceding list focused primarily on logistics. In addition to these considerations, the structure of your presentation is vital. As a simple example, which of the following two sentences is easier to understand?

Sentence A: While perambulating in the antithesis of the metropolis to evade the intemperate brouhaha thereof, my visual cortex perceived an ophidian.

Sentence B: I saw a snake while taking a relaxing walk in the woods.

Although sentence A may be phrased in a more intriguing manner, it tends to obscure the underlying meaning. This is perhaps desirable in poetry or fancy fiction but generally detrimental to a professional engineering presentation.

To help you avoid such pitfalls, we offer the 4-S Formula for structuring presentations.

Shortness + Simplicity + Strength + Sincerity = An Effective Presentation

The 4-S Formula
– Short
– Simple
– Strong
– Sincere

- **Shortness:** Use short sentences; avoid too many details, and don't talk too long.
- **Simplicity:** Avoid wordy, lengthy phrases. Remember the old saying: "Eschew obfuscation!"

- **Strength:** Use active voice and action verbs, not passive voice and "to be" verbs.
- **Sincerity:** Convey empathy, understanding, and respect for the audience.

Three Structural Parts

Keep in mind the purpose of discrete elements of a speech.

> Basic Elements
>
> – Introduction: capture the interest
> – Body: keep the interest
> – Conclusion: pull it all together

- **Introduction:** Purpose: to capture the interest of the audience. Your first task is to *hook your audience*. What is it about your subject that *they* (and not necessarily you) would find most interesting and relevant?
- **Body:** Purpose: to keep your audience interested. They will continue to pay attention if you keep the material interesting and relevant to them.
  - Divide the presentation into two or three main points, and organize your thoughts and ideas around them.
  - Use one or more simple examples to illustrate each major point.
  - Include slides, transparencies, or other visual aids.
- **Conclusion:** Purpose: to pull it all together.
  - Summarize major points.
  - Show appreciation for your audience's attention.
  - Allow for a few questions, but be sensitive to your audience and the other speakers.

Preparing Visual Aids

Most of you have seen presentations that used slides with unreadable text, incomprehensible graphics, or annoying special effects. Well-designed graphics can greatly enhance your presentation, not only making it easier for the audience to understand, but also keeping their attention focused. A picture really is worth about 2^{10} words! The guidelines below will help you design visual aids that are easily understood without being boring. Although our focus is on PowerPoint presentations, these suggestions apply to other media as well.

Helpful Hints

- Keep each slide simple, with one concept per slide. As a rule, use no more than *6 lines per slide*. Each slide should correspond to an average of *60 seconds* of speech.
- If possible, make slides in *landscape* format.
- Present data in simple graphs rather than in lists or tables. Avoid excessively complex graphs with extensive data. If you must present tables, divide them among several slides.
- Pictures, diagrams, and video simulations all may enhance your presentation. Be sure all are large enough to be seen by the audience, and have color schemes that do not appear washed out when projected. Often, such items are designed for viewing on a small screen and do not project well. Be sure to test them prior to your presentation.
- Use bullet points with important phrases to convey ideas. Avoid complete sentences.
- Large size text is best. A font size of at least 18 point and preferably no less than 24 point should be used. This includes *all* objects, such as axis and legend captions, table headings, figure symbols, and subscripts.
- Use high contrast colors. Avoid fancy fonts, such as cursive, or light colors, such as yellow or other pastels. Avoid using all capitals.
- Use a light background and dark print to keep the room brighter. Avoid black slides with white text.
- Keep background styles simple and minimize animation to avoid distracting from the presentation. Keep all the slide backgrounds the same throughout a single presentation.

4.2 SAMPLE PRESENTATIONS

To illustrate the visual aids caveats, we critiqued three student presentations.

Sample Student Presentation #1

Original Presentation: Critique:

- Slide 1 font difficult to read, poor choice of abbreviation for approximately
- Slide 2 dates and text appear disjointed due to text size and graphic; graphic too large
- Slide 3 graph difficult to read text, too many gridlines
- Slide 4 too many words; graphic too large

Improved Presentation:

Sample Student Presentation #2

Original Presentation: Critique:

- Different graphics on every page distracting
- White color is hard to project unless on black background
- Slide 4 graphic is difficult to read; yellow highlights make it worse

Improved Presentation:

Sample Student Presentation #3

Original Presentation: Critique:

- Green backgrounds, white text do not project well; blue text especially hard to read
- Slides 3 and 5 have too many words; should use bullets, not sentences
- No graphics; some pictures of acid rain damage would be nice

Improved Presentation:

Making the Presentation

Oral presentations present several challenges for effective communication. How many of you have had an instructor who simply reads the contents of the slides with no embellishment? How many have had an instructor who seems to be terrified of the audience, cowering in fear and trying to disappear into the wall? How many have tried desperately to read the contents of a slide containing hundreds of words in a minuscule font? If you have not yet suffered through one or more such painful presentations—you will! Don't inflict such things on your own audiences.

Presentation Do's and Don'ts

When delivering a presentation, do

- *Relax!*
- Speak slowly and clearly, making good eye contact.
- When your hands aren't busy, drop them to your sides.
- Rehearse your presentation *out loud* multiple times. If possible, have a friend critique.
- Arrive early enough to make sure all technology is present and working, and resolve any problems you may discover.

When delivering a presentation, don't

> The key to improving presentation skills is practice, practice, practice!

- Lean on your surroundings, turn your back to the audience, or cover your mouth while speaking.
- Read your presentation from a prepared text.
- Tell inappropriate jokes.
- Stammer, over use the pronoun "I," or repeatedly say "um" or "uh." Do not be afraid of a little silence if you need to glance at notes or collect your thoughts.
- Chew gum, remove coins from pockets, crack your knuckles, etc.
- Shuffle your feet or slouch; move repetitively, for example, pace back and forth or sway.
- Play with your notes.

4.3   BASIC TECHNICAL WRITING SKILLS

Although most of you probably consider written communications much easier than oral presentations (the fear factor is largely absent), technical documents you produce will often be far more important to your company and your career than a live presentation of the same information.

General Guidelines

In addition to many of the points made earlier, effective technical writing requires its own set of guidelines.

- **Be clear;** use precise language. Keep wording efficient without losing meaning. Don't exaggerate.
- Ensure that the finished copy logically and smoothly **flows** toward a conclusion. Beware of "choppiness" or discontinuity. Avoid extremely long sentences because they may confuse the reader.
- If possible, use **10-point font size and 1.5 line spacing.**
- **Generally, prefer past tense verbs.** Keep verb tenses in agreement within a paragraph.
- **Define any terms** that might be unfamiliar to the reader, including acronyms and symbols within equations.

Table 1. Example of a properly formatted table.

Current (I) [A]	2	6	10	14	16
Energy of Inductor #1 (E1) [J]	0.002	0.016	0.050	0.095	0.125
Energy of Inductor #2 (E2) [J]	0.010	0.095	0.250	0.510	0.675
Energy of Inductor #3 (E3) [J]	0.005	0.045	0.125	0.250	0.310

Figure 1. Example of a properly formatted figure.

- **Present facts or inferences** rather than personal feelings.
- **Maintain a professional tone.** Do not be emotional or facetious.
- **Number and caption all tables, figures, and appendices.** Refer to each from within the body of the text, numbering them in order of appearance within the text.
  - Tables are numbered and captioned *above* the table.
  - Figures are numbered and captioned *below* the figure.
- **Proofread and edit several times.**
  - **Remember to include** headings, figures, tables, captions, and references.
  - Do not assume the spell check on the computer will catch everything! It will not distinguish between such words as whether and weather or was and as.
- **Read it twice:** once for technical content and once for flow.
  - As you proofread, look for and *remove* the following: unnecessary words; sentences that don't add to the message; superfluous paragraphs.
  - Do one proofreading aloud. When you encounter commas, semicolons, colons, or periods, pause. Read a comma as a brief pause. Read a colon or semicolon as a longer pause. Read a period as a complete stop before the next sentence. Read what is actually written, not what you "think" it should say. If the text sounds stilted or blurred when read, you probably need to reconsider your use of these punctuation marks.
  - If possible, have someone not associated with the project (a roommate, a friend, a mentor) read it and ask that person for suggestions.
- **Spell out a number that starts a sentence**. If the number is large (e.g., a date), reword the sentence.
  - 23 points were outliers. (Unacceptable)
  - Twenty-three points were outliers. (OK)
- **Keep the leading zero with a decimal.**
  - The bridge cost .23 dollars per gram. (Unacceptable)
  - The bridge cost 0.23 dollars per gram. (OK)
- **For long numbers, do not spell out.**
  - The average was one thousand, two hundred fifty-five grams. (Unacceptable)
  - The average was 1,255 grams. (OK)
- **Use the dollar symbol.**
  - The bridge cost four thousand dollars. (Unacceptable)
  - The bridge cost $4,000. (OK)
- **Watch for significant figures. Keep it reasonable!**
  - The bolt is 2.5029 inches long. (Unacceptable)
  - The bolt is 2.5 inches long. (OK)

Proper Use of References

Contributed by: Ms. Jan Comfort—Engineering Reference Librarian, Clemson University Libraries

In today's wired age, most students immediately go to the Internet to find information. Although this can be an excellent source, particularly for preliminary research, there are definite risks associated with using online sources, since essentially anyone can put anything they want on the web. For example, type "flat earth society" into your favorite search engine and check out the "truth" concerning our home planet, or explore how you can save the endangered Pacific Northwest Tree Octopus.

When making presentations or writing reports, it is important to verify the veracity of any sources you consult. These guidelines will hopefully help you avoid egregious errors in your own technical communications.

The ABC's of evaluating information offer a useful start.

- **Authority:** Is it clear who is responsible for the site? What are the author's credentials? Is the author an expert in the field? Is it a .com or .gov or .edu site?
- **Bias:** What is the purpose of the article? Is it free of obvious bias? Is the author presenting an objective view of the subject matter?
- **Currency:** When was the information created or last updated?

But there's more to evaluating resources than that. Good students take it to the next level. Here's how you can, too.

- **Use sources that have been reviewed by experts.** Instead of searching for hours trying to find websites that meet stringent requirements, try using library sources to identify good quality sources that have already been through a review process.
- **Secure a peer review:** An expert in the appropriate field evaluates something proposed (as for research or publication).[1] Academic Search Premier and Expanded

WISE WORDS: HOW IMPORTANT ARE COMMUNICATION SKILLS AT YOUR JOB?

Projects are successful when the people who work together on them are able to communicate clearly with each other and work together to achieve a common goal. Misunderstanding and miscommunication leads to delays, poor quality and frustration.

A. Hu, EE

As an engineer you communicate at all levels from the least senior production employee on the factory floor to the president of the company. Tailoring the message to the audience is the difference in acceptance and rejection.

J. Huggins, ME

Writing clear and concise specifications for construction can make the difference between an under-budget, on-time project and an over-budget, late, and unsafe final product.

L. Johnson, CE

As a consultant, extremely. If clients can't get along with you, they won't hire you. Every job requires a proposal and an interview.

J. Meena, CE

Much of my work in my current job involves researching what is going on in the world, and then putting that information into a format that makes sense to people and helps other people draw conclusions from it. Good written communication skills are essential for what I do every day.

M. Peterson, EE

All the social skills are *extremely* important because of the different functions and technical levels that I interface with.

E. Styles, EE

Here's a joke: "How can you tell an extroverted engineer?" "He looks at *your* shoes when he talks to you." Certainly, engineers are notoriously bad at communicating, but communication— along with teamwork—really separates bad engineers from good ones. Someone could have the best idea in the world, but if he isn't able to describe the invention or provide reasons as to why it should be developed, the idea is useless. Plus, engineers are trained to be rational and thus perfect for managerial positions. If you have good communication skills, one can easily expect you to climb quickly up the corporate ladder.

A. Thompson, EE

[1] peer review. (2009). In Merriam-Webster Online Dictionary. Retrieved May 13, 2009, from http://www.merriam-webster.com/dictionary/peer review

Academic ASAP are the names of two very good multisubject databases that contain scholarly (peer-reviewed) as well as popular articles. One or both of them should be available at your library.

- **Compare the information found in your article or website with content from other websites, or from reviewed sources.** Comparing sources can also alert you to controversial information or bias that will need further study. Are facts from one website the same as those of another? How about depth of coverage? Maybe one site has better-quality information. Does the site have photos or other unique features that make it a good choice? Or perhaps a journal article from a library database is a better source. Until you compare several sources, you won't know what you're missing!

- **Corroborate the information.** Verify the facts from your source—regardless of where you found it—against one or more different sources. Don't take the word of one person or organization. A simple rule might be: "do not use information unless you have corroborated it. Corroboration with varied and reviewed sources increases the probability of success."[2]

4.4 COMMON TECHNICAL COMMUNICATION FORMATS

Technical communications can take on a variety of formats. Here we will specifically address email, memos, and short technical reports. Other, usually longer formats will probably be addressed later in your engineering career, but the same general guidelines apply regardless of form or length of content.

Email

Many students believe the rules they use for instant messaging, Twitter, etc., apply to email also. When sending informal messages to friends, this is acceptable. However, when using email in a professional context, (including email to professors!), more formal rules should be followed. The suggestions below will help you write email that is clear, concise, and appropriate for the recipient.

After you have composed your email, ask yourself if you would mind the president of the university, the CEO of your company, or your parents reading it. If the answer to any of these is no, then you probably should reword it.

Email Etiquette

- Be sure to correctly address the recipient. If you are unsure of a person's proper title (Dr., Mrs., Prof.), look it up!
- Use an appropriate subject line. Avoid silly subjects (Hey – Read This!) or omitting the subject line—this may cause the email to end up in the Junk Mail folder.
- Sign your *full* name and include contact information for email, phone, fax or mailing address if appropriate. When sending email about a class, including your course number and course day and time is often helpful.

---

[2]Meola, Marc. (2004). "Chucking the checklist: A contextual approach to teaching undergraduates web-site evaluation." portal: *Libraries and the Academy*, 4(3), 331–344.

Keep two email accounts: one for professional use, and one for personal use.

Choose email names carefully. Some of our favorites:

- sugarbritches
- guitarfreak
- cdbpimpman
- jessiethestudent
- Bombom bombombo
- YoltsPreston
- fatmarauder

- Change your sending name to your full name (such as Elizabeth Stephan) or an appropriate nickname (Beth Stephan). Do not leave your account as Student or the computer default setting (such as Noname Stephan). The easier it is for your reader to identify who you are, the better your chance for a response.
- Keep it brief. Do not use one continuous paragraph—make it easy to read.
- If you expect a response from your reader, be sure action items are clearly defined.
- Use correct capitalization and punctuation. Spelling does count—even in email! Avoid Instant Messaging (IM) speak (e.g., LOL, IMHO, IIRC).
- Avoid putting anything in email you would not say in person. Do not use email to "vent" or write anything that can be easily misinterpreted by the reader.
- To avoid sending an email before you have a chance to check over your work, fill in the To: and CC: lines last.
- When waiting for a reply, allow a grace period of 48 to 72 hours. If you have not received a reply after 48 hours and a deadline is approaching, you can resend your message, inquiring politely if it was received. Items do sometimes get lost in cyberspace! If the matter is critical, try the phone or request a face-to-face meeting if the first contact does not elicit a response.

Sample Email

To…	R. Swarthmore, Ph.D. [swart@reactorsealsrus.com]
cc…	C. Ohland [carson@reactorsealsrus.com]; K. Stephan [katie@reactorsealsrus.com]
Subject:	Leaky gel reactor seal

Dr. Swarthmore:

The gel reactor seals in B4L3 are leaking and causing production losses (over 200K for FY 2001). The Materials Engineering Lab was asked to test other seal materials in the laboratory and to arrive at a seal couple that has a longer, more efficient service life. Laboratory tests identified six material couples that produced better wear resistance than the current seal. A prototype seal was made with a new material, self-mated cemented carbide, but the carbide on the seal cracked during fabrication.

The purpose of this email is to request an additional $40,000 and four months' project time to fabricate and test another new seal configuration.

To proceed as quickly as possible, your approval of this program before Friday at noon will allow us to proceed with the project without delays. If you have further questions or would like more information, please contact me.

Sincerely,
J. Brock
Design Team Manager, Reactor Seals R Us
(123) 456-7890 × 1234
jbrock@reactorsealsrus.com

EXAMPLE 4-1	**Samples of Poorly Written (Actual Student) Email**

Subject: Carrier Fair

I hoping to be balle ot show up to the Friday class, or take the quiz early on Wed. to allow more time for myself at the Carrier Fair on Wed.

Subject: FW: John Doe fri 8

Im sorry for missing class this morning, I stayed up really lait working on an english paper. Im not really doing that well in any of my classes, but im confident that im going to pass the other ones. Thank god for that. Anyway last night i ate a bunch of fish too, and i think fish doesn't go to well on my stomach because i feel really sick. Could i meet with you sometime this weekend?

Subject: excel

I don't know why it sent 10 other e-mails but sorry. I'm writing to you about the excell test today. I've been sick for the last three days and i wasn't feeling great today either. Also i had a hard time getting started on the test because of downlad problems. Anyway by the time I had figured out the format of the test there were only 15 minutes left so if i could come and talk to you about that then that and one other thing then that would be great.

Subject: Hey, I need help

Hey,

I can't figue out how to do a complex interpolation problem. I understand how to do the simple one, and i used to no how to do the complex one but now everytime I do it i get it wrong. Do you have any office hours I can come in a lean how to do it? If you can't could you send my a worked example of how to do one?
Thank You

Subject: classes for next semester

from what I have heard the math classes we pick at our time pick classes are only for a time. and that we dont pick the teacher we want. is there any way where we can pick the teacher we want for next semester?

Memo (1 Page Limit)

The following is a template for a brief memo. Your professor may ask you to adhere to this format, or may suggest a different one.

General Information:
You should use a 10 point font such as Times New Roman or Verdana, with 1.5 line spacing.

Margins should be set to one inch all around.

Be sure to use correct spelling and grammar.

Include the headings given here, in bold.

To:	Dr. Engineering
From:	Ima Tiger, Section 000 (IMT@school.edu)
Subject:	Memo Guidelines
Date:	May 21, 2009

Introduction: The first 3 or 4 sentences should explain the purpose behind the memo. You should attempt to explain what you were asked to do, what questions you are trying to solve, what process you are attempting to determine, etc.

Results: Place any experimental results, in tabular and/or graphical format, here. As space is limited, this normally only includes two items, either two tables, or two figures, or one table and one figure. Be sure each is clear enough to stand alone, with one to two sentences of explanation. Include a table caption at the top of each table, and a figure caption at the bottom of each figure. The caption should include a number and a word description. When a figure is used within a document, a title is not necessary on the graph and is replaced by the caption. The two items should be pasted side by side from EXCEL into WORD using the **Paste Special > JPEG** command and then sized appropriately.

Discussion: In this section, discuss how you obtained your data, the meaning of any trends observed, and significance of your results. Refer to the tables or figures shown in results by name (Table 1 or Figure 1). Explain any errors in your data (if possible) and how your data differs from theory. If you are deciding between several alternatives, be sure to explain why you did NOT choose the other options.

References: List any sources you use here. You do not need to reference your textbook (Thinking Like an Engineer). You may use a new page for references if necessary. Any reference format is acceptable; Modern Language Association (MLA) citation style is preferred.

EXAMPLE 4-2	Sample of a Poorly Written Student Memo

To: Dr. Engineering
From: Ima Student
Subject: Memo
Date: April 1, 2009

Introduction: We are given the job to analyze the cost of upgrade a machine line, which produces widgets. We were given three companies to choose from, to figure out witch would be the best for the cost and its production. Just by graphing the variables would allow us to find our answer.

Results:

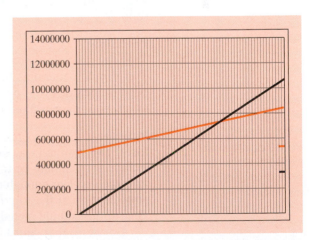

	Klein Teil	
Varible Cost	0.95	
Fixed Cost	5.00E+06	
Material Cost	0.75	
Energy Cost	0.15	
Labor Cost	0.05	
Selling Price	3	
apacity per da	6500	
antity Produc	ein Teil Total Co	Revenue
0.00E+00	5.00E+06	0.00E+00

Discussion: We got the data by taking all variables from the information provided, then graphing the results together. This allowed us to see which machine line would provide the better outcome for the situation at hand. Considering the cost of the machine, material, labor, and energy into consideration with what would produce the quantity and quality product we're striving for. Figure 1 display all three solutions': total cost, revenue for us, and our breakeven point to ensure us of our choice. From observation of the graph we see that the Klein Teil machine is better. It's breakeven is at $2,400,000 and the profit is twice as much.

Summary: So to answer the question, Klein Teil would be our best option. The results yielded that the Klein Teil machine would give us the most quality for its price, a better production rate, and more money in return. From this research I hope you choose to take the Klein Teil machine.

Comments on this memo.
There are so many problems with this submission that we address only the major problems.

- The subject line simply informs us that this is a memo.
- The introduction tells us very little about what the memo will address.
- The same data is presented in both the graph and table. The formatting of the table is very poor; the formatting of the graph is worse.
- The Discussion does not explain how the data was analyzed, and the justification of the final recommendation is essentially nonexistent.
- Similarly, the Summary says almost nothing. Although not an appropriate summary, the three words "Buy Klein Teil" would probably have been more effective.

How many more problems can you find in this sample memo?

Short Report (2–4 pages)

The following template is for a short reports. Again, your professor may ask you to follow these guidelines or provide a somewhat modified version.

Introduction
Type the introduction here. This should be 4 or 5 sentences. What is the problem that will be addressed in this memo? What can the reader expect to learn by reading your report?

Procedure
Type the procedure here. This should be at most ¾ page. It must be in bulleted format. You should generalize the procedure used to include the basic steps, but you do not need to include every detail. The reader should gain an understanding of how you collected your data and performed your analysis.

Results
Insert the results here, but do not discuss them or draw any conclusions. This may include a maximum of three illustrations, in a combination of figures and tables. Be sure each is clear enough to stand alone, with 1 or 2 sentences of explanation. Include a table caption at the top of the table, and a figure caption at the bottom of the figure. The caption should include a number and a word description.

　Be aware that your tables and figures should illustrate *different* ideas; they should not contain the same data. Do *not* include large tables of raw data or *every* graph generated. This section should be a *sample* of those items, used to illustrate the points of your discussion below.

Discussion of Model
Talk about your results here. This can be up to one page maximum. Refer to the table and figure shown in results by name. Be sure to include the items specifically requested in the original project description.

Table 4-1 Example of a properly formatted table

Section	Instructor	E-mail	Time
-030	Dr. Stephan	beths	M 8:00–9:55
-031	Dr. Park	wpark	M 12:20–2:15
-032	Dr. Sill	sillb	W 8:00–9:55
-033	Prof. Bowman	dbow	F 12:20–2:15
-034	Dr. Ohland	matto	T 9:00–10:55

Figure 4-1 Example of a properly formatted figure.

Summary

What is the final conclusion? As your boss, I may only have time to read the introduction (what the problem is) and the summary (what the answer is) and glance at the results (how you got the answer) before I head to a big meeting. This should be 4 to 5 sentences long, and answer the initial questions asked in the introduction and summarize any important findings.

Poster Presentation

The following template is for posters. Again, your professor may ask you to follow these guidelines or provide a somewhat modified version. The template provided online is a PowerPoint format, but is meant to be printed. The default is set to 8½ × 11 printing, which will allow you to submit this to your instructor without the need for a plotter. This could easily be changed, however, and this template be used to create a large poster.

LOGO

LOGO

Project Title

Course- Section
Team Names

BACKGROUND

What important information does the reader need to know before understanding your discussion? Have you built upon someone else's previous work? Why is this work important?

CHART or PICTURE
Illustrating the Purpose
or Background

PURPOSE

Type the introduction here. What is the problem that will be addressed? What can the reader expect to learn?

You may present this as a bulleted list or short sentences. Should be concise and easy to read.

RESULTS

Insert the results here, but do not discuss them or draw any conclusions. Include a table caption at the top of the table, and a figure caption at the bottom of the figure. The caption should include a number and a word description.

Be aware that your tables and figures should illustrate DIFFERENT ideas; they should not contain the same data. Do NOT include large tables of raw data. This section should be a SAMPLE of your work, used to illustrate the points of your discussion. Be sure to follow proper plot rules. Be sure your graphs can be easily read from a distance.

Replace these words with
FIGURE or TABLE

FIGURE or TABLE
of Results

DISCUSSION

Talk about your results here. Refer to the table and figure shown in results by name. Be sure to include the items specifically requested in the original project description.

You may present this as a bulleted list or short sentences. Should be concise and easy to read.

CONCLUSIONS

What is the final conclusion? I may only have time to read the purpose (what is the problem) and the conclusion (what is the answer) and glance at the results (how did you get the answer. This should be short, answer the initial questions asked in the introduction and summarize any important findings. Do NOT introduce any new information here.

REFERENCES (if needed)

(1)Put references here, if you used any

(2)If not, put something else creative here!

In-Class Activities

ICA 4-1

Critique the following student presentation, discussing improvement strategies.

Purpose

In this Business and Engineering Collaborative Project we were given to task to design a product that would make dorm life safer, more enjoyable, or more productive. Our design was driven by our desire to meet customers needs, which would correspond to the ability to make a profit. We will show you all the steps it took to make our product marketable. Finally, we would like to thank everyone for coming and we hope that you agree with us that our product is marketable and of course a masterpiece.

Idea Generation

Here are some of our best ideas....

1. Insulated rooms (sound proof)
2. Loft (size and cost efficient)
3. Air Freshener (Make dorms more livable)
4. Universal tack board (Tack board, phone holder, etc)
5. Fan and light that doubles as one
6. Improving air conditioning unit (Make rooms colder and warmer)
7. Universal Bathroom carrier (Holds toothpaste, vitamins, floss, etc.)
8. Fold down ironing board (Ironing board that folds down with cover behind it)
9. Fold up coffee table (Coffee table that folds up and stores easily)
10. Trash Compactor (Compacts trash so that students don't have to empty trash as often)
11. Black-out shades (Shades that blocks out all light)
12. Safe (Portable and light weight safe that makes storing valuables easy)
13. Better Lighting (More lights and in better locations)
14. Larger laundry bags (Some kind of bag that stores more clothes, but takes up less space)
15. Door Bell (Some kind of portable and cheap door bell)

ICA 4-2

Critique the following student presentation, discussing improvement strategies.

1| Task

Goals:

- Determine a perceivable hole in the college-age consumer market
- Find a marketable product that would bridge that gap
- Research customer needs
- Adhere to strict safety statutes
- Make dorm life more productive and enjoyable.

2| Team Structure

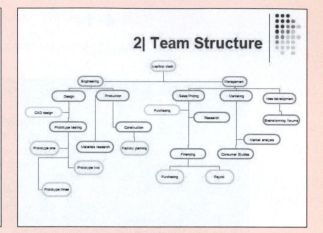

The following three ICAs are possible presentations you may be asked to make in this course. For your presentation, you may be required to choose one of the topics from the list provided, your instructor may specify different topics, or you may even be allowed to choose one of your own. Your instructor will make clear which options you have. Remember, you are giving a *technical* presentation. *Any discussion of the topic that is considered inappropriate by the instructor will be subject to a grade reduction.*

ICA 4-3

This ICA is meant to give you practice getting up in front of a group. The topics are informal, as the purpose is simply to help you gain confidence in public speaking. The presentations topic will be assigned to you at random shortly before your scheduled presentation; your instructor will determine how "shortly" before. The presentation should last *one minute*. If your speech is shorter or longer by more than 15 seconds, 10 points will be deducted from your presentation grade. For your presentation, here are some possible approaches you may want to take.

Make it informative.
Suppose your word was "WHALE." To give an informative talk, make a mental note of important things you know about whales. They are big. Some species are almost extinct. They are mammals. The blue whale is the largest animal that has ever lived on earth. They were hunted for their blubber and oil that was used in lamps before electric lights. They hold their breath while under the water.

Tell a story.
If you know *Moby Dick*, you could give a brief outline of the most important points or of a particular scene in the book. You could tell of a visit to Sea World where you saw the trained killer whales perform. You could make up a short bedtime story that you relate as one you would tell your child or that was told to you as a child.

Be innovative.
You could do something like this: Whale is spelled W-H-A-L-E. The W stands for "water" where the whales live (talk about the ocean for a while); H stands for "huge" which is what we think of when we say whale; A stands for "animal" and the whale is the largest; L stands for "large" which is a lot like huge; and E stands for "eating" since the whale has to eat a lot, or E could stand for "enormous" like huge and large.

or

You could do something like this: I dreamed about a whale that was in the parking lot at K-Mart. It kept smiling, and let out an enormous burp every few minutes. Being an animal lover, I went up next to the whale and stroked its sides. I was amazed at how smooth it was. . . .

or

You could do something like this: I have always wanted to scuba dive, and in particular to ride on the back of a killer whale. I know that it would be dangerous, but it would be something I could remember the rest of my life. . . .

ICA 4-3: Presentation Topic Suggestions

- Person Who Impressed You
- A Hobby
- Favorite Sport
- Favorite Team Mascot
- Favorite Course
- Importance of Cars
- Importance of Space Exploration
- Why Engineers Are Neat
- Why Calculus Is Important
- My Favorite Food
- My Favorite Dessert
- The Worst Insect
- How I Would Improve My School
- Coolest Animal
- Favorite Day of the Week
- A Childhood Memory

- My Favorite Relative
- Someone I Admire and Why
- The Best Car
- Favorite Month of the Year
- Why I Came to This College
- My Dream Job (After College)
- Why I Want to Be an Engineer
- Place I Would Like to Visit
- If I Had a Million Dollars
- The Ultimate Dorm Room
- Favorite Phone App
- Favorite Vacation
- Why I Would (Would Not) Like to be a Professor
- The Best Thing About Being in College
- The Best Thing About Computers

- What Kind of Boss I Would Make
- Why I Would (Would Not) Like to Be the President
- If I Weren't an Engineering Major, I Would Major in . . .
- My Opinion on Napster
- Something I Learned This Week
- The Best Thing About My School
- Advice for Incoming Freshman
- How to Improve This Classroom
- A Feature to Add to the Car
- A Good Book I've Read
- My Favorite Year in Grade School (K–12)
- The Farthest I've Ever Been From My Hometown

ICA 4-4

In this ICA, the speeches will discuss the graphical representation of various phenomena, such as "A total eclipse" or "Using a toaster." The presentations should last *one minute, thirty seconds*. If your speech is more than 15 seconds longer or shorter than this, 10 points will be deducted from your presentation grade.

Later in this text, much time is spent on graphing mechanics and interpretation. You are to choose a topic from the following list and represent it graphically. A variety of presentations of the available data that tell different things are possible. Don't forget to use all your senses and your imagination.

You must prepare a proper graph, incorporated in PowerPoint, and use it during your presentation. You are not allowed to copy a pre-made graph from the Internet! In your speech, you should discuss the process, explain how information is shown on the graph, and be prepared to answer questions about your process. YOU ARE LIMITED TO THREE SLIDES: (1) a title slide, (2) an introductory slide, and (3) a graph. If needed, you will be allowed a single 3″ × 5″ index card with notes on one side only.

ICA 4-4: Presentation Topic Suggestions

- Moving a desk down a set of stairs
- Letting go of a helium balloon
- Feedback from an audio system
- A cow being picked up by a tornado
- A kangaroo hopping along
- A glass of water in a moving vehicle
- People in Florida
- People in Michigan
- The flight of a hot air balloon
- Using a toaster
- A train passing
- Dropping ice in a tub of warm water
- Slipping on a banana peel
- Sound echoing in a canyon
- Detecting a submarine using sonar
- Pouring water out of a bottle
- Boiling water in a whistling teapot
- Hammering nails
- Spinning a Hula-Hoop™

- Playing with a yo-yo
- Snow blowing over a roof
- Climbing a mountain
- Pumping air into a bicycle tire
- The plume from a smokestack
- Formation of an icicle
- A rabbit family
- Firing a bullet from a rifle
- Hiking the Appalachian trail
- Eating a stack of pancakes
- A solar eclipse
- Taking a bath
- Throwing a ball
- Driving from Clemson to Greenville
- Diving into a swimming pool
- A person growing up
- Burning a pile of leaves
- Driving home from work
- An oak tree over the years

- Airplane from airport to airport
- Football game crowd
- Train passing through town
- A coastal river
- The moon
- Daily electric power consumption
- A typical day
- A thunderstorm
- A day in the life of a chicken
- Popping corn
- Feeding birds at a bird feeder
- Skipping a stone on water
- A burning candle
- Pony Express
- Talking on a cell phone
- Using Instant Messenger
- An engineer's salary
- Baseball in play
- Exercising
- "Running the bulls"
- Drag racing
- D.O.R. possum ("dead on the road")
- Political affiliations
- Traffic at intersections
- River in a rainstorm
- A tiger hunting

- Hair
- Grades in calculus class
- Forest fire
- Cooking a Thanksgiving turkey
- Brushing your teeth
- Baking bread
- Eating at a fast-food restaurant
- Studying for an exam
- Power consumption of your laptop
- Power usage on campus
- Oil supply
- Air temperature
- Student attention span during class
- Baseball
- A diet
- Sleeping
- Bird migration
- Strength of concrete
- Human population
- A mosquito
- A video game
- Taking a test
- A trip to Mars
- The North Pole
- Space elevators

ICA 4-5

This presentation covers topics related to your future major. The presentations should last *three minutes*. If your speech is more than 15 seconds longer or shorter than this, 10 points will be deducted from your presentation grade.

In 2000, the National Academy of Engineering compiled a list of "Greatest Engineering Achievements of the 20th Century." The following list contains some of the achievements listed, as well as some of the greatest failures of the past century. The full list can be found at www.greatestchievements.org.

It is your responsibility to research the topic you choose so that you can intelligently discuss it in reference to the engineering involved and how it affects everyday life, both then and now.

You must prepare a PowerPoint presentation and use it in this speech. You should have between 5 and 8 slides, of which the first must be a title slide, and the last must be a slide listing references. Other visual aids may be helpful and are encouraged. If needed, you will be allowed a single 3″ × 5″ index card with notes on one side only.

ICA 4-5: Presentation Topic Suggestions

Electricity
- Grand Coulee Dam (1942)
- Electric light bulbs

Automobiles
- Mass production (1901: Olds; 1908: Ford)
- Octane ratings (1929)
- Goodrich "tubeless" tires (1947)

Airplanes
- Hindenburg (1937)
- Concord

Water Supply and Distribution
- Purification of water supply (1915: Wolman)
- Cuyahoga River pollution (1970s)
- Hinkley, CA (Erin Brockovich)

Electronics
- Integrated circuits (1958: Kilby)
- Hand-held calculator (1967: TI)

Radio and TV
- Color television (1928; 1954)
- Phonograph (Edison)

Agricultural Mechanization
- Mechanical cotton picker (1949: Rust)
- Irrigation equipment (1948: Zybach)

Computers
- Language compiler (1952: Hopper)
- Transistor (1947)

Telephones
- Touch-tone dialing (1961)

Air conditioning and Refrigeration
- Air conditioning (1902: Carrier)
- Freezing cycle for food (1914: Birdseye)

Highways
- PA turnpike
- Failure of the Tacoma Narrows Bridge
- Golden Gate Bridge
- Chunnel

Spacecraft
- Space shuttle *Challenger* explosion
- Hubble telescope
- Apollo 1 and Apollo 13 failures
- Gemini spacecraft

Internet
- Email (1972: Tomlinson)
- WWW (1992)

Imaging
- Sonar (1915: Langevin)
- Radar (1940)
- Ultrasound (1958)
- MRI (1980)

Household Appliances
- Dishwasher (1932)
- Clothes dryer (1935: Moore)
- Microwave (1946)

Health
- Defibrillator (1932: Kauwenhoven)
- Artificial Heart (1980: Jarvik)

Petroleum
- Platforming (1982: Scotland platform)
- Distillation (1920s: Fischer & Tropsch)
- Ethyl gas (1921)
- Alaskan pipeline

Laser and Fiber Optics
- Chemical vapor deposition (1974)
- Optical fibers (1970: Maurer, Corning)

Nuclear Technologies
- Van de Graff generator (1937)
- Manhattan Project (1939–45)
- Three Mile Island power plant (1979)

High Performance Materials
- Vulcanization of rubber (1926)
- Polyvinyl chloride – "PVC" (1927)
- Teflon (1938: Plunkett)
- Kevlar (1971: Kwolek)

ICA 4-6 and 4-7 were written to address the question of what obligations humans may have to protect extraterrestrial life if it is discovered.

ICA 4-6

Critique the following writing assignment, discussing improvement strategies.

If life were to be discovered elsewhere in the universe, it would be our duty to protect that life only to the extent in which we do not compromise our own stability. The nation as a whole currently already has debt issues and is going through a depression. America could not afford to spend money on a life when they are struggling to take care of themselves. According to Hodges, "America has become more a debt 'junkie'—than ever before with total debt of $53 Trillion—and the highest debt ratio in history."(2) Globally there are people who are dying from starvation each day. "On Tuesday September 11, 2001, at least 35,615 of our brother and sisters died from the worst possible death, starvation. Somewhere around 85% of these starvation deaths occur in children 5 years of age or younger." (1) What would make this new life form a higher priority than the neighbors right around the corner? If they have managed to survive this long we shouldn't worry about helping them until we have taken care of ourselves. Even if they are extremely advanced with many discoveries we have not made, what good would it do to spend billions on reverse engineering but leave our own world in *desperation. Now if we were to reach a point in the future in which we could help them; we would naturally have to approach with caution. Politically one of the worst first impressions is when new people meet and one group starts a biological plague. In 1493 right after Columbus came to America he went back to Hispaniola, bringing livestock in order to start a colony there. "Influenza, probably from germs carried by the livestock, swept through the native people, killing many of them. Modern researchers believe that American Indian traders carried the disease to Florida and throughout the Caribbean." (3) We would not know what would hurt them as well as what could hurt us. In order to avoid this terrible introduction, we would need to do research with probes and satellites. These observations from afar would make a seamless introduction possible. After the introduction it would be top priority to keep diplomatic relations intact. Finding another life out in space would be great, but not if we made them our enemies.*

ICA 4-7

Critique the following writing assignment, discussing improvement strategies.

For centuries, people have wondered if we were all alone in our universe. Still today, even with our advancements in science we still have not been able to find life anywhere else. Recently the Phoenix Lander discovered what seems to be incontrovertible evidence that water exists on Mars. Since life is dependent on water eye brows have again been raised about the probability that life may exist, or has existed on Mars. So what if life is discovered somewhere other than on earth? Are we obligated to protect that life? Does that obligation depend on how advanced the life is? How do we stop ourselves from not destroying the non-terrestrial life? If life is discovered somewhere other than earth, some people will feel threaten, others will feel that maybe somehow human life could be supported by the non-terrestrial's planet and that life is elsewhere out there, maybe life like ours. As far as our obligation to protect that life, it is recommended that we do not interfere. We are having problems with our own planet, people pollute, litter, kill wildlife, start wildfires, cut down forests, and destroy whole eco-systems. Imagine what we will do to life if it doesn't matter to us if it lives or not. No matter how advance the life is we should leave the life to itself, unless the life was able to communicate with us somehow. The only way to keep ourselves from destroying the non-terrestrial life is to just to leave it alone. If we leave it, it wouldn't be our fault if it survives or eventually dies off. Yet still we haven't been able to find life anywhere besides earth so until we do we should try finding ways to turn our planet in the right direction before we lose what supports us.

ICA 4-8 and 4-9 were written to address the question of whether we should pursue manned exploration of space or restrict such activities to unmanned robotic devices.

ICA 4-8

Critique the following writing assignment, discussing improvement strategies.

Manned, unmanned space travel has been a big controversy for many years; it's too risky or space bots gets just as much done as humans. I agree 100% that robots are far better off into space than humans; manned space travel is too dangerous. Almost all the money for such a mission would be spent simply to keep the people alive. (John Tierney) Space bots are more reliable, don't get tired, and do what it's programed to do, but more proficient. Things humans need for survival includes: air, food, water, things we must take with us for space travel, which adds to the cost for each individual we fly out into space. While space bots don't need any of these sources for survive, which makes space travel less expensive and more productive. Also the bots can retrieve more data because there aren't any stopping periods for breaks or any other reasons for that matter. It cost about 1.3 billion dollars per shuttle launched out into orbit, while it cost far more per individual we send out into space. Manned space travel is not all bad; things humans can do like, making quick and on the spot decision or human senses to evaluate our surroundings are things robots can't imitate. It's nothing like being somewhere physically and knowing what's there instead of watching it through another pair of eyes. By using unmanned space travel, in reality we can only see what the robot sees, but is a human life worth the risk just for data. A human life is priceless, it like time when it's gone; it's gone, so why not send robotic material into space to collect data. It's only scrap metal, but the life of any human is far more valuable just for the expense of useless data.

ICA 4-9

Critique the following writing assignment, discussing improvement strategies.

Since the early 1960's, humans have been venturing beyond the Earth's atmosphere, a few times as far as the moon, a quarter of a million miles away. NASA currently plans to return humans to the moon by 2020, and plans Mars missions after that. Much criticism has been leveled at the entire idea of manned spaceflight, claiming that unmanned craft can do essentially all of the jobs people can with less cost and less risk. There are of course two sides to the situation. The idea for space exploration and those against it.

Space flight is a very integral part of exploration and adventure. Discovering the worlds around us I a very key part to discovering space. NASA's mission is to pioneer the future in space exploration, scientific discovery and aeronautics research. To do that, thousands of people have been working around the world – and off of it – for almost 50 years, trying to answer some basic questions. What's out there in space? How do we get there? What will we find? What can we learn there, or learn just by trying to get there, that will make life better here on Earth? (NASA). The ideal of many working together to the common goal of space exploration makes it a very important part. Manned spaceflight can be very dangerous. Many deaths and tragedies have occurred because of the exploration of space.

I think that space should still be explored by man kind. With technology advances mentioned inIn the first 25 years of its existence, NASA conducted five manned spaceflight programs: Mercury, Gemini, Apollo, Skylab, and Shuttle.

The latter four programs produced spacecraft that had on-board digital computers. The Gemini computer was a single unit dedicated to guidance and navigation functions. Apollo used computers in the command module and lunar excursion module, again primarily for guidance and navigation. Skylab had a dual computer system for attitude control of the laboratory and pointing of the solar telescope. NASA's Space Shuttle is the most computerized spacecraft built to date, with five general-purpose computers as the heart of the avionics system and twin computers on each of the main engines. The Shuttle computers dominate all checkout, guidance, navigation, systems management, payload, and powered flight functions. The computers helped and as they advance manned spaceflight gets better. In the long run the risk is worth the gain for knowledge.

ICA 4-10 – ICA 4-12 were written memos reporting the findings of experiments measuring the properties of cantilever beams.

ICA 4-10

Critique the following writing assignment, discussing improvement strategies.

Introduction: *This memo is to explain the relationship between the width and elasticity of a cantilever beam. To prepare for this memo, data had to be taken in a lab using mass weights. From there, it is my intention to plot the data on a graph to explain my results. I hypothesize that the wider the beam is, the less elastic it is. I believe that the results will prove the relationship to be exponential.*

Results:

Figure 1. (Force vs. Deflection)

Figure 2. (Stiffness vs. Width)

Discussion: *To acquire the data used, the group had to take measurements using many different Cantilever Beams. All the beams were 18 in. long but varied in width. We hung different masses from the end of the beams and measured the displacement. Figure 1 shows how as more mass was added to the beams, they bent more. This makes sense, as it seems pretty logical. The data was recorded and graphed. A line of best fit was added to the graphs, which gave us Young's Modulous. Young's Modulous was used to graph the data versus the beam width in Figure 2. The trendline for this shows how the beams were exponentially stiffer with wider widths than with narrower beams. Possible errors within our data would most likely result from the accuracy of measurement. The equation of the displacement of a beam is $\delta = FL^3/3EI$ where δ is the displacement, F is the Force, L is the Length of the beam, E is Young's Modulous, which for steel is about 205 GPa. And finally I is the moment of inertia for the beam. Using this equation, I calculated the slope equation for the .5 in beam to be 2.48 whereas we found the actual value to be 2.75. For a percent yield of 90%, this isn't tooo bad. The data in Figure 2 proved my hypotheses to be correct that the beams would bend exponentially more with narrower widths.*

ICA 4-11

Critique the following writing assignment, discussing improvement strategies.

Introduction: *The purpose of this memo is to evaluate the material of Aluminum for stiffness. This is achieved by looking at the relationship between the length of the bar and the deflection. Therefore, through our data we hope to discover the spring constant, k, of Aluminum.*

Results:

Figure 1. Beam Deflection of Various Lengths

Figure 2. Stiffness of Beams of Aluminum

Discussion: *The data was obtained through careful experimentation in which a mass was placed on a beam of Aluminum and the resultant deflection from the original height was measured. Figure 1 shows the originally obtained data for the four lengths of the beams. 4 separate masses were used for four lengths of the beams. The slope of the trend line of the data is the resultant stiffness of the beam with regard to the Length of the beam. It was discovered that the smaller the length of the beam, the stiffer the beam will be. The relationship between the spring constant and the elements of the beam can be defined through the equation and therefore the relationship between our data and the theoretical values of the spring constant gave us data that had an average percent error of 3.64%.*

$$\frac{F}{d} = \frac{Ewt^3}{4L^3} = k$$

ICA 4-12

Critique the following writing assignment, discussing improvement strategies.

Introduction: *The purpose of this memo is to demonstrate the relationship of the stiffness of a cantilever beam to its width. We did this by finding experimental data in the lab and then creating a series of graphs that help us draw conclusions. It turned out that the first graph actually led us to another graph, which we then drew a conclusion from. Here are the graphs:*

Results:

Figure #1: Different Width Beams with Varying Weights

Figure 2: Beam Width's direct effect on the stiffness.

Discussion: *The first chart is a representation of the amount each different beam flexed when we applied different weights (forces) to them. As you can see from the results, each beam deflects proportionately to how much weight was applied, discounting the first two, where I believe we made some errors. These may have arisen from measuring inaccuracies, or clerical errors. However, I looked up the supposed relationship online, and it appears that my original inference is correct; the deflection versus the force is linear. From this, we can then find the stiffness of each individual beam. The stiffness is just the slope of the line formed in figure #1. Then, if we plot it on the ordinate and the beam width on the abscissa, we can see that the relationship demonstrated in figure #2 is also roughly linear. This can also be confirmed on the internet site I found. The significance of this graph is apparent; there is a direct, linear correlation between a beams width and how stiff it is.*

Writing Assignments

For each question, write a response according to the directions given by your instructor. Each response should contain correct grammar, spelling, and punctuation. Be sure to answer the question completely, but choose your words carefully so as to not exceed the word limit if one is given. There is no right or wrong answer; your score will be based upon the strength of the argument you make to defend your position.

1. In the July 20, 2007 edition of the *Greenville News,* the following article appeared:

   NASA faces new worries on O-rings by Traci Watson (USA Today)
   In an eerie parallel to the problems that doomed space shuttle Challenger, *NASA is grappling with O-rings as the agency prepares to launch a second teacher into space. A NASA team is studying why recent batches of O-rings have a higher-than-usual number of specks of unmixed rubber, similar to pockets of flour in a partially mixed bowl of cake batter. If such specks are too large or too close together, they can make the rings stiffer. The specks, detected by X-ray, are about the size of a grain of salt.*

   The latest O-ring problem has been the subject of weeks of testing. Initial tests show the rings with the defects are as strong and resilient as they should be . . . Engineers started to raise questions three to four months ago, when they noticed that the number of unmixed specks had risen. The rings passed all tests for resilience, strength, and other quantities.

   What are the key factors in considering this issue, and what recommendation would you make?

2. In August, 2007, the space shuttle *Endeavor* suffered minor damage to the heat shield tiles during liftoff. After the *Columbia* burned up on reentry in 2003 due to damaged tiles, NASA developed a protocol for repairing damaged tiles while in orbit. After much consideration, NASA decided *not* to attempt to repair the tiles on *Endeavor* while in orbit on this mission.

   Find a minimum of three different references discussing this incident and NASA's decision. Summarize the reasons both for and against repairing the tile while in orbit. If you had been on the team making this decision, would you have argued for or against the repair and why?

3. Research the failure of the Teton Dam. Describe what you think are the fundamental ethical issues involved in the failure. In terms of ethics, compare the Teton Dam failure in 1976 to the failure of levees in New Orleans in 2005.

4. In recent years, many studies have been conducted on the use of cell phones while driving. University of Utah psychologists have published a study showing that motorists who talk on handheld or hands-free cellular phones are as impaired as drunken drivers.

   "We found that people are as impaired when they drive and talk on a cell phone as they are when they drive intoxicated at the legal blood-alcohol limit of 0.08 percent, which is the minimum level that defines illegal drunken driving in most U.S. states," says study

co-author Frank Drews, an assistant professor of psychology. "If legislators really want to address driver distraction, then they should consider outlawing cell phone use while driving."

[*Strayer and Drews, "Human Factors."* The Journal of the Human Factors and Ergonomics Society, *Summer 2006.*]

Another way to approach this problem is to force manufacturers of cell phones to create devices on the phones that would restrict usage if the user is driving, much like a safety on a gun trigger. Research and report current statistics on cell phone or other texting device usage during driving. Discuss the hazards (or lack of hazards) of driving while using a cell phone or other texting device. Discuss and justify whether cell phone manufacturers should be required to provide "safety" features on their devices to prevent usage while driving.

5. According to an article in the *Christian Science Monitor* (January 16, 2008 issue), environmentalists claim that a development around a remote lake 140 miles north of Augusta, Maine, would emit 500,000 tons of carbon dioxide over 50 years, including estimated emissions from cars travelling to and from the development. An environmental group has presented this "carbon footprint" to the state, and is requesting that the impact on the environment become part of the process for granting development permits.

As many as 35 states have adopted climate-action plans, but there are few cases like this in which environmental impact factors into government approval of land development. This could have a significant effect on engineers involved with land development, structures, land-use planning, or environmental impact assessments in the future. The original article can be found at: http://www.csmonitor.com/2008/0116/p01s04-wogi.html.

Discuss both sides of this issue, and take a stance for or against mandating carbon footprint assessment for new developments. Justify your position, including information from at least three sources. Ideas to include in your essay are cases in other states, climate-action plans, calculation of a "carbon footprint," or the land-development approval process. Most importantly, consider how an engineer would view this issue.

6. From 2005–2009, much of the southeast experienced a serious rainfall deficit. Lake levels were at record lows, water restrictions were debated and enacted, and the ecology of the area was changed. Amid all of this are nonessential services that are extremely heavy users of water. Perhaps the most egregious examples are sports fields, and in particular, golf courses. On average, an 18-hole golf course in the United States uses about 100 *million* gallons per year.

Summarize the arguments both in favor and against the heavy use of water for nonessential services, particularly in areas where water is in limited supply. Consider the following:

- If you were in a position to recommend legislation for such water use, what policies would you recommend and why? Consider this from an engineer's point of view.
- Consider the impact on local environments for building dams for water supplies and hydropower. As an engineer, what would you consider to be the benefits and drawbacks of this type of project, and why?

7. Since the early 1960s, humans have been venturing beyond Earth's atmosphere, a few times as far as the moon, a quarter of a million miles away. NASA currently plans to return humans to the moon by 2020, and plans Mars missions after that.

Much criticism has been leveled at the entire idea of manned spaceflight, claiming that unmanned craft can do essentially all the jobs people can with less cost and less risk.

Summarize the arguments on both sides of this issue. If you were in a position to recommend manned versus unmanned exploration of our solar system, which would you choose and why?

8. In summer 2008, the *Phoenix Lander* discovered what seems to be incontrovertible evidence that water exists on Mars. Since life as we know depends on water (for a variety of reasons), this discovery raises the probability that life may exist, or did exist in the past, on Mars.

If life is discovered somewhere other than on Earth, what are our obligations to protect that life? Does that obligation depend on how "advanced" we perceive that life to be? What steps, if any, should be taken to ensure that we do not decimate or destroy the nonterrestrial life?

9. Advances in genetic engineering may make it possible to bring extinct species back to life.

   (a) Should we attempt to restore populations of recently extinct animals, such as the Passenger pigeon, Carolina parakeet, ivory-billed woodpecker, Tasmanian tiger, or Formosan clouded leopard, for which there are many preserved specimens from which DNA could be acquired? There is currently an effort to obtain enough DNA to reestablish a population of Woolly mammoths. What are the ethical issues involved with bringing back a mammal extinct for 3500 years?

   (b) It was announced in February 2009 that a first draft of the genome of *Homo sapiens neanderthalensis* (Neanderthal), comprising about two-thirds of the base pairs, has been completed. If future advances make it possible, should we attempt to bring Neanderthals back to life?

10. If you could go back in time and be part of any engineering achievement in the past, what would you choose?

    Your choice must involve something that was accomplished before 1970, although it might still be in use today. You may go as far back in time as you wish, but not more recently than 1970.

    Your essay must include the following.

    ▪ Why did you choose this specific thing?
    ▪ Discuss the actual development of the item in question. Include issues such as why the item was desired, what problems the designers confronted, and specific design decisions that were made.
    ▪ What effects has the development of this device (or process) had on society and the planet, including positive, negative, and neutral effects.
    ▪ A bibliography with at least three distinct entries. All may be online references, but no more than one may be Wikipedia or similar sites.

11. Choose a bridge innovation to research from a list of potential topics given below, and write a memo or presentation that includes the following:

    ▪ The name and location of the structure, when it was constructed, and other significant attributes.
    ▪ If possible, a photo or sketch.
    ▪ A summary table of the design features. Potential features to consider including, if appropriate:

      • Type of bridge design (arch, beam, suspension, etc.)
      • Dimensions (span, height)
      • Unique design features
      • Cost
      • Awards or superlatives (longest, highest, most cost-effective, costliest).

    ▪ A discussion of the aspects of the bridge design that contributed to its success (materials, structural design, geographical, topographical or climatic challenges, aesthetic qualities, etc.).

    (a) Akashi Kaikyo Bridge, linking the islands of Honshu and Shikoku in Japan
    (b) Cooper River Bridge in Charleston, South Carolina
    (c) Forth Bridge, Scotland
    (d) Gateshead Millennium Bridge, spanning the River Tyne in England
    (e) Hanzhou Bay Bridge, China
    (f) Humber Bridge, England
    (g) Lake Pontchartrain Causeway, Louisiana
    (h) Mackinac Bridge, Michigan
    (i) Millau Viaduct, Millau, France
    (j) Natchez Trace Bridge in Franklin, Tennessee
    (k) Penobscot Narrows Bridge, Maine
    (l) Rio-Antirio Bridge, Greece
    (m) Sundial Bridge at Turtle Bay in Redding, California
    (n) Sunshine Skyway Bridge in Tampa, Florida

   (o) Sydney Harbor Bridge, Australia
   (p) Tower Bridge, London
   (q) Woodrow Wilson Bridge (proposed), Washington, DC
   (r) Zakim Bunker Hill Bridge in Boston in Massachusetts.

12. Choose a major transportation structure failure to research from a list of potential topics given below, and write a memo or presentation that includes the following:

- The name and location of the structure, when it was constructed, and other significant attributes.
- If possible, a photo or sketch.
- A summary table of the design features. Potential features to consider including, if appropriate:
  - Type of bridge design (arch, beam, suspension, etc.)
  - Dimensions (span, height)
  - Unique design features
  - Cost
  - Awards or superlatives (longest, highest, most cost-effective, costliest).
- A summary of the event(s) surrounding the structure's failure, including dates. Discuss aspects of the bridge design that contributed to its failure (materials, structural flaws, misuse, improper maintenance, climatic conditions, etc.). If it was rebuilt, repaired, or replaced, how was the design modified or improved?

   (a) Angers Bridge over the Maine River in Angers, France
   (b) Autoroute 19 Overpass, Quebec
   (c) Arroyo Pasajero Twin Bridges in Coalinga, California
   (d) Banqiao Dam, China
   (e) Charles de Gaulle Airport, France
   (f) Hartford Civic Center, Connecticut
   (g) Hyatt Regency hotel walkway in Kansas City, Missouri
   (h) Kemper Arena, Missouri
   (i) L'ambiance Plaza in Bridgeport, Connecticut
   (j) Loncomilla Bridge, Chile
   (k) Mianus River Bridge in Greenwich, Connecticut
   (l) Millenium Footbridge, London
   (m) Sampoong Department Store, South Korea
   (n) Seongsu Bridge, spanning the Han River in Seoul, South Korea
   (o) Sgt. Aubrey Cosens VC Memorial Bridge in Latchford, Ontario
   (p) Silver Bridge, between Point Pleasant, WV, and Kanauga, Ohio
   (q) The Big Dig in Boston, Massachusetts
   (r) West Gate Bridge in Melbourne, Victoria, Australia.

13. Choose an environmental issue to research from a list of potential topics given below, and write a memo or presentation that includes:

- A clear definition of the problem.
- If possible, a photo or sketch.
- Discussion of the issue; potential features to consider include, if appropriate:
  - Affected areas (local, region, continent, global)
  - Causes, sources
  - Effects (human, animal, vegetation, climate)
  - Prevention or reversal options (including feasibility and cost)
  - Important legislation (or lack thereof).
- The topic should be discussed from an *engineering* viewpoint.

   (a) Acid rain
   (b) Air purification
   (c) Bioremediation
   (d) Clean Air Act
   (e) Cuyahoga River pollution/Clean Water Act

    (f) Deforestation
    (g) Erosion control
    (h) Groundwater pollution
    (i) Kyoto Protocol
    (j) Lake Eutrophication/hypoxia
    (k) Loss of wetlands/ecosystems/pollinators
    (l) Ocean acidification/pollution
    (m) Purification/safe drinking water
    (n) RCRA (Resource Conservation and Recovery Act)
    (o) Smog/particle reduction
    (p) Soil contamination
    (q) Solar energy
    (r) Superfund/CERCLA
    (s) Water Pollution Control Act
    (t) Wind energy.

14. On a separate sheet, write a one-page résumé for one person from the following list. Include information such as education, job experience, primary accomplishments (inventions, publications, etc.), and references. If you want, you can include a photo or likeness no larger than 2 inches by 3 inches. You can add some "made up" material such as current address and references, but don't overdo this.

- Ammann, Othmar
- Ampere, Andre Marie
- Arafat, Yasser
- Archimedes
- Avogadro, Armedeo
- Bernoulli, Daniel
- Bessemer, Henry
- Bezos, Jeffrey P.
- Birdseye, William
- Bloomberg, Micheal
- Bohr, Niels
- Boyle, Robert
- Braun, Wernher von
- Brezhnev, Leonid
- Brown, Robert
- Brunel, Isambard
- Calder, Alexander
- Capra, Frank
- Carnot, Nicolas
- Carrier, Willis Haviland
- Carter, Jimmy
- Cauchy, Augustin Louis
- Cavendish, Henry
- Celsius, Anders
- Clausius, Rudolf
- Coriolis, Gaspard de
- Coulomb, Charles
- Cray, Seymour
- Crosby, Philip
- Curie, Marie
- Dalton, John
- Darcy, Henri
- Deming, W. Edwards
- Diesel, Rudolf
- Dunbar, Bonnie
- Eiffel, Gustave
- Euler, Leonhard
- Fahrenheit, Gabriel
- Faraday, Michael
- Fleming, Sandford
- Fleming, Sanford
- Ford, Henry
- Fourier, Joseph
- Fung, Yuan-Cheng
- Gantt, Henry
- Gauss, Carl Friedrich
- Gibbs, Josiah Willard
- Gilbert, William
- Gilbreth, Lillian
- Goizueta, Robert
- Grove, Andrew
- Hancock, Herbie
- Henry, Beulah
- Hertz, Heinrich Rudolf
- Hitchcock, Alfred
- Hooke, Robert
- Hoover, Herbert
- Hopper, Grace
- Iacocca, Lee
- Joule, James Prescott
- Juran, Joseph Moses
- Kelvin, Lord
- Kraft, Christopher Jr.
- Kwolek, Stephanie
- Landry, Tom
- Laplace, Pierre-Simon
- Leibniz, Gottfried
- LeMessurier, William
- MacCready, Paul
- Mach, Ernst
- McDonald, Capers
- Mestral, George de
- Midgley, Thomas Jr.
- Millikan, Robert
- Navier, Claude-Louis
- Newton, Isaac
- Nielsen, Arthur
- Ochoa, Ellen
- Ohm, Georg
- Pascal, Blaise
- Poiseuille, Jean Loius
- Porsche, Ferdinand
- Prandtl, Ludwig
- Rankine, William
- Rayleigh, Lord
- Resnik, Judith
- Reynolds, Osborne
- Rømer, Ole
- Sikorsky, Igor Ivanovich
- Stinson, Katherine
- Stokes, George
- Sununu, John
- Taguchi, Gen'ichi
- Taylor, Fredrick
- Teller, Edward
- van der Waals, Johannes
- Venturi, Giovanni
- Volta, Count Alessandro
- von Kármán, Theodore
- Watt, James
- Welch, Jack
- Wyeth, Nathaniel
- Yeltsin, Boris

Personal Reflections

15. Please address the following questions in approximately one page. You may write, type, draw, sketch your answers . . . whatever form you would like to use.

  (a) Where are you from?
  (b) What type of engineering are you interested in? Why that field specifically?
  (c) In the circle of your friends and family, are there any engineers?
  (d) How would you classify yourself as a student (new freshman, transfer, upperclassman)?
  (e) What are your activities and hobbies?
  (f) What are you most proud of?
  (g) What are you fantastic at?
  (h) What are you passionate about?
  (i) What has been the most difficult aspect of college so far?
  (j) What do you expect to be your biggest challenge this term?
  (k) Do you have any concerns about this class?
  (l) Anything else you would like to add.

16. In 2008, the National Academy of Engineers (NAE) met and determined the Grand Challenges for Engineering, a list of the 14 key problems for engineers to work toward solving throughout the next century. For any aspiring engineer, reading this list should feel like reading a description of the challenges you will face throughout your career. To read a description of each challenge, as well as a description of some connected areas within each grand challenge, visit the NAE's Engineering Grand Challenges website at http://www. engineeringchallenges.org/.

   After reading through the list of Grand Challenges, write a job description of your dream job. The job description should include the standard components of a job description: title, responsibilities (overall and specific), a description of the work hierarchy (whom you report to, whom you work with), as well as any necessary qualifications. Do not include any salary requirements, as this is a completely fictional position.

   After writing the description of your dream job, identify which of the 14 Grand Challenges this position is instrumental in working toward solving. Cite specific examples of the types of projects your fictional job would require you to do, and discuss the impact of that project on the Grand Challenge.

17. Choose and explain your choice of major, and the type of job you envision yourself doing in 15 years. Consider the following:

  (a) What skills or talents do you possess that will help you succeed in your field of interest?
  (b) How passionate are you about pursuing a career in engineering? If you do not plan on being an engineer, what changed your mind?
  (c) How confident are you in your choice of major? Are you still trying to decide among majors?
  (d) How long will it take you to complete your degree?
  (e) Will you obtain a minor?
  (f) Will you pursue Study Abroad, Co-op, or Internship?
  (g) Do you plan to pursue an advanced degree, or become a Professional Engineer (PE)?
  (h) What type of work (industry, research, academic, medical, etc.) do you see yourself doing?

18. Both the ICA completion record in your textbook (look in the back) and the Ethics Chapter list the ENGINEER'S CREED.

   If you are planning to pursue a career in engineering: Type the creed word for word, then write a paragraph (100–200 words) on what the creed means to you, in your own words, and how the creed make you feel about your chosen profession (engineering).

   If you are planning to pursue a career other than engineering, does your future discipline have such a creed? If so, look this up and type it, then write a paragraph (100–200 words) on what the creed means to you, in your own words, and how the creed makes you feel about your chosen profession. If not, write a paragraph (100–200 words) on what items should be included in a creed if your profession had one and how the lack of a creed makes you feel about your chosen profession.

19. In 2008, the University of Memphis made national headlines when the Memphis Tigers played in the NCAA basketball national championship game. When interviewed by a local newspaper about how he helped his team "own" the tournament, Coach Calipari revealed that before each game, he had his star player write an essay about how the game would play out. This particular player was prone to nervousness, so to help him focus, the coach told him to mentally envision the type of plays and how he himself would react to them.

In this assignment, write a short essay on how you will prepare for and take the final exam. Consider the following:

(a) What will you do to study? What materials will you gather, and how will you use them? Where will you study? Will it be quiet? Will you play music?

(b) What kinds of things could go wrong on the day of the exam, and how would you avoid them? (List at least three.)

(c) What will the exam look like, and how will you work through it?

20. Take a personality test to find your Myers-Briggs Type Indicator (MBTI) code.

(a) Find a description of your 4-letter code. List your MBTI code and four corresponding key terms that describe a person with this code type.

(b) How well do you think these keywords from part (a) describe you? Explain why and give specific examples.

(c) Determine the 4-letter code that is the exact opposite of your code. Find a description of this 4-letter code. List the 4-letter code and the keywords for the personality type that is exactly opposite of your personality.

(d) If you are placed on a team with someone who has the exact opposite personality code from yours, you will exhibit some characteristics that are exactly opposite from your team-mate. Based on your MBTI code, what is one weakness you might bring to a team?

(e) Based on the opposite MBTI code from your own, what is one strength trait that you are lacking and that another person might bring to the team?

(f) Describe two things you can do to work effectively with someone on a team who has an opposite MBTI code from yours.

21. Read the essay "Engineering is an . . . itch!" in the Engineering Essential introduction. Reflect on what it means to have performance-focused versus mastery-focused learning goals.

(a) Describe in your own words what it means to be a performance-based learner compared with a mastery-based learner.

(b) What learning goals do you have? Are these goals performance based or mastery based?

(c) Is it important to you to become more mastery focused?

(d) Do you have different kinds of learning goals than you had in the past, and do you think you will have different learning goals in the future?

22. *Never be afraid to sit awhile and think*. ~ Lorraine Hansberry

A recent article in *Science News* addresses the topics of nature and technology. In our electronic world, we are in constant contact with others, through our cell phones, Ipods, Facebook, email, etc. Researchers at the University of Washington have determined that this effect may create long-term problems in our stress levels and in our creativity.

Scenes of Nature Trump Technology in Reducing Low-level Stress

Technology can send a man to the moon, help unlock the secrets of DNA and let people around the world easily communicate through the Internet. But can it substitute for nature?

"Technology is good and it can help our lives, but let's not be fooled into thinking we can live without nature," said Peter Kahn, a University of Washington associate professor of psychology who led the research team.

"We are losing direct experiences with nature. Instead, more and more we're experiencing nature represented technologically through television and other media. Children grow up

watching Discovery Channel and Animal Planet. That's probably better than nothing. But as a species we need interaction with actual nature for our physical and psychological well-being."

Part of this loss comes from what the researchers call environmental generational amnesia. This is the idea that across generations the amount of environmental degradation increases, but each generation views conditions it grew up with as largely non-degraded and normal. Children growing up today in the cities with the worst air pollution often, for example, don't believe that their communities are particularly polluted.

"This problem of environmental generational amnesia is particularly important for children coming of age with current technologies," said Rachel Severson, a co-author of the study and a University of Washington psychology doctoral student. "Children may not realize they are not getting the benefits of actual nature when interacting with what we're calling technological nature."

[University of Washington (2008, June 16). "Scenes of Nature Trump Technology in Reducing Low-level Stress." *Science Daily*]

Complete the following steps:

Go someplace quiet and spend at least 10 minutes clearing your head. Only after these 10 minutes, get out a piece of paper and sketch something: you see; are thinking of; want to create or invent or imagine, etc. On the same sheet of paper, write a poem about your sketch; something you are thinking of; or quiet, wonder of the universe, lack of technology . . .

Rules:

- Must be done all by hand.
- Must be original work (no copied poems or artwork).
- Ability does not count—draw like you are 5 years old!
- Draw in any medium you want: use pencil, pen, colored pencils, markers, crayons, watercolors.
- Poetic form does not matter: use rhyme, no rhyme, haiku, whatever you want it to be.
- It does not need to be elaborate; simple is fine.

PART 2

PROBLEM PARADIGMS

It isn't that they can't see the solution. It is that they can't see the problem.

GILBERT KEITH CHESTERTON

Paradigm: **par·a·dime** ~ noun;
   definition – _____

Not only do engineers solve problems, but they must also communicate what they have done to others, whether it is their boss, their coworkers, or the general public. Part of clear communication is in presentation, but an equally important part is the proper documentation of the procedures that were followed in arriving at the solution.

Suppose that you have been hired by ExoTechnics Corporation to take over an important project from an engineer that unexpectedly left to follow his dream of herding yaks in Tibet. When you begin looking over the documentation for the project, you find the graph shown below.

Using only what you see on this graph, what can you deduce about the nature of the project and the parameters being presented? For any aspect that is unclear, make your best guess concerning the nature of the project and parameters. List your thoughts here or on a separate piece of paper.

1.

2.

3.

4.

You may find yourself wondering if the former engineer is better at herding yaks than he was at engineering.

Most of you probably quickly realized that even if you have what seems a reasonable guess for the nature of the variables T and/or P, you certainly cannot be sure of your deductions. Some of you may have guessed from the numbers on the horizontal axis, that T is time in years. You would be wrong. Without sufficient information given on a graph, or in the presentation of a project, the reader might as well begin the project from scratch.

To illustrate the value of a properly constructed graph, let's present the same information shown on the previous page, but following the rules we cover later in this section.

Although you may not be familiar with reactor vessels or incandescence (if you had been hired to work on this project, you would be), you now know far more about the project than you could possibly deduce from the previous graph. Make a new list of everything you can learn from this graph and compare it to your previous list.

1.

2.

3.

4.

WISE WORDS: WHAT WAS THE HARDEST ADJUSTMENT FROM HIGH SCHOOL TO COLLEGE?

The biggest adjustment was the overwhelming amount of responsibility that I had to take on. There was no longer anybody there to tell me what to do or when to do it. I had to rely on myself to get everything done. All the things I took for granted when I was at home—not having to do my own laundry, not preparing all of my meals, not having to rely on my alarm clock to wake me up, etc.—quickly became quite apparent to me after coming to college. I had to start managing my time better so that I would have time to get all of those things done.

T. Andrews, CE

For me, the most difficult adjustment from high school to college has been unlearning some of the study habits adopted early on. In high school, you can easily get by one semester at a time and just forget what you "learned" when you move into a new semester or a new chapter of the text. College is just a little bit different. To succeed, you have to really make an effort to keep up with your studies—even the classes you have finished already. If you don't, chances are that a topic mentioned in a prerequisite course is going to reappear in a later class, which requires mastery of the previous material in order to excel.

R. Izard, CpE

The hardest adjustment was learning how to study. I could no longer feel prepared for tests by simply paying attention in class. I had to learn to form study groups and begin studying for tests well in advance. You can't cram for engineering tests.

M. Koon, ME

The hardest adjustment for me, as clichéd as it may sound, was simply time management . . . which goes hand-in-hand with making wise decisions! Engineering demands so much focus and dedication that you really have to *want* it to make it through. The curriculum is tough and fast paced . . . and if you're not prepared for it, it can really take you by surprise.

M. Morgan, ME

Studying, networking, talking to my professors about my strengths and weaknesses, taking responsibility for my actions, just the whole growing up into an adult was tough.

C. Pringle, IE

The hardest adjustment I had to make going from high school to college was realizing that I was on my own—and not just for academics, either. I was responsible for making sure I remembered to eat dinner, for not eating candy bars for lunch every day, for balancing my social life with my studies, for managing my money . . . for everything.

J. Sandel, ME

The hardest adjustment from high school to college was changing my study habits. In high school, teachers coordinated their tests so we wouldn't have several on the same day or even in the same week. I had to learn how to manage my time more efficiently. Moreover, it was difficult to find a balance between both the social and academic aspects of college.

D. Walford

Since the tests cover more material and have more weight in college, I had to alter my study habits to make myself start studying more than a day in advance. It was overwhelming my first semester because there was always something that I could be studying for or working on.

A. Zollinger, CE

LEARNING OBJECTIVES

The overall learning objectives for this part include the following:

- Apply basic concepts of statistics to experimental data;
- Communicate technical information effectively by correctly apply graphing conventions;
- Identify process variability and measurement uncertainty associated with an experimental procedure, and interpret the validity of experimental results;
- Use graphical techniques to create "proper" plots, sketch functions, and determine graphical solutions to problems;
- Use 'practical' skills, such as visualizing common units and conducting simple measurements, calculations, and comparisons to make estimations.

CHAPTER 5
ESTIMATION

Enrico Fermi was a Nobel laureate and one of many brilliant scientists and engineers involved in the Manhattan Project, which developed the first nuclear weapons during World War II. He wrote the following after witnessing the first test of an atomic bomb, called the Trinity Test (see Figure 5-1). Of particular note is the final paragraph.

> On the morning of the 16th of July, I was stationed at the Base Camp at Trinity in a position about ten miles from the site of the explosion.
>
> The explosion took place at about 5:30 A.M. I had my face protected by a large board in which a piece of dark welding glass had been inserted. My first impression of the explosion was the very intense flash of light, and a sensation of heat on the parts of my body that were exposed. Although I did not look directly towards the object, I had the impression that suddenly the countryside became brighter than in full daylight.
>
> I subsequently looked in the direction of the explosion through the dark glass and could see something that looked like a conglomeration of flames that promptly started rising. After a few seconds the rising flames lost their brightness and appeared as a huge pillar of smoke with an expanded head like a gigantic mushroom that rose rapidly beyond the clouds probably to a height of the order of 30,000 feet. After reaching its full height, the smoke stayed stationary for a while before the wind started dispersing it.
>
> About 40 seconds after the explosion the air blast reached me. I tried to estimate its strength by dropping from about six feet small pieces of paper before, during and after the passage of the blast wave. Since at the time there was no wind I could observe very distinctly and actually measure the displacement of the pieces of paper that were in the process of falling while the blast was passing. The shift was about 2½ meters, which, at the time, I estimated to correspond to the blast that would be produced by ten thousand tons of T.N.T.

*Citation: U.S. National Archives, Record Group 227, OSRD-S1 Committee, Box 82 folder 6, "Trinity."

Figure 5-1 The first atomic bomb test, Alamogordo, New Mexico. Courtesy of: Truman Presidential Museum & Library.

Before the test, no one knew what would happen. Speculation among the many people involved concerning the results of the test ranged from nothing (no explosion at all) to setting the planetary atmosphere on fire and destroying all life on Earth. When all the data from the blast were analyzed, the true strength of the blast was calculated to be 19 kilotons. By simply observing the behavior of falling bits of paper

101

ten miles from ground zero, Fermi's estimation of 10 kilotons was in error by less than a factor of 2.

After the war, Fermi taught at the University of Chicago, where he was noted for giving his students problems in which so much information was missing that a solution seemed impossible. Such problems have been named **Fermi problems**, and in general they require the person considering them to determine an answer with far less information than would really be necessary to calculate an accurate value. Engineers are often faced with solving problems for which they do not have all the information. They must be adept at making initial estimates. This skill helps them identify critical information that is missing, develop their reasoning skills to solve problems, and recognize what a "reasonable" solution will look like.

Most practical engineering problems are better defined than Fermi problems and can be estimated more easily and accurately in general. The following are just a few examples of real engineering problems. See if you can estimate answers for these problems. These should be done without reference to the web or any other source of information.

- How many cubic yards of concrete are needed to pave one mile of interstate highway (two lanes each direction)?
- How many feet of wire are needed to connect the lighting systems in an automobile?
- What is the average flow rate in gallons per minute of gasoline moving from the fuel tank to the fuel injectors in an automobile cruising at highway speed?

An accomplished engineer knows the answer to most problems before doing any calculations. This does not mean an answer to three significant figures, but a general idea of the range of values that would be reasonable. For example, if you throw a baseball as high as possible, how long will it take for the ball to hit the ground? Obviously, an answer of a few milliseconds is unreasonable, as is several months. Several seconds seems more realistic. When you actually do a calculation to determine this time, you should ask yourself, "Is my answer reasonable?"

Sample Fermi Problems:
- Estimate the total number of hairs on your head.
- How many drops of water are there in all of the Great Lakes?
- How many piano tuners are there in New York?

EXAMPLE 5-1

Typical Fermi Problem

Every year, numerous people run out of fuel while driving their vehicle on the road. Determine how many gallons of gasoline are carried to vehicles with empty fuel tanks each year in the United States so that the vehicle can be driven to the nearest gas station. Do this without reference to any other material, such as the Internet or reference books.

Estimations

In almost all cases, the first step is to estimate unknown pieces of information that are not available. In general, there are numerous paths to a solution, and different people may arrive at different answers. Often, the answer arrived at is only accurate to within an order of magnitude (a factor of ten) or less. Nonetheless, such problems can provide valuable insight not only into the problem itself but also into the nature of problem solving in general.

Remember, these are estimates, not accurate values. Someone else making these estimates might make different assumptions.

- *Number of people in the United States: 500,000,000 persons*
- *Fraction of people in the United States that drive: 7 drivers per 10 persons*

- *Average times a person runs out of gas per year: 1 "out of gas" per 4 years per driver*
- *Fraction of "out of gas" incidents in which gas is brought to the car (rather than pushing or towing the car to a station): 24 "bring gas to car" per 25 "out of gas"*
- *Average amount of gas carried to car: 1.5 gallons per "bring gas to car"*

Calculation

This is where you combine your estimates to arrive at a solution. In the process, you may realize that you need further information to complete the computation.

$$\text{Drivers in United States} = \left(\frac{5 \times 10^8 \text{ people}}{} \right) \left(\frac{7 \text{ drivers}}{10 \text{ people}} \right) = 3.5 \times 10^8 \text{ drivers}$$

$$\text{Number out of gas per year} = \left(3.5 \times 10^8 \text{ drivers} \right) \left(\frac{1 \text{ out of gas}}{(4 \text{ years})(1 \text{ driver})} \right)$$

$$= 8.75 \times 10^7 \frac{\text{out of gas}}{\text{year}}$$

$$\text{Number bring gas to car per year} = \left(\frac{8.75 \times 10^7 \text{ out of gas}}{\text{year}} \right) \left(\frac{24 \text{ bring to car}}{25 \text{ out of gas}} \right)$$

$$= 8.4 \times 10^7 \frac{\text{bring gas to car}}{\text{year}}$$

$$\text{Amount of gas to cars} = \left(\frac{8.4 \times 10^7 \text{ bring gas to car}}{\text{year}} \right) \left(\frac{1.5 \text{ gallons of gas}}{\text{bring gas to car}} \right)$$

$$= 1.26 \times 10^8 \frac{\text{gallons}}{\text{year}}$$

Thus, we have estimated that about 125 million gallons of gas are taken to "out of fuel" vehicles each year in the United States. It would be perfectly valid to give the answer as "about 100 million" gallons, since we probably have only about one significant digit worth of confidence in our results, if that.

A few things to note about this solution:

- *"Units" were used on all numerical values, although some of these units were somewhat contrived (e.g., "out of gas") to meet the needs of the problem. Keeping track of the units is critical to obtaining correct answers and will be highly emphasized, not only in this text but also throughout your engineering education and career.*
- *The units combine and cancel according to the regular algebraic rules. For example, in the first computation, the unit "persons" appeared in both the numerator and the denominator, and thus canceled, leaving "drivers."*
- *Rather than the computation being combined into one huge string of computations, it was broken into smaller pieces, with the results from one step used to compute the next step. This is certainly not an immutable rule, but for long computations it reduces careless errors and makes it easier to understand the overall flow of the problem.*

Many cars today use a plastic fuel tank. It is more durable and lighter than a conventional steel tank. The picture below shows a blow-molded fuel tank with six layers of plastics. Six layers are required to eliminate any permeation of the fuel through the tank.

The picture on the right shows a model of how the melted plastic flows through the extruder head of a blow-molding machine. The six layers are combined at the bottom to form a parison, which is then blow molded to form the fuel tank.

[Photos courtesy of Kautex Machines]

5.1   GENERAL HINTS FOR ESTIMATION

Order of Magnitude: The term "orders of magnitude" is often used when comparing things of very different scales, such as a small rock and a planet. By far the most common usage refers to factors of ten — for example, three orders of magnitude refer to a difference in scale of $10^3 = 1,000$. If we wanted to consider the order of magnitude between 10,000,000 and 1,000, we would calculate the logarithm of each value ($\log(10,000,000) = 7$ and $\log(1,000) = 3$), thus there are $7 - 3 = 4$ orders of magnitude difference between 10,000,000 and 1,000.

As you gain more knowledge and experience, the types of problems you can estimate will become more complicated. Here we give you a few hints about making estimates.

- Try to determine the accuracy required. Is order of magnitude enough? How about ±25%?
  - What level of accuracy is needed to calculate a satellite trajectory?
  - What level of accuracy is needed to determine the amount of paint needed to paint a specified classroom?
- Remember that a "ballpark" value for an input parameter is often good enough.
  - What is the square footage of a typical house?
  - What is the maximum high temperature to expect in Dallas, Texas, in July?
  - What is the typical velocity of a car on the highway?
- Always ask yourself if it is better to err on the high side or the low side. There are two primary reasons for asking this question.
  - Safety and practical considerations. Will a higher or lower estimate result in a safer or more reliable result?
    - If estimating the weight a bridge can support, it is better to err on the low side, so that the actual load it can safely carry is greater than the estimate.

- For the bridge mentioned above, if estimating the load a single beam needs to support, it is better to err on the high side, thus giving a stronger beam than necessary. Be sure you understand the difference between these two points.
- Estimate improvement. Can the errors cancel each other?
  - If estimating of the product of two numbers, if one of the terms is rounded low, the other should be rounded high to counteract for the lower term.
  - If estimating a quotient, if you round the numerator term on the low side, should the denominator term be rounded low or high?
- Don't get bogged down with second-order or minor effects.
  - If estimating the mass of air in the classroom, do you need to correct for the presence of furniture?
  - In most instances, can the effect of temperature on the density of water be neglected?

5.2 SIGNIFICANT FIGURES

Significant figures or "sig figs" are the digits considered reliable as a result of measurement or calculation. This is not to be confused with the number of digits or decimal places. The number of **decimal places** is simply the number of digits to the right of the decimal point. Example 5.2 below illustrates these two concepts.

EXAMPLE 5-2

Decimal places is the number of digits to the right of the decimal point.

Significant figures are the digits considered reliable.

Number	Decimal Places	Significant Figures
376	0	3
376.0	1	4
376,908	0	6
3,760,000	0	3
3,760,000.	0	7
0.376	3	3
0.37600	5	5
0.0037600	7	5
376×10^{-6}	0*	3

*There is not universal agreement concerning whether numbers in scientific or engineering notation should be considered to have the number of decimal places indicated in the mantissa (as shown), or the number that would be present if the number were written out in standard decimal notation (6 in the last example above).

For those who did not run across the term "mantissa" in high school (or have forgotten it), the two parts of a number expressed in either scientific or engineering notation are the **mantissa** and the **exponent**. (The difference between scientific and engineering notation is discussed later in this chapter.) The mantissa is the part that gives the numerical values of the significant figures; the exponent specifies the location of the decimal point, thus the magnitude of the overall number. In the last row of the table above, the mantissa is 376, the exponent is −6. The mantissa can also contain a decimal point, for example, 3.76×10^{-4}: in this case the mantissa is 3.76 and the exponent is −4.

Mantissa: gives the numerical value of significant figures.

Exponent: specifies the location of the decimal point.

The Meaning of "Significant"

All digits other than zero are automatically considered significant. Zero is significant when:

- It appears between two non-zero numbers
  - 306 has three significant figures
  - 5.006 has four significant figures

- It is a "terminal" zero in a number with a decimal point
  - 2.790 has four significant figures
  - 2000. has four significant figures
  - 2000.0 has five significant figures
  - 2.9100 has five significant figures

Zero is not significant when:

- It is used to fix a decimal place
  - 0.0456 has three significant figures

- It is used in integers without a decimal point that could be expressed in scientific notation without including that zero

  - 2000 has one significant figure (2×10^3)
  - Carefully note the difference between "2000" (2×10^3, thus one sig fig) and "2000." (2.000×10^3, thus four sig figs). The decimal point makes a difference!
  - 35,100 has three significant figures (3.51×10^4).

Comprehension Check 5-1

Determine the number of significant figures and decimal places for each value.

(a) 0.0050
(b) 30,010
(c) 3.00
(d) 447×10^9
(e) 75×10^{-3}
(f) 7,790,200
(g) 8,000,000.
(h) 20.000

Calculation Rules

As an engineer, you will likely find that your job involves the design and creation of a product. It is imperative that your calculations lead to the design being "reasonable." It is also important that you remember that others will use much of your work, including people with no technical training.

   Engineers must not imply more accuracy in their calculations than is reasonable. To assist in this, there are many rules that pertain to using the proper number of "significant figures" in computations. These rules, however, are cumbersome and tedious. In your daily life as an engineer, you might use these rules only occasionally. The rules given below provide a reference if you ever need them; in this text, however, you are simply expected to be *reasonable,* the concept of which is discussed in Section 5.3.

In general, asking yourself if the number of significant figures in your answer is reasonable is usually sufficient. However, it is a good idea to be familiar with the rules, or at least know how to find them and use them if the need ever arises.

Multiplication and Division

A quotient or product should contain the same number of significant figures as the number with the fewest significant figures.

EXAMPLE 5-3

$(2.43)(17.675) = 42.95025 \cong 43.0$

- *(2.43) has three significant figures.*
- *(17.675) has five significant figures.*
- *The answer has three significant figures.*

EXAMPLE 5-4

$(2.479 \text{ hours})(60 \text{ minutes/hour}) = 148.74 \cong 148.7 \text{ minutes}$

- *(2.479 hours) has four significant figures.*
- *(60 minutes/hour) is an exact conversion.*
- *The answer has four significant figures.*

Addition and Subtraction

The answer resulting from an addition or subtraction should show significant figures only as far to the right as the least precise number in the calculation. For addition and subtraction operations, the "least precise" number should be that containing the lowest number of decimal places.

EXAMPLE 5-5

$1725.463 + 489.2 + 16.73 = 1931.393$

- *489.2 is the least precise.*
- *The answer should contain one decimal place: 1931.4*

EXAMPLE 5-6

$903,000 + 59,600 + 104,470 = 1,067,070$

- *903,000 is the least precise.*
- *The answer should be carried to the thousands place: 1,067,000*

Rounding

If the most significant figure dropped is 5 or greater, then increase the last digit retained by 1.

EXAMPLE 5-7

Quantity	Rounded to
43.48	43.5 if three significant figures
43.48	43. if two significant figures
0.0143	0.014 if two significant figures
0.0143	0.01 if one significant figure
1.555	1.56 if three significant figures
1.555	1.6 if two significant figures
1.555	2 if one significant figure

Some numbers, such as certain unit conversions, are considered "exact" by definition. Do not consider them in the determination of significant figures.

In calculations with a *known constant* (such as pi (π), which is defined to an infinite number of significant figures), *include at least two more significant figures* in the constant than are contained in the other values in the calculation.

At what point in a calculation should I round my values?

Calculators are quite adept at keeping track of lots of digits—let them do what they are good at. In general, it is neither necessary nor desirable to round intermediate values in a calculation, and if you do, maintain at least two more significant figures for all intermediate values than the number you plan to use in the final result. The following example illustrates the risk of excessively rounding intermediate results.

Evaluate $C = 10,000(0.6 - (5/9 + 0.044))$

- Using calculator with no intermediate rounding:
$$C = \mathbf{4.4}$$

- Rounding value in inner parenthesis to two sig figs:
$$C = 10,000(0.6 - (0.599555)) = 10,000(0.6 - (0.60)) = \mathbf{0}$$

- Rounding 9/5 to two sig figs:
$$C = 10,000(0.6 - (0.55555 + 0.044)) = 10,000(0.6 - (0.56 + 0.044))$$
$$= 10,000(0.6 - (0.604)) = \mathbf{-40}$$

As you can see, rounding to two significant figures at different points in the calculation gives dramatically different results: 4.4, 0, and −40.

Be very sure you know what effect rounding of intermediate values will have if you choose to do so!

Comprehension Check 5-2

Express the answer to the following, using the correct number of significant digits.

(a) $102.345 + 7.8 - 169.05 =$
(b) $20. * 3.567 + 175.6 =$
(c) $(9.78 - 4.352)/2.20 =$
(d) $(783 + 8.98)/(2,980 - 1,387.2) =$

Combined Arithmetic Operations

EXAMPLE 5-8

$$\frac{(68.71 - 41.9)}{(241.013 + 16.87)} = \frac{26.81}{257.88} = 0.103961874 \cong 0.104$$

5.3 REASONABLENESS

In the preceding discussion of estimation, the word *reasonable* was mentioned in several places. We consider two types of reasonableness in answers to problems in this section.

- **Physically reasonable.** Does the answer make sense in light of our understanding of the physical situation being explored or the estimates that we can make?
- **Reasonable precision.** Is the number of digits in the answer commensurate with the level of accuracy and precision available to us in the parameters of the problem?

When is Something Physically Reasonable?

Here are a few hints to help you determine if a solution to a problem is physically reasonable.

- First, ask yourself if the answer makes sense in the physical world.

  - You determine that the wingspan of a new airplane to carry 200 passengers should be four feet. This is obvious rubbish.
  - You determine that a sewage treatment plant for a community of 10,000 people must handle 100,000 pound-mass of sewage effluent per day. Since a gallon of water weighs about 8 pound-mass, this is about 1.25 gallons per person per day, which is far too low to be reasonable.
  - In an upper-level engineering course, you have to calculate the acceleration of a 1982 Volkswagen® Rabbit with a diesel engine. After performing your calculations, you find that the time required to accelerate from 0 to 60 miles per hour is 38 seconds. Although for a similarly sized gasoline engine, this is a rather low acceleration; for a small diesel engine, it is quite reasonable.
  - You are designing playground equipment, including a swing set. The top support (pivot point) for the swings is 10 feet above the ground. You calculate that a child using one of the swings will make one full swing (forward, then backward) in 3.6 seconds. This seems reasonable.

- If the final answer is in units for which you do not have an intuitive feel, convert to units for which you do have an intuitive feel.

  - You calculate the speed of a pitched baseball to be 2×10^{13} millimeters per year. Converting this to miles per hour gives over 1,400 miles per hour, obviously too fast for a pitched baseball.
  - You are interested in what angle a smooth steel ramp must have before a wooden block will begin to slide down it. Your calculations show that the value is 0.55 radians. Is this reasonable? If you have a better "feel" for degrees, you should convert the value in radians to degrees, which gives 32 degrees, this value seems reasonable.
  - You have measured the force of a hammer hitting a nail by using a brand-new sensor. The result is a value of 110 million dynes. Do you believe this value? Since few engineers work in dynes, it seems reasonable to convert this to units that are more familiar, such as pound-force. This conversion gives a value of 240 pound-force, which seems reasonable.

- You are told by a colleague that a ¾-inch pipe supplying water to a chemical process delivers 10 cubic meters of water per day. Converting to gallons per minute gives 1.8 gallons per minute, which seems completely reasonable.

■ If your solution is a mathematical model, consider the behavior of the model at very large and very small values.

- You have determined that the temperature (T in degrees Fahrenheit) of a freshly forged steel ingot can be described as a function of time (t in minutes) by this expression: $T = 2500 - 10t$. Using this equation to calculate the temperature of the ingot, you discover that after less than 300 minutes (6 hours), the temperature of the ingot will be less than absolute zero!

- You have determined the temperature (T in degrees Celsius) of a small steel rod placed over a Bunsen burner with a flame temperature of 1000 degrees Celsius is given by $T = 960 - 939e^{-0.002t}$, where t is the time in seconds from the first application of the flame to the rod. At time $t = 0$, $T = 21$ degrees Celsius (since the exponential term will become $e^0 = 1$). This seems reasonable, since it implies that the temperature of the rod at the beginning of the experiment is 21 degrees Celsius, which is about room temperature. As the value of time increases, the temperature approaches 960 degrees Celsius (since $e^{-\infty} = 0$). This also seems reasonable since the ultimate temperature of the rod will probably be a bit less than the temperature of the flame heating, because of inefficiencies in the heat transfer process.

- A large tank is filled with water to a depth of 10 meters, and a drain in the bottom is opened so that the water begins to flow out. You determine that the depth (D) of water in the tank is given by $D = 5t^{-0.1}$, where t is the time in minutes after the drain was opened at $t = 0$. As t increases, D approaches zero, as we would expect since all of the water will eventually drain from the tank. As t approaches zero, however, D approaches infinity, an obviously ridiculous situation; thus, the model is probably incorrect.

When Is an Answer Reasonably Precise?

First, we need to differentiate between the two terms accurate and precise. Although laypersons tend to use these words interchangeably, an aspiring engineer should understand the difference in meaning as applied to measured or calculated values.

Accuracy is a measure of how close a calculation or measurement (or an average of a group of measurements) is to the actual value. For measured data, if the average of all measurements of a specific parameter is close to the actual value, then the measurement is accurate, whether or not the individual measurements are close to each other. The larger the difference between the measured value and the actual value then the measurement is accurate, is the error in the measurement. Errors come about due to lack of accuracy of measuring equipment, poor measurement techniques, misuse of equipment, and factors in the environment (e.g., temperature or vibration).

Repeatability is a measure of how close together multiple measurements of the same parameter are, whether or not they are close to the actual value.

Precision is a combination of accuracy and repeatability, and is reflected in the number of significant figures used to report a value. The more significant figures, the more precise the value is, assuming it is also accurate.

To illustrate these concepts, consider the distribution of hits on a standard "bull's-eye" target. The figure shows all four combinations of accuracy and repeatable.

■ Neither repeatable nor accurate
■ Repeatable, but not accurate

Most common measuring devices can reliably measure the parameter in question to only three or four significant figures at best, and sometimes less than that. To obtain a really repeatable measurement (e.g., to seven significant figures) in general requires sophisticated (and expensive) equipment.

Not Repeatable | Repeatable

Not Accurate

Accurate

Precise

- Accurate, but not repeatable
- Both repeatable and accurate. This is called precise.

When considering if the precision of a numeric value is reasonable, always ask yourself the following questions:

- How many significant figures do I need in my design parameters? The more precision you specify in a design, the more it will cost and the less competitive it will be unless the extra precision is really needed.

  - You can buy a really nice 16-ounce hammer for about $20. If you wanted a 16 ± 0.0001-ounce hammer, it would probably cost well over a hundred dollars, possibly thousands.

- What are the inherent limitations of my measuring equipment? How much is the measurement affected by environmental factors, user error, etc.?

  - The plastic ruler you buy at the discount store for considerably less than a dollar will measure lengths up to twelve inches with a precision of better than 0.1 inch, but not as good as 0.01 inch. On the other hand, you can spend a few hundred dollars for a high-quality micrometer and measure lengths up to perhaps six inches to a precision of 0.0001 inches.

  - Pumps at gas stations all over the United States often display their gas price and the amount of gas pumped to three decimal places. When gas prices are high, it is extremely important to consumers that the pumps are correctly calibrated. The National Institute of Standards and Technology (NIST) requires that for every 5 gallons pumped, the amount must not be off by more than 6 cubic inches. To determine if a gas pump is calibrated correctly, you need to be able to see to three decimal places the amount pumped since 6 cubic inches is approximately 0.026 gallons.

You should report values in engineering calculations in a way that does not imply a higher level of accuracy than is known. Use the fewest number of decimal places without reducing the usefulness of the answer. Several examples, given below, illustrate this concept.

- We want to compute the area (A) of a circle. We measure the diameter (D) as 2.63 centimeters. We calculate

$$A = \frac{1}{4}\pi D^2 = \frac{1}{4}\pi(2.63)^2 = 5.432521 \text{ centimeters squared}$$

The value of π is known to as many places as we desire, and ¼ is an exact number. It seems reasonable to give our answer as 5.4 square centimeters or 5.43 square centimeters since the original data of diameter is given to two decimal places. Most of the time, reporting answers with two to four significant digits is acceptable and reasonable.

- We want to compute the area (A) of a circle. We measure the diameter (D) as 0.0024 centimeters. We calculate

$$A = \frac{1}{4}\pi D^2 = \frac{1}{4}\pi(0.0024)^2 = 0.0004446 \text{ centimeters squared}$$

If we keep only two decimal places, our answer would be 0.00 square centimeters, which has no meaning. Consequently, when reporting numerical results, particularly those with a magnitude much smaller than 1, we use significant figures, not decimal places. It would be reasonable to report our answer as 0.00044 or 4.4×10^{-4} square centimeters.

- We want to determine a linear relationship for a set of data, using a standard software package such as Excel®. The program will automatically generate a linear relationship based on the data set. Suppose that the result of this exercise is

$$y = 0.50236x + 2.0378$$

While we don't necessarily have proof that the coefficients in this equation are nice simple numbers or even integers, a look at the equation above suggests that the linear relationship should probably be taken as

$$y = 0.5x + 2$$

- If calculations and design procedures require a high level of precision, pay close attention to the established rules regarding significant digits. If the values you generate are small, you may need more significant digits. For example, if all the values are between 0.02 and 0.04 and you select one significant figure, *all* your values will read 0.02, 0.03 or 0.04; going to two significant figures gives values such as 0.026 or 0.021 or 0.034.
- Calculators are often set to show eight or more decimal places.
  - If you measure the size of a rectangle as 2¹⁄₁₆ inches by 5⅛ inches, then the area is calculated to be 6.4453125 square inches since the calculator doesn't care about how many significant digits result. It is unreasonable that we can determine the area of a rectangle to seven decimal places when we made two measurements, the most accurate of which was 0.0625 inches, or four decimal places.
  - If a car has a mass of 1.5 tons, should we say it has a mass of 3010.29 pound-mass?
- Worksheets in Excel often have a default of six to eight decimal places. Two important reasons to use fewer are that: (1) long decimal places are often unreasonable and (2) columns of numbers to this many decimal places make a worksheet difficult to read and unnecessarily cluttered.

> In general, it is a good idea to set your calculator to show answers in engineering format (or generally less desirable, scientific format) with two to four decimal places.

Comprehension Check 5-3

In each of the cases below, a value of the desired quantity has been determined in some way, resulting in a number displayed on a calculator or computer screen. Your task is to round each number to a *reasonable* number of significant digits—*up* if a higher value is conservative, *down* if a lower value is conservative, and to the *nearest* value if it doesn't make a difference. Specify why your assumption is conservative.

(a)	The mass of an adult human riding on an elevator	178.8 pounds
(b)	The amount of milk needed to fill a cereal bowl	1.25 cups
(c)	The time it takes to sing *Happy Birthday*	32.67 seconds

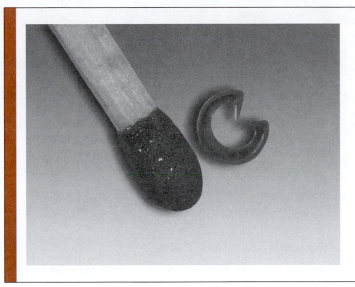

Increasingly, engineers are working at smaller and smaller scales. On the left, a vascular clamp is compared to the tip of a match. The clamp is made from a bio-absorbable plastic through the process of injection molding.

Photo courtesy of S. Stephan

5.4 NOTATION

When discussing numerical values, there are several different ways to represent the values. To read, interpret, and discuss values between scientists and engineers, it is important to understand the different styles of notation. For example, in the United States a period is used as the decimal separator and a comma is used as a digit group separator, indicating groups of a thousand (such as 5,245.25). In some countries, however, this notation is reversed (5.245,25) and in other countries a space is used as the digit group separator (5 245.25). It is important to always consider the country of origin when interpreting written values. Several other types of notations are discussed below.

Engineering Notation versus Scientific Notation

In high school, you probably learned to represent numbers in scientific notation, particularly when the numbers were very large or very small. Although this is indeed a useful means of representing numeric values, in engineering, a slight modification of this notation, called engineering notation, is often more useful. This is particularly true when the value of a parameter can vary over many orders of magnitude. For example, electrical engineers routinely deal with currents ranging from 10^{-15} amperes or less to 10^2 amperes or more.

> Scientific Notation:
> $\#.\#\#\# \times 10^N$
>
> N is an integer
>
> Engineering Notation:
> $\#\#\#.\#\#\# \times 10^M$
>
> M is an integer multiple of 3

Scientific notation is typically expressed in the form $\#.\#\#\# \times 10^N$, where the digit to the left of the decimal point is the most significant non-zero digit of the value being represented. Sometimes, the digit to the right of the decimal point is the most significant digit instead. The number of decimal places can vary, but is usually two to four. N is an integer, and multiplying by 10^N serves to locate the true position of the decimal point.

Engineering notation is expressed in the form $\#\#\#.\#\#\# \times 10^M$, where M is an integer multiple of 3, and the number of digits to the left of the decimal point is either 1, 2, or 3 as needed to yield a power of 10 that is indeed a multiple of 3. The number of digits to the right of the decimal point is typically between two and four.

EXAMPLE 5-9

Standard	Scientific	Engineering
43,480,000	4.348×10^7	43.48×10^6
0.0000003060	3.060×10^{-7}	306.0×10^{-9}
9,860,000,000	9.86×10^9	9.86×10^9
0.0351	3.51×10^{-2}	35.1×10^{-3}
0.0000000522	5.22×10^{-14}	52.2×10^{-15}
456200	4.562×10^8	456.2×10^6

Comprehension Check 5-4

Express each of the following values in scientific and engineering notation.

(a) 58,093,099
(b) 0.00458097
(c) 42,677,000.99

Calculator E-notation

Some computer programs like MATLAB give you an option to display numbers in scientific notation with an upper case or lower case "E."

Most scientific calculators use the capital letter E as shorthand for both scientific and engineering notation when representing numbers. To state the meaning of the letter E in English, it is read as "times 10 raised to the __." For example, 3.707 E −5 would be read as "3.707 times 10 to the negative 5." When transcribing numbers from your calculator, in general it is best *not* to use the E notation, showing the actual power of 10 instead. Thus 3.707 E −5 should be written as 3.707×10^{-5}.

Never use a lower case e for transcribing these values from the calculator, such as 3.707 e^{-5}, since this looks like you are multiplying by the number e ($\cong 2.717$) raised to the negative 5. But if you do use a capital E (which is occasionally, though rarely, justifiable), *do not* superscript the number following the E (e.g., 3.707 E^{-5}) since this looks like you are raising some value E (whatever it may be) to a power.

Situations for Use of an Exponential Notation

Use exponential notation when the magnitude greater than 10,000 or less than 0.0001.

In general, if the magnitude of a number is difficult to almost instantly determine when written in standard notation, use an exponential notation like scientific or engineering notation. Although there are no definite rules for this, if the magnitude is greater than 10,000 or less than 0.0001, you probably should consider using exponential notation. For larger numbers, using the comma notation can extend this range somewhat, for example, 85,048,900 is quickly seen to be 85 million plus a bit. However, there is no similar notation for very small numbers: 0.0000000483 is difficult to simply glance at and realize that it is about 48 billionths.

Note that it is never actually incorrect to use either exponential or standard notation; it is merely a matter of readability. To write 5×10^{20} as 500000000000000000000 is not wrong, but it may leave the readers' eyes vibrating trying to keep track of all the zeros. On the other hand, it is usually silly to write a number like 7 as 7×10^0.

Representation of Fractions and Use of Constants

Many of you have been previously taught that representing a numeric result exactly as a fraction is preferable to giving an approximate answer. This is seldom the preferred method of reporting values in engineering, for two reasons:

- Fractions are often difficult to glance at with instant comprehension of the actual value.

  - Quick! What does 727/41 equal? Did you immediately recognize that it is a little less than 18?
  - Is it easier to know the magnitude of 37/523 or 0.071?

- Seldom do engineers need a precision of more than 3 or 4 digits; thus, there is no need to try to represent values exactly by using the harder-to-read fractions.

Similarly, you may have learned earlier that when calculating with constants such as pi, it is better to leave answers in terms of that constant. For the same reasons cited above, it is generally better to express such values as a decimal number, for example, 27 instead of $8.6\,\pi$.

There are, of course, exceptions to these rules, but in general, a simple decimal number is more useful to engineers.

In-Class Activities

ICA 5-1

With your team, you are to determine common, readily available or understood quantities to help you estimate a variety of parameters. For example, a two-liter bottle of soda weighs about 4 pounds (this is an understood quantity since almost everyone has picked up one of these many times), or the end joint of your middle finger is about an inch long (this is a readily available quantity since it goes everywhere you do). To determine the benchmarks or "helpers," you may use whatever measuring tools are appropriate (rulers, scales, watches, etc.) to determine the values of the common objects or phenomena you choose. Try to determine at least two different estimation "helpers" for the following units:

- **Lengths:** millimeter, centimeter, inch (other than the example given above), foot, meter, kilometer
- **Areas:** square centimeter, square inch, square foot, square meter, acre
- **Volumes:** cubic centimeter, cubic inch, cubic foot, cubic meter, gallon
- **Weights and masses:** gram, newton, pound, kilogram, ton
- **Time:** second, minute, hour (the "helpers" you choose cannot be any form of device designed for measuring time).

ICA 5-2

Materials

Ruler	Tape Measure	Calipers

Procedure

The following measurements and estimations are to be completed individually, not using one set of measurements per team. This activity is designed to help you learn the size of common items to help you with future estimates. Be sure give both the value and the unit for all measurements.

Measure the following:

(a) Your height in meters
(b) Your arm-span (left fingertip to right fingertip) in meters,
(c) The length of your index finger, in centimeters
(d) The width of your thumb, in centimeters
(e) The width of the palm of your hand, in centimeters
(f) The length of your shoe, in feet
(g) The length of your "pace," in feet. A pace is considered the distance between the toe of the rear shoe and the toe of the lead shoe during a normal step. Think about how to make this measurement before doing it.

Determine the following relationships:

(a) How does your arm span measurement compare to your height?
(b) How does your knee height compare to your overall height?
(c) How does the length of your index finger compare to the width of your thumb?

Determine through estimation:

(a) The height of a door in units of feet
(b) The length of a car in units of yards
(c) The area of the floor of the classroom in units of square meters
(d) The volume of your body in units of gallons.

ICA 5-3

(a) If estimating the amount of time to design a new product, should you err on the high side or the low side?
(b) If estimating the switching speed of the transistors for a faster computer, is it better to err on the low side (slower switching) or high side?
(c) How many square yards of fabric are needed to cover the seats in a typical minivan? Is it better to err on the high side or the low side?
(d) Estimate the dimensions of the classroom, in feet. Using these values and ignoring the fact that you would not paint over the windows and doors, estimate the gallons of paint required to paint the classroom. A gallon of paint covers 400 square feet. Would it be better to round the final answer up or down to the nearest gallon?

ICA 5-4

In each of the cases below, a value of some desired quantity has been determined in some way, resulting in a number displayed on a calculator or computer screen. Your task is to

- Round each number to a *reasonable* number of significant digits—*up* if a higher value is conservative, *down* if a lower value is conservative, and to the *nearest* value if it doesn't make a difference.
- Specify why your assumption is conservative.

(a)	Paint needed to cover a single room	4.36 gallons
(b)	Paint needed to paint all rooms in a given building	1123.05 gallons
(c)	The distance from Tampa to New York	1484.2 miles
(d)	The flight time from Charlotte to Atlanta	3 hour, 56 minutes
(e)	The brightness of a light bulb	1690 lumens
(f)	The size of a bolt head	0.500 inch
(g)	The mass of a typical paperclip	1.125 grams

ICA 5-5

(a) How many toothpicks can be made from an 8-foot long 2-inch by 4-inch board?
(b) Who travels farther: a star player during a college basketball game or a golfer playing 18 holes? By how much farther? State all estimations that determine your answer.
(c) Noah's ark has been described as having the following dimensions: 300 cubits long × 50 cubits wide × 30 cubits high. If a cubit is 18 inches, how many people could fit into the ark?
(d) Due to drought, the water level of a 10-acre pond is five feet below normal. If you wanted to fill the pond to normal capacity by using a hose connected to your kitchen faucet, how long would it take to fill the pond? Select an appropriate unit of time to report your answer.
(e) A cubic meter of air has a mass of about 1.2 kilograms. What is the total mass of air in your home or the in a designated building? What is the total mass of air within 1000 feet of sea level on planet Earth?

ICA 5-6

(a) How many basketballs can be placed in the classroom?

(b) A portable device uses two D cells for power and can be operated for 50 hours before the batteries must be replaced. How long would the same device be able to operate if two AA cells were used for power instead of D cells?

(c) How many leaves are on the tree pointed out by the professor? What is the mass of that tree (aboveground portion only)?

(d) How many carrots are used to make all of the canned soup consumed in the United States in one year? (Alternatively, how many acres are used to grow these carrots?)

(e) Estimate the mass of planet Earth.

(f) How much water would be saved if everyone in the United States who does not turn off the faucet while they brush their teeth, did so?

(g) In 1978, cars that got about 40 miles per gallon were readily available. If the average fuel economy of all cars sold since then was 40 miles per gallon (instead of the lower average mileage of the cars that were actually sold), how many billions of gallons of gas would have been saved in the United States since 1978?

(h) If it is raining steadily at a rate of one inch per hour (a container with perpendicular sides and a flat bottom would accumulate water to a depth of one inch in one hour), how many raindrops are in each cubic yard of air approximately ten feet above the ground in an open field at any given instant?

CHAPTER 6
SOLVEM

Everyone solves problems in different ways. There are, however, some procedures and techniques that can help in developing a generally successful problem-solving approach. Although it is not possible to write down a specific recipe that will always work, some broad approaches will help.

6.1 DEFINING SOLVEM

Some problem-solving ideas have been assembled as an approach and given the acronym

SOLVEM:

Sketch
Observations or **O**bjectives
List
Variables and
Equations
Manipulation

SOLVEM:

Sketch
Observations or
Objectives
List
Variables and
Equations
Manipulation

Note that this approach is equally useful for problems involving estimation and more precise calculations. Each step is described below.

Sketch

Figure 6-1 illustrates how a drawing can help you visualize a problem. In sketching a problem, you are subconsciously thinking it. Be sure to draw the diagram large enough so everything is clear, and label the things that you know about the problem in the diagram. For some problems, a before-and-after set of diagrams may be helpful. In very complex problems, you can use intermediate diagrams or subdiagrams as well.

Comprehension Check 6-1

Create a sketch for the following problem. We use SOLVEM to complete this problem in the Comprehension Checks in this chapter.

Calculate the mass in kilograms of gravel stored in a rectangular bin 18.5 feet by 25.0 feet. The depth of the gravel bin is 15 feet, and the density of the gravel is 97 pound-mass per cubic foot.

Figure 6-1 Sketches for seven problems.

Observations, Objectives

An essential part of stating observations is to state the objectives.

These can be in the form of simple statements, questions, or anything else that might acquaint you with the problem at hand. It often helps to divide your observations and objectives into several categories. Some of the easiest to remember are:

- Objective to be achieved
- Observations about the problem geometry (size, shape, etc.)
- Observations about materials and material properties (density, hardness, etc.)
- Observations about parameters not easily sketched (temperature, velocity, etc.)
- Other miscellaneous observations that might be pertinent.

You will almost always find after writing down some observations that you actually know more about the problem than you originally thought.

Here are some typical examples:

When making observations, don't forget you have five senses.

Objective

- Find the velocity, force, flow rate, time, pressure, etc., for a given situation.
- Profitably market the device for less than $25.
- Fit the device into a 12-cubic-inch box.

Observations

PROBLEM GEOMETRY

- The liquid has a free surface.
- The submerged plate is rectangular.
- The support is vertical.
- The tank is cylindrical.
- The cross-sectional area is octagonal.
- The orbit is elliptical.

MATERIALS AND MATERIAL PROPERTIES

- The gate is steel.
- The coefficient of static friction is 0.6.
- The specific gravity is 0.65.
- Ice will float in water.
- The alloy superconducts at 97 kelvin.
- The alloy melts at 543 degrees Fahrenheit.

OTHER PARAMETERS

- If depth increases, pressure increases.
- If temperature increases, resistance increases.
- The flow is steady.
- The fluid is a gas and is compressible.
- The pulley is frictionless.
- The magnetic field is decreasing.
- Temperature may not fall below 34 degrees Fahrenheit.

MISCELLANEOUS

- The force will act to the right.
- Gravity causes the ball to accelerate.
- The sphere is buoyant.
- Drag increases as the speed increases.

> Quantities whose value is zero contain valuable information about the problem.

Remember to include those quantities whose value is zero! Often such quantities are hidden with terms such as:

- constant (implies derivative = 0)
- initially (at time = 0)
- at rest (no motion)
- dropped (no initial velocity)
- at the origin (at zero position)
- constant velocity (force = 0)
- melts (changes phase, temperature is constant).

Finally, one of the most important reasons to make *many* observations is that you often will observe the "wrong" thing. For example, write down things as you read this:

A bus contains 13 passengers.
At the first stop, 4 get off and 2 get on.
At the next stop, 6 get off and 1 gets on.
At the next stop, nobody gets off and 5 get on.
At the next stop, 8 get off and 3 get on.
At the next stop, 1 gets off.
At the last stop, 4 get off and 4 get on.

After putting your pencil down and without looking again at the list, answer the question given below textbox discussing "The Importance of Observations".

Comprehension Check 6-2

State the objective and any relevant observations for the following problem:

> Calculate the mass in kilograms of gravel stored in a rectangular bin 18.5 feet by 25.0 feet. The depth of the gravel bin is 15 feet, and the density of the gravel is 97 pound-mass per cubic foot.

THE IMPORTANCE OF OBSERVATIONS

An excerpt adapted from The Crooked Man *by Sir Arthur Conan Doyle*

Dr. Watson writes: I looked at the clock. It was a quarter to twelve. This could not be a visitor at so late an hour. A patient, evidently, and possibly an all-night sitting. With a wry face I went out into the hall and opened the door. To my astonishment, it was Sherlock Holmes who stood upon my step.

"Ah, Watson, I hoped that I might not be too late to catch you."

"My dear fellow, pray come in."

"You look surprised, and no wonder! Relieved, too, I fancy! Hum! You still smoke the Arcadia mixture of your bachelor days, then! There's no mistaking that fluffy ash upon your coat. It's easy to tell that you've been accustomed to wear a uniform, Watson; you'll never pass as a pure-bred civilian as long as you keep that habit of carrying your handkerchief in your sleeve. Could you put me up for the night?"

"With pleasure."

"You told me that you had bachelor quarters for one, and I see that you have no gentleman visitor at present. Your hat-stand proclaims as much."

"I shall be delighted if you will stay."

"Thank you. I'll find a vacant peg, then. Sorry to see that you've had the British work-man in the house. He's a token of evil. Not the drains, I hope?"

"No, the gas."

"Ah! He has left two nail marks from his boot upon your linoleum just where the light strikes it. No, thank you, I had some supper at Waterloo, but I'll smoke a pipe with you with pleasure."

I handed him my pouch, and he seated himself opposite to me, and smoked for some time in silence. I was well aware that nothing but business of importance could have brought him to me at such an hour, so I waited patiently until he should come round to it.

"I see that you are professionally rather busy just now."

"Yes, I've had a busy day. It may seem very foolish in your eyes, but I really don't know how you deduced it."

"I have the advantage of knowing your habits, my dear Watson. When your round is a short one you walk, and when it is a long one you use a hansom (a carriage). As I per-ceive that your boots, although used, are by no means dirty, I cannot doubt that you are at present busy enough to justify the hansom."

"Excellent!"

"Elementary. It is one of those instances where the reason can produce an effect which seems remarkable to his neighbor, because the latter has missed the one little point which is the basis of the deduction."

While we cannot all be as observant as Sherlock Holmes, we can improve our powers of observation through practice. This will pay dividends as we seek to be engineers with high levels of analytical skills.

QUESTION: For the bus problem, how many stops did the bus make?

The lesson here is that often we may be observing the wrong thing.

List of Variables and Constants

Go over the observations previously determined and list the variables that are important. It may help to divide the list into several broad categories—those related to the geometry of the problem, those related to the materials, and a properties category—although for some types of problems those categories may not be appropriate. Include in your list the written name of the variable, the symbol used to represent the quantity, and, if the value of the variable is known, list the numeric value, including units. If a value is a constant you had to look up, record where you found the information.

List of Variables and Constants should include:
  written name
  symbol
  value
  units
  record of
    information
  location if you
    looked up the
    value

INITIAL AND FINAL CONDITIONS
- initial temperature (T_0) 60 [°F]
- initial radius (r_i) 5 [cm]
- mass of the object (m) 23 [kg]

CONSTANTS
- acceleration of gravity (g) 32.2 [ft/s^2]
- ideal gas constant (R) 8314 [(Pa L)/(mol K)]
- Faraday's constant (F) 9.65×10^4 [C/mol]

GEOMETRY
- length of beam (L) [m]
- diameter of a wheel (d) [m]
- cross-sectional area of a pipe (A) [cm^2]
- volume of a reactor vessel (V) [gal]

MATERIALS

- steel
- polyvinyl chloride (PVC)
- plasma
- gallium arsenide
- medium-density balsa wood

PROPERTIES

- dynamic viscosity of honey (μ) 2,500 [cP]
- density of PVC (ρ) 1380 [kg/m^3]
- spring constant   (k)   0.05 [N/m]
- specific gravity   (SG)   1.34

Comprehension Check 6-3

Create a list of variables and constants for the following problem:

> Calculate the mass in kilograms of gravel stored in a rectangular bin 18.5 feet by 25.0 feet. The depth of the gravel bin is 15 feet, and the density of the gravel is 97 pound-mass per cubic foot.

Equations

Only after the steps above should you begin to think about equations that might govern the problem. It is useful to make a list of the pertinent equations in a broad sense before listing specific expressions. For example:

> For an equation, list the broad category of the equation (Hooke's Law) and then the actual expression (F = k x) to help with problem recognition.

- Conservation of energy
- Conservation of mass
- Conservation of momentum
- Frequency equations
- Ideal gas law

- Newton's laws of motion
- Stress-strain relations
- Surface areas of geometric solids
- Volumes of geometric solids
- Work, energy relations

You may need "subequations" such as:

- distance = (velocity) (time)
- energy = (power) (time)
- force = (pressure) (area)

- mass = (density) (volume)
- voltage = (current) (resistance)
- weight = (mass) (gravity)

Comprehension Check 6-4

Create a list of equations for the following problem:

> Calculate the mass in kilograms of gravel stored in a rectangular bin 18.5 feet by 25.0 feet. The depth of the gravel bin is 15 feet, and the density of the gravel is 97 pound-mass per cubic foot.

Manipulation

> *Do not substitute numerical values of the parameters into the equation right away.*
>
> Instead, manipulate the equation algebraically to the desired form.

Most of the time, you need to manipulate pertinent equations before you can obtain a final solution. *Do not substitute numerical values of the parameters into the equation right away.* Instead, manipulate the equation algebraically to the desired form. Often

---

NOTE
The SOLVEM acronym doesn't contain a word or step for "numerical solution." In fact, this process helps you analyze the problem and obtain an expression or procedure so that you can find a numerical answer. The thought here is that engineers need training to be able to analyze and solve problems. If you can do everything except "substitute numbers" you are essentially finished—as an engineer, you will "be paid the big bucks" for analysis, not for punching a calculator.

you will discover terms that will cancel, giving you a simpler expression to deal with. By doing this, you will:

- Obtain general expressions useful for solving other problems of this type
- Be less likely to make math errors
- Be able to judge whether your final equation is dimensionally consistent
- Better understand the final result.

6.2 REPRESENTING FINAL RESULTS

When you have completed all the steps to SOLVEM, plug in values for the variables and constants and solve for a final answer. Be sure to use reasonableness. The final answer should include both a numeric and unit portion. In addition, it is often useful to write a sentence describing how the answer meets the objectives. Box your final answer for easy identification.

> Don't plug values into the equation until the final step.

Repeated use of SOLVEM can help you develop a better "gut-level understanding" about the analysis of problems by forcing you to talk and think about the generalities of the problem before jumping in and searching for an equation into which you can immediately substitute numbers.

Comprehension Check 6-5

Manipulate and solve for the following problem, using the information from Comprehension Checks 1–4.

Calculate the mass in kilograms of gravel stored in a rectangular bin 18.5 feet by 25.0 feet. The depth of the gravel bin is 15 feet, and the density of the gravel is 97 pound-mass per cubic foot.

6.3 AVOIDING COMMON MISTAKES

Erroneous or argumentative thinking can lead to problem-solving errors. For example,

> Be sure your equation is the right one for the problem you are solving.

- **I can probably find a good equation in the next few pages.** Perhaps you read a problem and rifle through the chapter to find the proper equation so that you can start substituting numbers. You find one that looks good. You don't worry about whether the equation is the right one or whether the assumptions you made in committing to the equation apply to the present problem. You whip out your calculator and produce an answer. *Don't do this!*
- **I hate algebra, or I can't do algebra, or I've got the numbers, so let's substitute the values right in.** Many problems become much simpler if you are willing to do a little algebra before trying to find a numerical solution. Also, by doing some manipulation first, you often obtain a general expression that is easy to apply to another problem when a variable is given a new value. By doing a little algebra, you can also often circumvent problems with different sets of units. *Do some algebra!*

- **It's a simple problem, so why do I need a sketch?** Even if you have a photographic memory, you will need to communicate with people who do not. It is usually much simpler to sort out the various parts of a problem if a picture is staring you right in the face. *Draw pictures!*

- **I don't have time to think about the problem, I need to get this stuff finished.** Well, most often, if you take a deep breath and jot down several important aspects of the problem, you will find the problem much easier to solve and will solve it correctly. *Take your time!*

EXAMPLE 6-1 A Million Paper Clips

Estimate how many miles of wire stock are needed to make one million standard paper clips.

Sketch:

See the adjacent diagram.

Objective:

Determine the amount of wire needed to manufacture a million paper clips.

Observations:

- *Paper clips come in a variety of sizes.*
- *There are four straight segments and three semicircular sections in one clip.*
- *The three semicircular sections have slightly different diameters.*
- *The four straight sections have slightly different lengths.*

List of Variables and Constants:

- L   *Overall length of clip*
- W   *Overall width of clip*
- L_1, L_2, L_3, L_4 *Lengths of four straight sections*
- D_1, D_2, D_3 *Diameters of three semicircular sections*
- P_1, P_2, P_3 *Lengths of three semicircular sections*
- A   *Total amount (length) of wire per clip*

Estimations and Assumptions:

- *Length of clip:* L = 1.5 inch
- *Width of clip:* W = $\frac{3}{8}$ inch
- *Diameters from largest to smallest*
  - D_1 = W = $\frac{3}{8}$ inch
  - D_2 = $\frac{5}{16}$ inch
  - D_3 = $\frac{1}{4}$ inch
- *Lengths from left to right in sketch*
  - L_1 = To be calculated
  - L_2 = 0.8 inch
  - $L_3 \approx L4$ = 1 inch

Equations:

- *Perimeter of semicircle:* P = πD/2 *(half of circumference of circle)*
- $L_1 = L - D_1/2 - D_2/2$
- *Total length of wire in clip:* A = $L_1 + L_2 + L_3 + L_4 + P_1 + P_2 + P_3$

Manipulation:

In this case, none of the equations need to be manipulated into another form.

Length of longest straight side: $L_1 = 1.5 - \frac{3}{16} - \frac{5}{32} \approx 1.2$ inch

Lengths of semicircular sections: $P_1 = \pi \frac{3}{16} \approx 0.6$ inches

$P_2 = \pi \frac{5}{32} \approx 0.5$ inches

$P_3 = \pi \frac{1}{8} \approx 0.4$ inches

Overall length for one clip: $A = 1.2 + 0.8 + 1 + 1 + 0.6$
$+ 0.5 + 0.4 = 5.5$ inches per clip [in/clip]

Length of wire for one million clips: $(5.5 \text{ in/clip}) (1 \times 10^6 \text{clips}) = 5.5 \times 10^6$ inches

Convert from inches to miles: $(5.5 \times 10^6 \text{ inches}) (1 \text{ foot}/12 \text{ inches})$
$(1 \text{ mile}/5280 \text{ feet}) \approx 86.8$ miles

One million, 1.5-inch paper clips require about 87 miles of wire stock.

| EXAMPLE 6-2 | Inflating a Balloon |

A spherical balloon has an initial radius of 5 inches. Air is pumped in at a rate of 10 cubic inches per second, and the balloon expands. Assuming that the pressure and temperature of the air in the balloon remain constant, how long will it take for the surface area to reach 1000 square inches?

Sketch:

 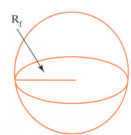

R_i R_f

Objective:

Determine how long it will take for the surface area of the balloon to reach 1000 in^2.

Observations:

- *The balloon is spherical.*
- *The balloon, thus its volume and surface area, gets larger as more air is pumped in.*
- *The faster air is pumped in, the more rapidly the balloon expands.*

List of Variables and Constants:

- *Initial radius: $R_i = 5$ in*
- *Final radius: R_f [in]*
- *Initial surface area: A_i [in^2]*
- *Final surface area: A_f [in^2]*
- *Change in volume: ΔV [in^3]*

- *Initial volume: V_i [in^3]*
- *Final volume: V_f [in^3]*
- *Fill rate: $Q = 10$ [in^3/s]*
- *Time since initial size: t [s]*

Equations:

- *Surface area of sphere:* $A = 4\pi R^2$
- *Volume of sphere:* $V = 4\pi R^3/3$
- *Change in volume:* $\Delta V = Qt$

Manipulation:

There are a few different ways to proceed. The plan used here is to determine how much the balloon volume changes as air is blown into the balloon and to equate this to an expression for the volume change in terms of the balloon geometry (actually the radius of the balloon).

Radius of balloon in terms of surface area:
$$R = \left(\frac{A}{4\pi}\right)^{1/2}$$

Final balloon radius in terms of surface area:
$$R_f = \left(\frac{A_f}{4\pi}\right)^{1/2}$$

Final volume of balloon in terms of surface area:
$$V_f = \left(\frac{4\pi}{3}\right)\left(\frac{A_f}{4\pi}\right)^{3/2}$$

Volume change in terms of air blown in:
$$\Delta V = V_f - V_i = Qt$$

Volume change in terms of geometry:
$$V_f - V_i = \left(\frac{4\pi}{3}\right)\left(\frac{A_f}{4\pi}\right)^{3/2} - \left(\frac{4\pi}{3}\right)R_i^3$$

Solve for time to blow up balloon:
$$t = \left(\frac{4\pi}{3Q}\right)\left(\frac{A_f}{4\pi}\right)^{3/2} - \left(\frac{4\pi}{3Q}\right)R_i^3$$

And simplifying:
$$t = \left(\frac{4\pi}{3Q}\right)\left\{\left(\frac{A_f}{4\pi}\right)^{3/2} - R_i^3\right\}$$

It takes just over 4 minutes to increase the volume to 1000 cubic inches.

Whenever you obtain a result in equation form, you should check to see if the dimensions match in each term. Dimensions are discussed in Part 3.

IMPORTANT

Recall that you should manipulate the equations *before* inserting known values. Note that the final expression for elapsed time is given in terms of initial radius (R_i), flow rate (Q), and final surface area (A_f). So you now have a general equation that can be solved for any values of these three parameters. If you had begun substituting numbers into equations at the beginning and then wanted to obtain the same result for different starting values, you would have to re-solve the entire problem.

In-Class Activities

ICA 6-1

Each of these items should be addressed by a team. List as many things about the specified items as you can determine by observation. One way to do this is to let each team member make one observation, write it down, and iteratively canvass the team until nobody can think of any more additions (it is fine to pass). Remember that you have five senses. Also note that observations are things you can actually detect during the experiment, *not* things you already know or deduce that cannot be observed.

   To help, Example A and Example B are given here before you do one activity (or more) on your own. Remember that not all observations will be important for a particular problem, but write them down anyway—they may trigger an observation that is important.

Example A: A loudspeaker reproducing music

- Electrical signal is sent to speaker.
- Speaker vibrates from electrical signal.
- Speaker motion imparts kinetic energy to the air.
- Diaphragm accelerates.
- Speaker is grey.
- A magnet is involved.
- Speaker is circular.
- Speaker is about 10 inches in diameter.
- Speaker diaphragm is made of paper; vibrating air creates sound.
- Gravity acts on speaker.
- Speaker is in a "box."

Example B: Drinking a soft drink in a can through a straw

- Liquid is cold.
- Straw is cylindrical.
- Can is cylindrical.
- Low pressure in your mouth draws the liquid up.
- Liquid assumes the shape of the container.
- Gravity opposes the rise of the liquid.
- Moving liquid has kinetic energy.
- As the liquid rises, it gains potential energy.
- Table supports the weight of the can, liquid, and straw.
- Can is opaque.
- Liquid is brown.
- Liquid is carbonated (carbonic acid).
- Liquid contains caffeine.
- Friction between straw and lips allows you to hold it.
- Plan view of the can is a circle.
- Silhouette of the can is the same from any direction.
- Silhouette of the can is a rectangle.
- Can is painted.
- Can is metal.
- Liquid surface is horizontal.
- Liquid surface is circular.

From the list below or others as selected by the instructor, list as many observations as you can about the following topic:

(a) An object provided by the professor (placed on the desk)
(b) Candle placed on the desk and lit
(c) Ball dropped from several feet, allowed to bounce and come to rest
(d) A weight on a string pulled to one side and released; watched until it comes to rest.
(e) Coin spinning on the desk
(f) Cup of hot coffee placed on desk
(g) A glass of cold water placed on desk
(h) Book pushed across the desk
(i) Large container guided smoothly up a ramp

(j) Ruler hanging over the desk
(k) Your computer (when turned off)
(l) Your chair
(m) Your classroom
(n) A weight tied on a string and twirled.

Final Assignment of this ICA: You have done several observation exercises. In these, you thought of observation as just a "stream of consciousness" with no regard to organization of your efforts. With your previous observation as a basis, generalize the search for observations into several (three to six or so) categories. The use of these categories should make the construction of a list of observations easier in the future.

ICA 6-2

A ball is dropped from the top of a tall building. A short time (t) later, a second ball is dropped. As the two balls fall, does the distance between them increase, decrease, or remain the same?

Use the SOLVEM method to determine the answer.

ICA 6-3

A hungry bookworm bores through a complete set of encyclopedias consisting of n volumes stacked in numerical order on a library shelf. The bookworm starts inside the front cover of volume 1, bores from page 1 of volume 1 to the last page of the last volume, and stops inside the back cover of the last volume. Note that the worm starts inside the front cover of volume 1 and ends inside the back cover of volume n.

Assume that each volume has the same number of pages. For each book assume that you know how thick the cover is, and that the thickness of a front cover is equal to the thickness of a back cover; assume also that you know the total thickness of all the pages in the book. How far does the bookworm travel? How far will it travel if there are 13 volumes in the set and each book has 2 inches of pages and a $\frac{1}{8}$-inch thick cover?

Use the SOLVEM method to determine the answer.

ICA 6-4

Two cargo trains each leave their respective stations at 1:00 pm and approach each other, one traveling west at 10 miles per hour and the other on separate tracks traveling east at 15 miles per hour. The stations are 100 miles apart. Find the time when the trains meet and determine how far the eastbound train has traveled.

Use the SOLVEM method to determine the answer.

ICA 6-5

Water drips from a faucet at the rate of three drops per second. What distance separates one drop from the following drop 0.65 seconds after the leading drop leaves the faucet? How much time elapses between impacts of the two drops if they fall onto a surface that is 6 feet below the lip of the faucet?

Your sketch should include the faucet, the two water drops of interest, and the impact surface. Annotate the sketch, labeling the each item shown and denote the relevant distances in symbolic form, for example, you might use d_1 to represent the distance from the faucet to the first drop.

Use the SOLVEM method to determine the answer.

ICA 6-6

You purchase a truck having wheels and tires such that each wheel-tire combination has a radius of 14 inches. The speedometer, which reads accurately, is calibrated for these tires. Inspired by a monster-truck event, you decide to jack up your truck such that the wheel-tire radius is 20 inches. Identify the effect this has on your speedometer reading. Specifically, if your speedometer indicates that you are traveling at 60 miles per hour, what is your actual speed?

Use the SOLVEM method to determine the answer.

ICA 6-7

You are trying to push a container of water across the floor. It is so heavy that you can't get it to move. To help, you open a valve at the bottom so that the water begins to drain out. Determine the depth of water at which you can begin to move the container.

To help here, you may need equation that relates to the force of friction that opposed movement. The friction force, F_f is given by: $F_f = \mu w$, where μ is the coefficient of friction and w is the weight of the container.

Use the SOLVEM method to determine the answer.

ICA 6-8

During rush hour, cars back up when the traffic signal turns red. When cars line up at a traffic signal, assume that they are equally spaced (Δx) and that all the cars are the same length (L). You do not begin to move until the car in front of you begins to move, creating a reaction time (Δt) between the time the car in front begins to move and the time you start moving. To keep things simple, assume that when you start to move, you immediately move at a constant speed (v). If the traffic signal stays green for some time (t_g), how many cars (N) will make it through the light?

After you have analyzed this problem with SOLVEM, answer the following:

(a) If the light remains green for twice the time, how many more cars will get through the light?
(b) If the speed of each car when it begins to move is doubled, will twice as many cars get through the light? If not, what variable would have to go to zero for this to be true?
(c) For a reaction time of zero and no space between cars, find an expression for the number of cars that will pass through the light. Does this make sense?

ICA 6-9

Suppose that the earth were a smooth sphere and you could wrap a 25,000-mile-long band snugly around it. Now let's say that you lengthen the band by 10 feet, loosening it just a little. What would be the largest thing that could slither under the new band (assume that it is now raised above the earth's surface equally all the way around so that it doesn't touch anywhere): an amoeba, a snake or an alligator?

Use the SOLVEM method to determine the answer.

Often, the best way to present technical data is through a "picture." But if not done properly, it is often the worst way to display information. As an engineer, you will have many opportunities to construct such pictures. *The Visual Display of Quantitative Information* by Edward Tufte contains one of the best discussions on this topic. If technical data are presented properly in a graph, it is often possible to explain a point in a concise and clear manner that is impossible any other way.

7.1   GRAPHING TERMINOLOGY

Abscissa is the horizontal axis; **ordinate** is the vertical axis. Until now, you have probably referred to these as "x" and "y." This book uses, the terms *abscissa* and *ordinate*, or *horizontal* and *vertical*, since x and y are only occasionally used as variables in engineering problems.

The **independent** variable is the parameter that is controlled or whose values are selected in the experiment; the **dependent** variable is the parameter that is measured corresponding to each set of selected values of the independent variable. Convention usually shows the independent variable on the abscissa and the dependent variable on the ordinate.

Data sets given in tabular form are commonly interpreted and graphed with the leftmost column or topmost row as the independent variable and the other columns as the dependent variable(s). For the remainder of this book, if not specifically stated, assume the abscissa variable is listed in the leftmost column or topmost row in a table of data values.

> **MNEMONIC**
>
> Within the alphabet:
>
> a comes before o
>
> h comes before v
>
> x comes before y
>
> Therefore:
>
> Abscissa = Horizontal axis = x-axis
>
> Ordinate = Vertical axis = y-axis

Time (t) [s]	Distance (D) [m]	
	Car No. 1	Car No. 2
Abscissa	Ordinate	Ordinate

Time (t) [s]		Abscissa
Distance (D) [m]	Car No. 1	Ordinate
	Car No. 2	Ordinate

7.2 PROPER PLOTS

We call graphs constructed according to the following rules **proper plots**.

- Caption with a **brief description.** The restating of "d versus t" or "distance versus time" or even "The relationship between distance and time" does not constitute a proper caption. The caption should give information about the graph to allow the graph to stand alone, without further explanation. It should include information about the problem that does not appear elsewhere on the graph. For example, instead of stating "distance versus time," better choices would be "Lindbergh's Flight Across the Atlantic" or "The Flight of Voyager I" or "Walking between Classes across Campus, Fall 2008." When including a graph as part of a written report, place the caption below the graph.
- **Label both axes clearly.** Three things are required unless the information is unavailable: category (e.g., Time), symbol used (t), and units [s]. Units should accompany all quantities when appropriate, enclosed in square brackets [ ].
- **Select scale increments (both axes) that are easy to read and interpolate between.** With a few exceptions, base your scale on increments of 1, 2, 2.5, and 5. You can scale each value by any power of 10 as necessary to fit the data. Avoid unusual increments (such as 3, 7, 15, or 6.5).

Increment	Sequence				
1	0	10	20	30	40
5	0.05	0.10	0.15	0.20	0.25
2.5	−2500	0	2500	5000	7500
2	6×1^{-5}	8×10^{-5}	1×10^{-4}	1.2×10^{-4}	1.4×10^{-4}

In this final case, reading is easier if the axis is labeled something like Time (t) [s] $\times 10^{-4}$ so that only the numbers 0.6, 0.8, 1.0, 1.2 and 1.4 show on the axis.

- **Provide horizontal and vertical gridlines** to make interpolation easier to aid the reader in determining actual numerical values from the graph.

When minor gridlines are present, the reader should be able to easily determine the value of each minor increment. For example, examine the graphs shown in Figure 7-1.

Figure 7-1

In which graph is it easier to determine the abscissa value for the black point? In the graph on the left, the increment can easily be determined as 0.1 meters. In the graph on the right, it is more difficult to determine the increment as 0.08 meters.

- **Provide a clear legend** describing each data set of multiple data sets shown. Do not use a legend for a single data set. Legends may be shown in a stand-alone box or captioned next to the data set. Both methods are shown in Figure 7-2.

Figure 7-2

Figure 7-3

- **Show measurements as symbols. Show calculated or theoretical values as lines.** Do not display symbols for calculated or theoretical values. A symbol shown on a graph indicates that a measurement has been made. See Figure 7-3.
- **Use a different symbol shape or color** for each experimental data set and a different line style and color for each theoretical data set. **Never use yellow and other light pastel colors** for either symbols or lines. Remember that when graphs are photocopied, all

Figure 7-4

colored lines become black lines. Some colors disappear when copied and are hard to see in a projected image. For example, in Figure 7-4, left, it is much easier to distinguish between the different lines than in the figure on the right.

■ **Produce graphs in portrait orientation** whenever possible within a document. Portrait orientation does not necessarily mean the graph is distorted to be taller than it is wide; it means that readers can study the graph without turning the page sideways.

■ **Be sure the graph is large enough to be easily read.** The larger the graph, the more accurate the extracted information.

Figure 7-5 Example of a proper plot, showing multiple experimental data sets.

Below is an example of a poorly constructed plot. Some problems with this plot are listed below:

■ It is a plot of distance versus time, but is it the distance of a car, a snail, or a rocket? What are the units of distance—inches, meters, miles? What are the units of time— seconds, days, years? Is time on the horizontal or vertical axis?

- Two data sets are shown, or are there three? Why is the one data set connected with a line? Is it a trendline? Is the same data set shown in the triangles? What do the shaded and open triangles represent—different objects, different trials of the same object, or modifications to the same object?
- Lack of gridlines and strange axis increments makes it difficult to interpolate between values.

Figure 7-6

Example of poorly constructed graph.

EXAMPLE 7-1

When attempting to stop a car, a driver must consider both her reaction time and the braking time. The data were taken from www.highwaycode.gov.uk. Create a proper plot of these data.

Vehicle Speed	Distance	
(v) [mph]	Reaction (d_r) [m]	Braking (d_b) [m]
20	6	6
30	9	14
40	12	24
50	15	38
60	18	55
70	21	75

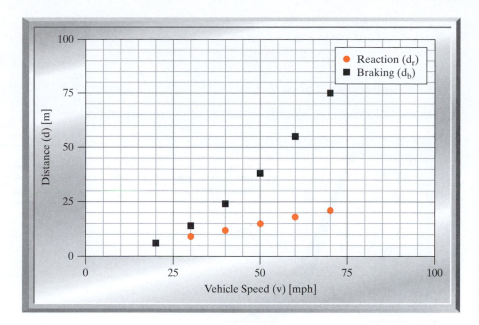

Figure 7-7 At various speeds, the necessary reaction time and braking time needed to stop a car.

EXAMPLE 7-2

Ohm's Law describes the relationship between voltage, current, and resistance within an electrical circuit, given by the equation $V = IR$, where V = voltage [V], I = current [A], and R = resistance [Ω]. Construct a proper plot of the voltage versus current, determined from the equation, for the following resistors: 3000 Ω, 2000 Ω, and 1000 Ω. Allow the current to vary from 0 to 0.05 amperes.

Note that while the lines were probably generated from several actual points along each line for each resistor, the points are not shown; only the resulting line is shown since the values

Figure 7-8 Ohm's law determined for a simple circuit to compare three resistor values within a current range of 0–0.05 amperes.

were developed from theory and not from experiment. If you create a plot like this by hand, you would first put in a few points per data set, then draw the lines and erase the points so they don't show on the final graph.

Comprehension Check 7-1

In the following plot, identify violations of the proper plot rules.

Comprehension Check 7-2

In the following plot, identify violations of the proper plot rules.

Comprehension Check 7-3

In the following plot, identify violations of the proper plot rules.

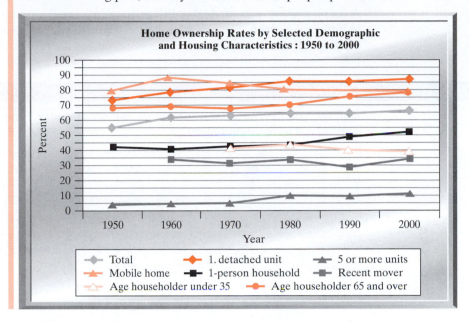

Comprehension Check 7-4

In the following plot, identify violations of the proper plot rules.

7.3   GRAPH INTERPRETATION

A graph conveys a great deal of information in a small amount of space. By being able to interpret a graph, you can infer the story behind the lines. In addition to the value of the slope of the line, the shape of the line contains useful information.

EXAMPLE 7-3

Suppose you have a 100 watt light bulb that is left on for 2 minutes. How much energy is delivered to the light bulb?

If we know that one watt is one joule per second, so for every second it is on, 100 joules are delivered to it. Since it is on for 120 seconds, it uses 12,000 joules.

$$(100 \text{ J/s})(120 \text{ s}) = 12{,}000 \text{ joules}$$

Let's consider the same problem graphically. The power is constant at 100 watts, so a graph of power versus time is simply a horizontal line. Now consider the area under the power line. The area of a rectangle is the width times the height.

Figure 7-9 Two-minutes-on light bulb.

If we make a point of using the units on each axis as well as the numeric values, we get

$$(100 \text{ W})(120 \text{ s})$$

Since a watt is a joule per second, this becomes

$$(100 \text{ J/s})(120 \text{ s}) = 12{,}000 \text{ joules}$$

This is exactly the same result we got above. In other words, the energy used is the area under the line.

This seems like much more effort than the straightforward calculation we did origi-nally, so why should we bother with the graph? Let's look at a slightly more compli-cated situation.

EXAMPLE 7-4

Assume a container is being filled with sand. Initially, the sand enters the container at 100 grams per second, but the rate of filling decreases linearly for 20 seconds, then stops. The final rate of sand into the container just before it stops is 25 grams per sec-ond. How much sand enters the container during the 20 seconds involved?

Let's compute the area under the line, being sure to include units, and see what we get. We can break this area into a rectangle and a triangle, which will make the calculation a bit easier.

- *The area of the rectangle at the base (below $\dot{m}$ = 25) is (25 grams/second) (20 seconds) = 500 grams.*
- *The area of the triangle is 0.5 (75 grams per second) (20 seconds) = 750 grams.*
- *The total area is 1250 grams, the total mass of sand in the container after 20 seconds.*

Again, many of you have realized that there is a much easier way to obtain this result. Simply find the average flow rate (in this case: 62.5 grams per second) and multiply by the total time.

However, what if the parameter plotted on the vertical axis was not a simple straight line? Consider the following example.

EXAMPLE 7-5

What if the parameter plotted on the vertical axis was not a simple straight line? For example, what if the power in the first example was defined by an exponential function: $P = 100\,e^{-0.02t}$? How would you determine the total energy in this case? A graph of this situation is shown in Figure 7-10.

Figure 7-10

Although we might be able to make a reasonable estimate of the area under the curve (the total energy), simple algebra will not suffice to arrive at an accurate value. Those of you who have already studied integral calculus should know how to solve this problem. However, some students using this text may not yet have progressed this far in math, so we will have to leave it at that. It is enough to point out that there are innumerable problems in many engineering contexts that require calculus to solve. To succeed in engineering, you must have a basic understanding of calculus.

EXAMPLE 7-6

From past experience of driving an automobile down a highway, you should understand the concepts relating acceleration, velocity, and distance. As you slowly press the gas pedal toward the floor, the car accelerates, causing both the speed and distance to increase. Once you reach a cruising speed, you turn on the cruise control. Now, the car is no longer accelerating and travels at a constant velocity while increasing in distance. These quantities are related through the following equations:

$$\text{velocity} = (\text{acceleration})(\text{time}) \qquad v = (a)(t)$$

$$\text{distance} = (\text{velocity})(\text{time}) \qquad d = (v)(t)$$

If we create a graph of velocity versus time, the form of the equation tells us that acceleration is the slope of the line. Likewise, a graph of distance versus time has velocity as the slope of the line.

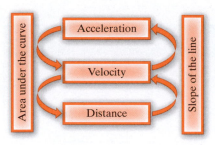

However, if we had a graph of velocity versus time and we wanted to determine distance . . . how can we do this? The distance is determined by how fast we are traveling times how long we are traveling at that velocity; we can find this by determining the area under the curve of velocity versus time. Likewise, if we had a graph of acceleration versus time, we could determine the velocity from the area under the curve. In technical terms, the quantity determined by the slope is referred to as the **derivative**; the quantity determined by the area under the curve is referred to as the **integral**.

In the graph shown Figure 7-11, we drive our car along the road at a constant velocity of 60 miles per hour [mph]. After 1.5 hours, how far have we traveled? The area under the curve, shown by the rectangular box, is

$$\text{Area of the rectangle} = (\text{height of rectangle})(\text{width of rectangle})$$
$$= (60 \text{ mph} - 0 \text{ mph})(1.5 \text{ h} - 0 \text{ h})$$
$$= 90 \text{ miles}$$

Can you explain, in your own words and *without* resorting to calculus terminology, *why* the area under the graph of acceleration and time gives velocity?

Figure 7-11
Example of distance calculation from area under velocity, time graph.

7.4   MEANING OF THE LINE SHAPES

In addition to the value of the slope of the line, the shape of the line contains useful information. In Figure 7-11 above, the speed is shown as a horizontal line. This implies it has a constant value; it is not changing over time. The slope of this line is zero, indicating that the acceleration is zero. Table 7-1 contains the various types of curve shapes and their physical meanings.

Table 7-1 What do the lines on a graph mean?

If the graph shows a . . .	. . . it means that the dependent variable . . .	Sketch
horizontal line	is not changing	
vertical line	has changed "instantaneously"	
straight line positive or negative slope neither horizontal nor vertical	is changing at a constant rate	
curved line concave up, increasing	is changing at an increasing rate	
curved line concave down, increasing	is changing at a decreasing rate	
curved line concave up, decreasing	is changing at a decreasing rate	
curved line concave down, decreasing	is changing at an increasing rate	

In Figure 7-12, the voltage is constant, as indicated by the horizontal line at 23 volts, from time = 0 to time = 8 seconds. At time = 8 seconds, the voltage changes instantly to 15 volts, as indicated by the vertical line. Between time = 8 seconds and time = 20 seconds, the voltage decreases at a constant rate, as indicated by the straight line, and reaches zero volts at time = 20 seconds, where it remains.

Figure 7-12

Figure 7-13

In Figure 7-13, the force on the spring increases at an increasing rate from time = 0 until time = 2 minutes, then remains constant for one minute, after which it increases at a decreasing rate until time = 5 minutes, after which it remains constant at about 6.8 newtons.

The height of a blimp is shown in Figure 7-14. The height decreases at an increasing rate for 5 minutes, then remains constant for 2 minutes, after which its height decreases at a decreasing rate until t = 10 minutes, after which its height remains constant at 10 meters.

Figure 7-14

Comprehension Check 7-5

Use the graph to answer the following questions. Choose from the following answers:

- Increasing at a constant rate
- Equal to a constant, positive value
- Equal to zero
- Equal to a constant, negative value
- Decreasing at a constant rate
- Increasing at a constant rate

- Increasing at a decreasing rate
- Increasing at an increasing rate
- Decreasing at a decreasing rate
- Decreasing at an increasing rate

(a) Between points (a) and (b), the acceleration is:
(b) Between points (b) and (c), the acceleration is:
(c) Between points (c) and (d), the acceleration is:
(d) Between point (d) and (e), the distance is:
(e) Between points (f) and (g), the distance is:
(f) Between points (g) and (h), the distance is:

EXAMPLE 7-7

As a club effort to raise money for charity, we are going to push a university fire truck along a flat section of highway. The fire truck weighs 29,400 newtons. Using teams, we will exert a force of 10 newtons for the first two minutes, 20 newtons for the next three minutes and 5 newtons for the next four minutes. We notice a billboard nearby that reads, "Newton says that 'FORCE = MASS × ACCELERATION'!"

Graph the acceleration and speed versus time on separate graphs, using the same time scale for each. In your analysis, ignore the effect of friction and assume the acceleration changes instantaneously.

Step One: Calculate acceleration based on Newton's Law. Plot acceleration versus time; the results are shown in Figure 7-15a. The details of this calculation are left for the reader . . . but as a hint, for the first two minutes, a force of 10 newtons is applied to the 3,000 kilogram mass truck, giving an acceleration of 0.0033 meters per second squared. Note that in the graph, the horizontal lines indicate that acceleration is constant.

The vertical lines imply that the acceleration instantaneously increased in value. While this is not exactly accurate, no information was given on how the acceleration changes between values. In such cases, an instantaneous change is often assumed.

Step Two: Calculate speed based on the area under the curve generated in Step One. Plot speed versus time. The details are left to the reader. Results shown in Figure 7-15b.

Figure 7-15a
Acceleration of a Fire Truck.

Figure 7-15b
Velocity determined from area under the acceleration curve.

Comprehension Check 7-6

The graph gives an acceleration profile for a car that begins to move from an initial speed of zero. Draw the corresponding velocity and the corresponding distance profile.

7.5 | GRAPHICAL SOLUTIONS

When you have two equations containing the same two variables, it is sometimes desirable to find values of the variables that satisfy both equations. Most of you have studied methods for solving simultaneous linear equations (there are many techniques); however, most of these methods apply only to linear equations and do not work if one or both of the equations is nonlinear. It also becomes problematic if you are working with experimental data.

For systems of two equations (or data sets in two variables), you can use a graphical method to determine the value or values that satisfy both. Essentially, graph the two equations and visually determine where the curves intersect. This may be nowhere, at one point, or at several points.

EXAMPLE 7-8

We assume that the current through two electromagnets is given by the following equations

$$\text{Electromagnet A:} \quad I = 5t + 6$$
$$\text{Electromagnet B:} \quad I = -3t + 12$$

We want to determine when the value of the current through the electromagnets is equal.

Graphing both equations gives the following. We know not to show the points when plotting data derived from equations.

The two lines cross at time 5 0.75 seconds (approximately), and the current at this time is approximately 9.7 amperes. The larger we make this graph and the more gridlines we include, the more accurately we can determine the solution.

Solution: t = 0.75 seconds, I = 9.7 amperes.

EXAMPLE 7-9

> - Total cost = Fixed cost + Variable cost * Amount produced
> - Revenue = Selling price * Amount sold
> - If Revenue − Total Cost > 0, result = Profit
> - If Revenue − Total Cost < 0, result = Loss
> - If Revenue − Total Cost = 0, result is Breakeven Point

Breakeven analysis determines the quantity of product a company must make before they begin to earn a profit. Two types of cost are associated with manufacturing: fixed and variable. **Fixed costs** include equipment purchases, nonhourly employee salaries, insurance, mortgage or rent on the building, etc., or "money we must spend just to open the doors." **Variable costs** depend on the production volume, such as material costs, hourly employee salaries, and utility costs. The more product produced, the higher the variable costs become. The product is sold at a **selling price,** creating **revenue.** Any excess revenue remaining after all production costs have been paid is **profit.**

Let the amount of product we produce be G [gallons per year]. Consider the following costs:

- Fixed cost: $1 million/year
- Variable cost: 10 cents/gallon of G
- Selling price: 25 cents/gallon of G.

Plot the total cost and the revenue versus the quantity produced. Determine the amount of G that must be produced to break even. Assume we sell everything we make.

There will be two curves on the plot: (1) total cost versus G; (2) revenue versus G.

- *The first curve:* total cost = fixed cost + variable cost × gallons of G produced, or TC = FC + VC × G.
- *The second curve:* revenue = selling price × gallons of G sold, or R = SP × G

The plot of these two functions is shown in Figure 7-16. The breakeven point occurs when the two graphs cross, at a production capacity of 6.7 million gallons of G.

Figure 7-16
Breakeven analysis definitions.

Comprehension Check 7-7

You are working for a tire manufacturer, producing wire to be used in the tire as a strengthening agent. You are considering implementing a new machining system, and you must present a breakeven analysis to your boss. You develop the graph, showing two possible machines that you can buy.

(a) Which machine has a higher fixed cost?
(b) Which machine has a lower variable cost?
(c) How much wire must be produced on Machine #1 to break even?
(d) If you make 3 million feet of wire, which machine will yield the highest profit?
(e) Which machine has the lower breakeven point?

EXAMPLE 7-10

NOTE

This example demonstrates the graphical solution of simultaneous equations when one of the equations is nonlinear. We do not expect you to know how to perform the involved circuit analyses. Those of you who eventually study electronics will learn these techniques in considerable detail.

The semiconductor diode is sort of like a one-way valve for electric current: it allows current to flow in one direction, but not the other. In reality, the behavior of a diode is considerably more complicated. In general, the current through a diode can be found with the **Shockley equation**,

$$I = I_0 \left(\frac{V_D}{e^{nV_t} - 1} \right)$$

where I is the current through the diode in amperes; I_0 is the saturation current in amperes, constant for any specific diode; V_D is the voltage across the diode in volts; and V_T is the thermal voltage in volts, approximately 0.026 volts at room temperature. The emission coefficient, n, is dimensionless and constant for any specific diode; it usually has a value between 1 and 2.

The simple circuit shown has a diode and resistor connected to a battery. For this circuit, the current through the resistor can be given by

$$I = \frac{V - V_D}{R}$$

where I is the current through the resistor in milliamperes [mA], V is the battery voltage in volts [V], V_D is the voltage across the diode in volts, and R is the resistance in ohms [Ω].

In this circuit, the diode and resistor are in series, which implies that the current through them is the same; this concept is covered later in the book. We have two equations for the same parameter (current), both of which are a function of the same parameter (diode voltage). We can find a solution to these two equations, and thus the current in the circuit, by graphing both equations and finding the point of intersection. For convenience of scale, the current is expressed in milliamps rather than amperes.

Plot these two equations for the following values and determine the current.

$$I_0 = 0.01 \text{ mA}$$
$$V = 3 \text{ V}$$
$$R = 24 \text{ Ω}$$
$$nV_T = 0.04 \text{ V}$$

The point of intersection is at $V_D = 0.64$ V and $I = 100$ mA; thus, the current in the circuit is 100 mA or 0.1 A.

In-Class Activities

ICA 7-1

Using the list provided, you may be assigned a topic for which to create a graph. You must determine the parameters to graph and imagine a set of data to show on the chart below (or a copy of the chart).

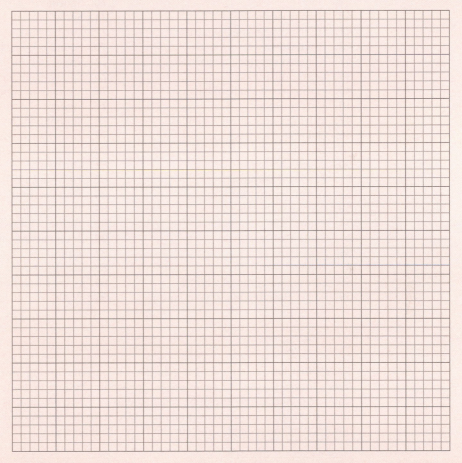

1. Air temperature
2. Airplane from airport to airport
3. Baking bread
4. Baseball
5. Bird migration
6. Birthday party
7. Boiling water in a whistling teapot
8. Bouncing a basketball
9. Brushing your teeth
10. Burning a pile of leaves
11. Burning candle
12. Climbing a mountain
13. Coastal river
14. Cooking a Thanksgiving turkey
15. Cow being picked up by a tornado
16. D.O.R. possum ("dead on the road")
17. Daily electric power consumption
18. Day in the life of a chicken
19. Detecting a submarine by using sonar
20. Diet of a parakeet
21. Diving into a swimming pool
22. Drag racing
23. Driving home from work
24. Dropping ice in a tub of warm water
25. Eating a stack of pancakes
26. Eating at a fast food restaurant
27. Email messages
28. Engineer's salary

29. Exercising
30. Feedback from an audio system
31. Feeding birds at a bird feeder
32. Firing a bullet from a rifle
33. Fishing
34. Flight of a hot air balloon
35. Football game crowd
36. Forest fire
37. Formation of an icicle
38. Glass of water in a moving vehicle
39. Hair
40. Hammering nails
41. Hiking the Appalachian trail
42. Homework
43. Kangaroo hopping along
44. Leaves on a tree
45. Letting go of a helium balloon
46. Marching band
47. Moving a desk down a staircase
48. Oak tree over the years
49. Oil supply
50. People in Florida
51. People in Michigan
52. Person growing up
53. Playing with a yo-yo
54. Plume from a smokestack
55. Political affiliations
56. Pony Express
57. Popping corn
58. Pouring water out of a bottle
59. Power consumption of your laptop

60. Power usage on campus
61. Pumping air into a bicycle tire
62. Rabbit family
63. Rain filling a pond
64. Recycling
65. River in a rainstorm
66. Running of the bulls
67. Skipping a stone on water
68. Sleeping
69. Slipping on a banana peel
70. Snoring
71. Snow blowing over a roof
72. Solar eclipse
73. Sound echoing in a canyon
74. Space station
75. Spinning a hula-hoop
76. Strength of concrete
77. Student attention span during class
78. Studying for an exam
79. Taking a bath in a tub
80. Talking on a cell phone
81. The moon
82. Throwing a ball
83. Thunderstorm
84. Tiger hunting
85. Traffic at intersections
86. Train passing through town
87. Typical day
88. Using a toaster
89. Using Instant Messenger
90. Washing clothes

ICA 7-2

Joule's first law relates the heat generated to current flowing in a conductor. It is named for James Prescott Joule, the same person for whom the unit of Joule is named. The Joule effect states that electric power can be calculated as $P = I^2R$, where R is the resistance [in ohms, Ω] and I is the electrical current [in amperes, A]. Use the following data to create a graph of the power (P, on the ordinate) and current (I, on the abscissa).

Current (I) [A]	0.50	1.25	1.50	2.25	3.00	3.20	3.50
Power (P) [W]	1.20	7.50	11.25	25.00	45.00	50.00	65.00

ICA 7-3

When an airplane is in the air, the temperature of the outside atmosphere changes as the altitude of the airplane changes. Use the following data to create a graph of temperature (T, on the ordinate) and time (t, on the abscissa).

Time (t) [h]	0.00	0.25	0.50	1.00	1.50	2.00
Temperature (T) [°C]	60	22	5	4	5	25

ICA 7-4

You may be assigned several of the following functions. Plot them, all on the same graph. The independent variable (angle) should vary from 0 to 360 degrees.

(a) $\sin(\theta)$

(b) $2\sin(\theta)$

(c) $-2\sin(\theta)$

(d) $\sin(2\theta)$

(e) $\sin(3\theta)$

(f) $\sin(\theta) + 2$

(g) $\sin(\theta) - 3$

(h) $\sin(\theta + 90)$

(i) $\sin(\theta - 45)$

(j) $2\sin(\theta) + 2$

(k) $\sin(2\theta) + 1$

(l) $3\sin(2\theta) - 2$

ICA 7-5

You may be assigned several of the following functions. Plot them, all on the same graph. The independent variable (angle) should vary from 0 to 360 degrees.

(a) $\cos(\theta)$

(b) $2\cos(\theta)$

(c) $-2\cos(\theta)$

(d) $\cos(2\theta)$

(e) $\cos(3\theta)$

(f) $\cos(\theta) + 2$

(g) $\cos(\theta) - 3$

(h) $\cos(\theta + 90)$

(i) $\cos(\theta - 45)$

(j) $2\cos(\theta) + 2$

(k) $\cos(2\theta) + 1$

(l) $3\cos(2\theta) - 2$

ICA 7-6

You need to create a graph showing the relationship of an ideal gas between pressure (P) and temperature (T). The ideal gas law relates the pressure, volume, amount and temperature of an ideal gas using the relationship: $PV = nRT$. The ideal gas constant, R, is 0.08207 atmosphere liter per mole kelvin. Assume the tank has a volume of 12 liters and is filled with nitrogen (formula: N_2, molecular weight: 28 grams per mole). Allow the initial temperature to be 270 kelvin at a pressure of 2.5 atmospheres. Create the graph, showing the temperature on the abscissa from 270 kelvin to 350 kelvin.

ICA 7-7

The decay of a radioactive isotope can be modeled with the following equation, where C_0 is the initial amount of the element at time zero and k is the decay rate of the isotope.

$$C = C_0 e^{-t/k}$$

Create a graph of the decay of Isotope A [k = 1.48 hours]. Allow time to vary on the abscissa from 0 to 5 hours with an initial concentration of 10 grams of Isotope A.

ICA 7-8

Myrmie Solenops, a biosystems engineer, has discovered that ultrasound at a specific frequency causes fire ants to go berserk and kill one another, including the queen ant. Solenops did breakeven analysis as part of a proposal to her company to manufacture her Fire Ant Destroyer. The graph below shows the results of this analysis.

(a) How many Fire Ant Destroyers must the company produce and sell to break even? Answer to the nearest 10,000 (the last four digits should be zero).

(b) How many Fire Ant Destroyers must the company produce and sell to make a profit of $500,000? Answer to the nearest 10,000 (the last four digits should be zero).

(c) What is the sales price of one Fire Ant Destroyer?

(d) If the sales price of one Fire Ant Destroyer was increased by $1, how many units must the company produce and sell to break even?

(e) If the sales price of one Fire Ant Destroyer was decreased by $1, how many units must the company produce and sell to break even?

ICA 7-9

You are working for a chemical manufacturer, producing solvents used to clean lenses for microscopes. You are working on determining the properties of three different solvent blends. You develop the following chart, showing the evaporation of the three blends.

(a) Which solvent evaporates at the slowest rate?
(b) Which solvent evaporates at the fastest rate?
(c) What is the initial mass of Solvent A? Be sure to include units.
(d) What is the rate of solvent evaporation of Solvent A? Be sure to include units.
(e) What is the rate of solvent evaporation of Solvent B? Be sure to include units.
(f) What is the rate of solvent evaporation of Solvent C? Be sure to include units.

ICA 7-10

You, with an assistant, will investigate the effect of varying (separately) the mass and the length of a pendulum on the period of oscillation. As in all experiments, the more carefully you make your measurements, the better your results will be.

Materials

Masses (5, ranging 10 g–200 g)	Stopwatch (2)	Steel Rods
Fishing line (2, 40″ pieces)	Tape Measure (2)	Table Clamps

Part I: Effect of Length on Period of Oscillation

Hang a 50-gram mass from the loop at the end of a piece of fishing line. Wrap the other end of the line around one of the steel rods such that the distance from the mass to the steel rod is about 10 inches. Carefully measure and record in a worksheet the distance from the center of the rod to the center of the mass.

Release the pendulum and use the stopwatch to determine the time required for five full swings, with the motion of the pendulum to the right and back to the left counting as one full swing. To release the pendulum, pull the mass to the side about 15° from the vertical, keeping the body of the mass in line with the fishing line.

After the pendulum is released, wait for at least one full swing and then start the stopwatch when the mass is at one side or the other and just beginning to reverse direction. After five full swings, stop the watch when the mass returns the fifth time to the point at which the watch was started. Record the time in the worksheet. Keeping the length constant, repeat this procedure for a total of three trials.

Repeat the entire procedure above for lengths of approximately 15, 20, 25, and 30 inches. Remember to carefully measure and record the actual distance from center of rod to center of the mass.

Calculate the average of the three times and record the average time. Divide each average time by 5 (you timed five complete swings, not just one) to obtain the value for the period corresponding to each condition.

Length (L) [in]	Time (t) [s]				Period (T) of one cycle [s]
	Trial #1	Trial #2	Trial #3	Average	

Part II: Analysis of Length

Draw a proper plot of period (T, on the ordinate) and length (L, on the abscissa) on the graph below or on a copy of it. Make the period scale of this graph identical to the one in Part IV so that it is easier to compare the two graphs.

Use Table 7-1 to describe the relationships shown on the graph from the shape of the line.

Part III: Effect of Mass on Period of Oscillation

Hang a 200-gram mass from the loop at the end of the shortest piece of fishing line. Wrap the other end of the line around one of the steel rods such that the distance from the mass to the steel rod is about 20 inches. Release the pendulum and use the stopwatch to determine the time required for five full swings, with the motion of the pendulum to the right and back to the left counting as one full swing. To release the pendulum, pull the mass to the side about 15° from the vertical, keeping the body of the mass in line with the fishing line.

After the pendulum is released, wait for at least one full swing and then start the stopwatch when the mass is at one side or the other and just beginning to reverse direction. After five full swings, stop the watch when the mass returns the fifth time to the point at which the watch was started. Record the time. Keeping the mass constant, repeat this procedure for a total of three trials.

Repeat the entire procedure above for masses of 100, 50, 20, and 10 grams.

Since the masses are different sizes, you will have to adjust the length from center of rod to center of mass each time you change masses. This length should be the same for all five masses.

Calculate the average of the three times and record the average time. Divide each average time by 5 (you timed five complete swings, not just one) to obtain the value for the period corresponding to each condition.

Mass (m) [g]	Time (t) [s]				Period (T) of one cycle [s]
	Trial #1	Trial #2	Trial #3	Average	
200					
100					
50					
20					
10					

Part IV: Analysis of Mass

Draw a proper plot of period (T, on the ordinate) and mass (m, on the abscissa) on a graph like the one below. Make the period scale of this graph identical to the one in Part II so that you can more easily compare the two graphs.

Use Table 7-1 to describe the relationships shown on the graph from the shape of the line.

ICA 7-11

This experiment investigates the oscillation frequency of a mass suspended from a spring. When a mass suspended from a spring is allowed to come to rest, the spring will be stretched so that the upward spring force balances the downward force (weight) of the mass. Call the location of the mass suspended in this way the rest position.

NOTE

To model the oscillating mass–spring system mathematically, differential equations are required. Solving such equations is beyond the scope of this book.

WARNING

If you are not careful, you will permanently damage the spring. Damaged springs will result not only in poor data but a grade penalty as well. Do not hook the masses together at any point during this experiment! Use each mass individually.

If the mass is raised above the rest position, the spring is less stretched and the spring force on the mass is less than the weight of the mass. If the mass is then released, it will move downward because of the imbalance in the forces. When the mass reaches the location of the rest position, its velocity is not zero. For the mass to be slowed to a stop, an upward force must be exerted on the mass. As the mass descends beyond the rest position, the spring is stretched more, increasing the upward spring force to a point where it is greater than the force (weight) of the mass.

Eventually the mass slows to a stop, but then a net upward force causes the mass to begin rising again. This continues over and over with the mass oscillating up and down with the rest position in the middle of the oscillation. Eventually, frictional forces bring the mass back to the rest position at a complete stop.

Materials

Mass (20 g, 50 g, 100 g)	Stopwatch	Yardstick
Spring	Wooden block with hook	Table Clamp

Part I: Initial Setup

1. Clamp the wooden block to the table securely with the hook on the underside of the block and positioned so that the hook is well beyond the edge of the table.
2. Hang the spring from the hook.
3. Measure the distance from the underside of the block to the lowest point on the spring.

   Record the initial spring length here or on a separate worksheet = _____

4. Read this entire step before doing it!
   - For each of the three masses provided, *carefully* hang the mass on the bottom of the spring, then slowly and gently lower it to the rest position.
   - Do not simply let go of the mass or you may damage the spring.
   - With each mass in the rest position, measure the distance from the underside of the block to the lowest point of the spring.
   - Record the values of the mass and the distance to the rest position in the table below or on your worksheet.

5. For each mass, calculate how much the spring was stretched by each mass (rest position minus initial spring length) and record in the table below or on a separate worksheet as the spring extension.

Mass [g]	Rest Position [in]	Spring Extension [in]

Part II: Effect of Mass on Period of Oscillation

1. Attach the smallest mass to the spring and carefully lower it to the rest position. Re-measure the rest position and verify that it agrees with the value recorded in Part I.
2. Raise the mass vertically until the spring is almost, but not quite, back to the unstretched length. The distance from the underside of the block to the lowest point on the spring will be slightly more than the initial spring position recorded during Part I. Determine the amount you raised the spring from the rest position. This will probably be less than an inch.

   Record the amount raised here or on a separate worksheet = _____

3. Read this entire step before doing it!
   - Reset the stopwatch.
   - Carefully release the mass from the raised position so that it oscillates up and down with no horizontal motion.

- Wait for three oscillations, then, while watching the mass, start the stopwatch when the mass reaches its lowest point.
- Count twenty full oscillations, and when the mass reaches the lowest point on the twentieth cycle, stop the watch.
- Record the time for twenty oscillations in the table below or on a separate worksheet.

4. Repeat Step 3 for the remaining masses, raising each the same amount in Step 2.
5. Calculate the period and frequency for each mass.

- The period is the length of time required for one oscillation.
- The frequency is the number of oscillations per second, which is simply the inverse of the period.

Mass [g]	Time for 20 oscillations [s]	Period [s]	Frequency [s^{-1}]

Part III: Analysis of Effect of Mass on Frequency

Draw a proper plot of frequency (f, on the ordinate) and mass (m, on the abscissa) on the graph below or on a copy. Make the ordinate scale of this graph identical to the one in Part V so you can easily compare the two graphs.

Use Table 7-1 to describe the relationships shown on the graph from the shape of the line.

Part IV: Effect of Release Position on Period of Oscillation

1. Attach the 50-gram mass to the spring, and carefully lower it to the rest position.
2. Raise the mass ¾ inch vertically.
3. Read this entire step before doing it!

   - Reset the stopwatch.
   - Carefully release the mass from the raised position so that it oscillates up and down with no horizontal motion.
   - Wait for 3 oscillations, then, while watching the mass, start the stopwatch when the mass reaches its lowest point.
   - Count 20 full oscillations, and when the mass reaches the lowest point on the 20th cycle, stop the watch.
   - Record the time for 20 oscillations in the table below or a copy of it.

4. Repeat Step 3 by raising the mass a distance of 1 inch, 1¼ inches, 1½ inches, and 1¾ inches.
5. Calculate the period and frequency for each mass.

   - The period is the length of time required for one oscillation.
   - The frequency is the number of oscillations per second, which is simply the inverse of the period.

Length [in]	Time for 20 oscillations [s]	Period [s]	Frequency [s^{-1}]

Part V: Analysis of Effect of Release Position on Frequency

Draw a proper plot of frequency (f, on the ordinate) and initial displacement (H, on the abscissa) on the graph below or on a copy. Make the ordinate scale of this graph identical to the one in Part III so that you can more easily compare the two graphs.

Use Table 7-1 to describe the relationships shown on the graph from the shape of the line.

ICA 7-12

Materials

Balloons (2)	Stopwatch (2)	String (40")	Tape Measure

Part I: Blowing Up a Balloon

One team member is to inflate one balloon, a second team member is to time the inhalation stage (how long it takes to inhale a single breath), and a third team member is to time the exhalation stage (how long it takes to exhale a single breath into the balloon). A fourth team member is to measure the balloon size at the end of each inhale/exhale cycle, using the string to measure the balloon circumference.

Record the observations on a worksheet similar to that shown below for three complete inhale/exhale cycles or until the balloon appears to be close to maximum volume, whichever occurs first. Repeat the entire balloon inflation process for a second balloon; average the times from the balloons to obtain the time spent at each stage and the average circumference at each stage. Calculate the balloon volume at each stage, assuming the balloon is a perfect sphere.

Balloon	Stage	Inhale Time	Exhale Time	Circumference
1	1			
	2			
	3			
2	1			
	2			
	3			

	Stage	Inhale Time	Exhale Time	Circumference	Volume
Average Balloon	1				
	2				
	3				

Part II: Analysis

Graph the balloon volume (V, ordinate) versus time (t, time) on the graph below or a copy. Allow the process to be continuous, although in reality it was stopped at various intervals for measurements. The resulting graph should contain only the time elapsed in the process of inhaling and exhaling, not the time required recording the balloon size. For this procedure, assume the air enters the balloon at a constant rate and the balloon is a perfect sphere.

(a) What does the assumption of the air entering the balloon at a constant rate indicate about the slope?

(b) Calculate the following graphically.

- The rate at which the air enters the balloon in the first stage.
- The rate at which the air enters the balloon in the third stage.

(c) On the same graph, sketch the balloon volume (V, ordinate) versus time (t, time) if you were inflating a balloon which contained a pinhole leak.

(d) On the same graph, sketch the balloon volume (V, ordinate) versus time (t, time) if you were inflating a balloon from a helium tank.

CHAPTER 8
INTERPOLATION

A small history lesson: when two of the authors of this book (Sill and Park) were in high school, pocket-sized electronic calculators did not exist. For scientific calculations, one used a mechanical gizmo called a slide rule. When these two authors started college (1963 and 1968), the mark of the engineering student was a hard-leather pouch about a foot long and a couple of inches wide hanging from the belt; this contained the student's slide rule.

> The slide rule was developed in the 1600s, shortly after John Napier invented logarithms. Edmund Gunter is credited with developing the logarithmic scale. The slide rule, using a combination of log scales, is credited to William Oughtred.

Photo provided by: Dr. William Park

Since slide rule use was seldom taught in high school, when numerical values for transcendental functions such as logarithms or trigonometric functions were needed before college, students turned to the appendices in the back of their math book, where multipage tables of values for common transcendental functions were always found. To obtain a specific value, students often had to interpolate between the values actually found in the tables. A very small portion of just such a logarithm table is shown below in Table 8-1. The two-digit number in the leftmost column indicates the first two digits of the value for which the logarithm was desired, and the topmost row gave the third significant figure. The other numbers gave the logarithms. It was up to each student to get the decimal point in the correct place.

As examples (see highlighted cells):

Log (3.74) = 0.5729
Log (368) = 2.5658
Log (0.399) = −0.6010

Table 8-1 Sample logarithmic values

x	0	1	2	3	4	5	6	7	8	9
3.5	0.5441	0.5453	0.5465	0.5478	0.5490	0.5502	0.5514	0.5527	0.5539	0.5551
3.6	0.5563	0.5575	0.5587	0.5599	0.5611	0.5623	0.5635	0.5647	0.5658	0.5670
3.7	0.5682	0.5694	0.5705	0.5717	0.5729	0.5740	0.5752	0.5763	0.5775	0.5786
3.8	0.5798	0.5809	0.5821	0.5832	0.5843	0.5855	0.5866	0.5877	0.5888	0.5899
3.9	0.5911	0.5922	0.5933	0.5944	0.5955	0.5966	0.5977	0.5988	0.5999	0.6010

All of these values can be directly determined from the table. What if the question asked the value of the log (3.815)? To obtain an answer, we must use the method of **interpolation**. For interpolation to work well, the behavior of the data between adjacent data points in the table needs to have a constant or nearly constant slope. That is, it should be roughly a straight line. This method is especially effective for data points given at equally spaced intervals, as is often found in reference materials. Furthermore, interpolation is most effective with less "noisy" data.

8.1   SINGLE INTERPOLATION

To understand how interpolation works, look at the previous chart. In the specific example used above, we wanted to find the log (3.815). Note that the two data points closest to 3.815 are 3.81 and 3.82. Also note that 3.815 is halfway between 3.81 to 3.82. Since we are assuming the data is linear between these points, then the corresponding logarithm should be 50% of the way between the log at these two values: 0.5809 and 0.5821, respectively.

To place this on a more solid mathematical footing, we can assign variable names to the values bracketing the desired value, calling them point 1 and point 2.

$A_1 = 3.81$	$A = 3.815$	$A_2 = 3.82$
$B_1 = 0.5809$	$B = ?$	$B_2 = 0.5821$

First, let's consider the fraction of the distance that A (3.815) is between A_1 (3.81) and A_2 (3.82). We can do this by dividing the numerical distance from A to A_1 by the numerical distance from A_2 to A_1. Mathematically, this is

$$\frac{(A - A_1)}{(A_2 - A_1)}$$

Since there is a one-to-one correspondence between the variable A and the variable B, we can write an analogous expression for B and set both expressions equal to each other:

$$\frac{(A - A_1)}{(A_2 - A_1)} = \frac{(B - B_1)}{(B_2 - B_1)}$$

To solve for B, we begin by multiplying both sides of the equation by the distance between the corresponding B values given by $B_2 - B_1$. Thus, the distance from B_1 to B is

$$B - B_1 = \frac{(A - A_1)}{(A_2 - A_1)}(B_2 - B_1)$$

Finally, we have to add the starting value B_1 to the numerical distance we travel from B_1 to B to get the actual value of B. Putting this all together gives an alternative interpolation formula:

$$B = B_1 + \frac{(A - A_1)}{(A_2 - A_1)}(B_2 - B_1)$$

Applying either interpolation formula to the specific problem solved above gives the **ratio interpolation formula**:

$$\frac{(3.815 - 3.81)}{(3.82 - 3.81)} = \frac{(B - 0.5809)}{(0.5821 - 0.5809)}$$

Solving for B:

$$B = 0.5809 + \frac{(3.815 - 3.81)}{(3.82 - 3.81)}(0.5821 - 0.5809) = 0.5815$$

INTERPOLATION PROCEDURE

1. Write the known value of the first variable "A," corresponding to the unknown desired value of the second variable "B."

A_1	A	A_2
B_1	B	B_2

2. Write the data point whose value occurs below the known value "A" in the table, calling it Point 1 (A_1, B_1).
3. Write the data point whose value occurs above the known value of "A" in the table, calling it Point 2 (A_2, B_2).
4. Calculate, using the interpolation formula and express your answer in "reasonable" terms. Don't forget to include units if needed with your final answer.
5. Check that the result occurs between the values of Point 1 and Point 2.

- Note: the roles of A and B could be reversed.
- Note: the calculated B value must *always* be between the two bracketing values of B. If not, we made an error somewhere. This is an easy check of our work.
- One more note: The equation used for interpolation applies to parameters that have a linear variation. An everyday example of this is the temperature scales, particularly Fahrenheit [°F] and Celsius [°C]. If we "interpolate" between the boiling and freezing points of water, we can write the relationship between these scales as

$$\frac{°F - 32}{212 - 32} = \frac{°C - 0}{100 - 0}$$

This expression is easy to remember and can be used (even for values outside the freezing and boiling points of water) to convert units of Fahrenheit to Celsius, and vice versa.

EXAMPLE 8-1

A spacecraft with an experimental ion engine is able to achieve a constant acceleration of one one-hundredth g, or just under 0.1 meters per second squared [m/s²]. The velocity of the craft is measured in miles per second [mi/s] every 10 hours [h], and the resulting information is shown.

Elapsed Time (t) [h]	10	20	30	40	50	60	70
Velocity (v) [mi/s]	2.2	4.4	6.6	8.8	11.0	13.2	15.3

(a) What is the velocity of the spacecraft after a day and a half?
(b) How long is required for the spacecraft to reach 14 miles per second?

Solution (a)

A day and a half is 36 hours, so the velocity will be between 6.6 miles per second [mi/s] at 30 hours and 8.8 mi/s at 40 hours. Applying the interpolation formula gives

$A_1 = 30$ h	time = $A = 36$ h	$A_2 = 40$ h
$B_1 = 6.6$ mi/s	velocity = $B = ?$ mi/s	$B_2 = 8.8$ mi/s

$$\frac{(36 - 30)}{(40 - 30)} = \frac{(v - 6.6)}{(8.8 - 6.6)} \qquad v = 6.6 + \frac{(36 - 30)}{(40 - 30)}(8.8 - 6.6) = 7.9 \text{ mi/s}$$

Solution (b)

When the craft is traveling 14 miles per second [mi/s], the elapsed time will be between 60 hours at 13.2 mi/s and 70 hours at 15.3 mi/s. Applying the interpolation formula yields

$A_1 = 13.2$ mi/s	velocity = $A = 14$ mi/s	$A_2 = 15.3$ mi/s
$B_1 = 60$ h	time = $B = ?$ h	$B_2 = 70$ h

$$\frac{(t - 60)}{(70 - 60)} = \frac{(14 - 13.2)}{(15.3 - 13.2)} \qquad t = 60 + \frac{(14 - 13.2)}{(15.3 - 13.2)}(70 - 60) = 63.8 \text{ h}$$

Comprehension Check 8-1

The temperature of air in a "standard atmosphere" is given below as a function of altitude. Determine the temperature in units of degrees Celsius [°C] at 23,500 feet [ft] using interpolation.

Altitude (A) [ft]	Temperature (T) [°C]
0	59.0
5,000	41.2
10,000	23.3
15,000	5.5
20,000	−12.3
25,000	−30.2
30,000	−48.0

Comprehension Check 8-2

The temperature of air in a "standard atmosphere" is given below as a function of altitude. Determine the altitude in units of feet [ft] at –20 degrees Celsius [°C] using interpolation.

Altitude (A) [ft]	Temperature (T) [°C]
0	59.0
5,000	41.2
10,000	23.3
15,000	5.5
20,000	−12.3
25,000	−30.2
30,000	−48.0

8.2 COMPLEX INTERPOLATION

Sometimes a parameter of interest depends on two variables instead of just one, as in the preceding examples. Wind chill is an example of such a phenomenon, depending on both temperature and wind speed. Table 8-2 shows wind chill factors over a wide range of temperatures and wind speeds. We can use this chart to interpolate wind chill factors for combinations of temperature and wind speed not listed.

If our desired temperature for wind chill is in the table but the wind speed is not (or vice versa), a simple single interpolation is all we need.

Table 8-2 Wind chill temperature

Actual Temp (AT) [°F]	Wind Chill Temperature (WCT) [°F]				
	Wind speed (WS) [mph]				
	0	10	20	30	40
40	40	34	30	28	27
30	30	21	17	15	13
20	20	9	4	1	−1
10	10	−4	−9	−12	−15
0	0	−16	−22	−26	−29
−10	−10	−28	−35	−39	−43
−20	−20	−41	−48	−53	−57

EXAMPLE 8-2

For a wind speed of 20 miles per hour [mph], what is the wind chill temperature at an actual temperature of 17 degrees Fahrenheit [°F]?

$A_1 = 20$ °F	$AT = A = 17$ °F	$A_2 = 10$ °F
$B_1 = 4$ °F	$WCT = B = ?$ °F	$B_2 = -9$ °F

$$\frac{(17 - 20)}{(10 - 20)} = \frac{(WCT - 4)}{(-9 - 4)} \qquad WCT = 4 + \frac{(17 - 20)}{(10 - 20)}(-9 - 4) = -1.2\,°F$$

EXAMPLE 8-3

What is the wind chill temperature for an actual temperature of –20 degrees Fahrenheit [°F] at a wind speed of 8.5 miles per hour [mph]?

$A_1 = 0$ mph	$WS = A = 8.5$ mph	$A_2 = 10$ mph
$B_1 = -20$ °F	$WCT = B = ?$ °F	$B_2 = -41$ °F

$$\frac{(8.5 - 0)}{(10 - 0)} = \frac{(WCT - (-20))}{(-41 - (-20))} \quad WCT = -20 + \frac{(8.5 - 0)}{(10 - 0)}(-41 - (-20)) = -38.9\,°F$$

However, what if we need a wind chill determination for a temperature and wind speed when neither is given directly in the table? Here, we have to do more than a single interpolation.

EXAMPLE 8-4

Determine the wind chill temperature at an actual temperature of 22.3 degrees Fahrenheit [°F] and a wind speed of 8.9 miles per hour [mph].

First find the bracketing values for both wind speed and temperature, and insert both a row and a column between them. Place the specified temperature (22.3 °F in this case) in a new column (between 20 and 30 °F) and place the specified wind speed (8.9 mph) in a new row (between 0 and 10 mph). This procedure yields a table that looks like this:

Temperature (T) [°F]	Wind Speed (WS) [mph]		
	0	8.9	10
30	30		15
22.3	22.3	desired	11.8
20	20		9

We can now proceed in one of two ways:

1. *Interpolate values for wind chills at 22.3 °F for wind speeds of both 0 mph and 10 mph.*
2. *Interpolate values for wind chill at 8.9 mph for temperatures of both 30 °F and 20 °F.*

Once we have either of those pairs of numbers, we interpolate between those two new values to obtain the desired value. As an example, we will follow plan 1 above.

At 0 mph and 22.3 °F, we get

$$\frac{(WCT - 20)}{(30 - 20)} = \frac{(22.3 - 30)}{(20 - 30)} \qquad WCT = 22.3\ °F$$

At 10 mph and 22.3 °F, we get

$$\frac{(WCT - 15)}{(9 - 15)} = \frac{(22.3 - 30)}{(20 - 30)} \qquad WCT = 11.8\ °F$$

Inserting these into the table above yields

Temperature (T) [°F]	Wind Speed (WS) [mph]		
	0	8.9	10
30	30		15
22.3	22.3	desired	11.76
20	20		9

Finally, we interpolate between these two "new" values for 22.3°F:

$$\frac{(WCT - 22.3)}{(11.8 - 22.3)} = \frac{(8.9 - 0)}{(10 - 0)} \qquad WCT = 12.9\ °F$$

Thus, the wind chill at 22.3 °F and wind speed 8.9 mph is 12.9 °F. Checking the answer below, it seems reasonable.

Temperature (T) [°F]	Wind Speed (WS) [mph]		
	0	8.9	10
30	30		15
22.3	22.3	**12.9**	11.8
20	20		9

As a practice exercise, calculate this same wind chill using plan 2 above; you should get the same value.

Comprehension Check 8-3

Determine the wind chill temperature at an actual temperature of -12 degrees Fahrenheit [°F] and a wind speed of 22 miles per hour [mph].

Comprehension Check 8-4

Determine the wind chill temperature at an actual temperature of -18 degrees Fahrenheit [°F] and a wind speed of 12 miles per hour [mph].

In-Class Activities

ICA 8-1

The following chart gives values of logarithms. The two-digit number in the leftmost column indicates the first two digits of the value for which the logarithm was desired, and the topmost row gives the third significant figure. The other numbers gives the logarithms. It is up to you to get the decimal point in the correct place.

x	0	1	2	3	4	5	6	7	8	9
3.5	0.5441	0.5453	0.5465	0.5478	0.5490	0.5502	0.5514	0.5527	0.5539	0.5551
3.6	0.5563	0.5575	0.5587	0.5599	0.5611	0.5623	0.5635	0.5647	0.5658	0.5670
3.7	0.5682	0.5694	0.5705	0.5717	0.5729	0.5740	0.5752	0.5763	0.5775	0.5786
3.8	0.5798	0.5809	0.5821	0.5832	0.5843	0.5855	0.5866	0.5877	0.5888	0.5899
3.9	0.5911	0.5922	0.5933	0.5944	0.5955	0.5966	0.5977	0.5988	0.5999	0.6010

Using the values in this table, determine the following logarithms by interpolation:

(a) log (3.8)
(b) log (3.77)
(c) log (3.816)
(d) log (3.5009)

(e) log (3876)
(f) log (360.4)
(g) log (0.37)
(h) log (0.003842)

Using the values in this table, determine the following powers of 10:

(i) $10^{0.5843}$
(j) $10^{0.5441}$
(k) $10^{0.561}$
(l) $10^{0.545}$

(m) $10^{0.5729}$
(n) $10^{0.5905}$
(o) $10^{-0.4498}$
(p) $10^{-0.43}$

ICA 8-2

The following chart gives the relationship between actual temperature (symbolized by AT, given down the column on the far left in degrees Fahrenheit [°F]), wind speed (symbolized by WS, given across the row on the top in miles per hour [mph]), and the wind chill temperature (the temperature it "feels" like, symbolized by WCT, given by the crossing of the rows and columns in degrees Fahrenheit [°F]).

Wind Chill Temperature (WCT) [°F]					
Actual Temp (AT) [°F]	Wind speed (WS) [mph]				
	0	10	20	30	40
40	40	34	30	28	27
30	30	21	17	15	13
20	20	9	4	1	−1
10	10	−4	−9	−12	−15
0	0	−16	−22	−26	−29
−10	−10	−28	−35	−39	−43
−20	−20	−41	−48	−53	−57

Use interpolation to determine the following.

(a) Wind chill temperature at an actual temperature of −20 °F and a wind speed of 13 mph.
(b) Wind chill temperature at an actual temperature of −20 °F and a wind speed of 38 mph.
(c) Wind chill temperature at an actual temperature of 8 °F and a wind speed of 30 mph.
(d) Wind chill temperature at an actual temperature of −18 °F and a wind speed of 10 mph.
(e) Wind chill temperature at an actual temperature of 3 °F and a wind speed of 37 mph.
(f) Wind chill temperature at an actual temperature of −7 °F and a wind speed of 22 mph.
(g) Wind speed at an actual temperature of 10 °F and a wind chill temperature of −7 °F.
(h) Wind speed at an actual temperature of −10 °F and a wind chill temperature of −40 °F.
(i) Actual temperature at a wind speed of 20 mph and a wind chill temperature of −30 °F.
(j) Actual temperature at a wind speed of 40 mph and a wind chill temperature of −20 °F.

ICA 8-3

The table below shows the velocity of a constant boost spacecraft for various values of acceleration after boosting for several different times. Answer the following questions using interpolation.

Acceleration (a) [g]	Velocity of constant boost spacecraft (v) [mi/s] Elapsed time (t) [h]							
	15	30	45	60	75	90	105	120
0.004	1.3	2.6	3.9	5.3	6.6	7.9	9.2	10.5
0.008	2.6	5.3	7.9	10.5	13.2	15.8	18.4	21.1
0.012	3.9	7.9	11.8	15.8	19.7	23.7	27.6	31.6
0.016	5.3	10.5	15.8	21.1	26.3	31.6	36.9	42.1
0.020	6.6	13.2	19.7	26.3	32.9	39.5	46.1	52.7
0.024	7.9	15.8	23.7	31.6	39.5	47.4	55.3	63.2
0.028	9.2	18.4	27.6	36.9	46.1	55.3	64.5	73.7
0.032	10.5	21.1	31.6	42.1	52.7	63.2	73.7	84.2
0.036	11.8	23.7	35.5	47.4	59.2	71.1	82.9	94.8
0.040	13.2	26.3	39.5	52.7	65.8	79	92.1	105.3

(a) What is the velocity in unit of miles per second [mi/s] of a spacecraft capable of a 0.02 g constant boost after accelerating for four days?
(b) What is the velocity in unit of miles per second [mi/s] of a spacecraft capable of 0.033 g constant boost after accelerating for 120 hours [h]?
(c) What is the velocity in unit of miles per second [mi/s] of a spacecraft capable of 0.015 g constant boost after accelerating for three days?
(d) How long in units of hours would a spacecraft capable of 0.032 g constant boost have to accelerate to attain a speed of 50 miles per second [mi/s]?
(e) At what acceleration rate [g] would a spacecraft have to travel to achieve a velocity of 30 miles per second [mi/s] after accelerating for 75 hours [h]?

ICA 8-4

We have made many measurements of coffee cooling in a ceramic coffee cup. We realize that as the coffee cools, it gradually reaches room temperature. Consequently, we report the value of the coffee temperature in degrees *above* room temperature (so after a long time, the temperature rise will be equal to 0). Also, we realize that the hotter the coffee is initially (above room temperature) the longer it will take to cool. The values presented here are in degrees Fahrenheit [°F].

Temperature Rise (T) [°F]						
Initial Temp Rise (T_0) [°F]	Cooling time elapsed (t) [min]					
	0	10	20	30	40	50
20	20	13	9	6	4	3
40	40	27	18	12	8	5
60	60	40	27	18	12	8
80	80	54	36	24	16	11

(a) What is the temperature (rise) of the cup of coffee after 37 minutes [min] if the initial rise of temperature is 40 degrees Fahrenheit [°F]?

(b) If the coffee cools for 30 minutes [min] and has risen 14 degrees Fahrenheit [°F] at that time, what was the initial temperature rise?

(c) Find the temperature rise of the coffee at 17 minutes [min] if the initial rise is 53 degrees Fahrenheit [°F].

(d) What is the temperature (rise) of the cup of coffee after 23 minutes [min] if the initial rise in temperature is 80 degrees Fahrenheit [°F]?

(e) If the coffee cools for 10 minutes [min] and has risen 32 degrees Fahrenheit [°F] at that time, what was the initial temperature rise?

(f) Find the temperature rise of the coffee at 34 minutes [min] if the initial rise is 45 degrees Fahrenheit [°F].

ICA 8-5

A bipolar junction transistor can be used in a variety of ways. When used as an amplifier, the collector current, I_C, depends both on the base current, I_B, and the voltage from collector to emitter, V_{CE}.

The table below shows typical values for a 2N2222 NPN transistor.

Collector Current (I_c) [mA]					
Base Current (I_B) [μA]	Collector-Emitter Voltage (V_{CE})[V]				
	2	4	6	8	10
40	8.0	8.2	8.4	8.6	8.8
80	16.0	16.3	16.6	16.9	17.2
120	23.0	23.7	24.4	25.1	25.8
160	32.0	32.8	33.6	34.4	35.2
200	40.0	40.9	41.8	42.7	43.6

(a) What is the collector current (I_C) in units of milliamperes [mA] when the collector-emitter voltage (V_{CE}) is 4 volts [V] and the base current (I_B) is 97 microamperes [μA]?

(b) What is the collector current (I_C) in units of milliamperes [mA] when the collector-emitter voltage (V_{CE}) is 7.2 volts [V] and the base current (I_B) is 200 microamperes [μA]?

(c) What is the base current (I_B) in units of microamperes [μA] when the collector-emitter voltage (V_{CE}) is 6 volts [V] and the collector current (I_C) is 28 milliamperes [mA]?

(d) What is the collector current (I_C) in units of microamperes [μA] when the collector-emitter voltage (V_{CE}) is 8.7 volts [V] and the base current (I_B) is 149 microamperes [μA]?

(e) What is the collector-emitter voltage (V_{CE}) in units of volts [V] when the base current (I_B) is 60 microamperes [μA] and the collector current (I_C) is 12.2 milliamperes [mA]?

ICA 8-6

The table below relates three quantities:

- Pressure (P): given as column of numbers on the left-hand side, in units of pound-force per square inch [psi]
- Pipe diameter (D): given as a row of numbers at the top, in units of inches [in]
- Mass flow rate ($\dot{m}$): shown as the intersecting cells, given as units of grams per hour [g/h]

Use interpolation to determine the required values.

Pressure (P) [psi]	Mass Flow Rate ($\dot{m}$) [g/h] Pipe Diameter (D) [in]					
	½	1	1½	2	2½	3
5	19	75	170	300	490	710
10	24	95	210	380	600	900
20	30	120	260	520	810	1100
30	42	155	360	655	950	1460
40	51	195	445	840	1250	1800
60	65	250	600	1000	1650	2400

(a) If the pipe diameter is 2 inches [in], what is the mass flow rate at a pressure of 18 pounds-force per square inch [psi]?

(b) If the pipe diameter is 3 inches [in], what is the mass flow rate at a pressure of 33 pounds-force per square inch [psi]?

(c) If the pipe diameter is 1 inch [in], what is the pressure if the mass flow rate is 170 grams per hour [g/h]?

(d) If the pipe diameter is 2 inches [in], what is the pressure if the mass flow rate is 450 grams per hour [g/h]?

(e) If the pipe diameter is 1¼ inches [in], what is the mass flow rate at a pressure of 24 pounds-force per square inch [psi]?

(f) If the pipe diameter is 2¾ inches, what is the pressure if the mass flow rate is 1000 grams per hour [g/h]?

CHAPTER 9
STATISTICS

Probability is associated with assessing the likelihood that an event will or will not occur. For example:

airplane crash   earthquake   nuclear reactor accident
tornado   failure of equipment   terminal cancer
river breaching a levee   space probe data reception   microprocessor failure

Statistics are used for design-concept evaluation because they provide quantitative measures to "things" that behave in a random manner. This evaluation helps us make rational decisions about events and manufactured products. Statistics, as well as probability, use numerical evidence to aid decision-making in the face of uncertainty. Roles of statistics in engineering include:

- Evaluation of new or alternative designs, concepts, procedures
- Estimation of amount to bid on projects
- Management (human uncertainty, economic uncertainty, others)
- Determination of degree of acceptable item-to-item variation (quality control).

Often, the best way to analyze an engineering problem is to conduct an experiment. When we take this approach, we face several questions:

- How many tests do we need?
- How confident are we in the results?
- Can we extrapolate the results to other conditions?
- Can we estimate how often the result will lie within a specified range?

There are many other related issues, but addressing all of them requires a separate book. Many readers will take or have taken an entire course in probability or statistics, so for now we just touch on some of the important fundamentals.

- Repeated tests: When a test is conducted multiple times, we will not get the same (exact) result each time. For example, use a ruler to measure the length of a particular brand of shoe manufactured by company X. You can produce a table of values, all of which will be nearly, but not exactly, the same. This is because you make slight errors in measurement, so even if every shoe is identical, there will always some errors in your measurement. Moreover, every shoe is not identical.
- Differences in a population: What is the heart rate for all students in a class? Obviously, not everyone will have the same heart rate. We do expect, however, that everyone's rate will lie between, say, 40 and 140 beats per minute. Through measurement, we can determine this variation. In fact, we may find that the average rate for females and males differs. Statistical procedures help us analyze situations such as this.

- Manufacturing errors: Suppose you are manufacturing a run of widgets and a buyer wants all of them to be exactly alike. Obviously, this is impossible, but you can make them almost alike and then tell the buyer how much variation to expect. If you measure each widget as it comes off the assembly line, there will certainly be some variation.
- Design criteria for products: When you build a house, you would like it to stand safely for some period of time. For example, you might specify that the house be designed to withstand a windstorm that would occur, on average, every 50 years. For that case, you must be able to calculate the wind speeds associated with such a storm. Statistical methods allow you to do this.

9.1 HISTOGRAMS

To illustrate several common statistical concepts, we use data representing the height of several freshman engineering students. Table 9-1 shows the height, to the nearest inch, of each student in a typical class. Table 9-2 shows the same data, summarized by number of students at each height.

Table 9-1 Student height

Student ID	Height (H) [in]
A	67
B	73
C	71
D	69
E	68
F	64
G	70
H	72
I	67
J	71
K	70
L	68
M	66
N	71
O	74
P	71
Q	68
R	72
S	67
T	64
U	75
V	74
W	72

Table 9-2 Summary of height data

Height (H) [in]	Number of Students
62	0
63	0
64	2
65	0
66	1
67	3
68	3
69	1
70	2
71	4
72	3
73	1
74	2
75	1
76	0
77	0
Total	23

When we graph the values shown in Table 9-1, we end up with a scatter plot with data that is exactly that: scattered, as shown in Figure 9-1.

Instead of using a scatter plot, we can group the data and plot the group values in a chart similar to a column chart, shown in Figure 9-2. Using the summarized data shown in Table 9-2, we will place two height ranges into a single column or **bin**. The first bin will contain all student-height values less than 62 inches. The next bin will contain student-height values of 62 and 63 inches. The next bin will contain student-height values of 64 and 65 inches, and so on. The abscissa of the graph is the height values; the ordinate is the number of students measured. Graphs of this nature are called **histograms**. By counting the number of blocks, we find the area under the curve represents the total number of samples taken, in this case, the total number of students (23) observed.

Figure 9-1 Example of student height, shown on scatter plot.

Figure 9-2 Example of student height, shown on a histogram.

Technically, before beginning this example we should have mathematically determined a bin size, rather than arbitrarily grouping the measurements in pairs (62 and 63 in one bin, 64 and 65 in the next bin, etc.) There are several ways to calculate the bin size that will best display the information. Here is one example.

DETERMINATION OF BIN SIZE

Step One: Determine the number of bins needed.

Number of Bins = Square root of number of data points, rounded to nearest whole number

Step Two: Determine the range of the data.

Range = $X_{max} - X_{min}$

Step Three: Determine the number of items in each bin.

Bin Size = Range divided by Number of Bins, rounded to nearest whole number

Let's apply this to our example.

Step 1: As shown, we have a class of 23 students, so we would need about five bins, since the square root of 23 is about 4.8, which rounds to 5. Four would probably also work fine, as would 6. Remember that this is just a rule of thumb.

Step 2: The shortest person in the class is 64 inches tall and the tallest is 75, so the range is 75–64, or 11 inches.

Step 3: Dividing the range determined in Step 2 by the number of bins determined in Step 1, we get 2.2, or about 2 inches per bin, the value used in our example. On the other hand, we might instead decide to have four bins or six bins. If we divide the range by 4, we have 2.75 or 3 inches per bin, whereas if we divide by 6, we have 1.83 or 2 inches per bin.

Depending on the number of bins, we sometimes get two different, but acceptable, bin sizes. By changing the bin size, we can change the appearance of the data spread, or the data **distribution**.

What happens to the student height data if we alter the bin size? The plot on the preceding page shows a 2-inch bin interval, and Figure 9-3 (a) below shows a 3-inch interval.

In Figure 9-3 (b), we have used a 4-inch bin interval, and while it is not what we obtained from the "rule of thumb" (2- or 3-inch intervals), it is still mathematically correct but not as informative as the other two.

Figure 9-3(a) Bin size of 3.

Figure 9-3(b) Bin size of 4.

STATISTICAL BEHAVIOR

When we have gathered the data and plotted a distribution, the next step is to explain the outcome to others. For convenience, we set of parameters to describe distributions.

One parameter of a distribution is the average value. The **average**, or **mean**, is an estimate of the value most representative of the population. This is often called the central tendency of the data. The computation of the mean ($\overline{X}$) of a data set containing N values, given in the equation below.

$$\text{Mean} = \overline{X} = \frac{1}{N}(X_1 + X_2 + \cdots + X_N) = \frac{1}{N}\sum_{j=1}^{N} X_j$$

In other words, the mean is simply the sum of all of the values divided by the total number of values.

The **median**, another measure of central tendency, is the value between the lower half and upper half of the population. In other words, if all data points are listed in numerical order, the median is the value exactly in the middle of the list. If the number of data points is odd, the median will be the middle value of the population. If the number of data points is even, however, the median will be the average of the two values at the center. A few examples should clarify this.

Set #	Data	Mean	Median
1	1, 2, 3, 4, 5, 6, 7	4	4
2	1, 50, 70, 100	55	60 (average of 50 & 70)
3	5, 10, 20, 40, 80	31	20
4	50, 50, 50, 50, 50, 1000	208	50 (average of 50 & 50)

Review the data in set #4 in shown above. It would seem logical if every data point has a value of 50 except one, the average of the data should be about 50; instead, it is 208! This illustrates the sensitivity of the mean to **extreme values**, or **outliers**. Note that the median is unaffected or only slightly affected. It is for this reason that the mean is insufficient to describe the central tendency of all distributions.

Two other terms are useful in describing a distribution: **variance** and **standard deviation**. Both of these terms quantify how widely a set of values is scattered about the mean. To determine the variance (V_x^2), the difference between each point and the mean is determined, and each difference is squared to keep all terms positive. This sum is then divided by one less than the number of data points.

$$\text{Variance} = V_x^2 = \frac{1}{N-1}\left((\overline{X} - X_1)^2 + (\overline{X} - X_2)^2 + \cdots + (\overline{X} - X_N)^2\right)$$

$$= \frac{1}{N-1}\sum_{j=1}^{N}(\overline{X} - X_j)^2$$

The standard deviation (SDx) is found by taking the square root of the variance:

$$\text{Standard deviation} = SD_x = \sqrt{V_x^2}$$

If we again examine the data found in Table 9-2, we can calculate the mean, median, variance, and standard deviation for our height data.

*Sidebar:*

Average or Mean = typical, expected value of the data set; sensitive to outliers.

Median = value representing the exact middle value of the list; typically unaffected by outliers. Data must be in ascending order to determine!

Variance = measure of data scatter; has SQUARED UNITS of the original data set.

Standard deviation = square root of the variance; has units of the original data set.

Height (H) [in]	Number of Students
62	0
63	0
64	2
65	0
66	1
67	3
68	3
69	1
70	2
71	4
72	3
73	1
74	2
75	1
76	0
77	0
Total	23

Calculation of the Mean:

Total number of points (N)
= 23 students

The sum of all heights
= (2 students * 64 inches/student) + (1 * 66) + (3 * 67) + (3 * 68) + (1 * 69)
 + (2 * 70) + (4 * 71) + (3 * 72) + (1 * 73) + (2 * 74) + (1 * 75)
= 1604 inches

Mean
= 1604 inches/23 students
= 69.7 inches/student

Calculation of the Median:

Put data in order of value, listing each entry once
 64, 64, 66, 67, 67, 67, 68, 68, 68, 69, 70, 70, 71, 71, 71, 71, 72, 72, 72, 73, 74, 74, 75

Find the center value since the total number of students is odd (23; center value at entry #12)
 64, 64, 66, 67, 67, 67, 68, 68, 68, 69, 70, **70**, 71, 71, 71, 71, 72, 72, 72, 73, 74, 74, 75

Median = 70 inches

Calculation of Variance: Note that the variance will have the same units as the variable in question, in this case, "inches^2."

$$\text{Variance} = \frac{1}{23-1}((69.7-64)^2 + (69.7-64)^2 + (69.7-66)^2 + \cdots$$
$$+ (69.7-75)^2) = 9.5 \text{ inches}^2$$

Calculation of Standard Deviation: The standard deviation has the same units as the variable in question, in this case, "inches."

$$\text{Standard deviation} = \sqrt{9.5} = 3.08 \text{ inches}$$

EXAMPLE 9-1

Consider the following data. Determine the mean, median, variance, and standard deviation of the data.

1	28	14	32	35	25	14	28	5
16	42	35	26	5	33	35	16	14

Calculation of the Mean:

Total number of points (N) = 18 entries

Sum of all data (ΣX_i) = (1) + (2 * 5) + (3 * 14) + (2 * 16) + (25) + (26)
 + (2 * 28) + (1 * 32) + (1 * 33) + (3 * 35) + (1 * 42)
= 404

Mean = 404/18 = 22.4

Calculation of the Median:

Put data in order, listing each entry once.
 1, 5, 5, 14, 14, 14, 16, 16, 25, 26, 28, 28, 32, 33, 35, 35, 35, 42

Find the center two values and average them, since total number of entries is even (18).
 1, 5, 5, 14, 14, 14, 16, 16, 25, 26, 28, 28, 32, 33, 35, 35, 35, 42

Median = (25 + 26)/2 = 25.5

Calculation of Variance:

$$\text{Variance} = \frac{1}{18 - 1}((22.4 - 1)^2 + 2*(22.4 - 5)^2 + 3*(22.4 - 14)^2 + \cdots$$
$$+ (22.4 - 42)^2) = 147$$

Calculation of Standard Deviation:

$$\text{Standard Deviation} = \sqrt{147} = 12.1$$

EXAMPLE 9-2

Consider the following data. Determine the mean, median, variance, and standard deviation of the data.

159	837	618	208	971	571	379	220	31

Calculation of the Mean:

Total number of points (N) = 9

Sum of all heights ($\Sigma\,X_i$) = 159 + 837 + 618 + 208 + 971 + 571 + 379 + 220 + 31
= 3994

Mean = 3994/9 = 443.7 = 444

Calculation of the Median:

Put data in order, listing each entry once.

31, 159, 208, 220, 379, 571, 618, 837, 971

Find the center value since the total number of students is odd (9; center value at entry #5).

31, 159, 208, 220, 379, 571, 618, 837, 971

Median = 379

Calculation of Variance:

$$\text{Variance} = \frac{1}{9 - 1}((444 - 31)^2 + (444 - 159)^2 + \cdots + (444 - 971)^2) = 105.059$$

Calculation of Standard Deviation:

$$\text{Standard Deviation} = \sqrt{105.059} = 324$$

Comprehension Check 9-1

For the following data, determine the mean, median, variance, and standard deviation.

8	7	9	11	16
12	2	9	10	9

Comprehension Check 9-2

For the following data, determine the mean, median, variance, and standard deviation.

105	120	110	100	102
103	58	110	100	118

9.3 DISTRIBUTIONS

From the charts on the preceding pages discussing histograms, you can see a similarity in the shape of all three plots. The values start small, increase in size, and then decrease again. In the case of student height, this means that a few people are short, most people have some "average" height, and a few people are tall. This same conclusion is true in many places in our world. For example:

- If we weigh many standard-size watermelons (neither miniature nor giant), we will find that most weigh between 20 and 30 pounds. A few weigh less than 20 pounds, and a few weigh more than 30 pounds.
- As we look through a dictionary, we find that there are many words with between four and six letters. There are a few with one, two, or three letters and a few with more than six, but clearly most have between four and six letters.
- To improve efficiency in the office, we had an expert monitor the length of phone calls made by the staff. The expert found that most of the time, phone calls lasted between three and five minutes, but a few were longer than five and some others lasted only a minute or two.

Normal Distributions

We wanted to know how many "flexes" it takes to cause a paper clip to fail, so we asked volunteers to test the bending performance of paper clips by doing the following:

- Unfold the paper clip at the center point so that the resulting wire forms an "S" shape.
- Bend the clip back and forth at the center point until it breaks.
- Record the number of "flexes" required to break the clip.

Using these data, we created Figure 9-4. This is the same as the earlier histogram, but the "boxes" are replaced by a smooth curve through the values.

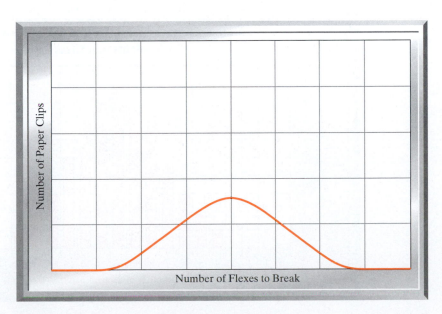

Number of Paper Clips

Number of Flexes to Break

Figure 9-4
Distribution of paper clip failure.

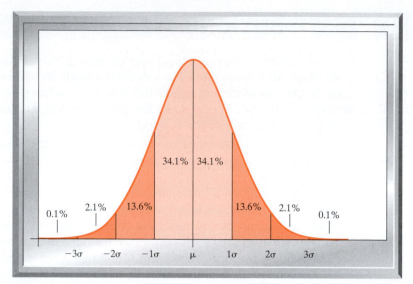

Figure 9-5 "Normal" distribution, showing the 68–95–99.7 rule.

Sketch from: wikipedia.org/wiki/Standard_deviation by Pat Beirne

When you are interested in the *shape* of the curve rather than the exact data values, you can replace the bars of the histogram with a smooth curve and rename the graph *distribution*. A distribution is considered *normal* if the following rules hold true. This is known as the *68-95-99.7 rule*, shown in Figure 9-5.

- 68 percent of values are within one standard deviation (1σ) of mean (μ).
- 95 percent of values are within two standard deviations (2σ).
- 99.7 percent of values are within three standard deviations (3σ).

EXAMPLE 9-3

Suppose we ask a class of students how many states they have visited. The results might appear as shown below.

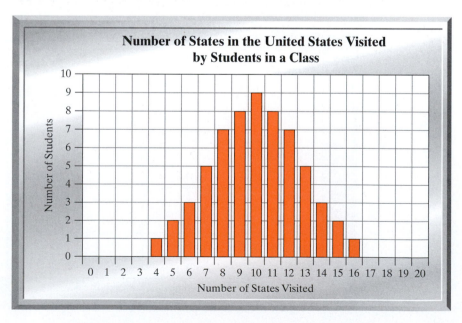

It seems that most have visited between 8 and 12, and that as many have visited more than 10 as have visited fewer than 10. A few have visited as many as 16 states, and all the students have visited at least 4. Let's calculate some values pertinent to this situation.

How many students are in the class?

To do this we simply add the number of students represented by each bar, or

$$1 + 2 + 3 + 5 + 7 + 8 + 9 + 8 + 7 + 5 + 3 + 2 + 1 = 61$$

What is the cumulative number of state visits?

We answer this by totaling the product of the bar height with the number of states represented by the bar. For example, 5 students have visited 7 states, so those 5 students have visited a total of 5 3 7 5 35 states. Or, 8 students have visited 11 states, so those students have visited a total of 88 states. We calculate

$$1*4 + 2*5 + 3*6 + 5*7 + 7*8 + 8*9 + 9*10 + 8*11 + 7*12 + 5*13$$
$$+ 3*14 + 2*15 + 1*16 = 610$$

What is the average number of states visited by a student?

Once we have the values from our first two answers, this is straightforward division: the total number of visits divided by the total number of students.

$$610/61 = 10$$

Notice that the value 10 is in the center of the distribution. For distributions that are symmetrical (such as this one), the average value is the one in the center, the one represented by the largest number of occurrences.

Decrease in Variance

Let's examine the shape of the distribution of paper clip failures discussed earlier. How would the distribution change if we brought in a machine that did it "exactly" the same way each time? Both the distribution from the data class and the distribution of the machine are shown below. The same number of clips was tested in each case, so the areas under each curve must be the same. The comparison is shown in Figure 9-6.

Figure 9-6 Distribution of paper clip failure after a decrease in variance.

This exercise illustrates that distributions that have the same mean (and median) can look very different. In this case, the difference between these two distributions is in their "spread," or their variation about the mean.

Shift in Mean

Redraw the paper clip distribution; then on the same plot, sketch the distribution if each volunteer tested clips that were manufactured by the same manufacturer as before but were stronger and typically required 10 more flexes to fail. The result is shown in Figure 9-7.

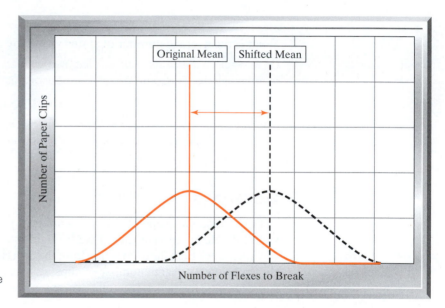

Figure 9-7
Distribution of paper clip failure with a shift.

Restriction of Range

How would the distribution change if you flexed the paper clips as usual but heated the ones that were still in one piece after the 13th flex (which weakens the clip) and then continued to test them with more flexes?

Figure 9-8 illustrates another factor that affects the shape of a distribution—the **restriction of range**. Similarly, range expansion changes the shape of a distribution. For example, mortgage payments among families with household incomes between $40,000 and $70,000 will have a certain distribution. If the limits are changed to include those families with household incomes from $30,000 to $70,000, the distribution will change—the mean will probably be lowered and the standard deviation increased. We could restrict a range by establishing a minimum possible outcome as well, such as a minimum or maximum score on an exam.

Figure 9-8 Distribution of paper clip failure with a restricted range.

Skewed Data

It is often easy to place an upper limit on the value of the possible outcome. In these cases, the distribution is no longer symmetric—it is **skewed**. A population is **positively skewed** if the mean has been pulled higher than the median, and **negatively skewed** if the mean has been pulled lower than the median (see Figure 9-9). You have probably heard news reports that use the median to describe a distribution of income in the United States. The median is used in this case because the distribution is positively skewed. This skew is caused by two factors—the presence of extreme values and the range restriction, the latter because income cannot be lower than $0. The extreme values causing the positive skew are not shown on the graph. Most of these would be far off the page to the right.

Figure 9-9 Distribution of positively skewed data.

Comprehension Check 9-3

For each graph shown below, decide if the mean, variance, or population size has changed.

(a)

(b)

(c)

EXAMPLE 9-4

For each scenario, identify the one graph from the following that best illustrates how the baseline curve would change under the conditions of that scenario. Each graph shows the usual distribution (labeled baseline) and the way the distribution would be modified from the baseline shape (labeled modified) under certain conditions.

The graphs show SAT composite (verbal + quantitative) scores, for which 400 is generally consider to be the minimum possible score and 1600 is considered to be the maximum possible score.

(a) The designers of the SAT inadvertently made the test more difficult.
This is shown by Curve F: Area is the same; mean shifted to left.

(b) The variability of scores is reduced by switching to true/false questions.
This is shown by Curve D: Area is the same; distribution is narrower.

(c) A population boom increases the number of students seeking college admission.
This is shown by Curve B: Area increases, distribution stays the same.

(a)

(b)

(c)

(d)

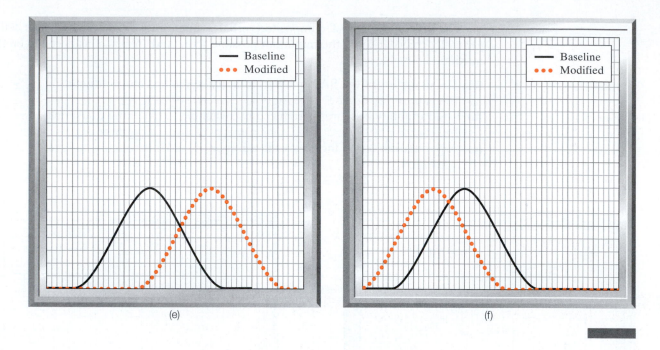

9.4   CUMULATIVE DISTRIBUTION FUNCTIONS

For the earlier student height plot using two heights per bin, we will graph the bin data but now show the values on the ordinate as a fraction rather than a whole number. To do this, we divide the number of students in each bin of the histogram by the total number of students. If we now add the heights of all the bars in the new plot, they should equal 1. This is called a **normalized plot**, shown in Figure 9-10. We "normalized" the values by dividing by the total number of data points. This graph holds no new information; it is simply a rescaling of the histogram we drew earlier.

Figure 9-10 Normalized plot of student height, originally shown in Figure 9-2.

This plot can be used as an intermediate step to obtain a final plot called a **cumulative distribution function**, or **CDF**. We derive this plot by summing the values for each bin on the normalized plot from the first bin up to each individual bin. For example, suppose the values in the first three bins were 0, 0.08, and 0.18. By adding the values, we get new "cumulative" values: bin 1 = 0; bin 2 = 0 + 0.08 = 0.08; and bin 3 = 0 + 0.08 + 0.18 = 0.26. It should be obvious that we can obtain the CDF value for each bin by adding the normalized value of that bin to the CDF value of the bin before it. The CDF values are usually shown as percentages rather than fractions, for example, 50% instead of 0.5.

As we move across the plot, the values should go from 0 to 1. Using the height data in the normalized plot above, we have produced a cumulative distribution shown in Figure 9-11. Sometimes, the CDF is shown as a continuous, curved line rather than a column chart. Both the original histogram and the cumulative distribution plot are useful tools in answering questions about the composition of a population.

Figure 9-11 CDF of student height, originally shown in Figure 9-2.

EXAMPLE 9-5

Consider the following data. Draw the histogram and CDF of the data.

36	9	33	11	23	3
34	39	56	51	39	1
27	25	2	1	53	32
14	41	55	28	29	19
51	15	25	10	35	38

(a)

(b)

Comprehension Check 9-4

Consider the following data. Draw the histogram and CDF of the data.

38	103	20	42	16
20	74	63	90	61
114	79	61	50	64

9.5 STATISTICAL PROCESS CONTROL (SPC)

We showed that a histogram such as one shown in Figure 9-12 visually summarizes how a set of values is distributed. Sometimes, however, we are not only interested in the values themselves but also in how the distribution changes over time.

For example, as a machine in a factory operates, it may slowly (or occasionally quickly) lose proper alignment or calibration due to wear, vibrations, etc. If a machine was making bolts with a mean length of one inch and a standard deviation of 0.01 inch when it first began operating, after it had made 100,000 bolts, the alignment may have drifted so that the mean was only 0.95 inches with a standard deviation of 0.02 inch. This may be unacceptable to the customer purchasing the bolts, so the parameters of the process need to be monitored over time to make sure the machine is readjusted as necessary.

A graph called a **quality chart** is often used to show how close to the mean the results of a process are when measured over time. The graph is usually a scatter plot, with the abscissa shown as time or another indicator that would change with time, such as batch number. Figure 9-13 shows a sample control chart, the mean, and the standard deviation.

When tracking a manufacturing process, engineers are often concerned with whether the process is "in control," or behaving as expected. **Statistical Process Control (SPC)** is a method of monitoring, controlling, and improving a manufacturing process. In some situations, the desired mean and acceptable deviation limits may be preset for

Figure 9-12 Sample histogram.

Figure 9-13 Sample quality chart.

a variety of reasons (chemistry, safety, etc.). Often, the upper or lower limits of control are determined by the desired end result.

- The reactor temperature must not rise above 85 degrees Celsius or the reactant will vaporize.
- The injection pressure should be between 50 kilopascals and 75 kilopascals to ensure that the part is molded properly.
- A bolt must be machined to ± 0.02 inches to fit properly in a chair leg.

An engineer will study how the process relates to the control limits and will make adjustments to the process accordingly. To discuss whether a process is in control, we can divide a chart into the following zones, shown in Figure 9-14, to create a control chart: The mean is determined either by the desired end result (the iron content of the product

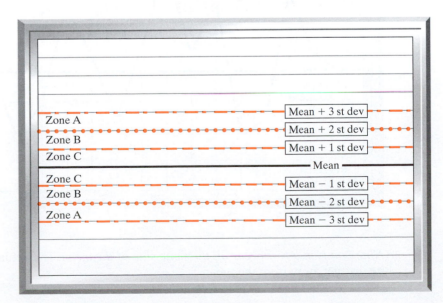

Figure 9-14 Standard deviation ranges.

must average 84%) or by the process itself (the reactor temperature should average 70 degrees Fahrenheit for an optimum reaction to occur). The standard deviation is most often determined by experimentation.

- The purity range for this product is ±0.0005%.
- The standard deviation for the reactor temperature must not exceed 5 degrees Fahrenheit or the reaction will create unwanted byproducts.
- The standard deviation of the current gain of the transistors being produced must be less than 15.

Eight Ways to Be Out of Control

A variety of conditions can indicate that a process is out of control. First published by Lloyd S. Nelson in the October 1984 issue of the *Journal of Quality Technology*, the **Nelson Rules** are listed below, with examples and graphs. In the graphs, solid points indicate the rule violations. The actual conditions may vary slightly from company to company, but most take the same standard form. For example, a company may operate with Rule 3 stated as seven or eight points in a row instead of six.

1. *A point falls anywhere beyond Zone A.* The value is more than three standard deviations away from the mean. May occur on either side of the mean.

   Ex: The mean temperature of a reactor is 85 degrees Celsius [°C] with a standard deviation of 5 °C. If the temperature exceeds 100 °C, the reactor vessel may explode. If the temperature falls below 70 °C, the reaction cannot proceed properly.

2. *Nine points in a row occur on the same side of the mean.* The actual value seems to be drifting away from the mean.

Ex: The percentage of boron in a semiconductor should be 250 parts per billion. Nine consecutive samples have a boron content less than this value. The machine incorporating the boron into the semiconductor material may need to be cleaned or recalibrated.

3. *Six points occur with a consistently increasing or decreasing trend.* If this pattern continues, the values will eventually become unacceptable.

Ex: The shaft length of a part is increasing with each successive sample; perhaps the grinding wheel needs to be changed.

4. *Fourteen points in a row alternate from one side of the mean to the other.* The process is unstable.

Ex: The control system for a crane errs from one side to the other. This may indicate a sensor failure or the need to reprogram the controller.

5. *Two out of three points in a row occur in Zone A; all three points occur on the same side of the mean.* The process is close to the upper limit; take preventive measures now.

Ex: A robot that is spot-welding parts in an automobile is coming close to the edge of the material being welded. It probably needs attention.

6. *Four out of five points in a row occur in Zone B; all five points occur on the same side of the mean.* The process is very close to the upper limit; take preventive measures now.

   Ex: Four out of five customers in the bank teller queue have waited more than one standard deviation to be helped. Perhaps another teller is needed.

Rule 6

7. *Fifteen points in a row occur in Zone C; points can occur on either side of the mean.* The process is running too perfectly; in many applications the restrictions can be loosened to save money.

   Ex: The thickness of all washers being manufactured for quarter-inch bolts is within 0.0005 inch of the desired mean. Very few applications require washers with such close tolerances.

Rule 7

8. *Eight points in a row occur beyond Zone C; points can occur on either side of the mean.* The process does not run close enough to the mean; the parts are never quite on target; may indicate a need for a process adjustment.

Ex: The postmark machine in a regional postal distribution center is stamping the envelopes too high or too low; it probably needs attention.

Rule 8

EXAMPLE 9-6

Part #	Length (L) [in]
1	10.00
2	10.25
3	10.65
4	9.50
5	9.36
6	9.00
7	10.50
8	10.20
9	9.80
10	10.00
11	10.25
12	10.65
13	9.50
14	9.36
15	9.25
16	10.50
17	10.20
18	9.80
19	10.45
20	10.10

The data shown in the table were collected from a manufacturing process that makes bolts.

Assume the process specifies an average bolt length of 10 inches, with a standard deviation of 0.25 inches. Is this process under statistical control?

The control chart for these data is shown below. Rule 1 is violated since a point falls outside of Zone A (part #6); therefore, the process is not in statistical control.

Comprehension Check 9-5

The pressure in a water filter is monitored in a chemical plant. The filter should operate at 18 pounds-force per square inch [psi], with a standard deviation of ±2 psi. Analyze the data shown to determine if the filter is behaving as expected (the process is in control; the filter does not require any attention) or if the filter required attention (the process is out of control; the filter should be cleaned). Refer to the Nelson Rules to explain your conclusion.

Comprehension Check 9-6

The following data were collected from the machines over the weekend. Using the 8 SPC rules, determine whether each machine is in statistical control. If the machine is not in statistical control, indicate which rule was violated.

(a)   (b)

In-Class Activities

ICA 9-1

Ten students are asked to participate in a golf-putting contest. Each student is given five balls, placed a distance of 60 feet from the hole. They are to try to get each ball in the hole with a single putt. The number of feet that each ball is from the hole when it comes to rest is measured. For simplicity, the assumption is that no students make their putt.

What do you think the values for the students in a class might look like? Sketch a histogram of this situation (number of students on the ordinate and distance the putt stops from the hole on the abscissa) using the data given.

Distance from hole [ft]	0	1	2	3	4	5	6	7	8	9	10	11	12	13	14
Number of Balls	0	1	3	2	3	5	4	6	9	7	5	3	2	0	0

ICA 9-2

This exercise includes the measurement of a distributed quantity and the graphical presentation of the results. You are to determine how many flexes it takes to cause a paper clip to fail.

Test the bending performance of 20 paper clips by doing the following:

- Unfold the paper clip at the center point so that the resulting wire forms an "S" shape.
- Bend the clip back and forth at the center point until it breaks.
- Record the number of flexes required to break the clip.

On a copy of the table below, record the raw data for the paper clips you break. Then, summarize the data for the team by adding up how many clips broke at each number of flexes. Each team member should contribute 20 data points.

Use the data to plot a histogram with an appropriate bin size, the normalized plot, and cumulative distribution function.

Paper clip flexing data

Paper Clip #	Flexes to Break
1	
2	
3	
4	
5	
6	
7	
8	
9	
10	
11	
12	
13	
14	
15	
16	
17	
18	
19	
20	

Summary of data

No. of Flexes	No. of Clips

ICA 9-3

For the following velocity data, recorded in units of feet per second, answer the questions below.

4	9	12	1	7
14	11	5	8	5
7	10	2	16	9

(a) What is the mean of the data?
(b) What is the median of the data?
(c) What is the variance of the data?
(d) What is the standard deviation of the data?

ICA 9-4

For the following pressure data, recorded in units of pound-force per square inch, answer the questions below.

1	14	2	15	10
6	3	1	18	

(a) What is the mean of the data?
(b) What is the median of the data?
(c) What is the variance of the data?
(d) What is the standard deviation of the data?

ICA 9-5

A technician tested a temperature probe by inserting it in boiling acetic acid (theoretical boiling point = 118 degrees Fahrenheit [°F]), recording the readings, removing and drying the probe, and repeating the process. The data are shown in the following table.

Temperature, Probe #1 (T) [°F]	120	118	105	115	105	120	125
Temperature, Probe #2 (T) [°F]	100	90	95	105	90		

(a) Determine the mean of Probe #1.
(b) Determine the median of Probe #1.
(c) A second probe was tested, yielding a mean of 95 degrees Fahrenheit [°F] and a median of 92.5 degrees Fahrenheit [°F]. If the data from Probe #2 are as shown above, determine the missing data point.
(d) If a probe has a standard deviation of 8 degrees Fahrenheit [°F], what is the variance of the probe?

ICA 9-6

A technician tested two temperature probes by inserting them in boiling water (theoretical boiling point = 100 degrees Celsius [°C]), recording the readings, removing and drying the probe, and repeating the process. The CDF for both probes is shown below.

(a) Using this information, which probe would you choose to use?
(b) How would you use the probe you chose above to ensure that you found the correct boiling point?
(c) Which probe has the higher standard deviation?

ICA 9-7

You are assigned to inspect metal-composite beam trusses for a new bridge being built over a nearby lake. The manufacturer has run a prototype set of 500 beams and conducted strength tests, which you consider to be the baseline case, shown by the solid line in all graphs. Examine the graphs on the following page, and explain the changes to the baseline curve observed in the dashed line by choosing a cause from the following list. The strength of the beam is shown on the abscissa.

(a) The manufacturer tested 1000 beams instead of 500 beams.
(b) The manufacturer tested 200 beams instead of 500 beams.
(c) A reinforcing coating was used on a sample of 500 beams.
(d) An impurity, which caused the beams to weaken, was discovered in a sample of 500 beams.
(e) A data entry clerk accidentally combines the data for 300 metal-composite beams with 200 wooden beams.
(f) The manufacturer upgrades processing equipment to lower the variability of the metal-composite strength.
(g) During a plant strike by union workers, substitute workers manufacture the beams.

ICA 9-8

The table below lists the number of computer chips rejected for defects during random testing over the course of a week on a manufacturing line. Four samples of 20 parts are pulled each day. Use the following data to generate a histogram and CDF.

1	1	8	0	2	0
0	2	10	1	3	2
0	1	12	0	2	1
1	6	15	0	0	
3	8	1	2	5	

ICA 9-9

We are interested in the population and health of chorous frogs as an indicator of the pollution affecting a lake. Eggs hatch into tadpoles in March. Tadpoles transform into frogs in May and mature by September. The baseline graph shows a sample of 100 frogs measured in July. Examine the graphs on the following page, and explain the changes to the baseline curve observed in the dashed line by choosing a cause from the list provided. The length of the frog is shown on the abscissa.

(a) A sample of 100 frogs is taken in May.
(b) A sample of 200 frogs is taken in July.
(c) A sample of 100 frogs is taken in July, after a growth hormone has been introduced into the water supply in May, causing the male frogs to be double the size of the female frogs by July. It can be assumed that 50 of the frogs sampled were male, and 50 were female.
(d) A new predator is introduced into the ecosystem in May, consuming many of the smaller frogs. In June, a drought occurs, killing many of the larger frogs. A sample of frogs is taken in July, with obvious reduction in variability.

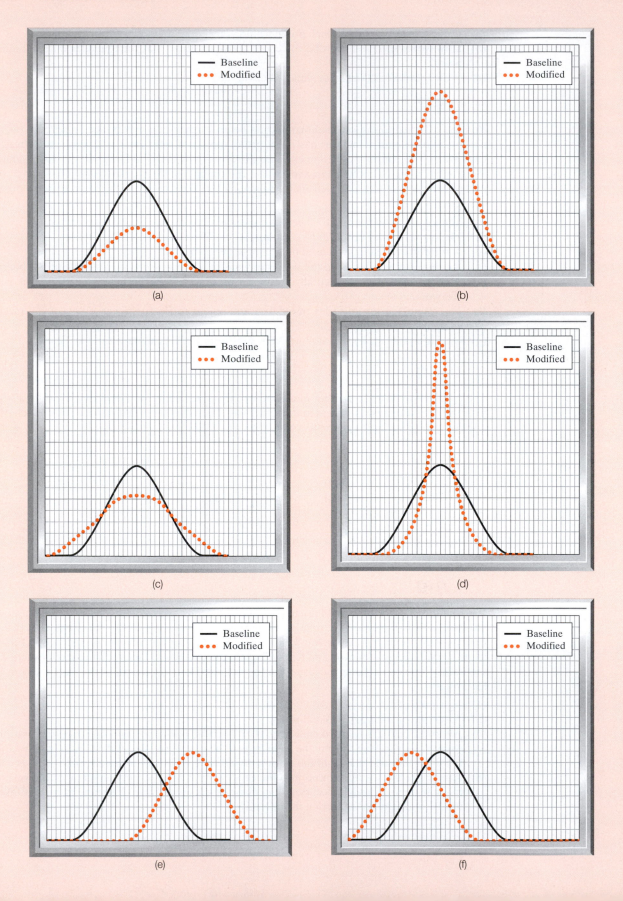

ICA **9-10**

From the CDF shown, draw the histogram that was used to create the CDF. Assume there are 20 points, with a minimum of 2 and a maximum of 50.

ICA **9-11**

The data below were collected from a manufacturing process for making plastic cylinders. According to the specifications, the cylinder diameter should be 100 inches (the average diameter is 100 inches) and the standard deviation is ± 5 inches.

Graph the data on a control chart. Be sure to clearly indicate the "Zones" of control.

Using the eight SPC rules, determine whether the process is in statistical control. If it is not in statistical control, indicate which rule or rules were violated.

Part #	Diameter (D) [in]	Part #	Diameter (D) [in]
1	100	10	96
2	106	11	89
3	103	12	89
4	99	13	87
5	90	14	92
6	95	15	94
7	105	16	87
8	107	17	96
9	97	18	98

1. A motorcycle weighing 500 pounds-mass plus a rider weighing 300 pounds-mass produces the following chart. Predict a similar table if a 50-pound-mass dog is added as a passenger. Use SOLVEM.

Velocity (v) [mi/h]	Time (t) [s]
0	0.0
10	2.3
20	4.6
30	6.9
40	9.2

2. A circus performer jumps from a platform onto one end of a seesaw, while her partner, a child of 12, stands on the other end. How high will the child "fly"? Analyze this problem with SOLVEM.

3. Your college quadrangle is 85 meters long and 66 meters wide. When you are late for class, you can walk (well, run) at 7 miles per hour. You are at one corner of the quad and your class is at the directly opposite corner. How much time can you save by cutting across the quad rather than walking around the edge? Analyze this problem with SOLVEM.

4. I am standing on the upper deck of the football stadium. I have an egg in my hand. I am going to drop it and you are going to try to catch it. You are standing on the ground. Apparently, you don't want to stand directly under me; in fact, you would like to stand as far to one side as you can so that if I accidentally release it, it won't hit you on the head. If you can run at 20 feet per second and I am at a height of 100 feet, how far away can you stand and still catch the egg if you start running when I let go? Analyze this problem with SOLVEM.

5. A one-kilogram mass has just been dropped from the roof of a building. I need to catch it after it has fallen exactly 100 meters. If I weigh 80 kilograms and start running at 7 meters per second as soon as the object is released, how far away can I stand and still catch the object? Analyze this problem using SOLVEM.

6. Neglect the weight of the drum in the following problem. A sealed cylindrical drum has a diameter of 6 feet and a length of 12 feet. The drum is filled exactly half-full of a liquid having a density of 90 pound-mass per cubic foot. It is resting on its side at the bottom of a 10-foot deep drainage channel that is empty. Suppose a flash flood suddenly raises the water level in the channel to a depth of 10 feet. Determine if the drum will float. The density of water is 62 pound-mass per cubic foot. Analyze this problem with SOLVEM.

7. Create a proper plot of the following data set showing the effect of the percent glycerol in water on the viscosity of the solution.

Glycerol (w) [% weight]	Viscosity (μ) [cP]
0	1.01
10	1.31
20	1.77
30	2.50
40	3.75
50	6.05
60	10.96

8. Create a proper plot of the following data set showing the effect of the percent glycerol in water on the freezing temperature of the solution.

Glycerol (w) [% weight]	Temperature (T) [°C]
10	−1.6
30	−9.5
50	−22.0
60	−33.6
70	−37.8
80	−19.2
100	17.0

9. A biosystems engineer has discovered a relationship between the population density of broomsedge (*Andropogon virginicus*) and levels of sulfur dioxide (SO_2) in the air, which is measured in parts per million [ppm]. The data table gives average kilograms of broomsedge per hectare for several values of SO_2 concentration. Draw a proper plot for the data. Assume that broomsedge biomass is the dependent variable and thus will be plotted on the ordinate.

Sulfur Dioxide Concentration	Average Broomsedge Biomass
75	900
120	2400
190	3500
275	4750
350	5300

10. The number of mosquitoes found in one square meter of the average South Carolina back-yard varies according to the total rainfall measured in inches during the summer months of June, July, and August. Draw a proper plot for the data.

Rainfall	Mosquito Count
6	5
11	8
18	17
24	20
31	15

11. Several reactions are carried out in a closed vessel. The following data are taken for the concentration (C) in units of grams per liter of solvent processed for compounds A and B as a function of time (t). Draw a proper plot for the data.

Time (t) [min]	Concentration [g/L]	
	of A (C_A)	of B (C_B)
36	0.145	0.160
65	0.120	0.155
100	0.100	0.150
160	0.080	0.140

12. Your team has designed three tennis ball launchers, and you have run tests to determine which launcher best meets the project criteria. Each launcher is set to three different launch angles, and the total distance the ball flies through the air is recorded. These data are summarized in the following table. Plot all three sets of data (one for each of the three launchers) on a single graph. Launch angle should be plotted on the abscissa.

Launcher #1		Launcher #2		Launcher #3	
Launch Angle (θ) [°]	Distance (D) [ft]	Launch Angle (θ) [°]	Distance (D) [ft]	Launch Angle (θ) [°]	Distance (D) [ft]
20	5	10	10	20	10
35	10	45	25	40	20
55	12	55	18	50	15

13. The following experimental data were collected on the current (symbolized by I) in the positive direction and voltage (symbolized by V) across the terminals of two different

thermionic rectifiers. Create a proper plot of the data. Use the grid provided, and place voltage on the abscissa.

Voltage (V) [volts]	Current (I) [milliamperes]	
	Rectifier A	Rectifier B
18	5	15
30	18	26
40	24	34
45	30	50

14. Below is a graph of the vertical position of a person bungee jumping, in meters.

(a) What is the closest this person gets to the ground?

(b) When this person stops bouncing, how high off the ground will the person be?

(c) If the person has a mass of 70 kilograms, how would the graph change for a jumper of 50 kilograms? Approximately sketch the results on the graph provided (or on a copy of it).

(d) If the person has a mass of 70 kilograms, how would the graph change for a jumper of 80 kilograms? Approximately sketch the results on the graph provided (or on a copy of it).

15. Shown are graphs of the altitude in meters, and velocity in meters per second, of a person skydiving.

(a) When does the skydiver reach the ground?

(b) How fast is he moving when he reaches the ground?

(c) What is the fastest he ever falls, and when does this occur?

(d) At what altitude does he open the parachute?

(e) Terminal velocity is the velocity at which the acceleration of gravity is exactly balanced by the drag force of air. How long does it take him to reach his terminal velocity without the parachute open?

(f) How long after his parachute opens does it take him to reach open parachute terminal velocity?

16. Generally, when a car door is opened, the interior lights come on and turn off again when the door is closed. Some cars turn the interior lights on and off gradually. Suppose that you have a car with 25 watts of interior lights. When a door is opened, the power to the lights increases linearly from 0 watts to 25 watts over 2 seconds. When the door is closed, the power is reduced to zero in a linear fashion over 5 seconds. If the door to the car is opened and then closed again 10 seconds later, how much total energy was delivered to the interior lights?

17. One of the 22 named, derived units in the metric system is the volt, which can be expressed as one joule per coulomb (V = J/C). A coulomb is the total electric charge on approximately 6.24×10^{18} electrons. The voltage on a capacitor, an electrical device that stores electric charge, is given by V = 0.2 Q + 5 volts, where Q is the total charge in units of coulombs stored. Create a proper plot of voltage (V, on the ordinate) and total charge (Q, on the abscissa) and use this plot to determine the total energy stored in the capacitor.

18. In a simple electric circuit, the current (I) must remain below 40 milliamps [mA] (I < 40 mA), and must also satisfy the function $I > 10^{-6} e^{25V}$, where V is the voltage across a device called a diode.

    (a) Sketch a graph of these two inequalities with current on the ordinate and voltage on the abscissa. The values on the vertical axis should range from 0 to 50 milliamperes, and the values on the horizontal axis should range from 0 to 1 volt. Shade the region of the graph where *both* inequalities are satisfied.
    (b) Graphically determine the maximum allowable voltage across the diode.

19. In a hard drive design, the faster the disk spins, the faster the information can be read from and written to the disk. In general, the more information to be stored on the disk, the larger the diameter of the disk must be. Unfortunately, the larger the disk, the lower the maximum rotational speed must be to avoid stress-related failures. Assume the minimum allowable rotational speed (S) of the hard drive is 6000 revolutions per minute [rpm], and assume that the rotational speed must meet the criterion $S < 12000 - 150 D^2$, where D is the diameter of the disk in inches.

    (a) Sketch a graph of these two inequalities with rotational speed on the ordinate and diameter on the abscissa. The values on the vertical axis should range from 0 to 12000 rpm, and the values on the horizontal axis should range from 0 to 7 inches. Shade the region of the graph where *both* inequalities are satisfied.
    (b) Graphically determine the range of allowable rotational speeds for a 4-inch diameter disk.
    (c) Graphically determine the largest diameter disk that meets the design criteria.

20. Your company has developed a new high-mileage automobile. There are two options for manufacturing this new vehicle.

    ■ Process A: The factory can be completely retooled and workers trained to use the new equipment.
    ■ Process B: The old equipment can be modified.

    A graph of the costs of each process and the revenues from sales of the vehicles is shown. Use the chart to answer the following questions.

(a) What is the sales price per vehicle?

(b) What is the break-even point (number of vehicles) for each of the two processes?

(c) Which process yields the most profit if 18,000 vehicles are sold? How much profit is made in this case?

(d) If the sales price per vehicle is reduced by $2,000 with a rebate offer, what is the new break-even point (number of vehicles) for each of the two processes?

21. You are an engineer for a plastics manufacturing company. In examining cost-saving measures, your team has brainstormed the following ideas. It is your responsibility to evaluate these ideas and recommend which one to pursue. You have been given a graph of the current process.

(a) What is the selling price of the product?

   Current Operations: The current process has been running for a number of years, so there are no initial fixed costs to consider.

   In the operating costs, the process requires the following:

   - Material cost: $2.00/pound of resin
   - Energy cost: $0.15/pound of resin
   - Labor cost: $0.10/pound of resin

(b) There is also a cost associated with taking the scrap material to the landfill. Using the total cost determined from the graph, find the cost of landfill, in dollars per pounds of resin.

   Idea (A): Use 10% scrap: Your customer will allow you to use up to 25% regrind (reprocessed plastic) in the parts instead of 100% virgin plastic. Your process generates 10% scrap. Evaluate using all your scrap material as regrind, with the regrind processed at your plant.

(c) You will need to purchase a regrind machine to process the plastic, estimated at a cost of $100,000. Using the regrind will also alter the following costs:

   - Material cost: $1.80/pound of resin
   - Energy cost: $0.16/pound of resin
   - Labor cost: $0.11/pound of resin

This idea will eliminate the landfill charge required in the current process (see part b above). Draw the total cost curve for Idea (A) on the graph (or on a copy).

(d) How long (in pounds of resin processed) before the company breaks even on the regrind machine?

(e) At what minimum level of production (in pounds of resin processed) will Idea (A) begin to generate more profit than the current process?

Idea (B): Use 25% scrap: Your customer will allow you to use up to 25% regrind (reprocessed plastic) in the parts instead of 100% virgin plastic. Your process generates 10% scrap. Evaluate using 25% regrind purchased from an outside vendor and selling your scrap to them in return.

(f) Using the regrind from the other company will also alter the following costs:

- Material cost: $1.85/pound of resin
- Energy cost: $0.15/pound of resin
- Labor cost: $0.11/pound of resin

This idea will eliminate the landfill charge required in the current process (see part b above). Draw the total cost curve for Idea (B) on the graph (or on a copy).

(g) At what minimum level of production (in pounds of resin processed) will Idea (B) begin to generate more profit than the current process?

(h) At a production level of 500,000 pounds of resin, which Idea (A, B, or neither one) gives the most profit over the current process?

(i) If the answer to (h) is neither machine, list the amount of profit generated by the current process at 500,000 pounds of resin. If the answer to (h) is A or B, list the amount of profit generated by that idea at 500,000 pounds of resin.

22. We have decided to become entrepreneurs by raising turkeys for the Thanksgiving holiday. We already have purchased some land in the country with buildings on it, so that expense need not be a part of our analysis. A study of the way turkeys grow indicates that the weight of a turkey (w) from the time it hatches (at time zero) until it reaches full weight is

$$w = K(1 - e^{-bt})$$

Here, we select values of K and b depending on the breed of turkey we decide to raise. The value (V) of our turkey is simply the weight of the turkey times the value per pound (S) when we sell it, or

$$V = Sw$$

Here, S is the value per pound (in dollars). Finally, since we feed the turkey the same amount of food each day, the cumulative cost (C) to feed the bird is

$$C = Nt$$

Here, N is the cost of one days supply of food [$/day].

Create a graph of this situation, showing three lines: cumulative food cost, bird value, and profit on a particular day. Plot all three of these parameters versus time (in days). For the graph, show the point after which you begin to lose money, and show the time when it is most profitable to sell the bird, indicating the day on which that occurs. Use values of $K = 21$ pounds, $b = 0.03$/day, $S = 1/pound, and $N = 0.12/day.

23. As an engineer, suppose you are directed to design a pumping system to safely discharge a toxic industrial waste into a municipal reservoir. The concentrated wastewater from the plant will be mixed with fresh water from the lake, and this mixture is to be pumped into the center of the lake. You realize that the more water you mix with the waste, the more dilute it will be and thus will have a smaller impact on the fish in the lake. On the other hand, the more water you use, the more it costs in electricity for pumping. Your objective is to determine the optimum amount of water to pump so the overall cost is a minimum.

- Assume that the cost of pumping is given by the expression $C_{pump} = 10\,Q^2$. The cost C_{pump} [$/day] depends on the pumping rate Q [gallons per minute, or gpm] of the water used to dilute the industrial waste.
- Now suppose that some biologists have found that as more and more water dilutes the waste, the fish loss C_{fish} [$/day] can be expressed as $C_{fish} = 2{,}250 - 150\,Q$.

   (a) With this information, construct a graph, with pumping rate on the abscissa showing the pumping cost, the fish-loss cost, and total cost on the ordinate. For the scale, plot 0 to 15 gallons per minute for flow rate.
   (b) Determine both the minimum cost and the corresponding flow rate.

24. Below is a set of data taken of catapult use. Use interpolation to determine the following:

Range (R) [m], arm released from horizontal		
Projectile Mass (m) [g]	Angle (θ) [°]	
	30	60
5	7.6	5.0
10	7.4	4.8
25	7.2	4.5
40	5.9	3.8
50	5.2	2.8

   (a) What would the mass of a projectile need to be at 60° to achieve a range of 3 meters?
   (b) What would the angle need to be for a 10-gram projectile to achieve a range of 6 meters?
   (c) What would the mass of a projectile need to be at 40° to achieve a range of 6 meters?
   (d) What would the angle need to be for a 35-gram projectile to achieve a range of 5 meters?

25. The heat index temperature is the temperature in degrees Fahrenheit [°F] that it "feels like" when the actual temperature [°F] and relative humidity [%] are taken into account.

Heat Index Temperature (HIT) [°F]				
Relative Humidit (RH) [%]	Actual Temperature (AT) [°F]			
	70	80	90	100
30	67	78	90	104
40	68	79	93	110
50	69	81	96	120
60	70	82	100	132
70	70	85	106	144

(a) What is the heat index temperature at an actual temperature of 80 degrees Fahrenheit, relative humidity of 57%?

(b) What is the heat index temperature at an actual temperature of 78 degrees Fahrenheit, relative humidity of 48%?

(c) What is the relative humidity at an actual temperature of 90 degrees Fahrenheit, heat index temperature of 102 degrees Fahrenheit?

(d) What is the actual temperature at a relative humidity of 60%, heat index temperature of 90 degrees Fahrenheit?

26. Glycerol and water solutions have numerous applications in the food-processing industry. The following table shows the effect of percent glycerol (given down the column on the far left) and temperature in degrees Celsius [°C] (given across the top row) on the viscosity (given by the crossing of the rows and columns) in units of centipoise [cP]. (Note: Glycerol is the technical term for glycerin or glycerine.)

Viscosity of Aqueous Glycerol Solutions (μ) [cP]											
Glycerol (w) [% by weight]	Temperature (T) [°C]										
	0	10	20	30	40	50	60	70	80	90	100
50	14.6	9.01	6	4.21	3.1	2.37	1.86	1.53	1.25	1.05	0.91
60	29.9	17.4	10.8	7.19	5.08	3.76	2.85	2.29	1.84	1.52	1.28
70	76	38.8	22.5	14.1	9.4	6.61	4.86	3.78	2.9	2.34	1.93
80	255	116	60.1	33.9	20.8	13.6	9.42	6.94	5.13	4.03	3.18
90	1310	498	219	109	60	35.5	22.5	15.5	11	7.93	6
100	12070	3900	1410	612	284	142	81.3	50.6	31.9	21.3	14.8

(a) Find the viscosity at a temperature of 57 degrees Celsius and glycerol percentage of 60%.

(b) Find the glycerol percentage at a temperature of 80 degrees Celsius and a viscosity of 8 centipoise [cP].

(c) Find the temperature if the viscosity of a 90% glycerol solution is 174 centipoise [cP].

(d) Find the viscosity at a temperature of 22 degrees Celsius and a glycerol percentage of 74%.

(e) For glycerol percentages between 50% and 100% and temperatures between 0 degrees Celsius and 100 degrees Celsius, how many possible combination of temperature and glycerol percentage would give a viscosity of 4.0 centipoise [cP]?

(f) For glycerol percentages between 50% and 100% and temperatures between 0 degrees Celsius and 100 degrees Celsius, what is the lowest possible temperature that would yield a viscosity of 4.0 centipoise [cP]?

(g) For glycerol percentages between 50% and 100% and temperatures between 0 degrees Celsius and 100 degrees Celsius, what is the highest percentage of glycerol that would yield a viscosity of 4.0 centipoise [cP]?

27. Below are two partial tables of the cosine function.

Cosine Table A						
Angle (θ) [°]	0	5	10	15	20	25
cos (θ)	1.0000	0.9962	0.9848	0.9659	0.9397	0.9063

Cosine Table B						
Angle (θ) [°]	10	11	12	13	14	15
cos (θ)	0.9480	0.9816	0.9781	0.9744	0.9703	0.9659

(a) Using Cosine Table A, determine the following:

    (i) cos 10°   (ii) cos 12°   (iii) cos 13.7°

(b) Using Cosine Table B, determine the following:

    (i) cos 10°   (ii) cos 12°   (iii) cos 13.7°

(c) In your own words, explain any differences in the values derived from the two tables. Which table will provide the most accurate interpolated values? Explain.

28. Since the magnetic north pole is not at the true North Pole, at most locations on the planet a compass will not point to true north. The difference between true north and magnetic north is called magnetic declination. Anyone navigating by compass needs to know the magnetic declination to stay on course, whether that person is a backpacker, a mariner, or someone exploring the Amazon rain forest. The table below lists magnetic declinations by latitude and longitude, all in degrees, for much of North America.

Magnetic Declination [°]				
	Longitude [°]			
Latitude (°)	80	90	100	110
20	−4.83	1.27	5.95	8.83
30	−6.72	0.12	6.00	10.00
40	−8.88	−1.10	6.47	11.60
50	−12.50	−3.03	6.35	13.00

(a) What is the magnetic declination at latitude of 47° and longitude of 80°?
(b) At latitude of 50°, what is the longitude if the magnetic declination is 0°?
(c) At longitude of 100°, what is the latitude if the magnetic declination is 6.4°?
(d) What is the magnetic declination at latitude of 22° and longitude of 105°?

29. Although vacuum tubes are largely a relic of the past, they still find applications in high-end audio equipment and a few other niche applications. The following table shows typical values for a 6AS7 triode tube. Note: OL means Overload. The device should not be subjected to this combination of parameters.

6AS7 Plate Current (Ip) [mA]					
Grid Voltage (V$_G$) [V]	Plate Voltage (V$_P$) [V]				
	11	13	15	17	19
0	290	350	OL	OL	OL
−20	162	210	255	310	OL
−40	80	120	160	210	256
−60	33	60	95	130	170
−80	11	23	45	70	96

   (a) What is the plate current (I$_P$) when the plate voltage (V$_P$) is 13 volts and the grid voltage (V$_G$) is −47 volts?
   (b) What is the plate current (I$_P$) when the plate voltage (V$_P$) is 16.4 volts and the grid voltage (V$_G$) is −80 volts?
   (c) What is the grid voltage (V$_G$) when the plate voltage (V$_P$) is 15 volts and the plate current (I$_P$) is 200 milliamperes?
   (d) What is the plate current (I$_P$) when the plate voltage (V$_P$) is 15.7 volts and the grid voltage (V$_G$) is −64 volts?
   (e) What is the plate voltage (V$_P$) when the grid voltage (V$_G$) is −35 volts and the plate current (I$_P$) is 200 milliamperes?

30. Over twelve school terms, Clemson University recycled the following amount of waste (www.clemson.edu/facilities/recycling/results). For each set of six academic years ('94 – '95 through '99 – '00 and '00 – '01 through '05 – '06), determine the mean, median, variance, and standard deviation by hand.

Year	'94 − '95	'95 − '96	'96 −'97	'97 − '98	'98 − '99	'99 − '00
Waste in Tons	324.5	204.5	413.7	192.5	280.9	536.9

Year	'00 − '01	'01 − '02	'02 − '03	'03 − '04	'04 − '05	'05 − '06
Waste in Tons	671.3	705.9	784.7	681.5	750.7	811.1

31. A technician tested two temperature probes by inserting their probes in boiling water, recording the readings, removing and drying the probes, and repeating the process. The results are shown below, giving temperature reading in degrees Celsius [°C]. For each probe, determine the mean, median, variance, and standard deviation by hand.

Probe #1	87.5	86.5	88	89.5	87	88.5	89
Probe #2	95.5	100	101.5	97.5	90.5	91.5	103.5

32. Metal plates and screws are used to fix complex bone fractures, such as in cases with multiple bone fragments or a large gap in the bone. There are many plate and screw designs with different features, and surgeons need information on how these designs perform in order to choose the correct designs for their cases. For example, Limited Contact Dynamic Compression (LCDC) plates have spherical screwheads that can pivot within the holes in the plate. This allows the bone fragments to compact when they are loaded, which can promote healing. Locking Compression (LC) plates have screws with threaded heads that

actually lock into the holes in the plate when they are tightened down. This makes the plate-screw combination stiffer and stronger than the traditional LCDC design and holds the fragments in place when the bone is loaded.

Biomechanical testing mimics the types of loads that are applied to the plates and screws when they are implanted and measures the amount of bending or deformation that a plate undergoes while loaded. The following table shows results from biomechanical testing for the two types of plates described above, under four different types of loading: compression along the long axis of the bone, torsion (or twisting) along the long axis of the bone, 4-point bending in the anterior-posterior (A-P, or forward-backward) direction, and 4-point bending in the medial-lateral (M-L, or side-to-side) direction. The plates and screws were fastened to composite resin models of the humerus, with a gap between the bone fragments.

Data and photos provided by Dr. L. Benson   The bone plate test setup is the same for loading in compression and torsion, shown in the adjacent figure. The bone is clamped at both ends in a mechanical testing machine, and loads are applied along the long axis of the bone. The test setup for the 4-point bending tests is shown below.

3.5-mm Locking Compression Plate Stiffness				
Test #	Compression	Torsion	A-P Bending	M-L Bending
1	400.18	1.05	419.45	587.94
2	863.95	0.96	427.31	435.99
3	526.52	1.03	408.22	533.99
4	596.58	1.03	388.35	543.57

4.5-mm Limited Contact Dynamic Compression Plate Stiffness				
Test #	Compression	Torsion	A-P Bending	M-L Bending
1	556.5	1.08	422.27	868.76
2	441.2	1.39	435.63	716.2
3	545.87	1.35	576.77	794.48
4	487.95	1.38	413.1	1025.75

(a) For each of the four loading conditions, calculate the mean and standard deviation for the stiffness of both plate designs.

(b) The most common mode of fracturing the humerus is a fall where the person puts an arm out to break the fall. This could load the humerus in the medial-lateral direction. Which of these plate designs would you recommend to withstand medial-lateral loading? Justify your answer.

(c) Test results are not considered "significantly different" if the ranges of one standard deviation overlap. For example, if Product A has a mean of 10 and a standard deviation of 5, and Product B has a mean of 14 and a standard deviation of 2, the products are not significantly different since Product A can range from 5 to 15 and Product B from 12 to 16; the two ranges overlap. Compare the two designs in the anterior-posterior bending and in torsional loading. Can one design be considered superior to the other design in these modes of loading?

33. You test several temperature probes by inserting them in boiling ethanol (theoretical boiling point = 78.4 °C), recording the readings, removing and drying the probe, and repeating the process 20 times. The distribution curves for the probes are shown on the following page. The solid line "baseline" curve in every graph is the same curve, for a previous probe tested 20 times in boiling ethanol.

(a) Which probe was tested 40 times instead of 20 times?

(b) Which probe has the highest standard deviation?

(c) During the testing of one probe, you suspect your assistant of using formic acid (which boils at 101 °C) instead of ethanol. Which probe did your assistant incorrectly test?

(d) Which probe has the lowest standard deviation?

(e) Which probe was tested 10 times instead of 20 times?

(f) During the testing of one probe, you suspect your assistant of using chloroform (which boils at 61 °C) instead of ethanol. Which probe did your assistant incorrectly test?

(g) If you could choose between Probe C or Probe F, which probe would you choose to use? In a single sentence, describe how you would use the probe you chose to ensure that you found the correct boiling point.

34. You work for a company developing a new hybrid car. Before releasing it to the market, the company has conducted many, many tests, in several distinct test programs.

■ Test Program #1: This is a test of 500 non-hybrid cars of the same basic design as the new hybrids.

■ Test Program #2: This is a test of 500 hybrid cars.

You are given six plots for different test programs below. Depending on the questions below, the abscissa variable may vary.

(a) The hybrid cars are supposed to have better gas mileage than the nonhybrids. Compare Test Program #1 (as the baseline) with Test Program #2 with respect to gas mileage.

(b) To ensure engine efficiency, the computer-controlled ignition sequence for the hybrids must have less car-to-car variation than the nonhybrids, thus providing better performance (horsepower). Compare Test Program #1 (as the baseline) and #2 for horsepower.

(c) For the nonhybrid cars, the average 0 to 60 mph acceleration time is 8 seconds. From the Test Program #1 results, it was determined that only 3 of the 500 cars had times of more than 11 seconds. Since the complex manufacturing process for the new hybrids still has some kinks in it, some cars don't perform as consistently as the nonhybrids. If the hybrid cars had the same average acceleration time, compare Test Program #1 for the nonhybrids (as the baseline) and Test Program #2 for the hybrids with respect to acceleration times.

(e)

(f)

35. Polyetheretherketone (PEEK)™ are polymers that are resistant to both organic and aqueous environments; they are used in bearings, piston parts, and pumps. Several tests were conducted to determine the ultimate tensile strengths in units of Megapascals [MPa]. The following CDF shows results from 320 points.

   (a) What is the frequency value of A on the chart?
   (b) What is the frequency value of B on the chart?
   (c) What is the frequency value of C on the chart?
   (d) What is the frequency value of D on the chart?
   (e) What is the frequency value of E on the chart?

36. The data below were collected from a manufacturing process involving reactor temperature measured in degrees Celsius [°C]. The following values are desired: Average = 100 °C; Standard Deviation = ±10 °C.

    Graph the data on a control chart. Be sure to clearly indicate the "Zones" of control.

    Using the eight SPC rules, determine whether the process is in statistical control. If it is not in statistical control, indicate which "rule" or "rules" were violated.

Reading No.	Temperature (T) [°C]	Reading No.	Temperature (T) [°C]
1	100	11	103
2	105	12	101
3	106	13	100
4	97	14	98
5	98	15	97
6	95	16	96
7	101	17	104
8	100	18	102
9	96	19	95
10	105	20	101

37. The data below were collected from a manufacturing process involving reactor temperature measured in degrees Celsius °C. The following values are desired: Average = 100 °C; Standard Deviation = ±5 °C.

    Graph the data on a control chart. Be sure to clearly indicate the "Zones" of control.

    Using the eight SPC rules, determine whether the process is in statistical control. If it is not in statistical control, indicate which "rule" or "rules" were violated.

Reading No.	Temperature (T) [°C]	Reading No.	Temperature (T) [°C]
1	101.0	11	97.5
2	103.5	12	100.0
3	98.5	13	95.0
4	100.5	14	97.0
5	96.5	15	103.0
6	102.5	16	103.0
7	105.0	17	105.5
8	100.0	18	100.5
9	102.0	19	102.5
10	98.0	20	98.5

PART 3

UBIQUITOUS UNITS

Before I came here, I was confused about this subject. Having listened to your lecture, I am still confused . . . but on a higher level.

ENRICO FERMI

You may not be sure what the word "ubiquitous" means . . . we suggest you look it up!

Ubiquitous: yoo·**bik·we·teous** ~ adjective;
definition - _____

Imagine you are in a small boat with a large stone in the bottom of it. The boat is floating in the swimming pool in the campus recreation center. What happens to the level of water in the pool if you throw the stone overboard? Assume no water splashes out of the pool or into the boat.

Archimedes was a Greek scientist and mathematician. Most people know Archimedes for his discovery of buoyancy. According to legend, the king asked Archimedes to determine if his new crown was made of pure gold. Before this, no method had been developed for measuring the density of irregularly shaped objects. While taking a bath, Archimedes noted that the water rose in proportion to the amount of his body in the tub. He shouted "Eureka (I have found it)!" and ran though the streets naked because he was so excited he forgot to get dressed. While Archimedes never recounts this tale himself, he does outline Archimedes' principle in his treatise *On Floating Bodies: A body immersed in a fluid is buoyed up by a force equal to the weight of the displaced fluid.* Before we can begin to answer the question of the boat and the stone, we need to understand the principles of dimensions and units.

223

Engraving from Mechanics Magazine published in London in 1824

"Give me a place to stand on, and I will move the Earth." ~ *Archimedes*

In addition to buoyancy, Archimedes made many contributions to science, including the explanation of the lever, and is considered one of the greatest mathematicians.

HOW PERSONALITY INFLUENCES SUCCESS IN COLLEGE

The Myers-Briggs Type Indicator, often referred to as the MBTI, is a method for assessing individuals—their attitudes and their modes of actions, learning, and interacting with others. By answering a series of questions, an individual is categorized according to four dichotomies. You can think of these as preferences for particular ways of conducting your life. This brief introduction is by no means intended to acquaint you with the intricacies of the MBTI, but simply to help you gain insight that will help you study, learn, and interact with others more effectively. An individual's preference relative to each of the four dichotomies is indicated by a single letter. When combined, these create a four-letter code that describes your preferences. The four dichotomies are:

Cognitive Attitude:	Extrovert or Introvert	E vs I
Mode of Data-Gathering:	Sensing or Intuition	S vs N
Mode of Decision-Making:	Thinking or Feeling	T vs F
Lifestyle Preference:	Judging or Perceiving	J vs P

Cognitive Attitude (or how one directs one's energy)
The Extrovert (E) prefers action, and focuses on people and objects.
The Introvert (I) prefers reflection, and focuses on concepts and ideas.

Mode of Data-Gathering (or how one perceives things)
The Sensing (S) individual tends to trust data gathered by the five senses and is most interested in details and facts.
The Intuitive (N) individual tends to trust information that is theoretical or abstract and is most interested in how data fit a theory or a pattern.

Mode of Decision-Making (or how one judges things)
The Thinking (T) person makes decisions with an emotionally detached attitude, being concerned with what is most logical and what established rules dictate.
The Feeling (F) person makes decisions by empathizing with the situation, being concerned with what will promote the most harmonious outcome.

Lifestyle Preference (or how one relates to the outside world)
The Judging (J) individual prefers making decisions to obtaining information, whether gathering that information by thinking or feeling.
The Perceiving (P) individual prefers gathering data to making decisions, whether gathering that information by sensing or intuition.

Estimated percentages of the 16 types in the American population.
http://www.infj.org/archive/typestats.html

ISTJ	ISFJ	INFJ	INTJ
11.6%	13.8%	1.5%	2.1%
ISTP	**ISFP**	**INFP**	**INTP**
5.4%	8.8%	4.3%	3.3%
ESTP	**ESFP**	**ENFP**	**ENTP**
4.3%	8.5%	8.1%	3.2%
ESTJ	**ESFJ**	**ENFJ**	**ENTJ**
8.7%	12.3%	2.4%	1.8%

Before proceeding further, it would be helpful if you took the MBTI yourself. Many universities offer this as a free service, usually through the career center. Ask your professor or advisor for information about this service at your school. If none is available through your school, try one of the numerous sites online that offer the MBTI for free.

HOW ATTITUDES ARE ASSESSED: FOUR PROFILES

Once you have identified your four-letter code, how will this help you in your studies, in your interactions with your professors and fellow students, and eventually with your coworkers on the job?

To simplify things a little, we focus here on only two of the dichotomies, how you gather information (sensing or intuition) and how you make decisions (thinking or feeling). This limitation reduces the number of possible dichotomy combinations to four instead of sixteen, which we will call profiles: ST, SF, NT, and NF.

A summary of typical characteristics of people from each of the four profiles is shown below. Remember that people who are dominant in a particular quadrant will certainly have talents that draw from the other quadrants as well, since essentially everyone has some of the characteristics of the other types. It is critical that you recognize that you do not act, think, and learn solely within your dominant quadrant. The dominant quadrant is likely, however, to describe the areas in which you are most comfortable and talented.

Typical Characteristics

ST	SF
Analytical thinkers	Concrete learner
Organized and methodical	Suited to cooperative learning
Like timely feedback	Processes information verbally
Answers are either right or wrong	Learns best by relating to content
Prefer clearly defined procedures	Talks through complex content
Memorize easily and well	Break large tasks into small steps
Repeat complex tasks	Talent for relating to people
Attentive to detail	Sequential
Sequential	Relate on a personal level
Intuitive	Creative learner
Logical thinking	Dislike routine work
Perceive patterns	Comfortable with disorganization
A strong need to understand	Do not like to memorize
Process material before discussion	Abstract
Must see the overall picture	Daydreamers
Do not memorize well	Learn through metaphors
Need to understand concepts	Build new learning by comparison
See images in 3 dimesions	Like to see the big picture
NT	**NF**

Study Strategies for Each of the Four Profiles

By now you've realized that some teachers present information in ways that you grasp easily, while other instructors are hard for you to follow. In college, you can get lost much faster because the pace and expectations are considerably higher than in high school. One way to prevent this is to know how you learn.

Dr. Lois Krause has developed a streamlined version of the MBTI called the Cognitive Profile Inventory (CPI) specifically designed to address the issue of helping students understand the study strategies that work best for them personally. The CPI determines which of the four subcategories mentioned above—ST, SF, NT, or NF—describes you best. In her book and the associated website, *How We Learn and Why We Don't*, Dr. Krause specifically discusses the following topics:

- Successful study strategies for each type
- How to learn from a professor whose teaching style does not match your learning style
- How to deal with and benefit from classmates with different learning styles

It is critical that you recognize that you do not act, think, and learn solely within your dominant quadrant. The dominant quadrant is likely, however, to describe the areas in which you are most comfortable and talented. The table below summarizes the learning strategies for each type.

Study Strategies

ST	SF
Study alone, free from distractions	Talk things through; if studying alone, talk to yourself
Conventional table and chair	Read chapters in order, take notes in outline or numbered form
Area well-organized	Review aloud every 5-10 minutes
Read chapters in order, take notes in outline or numbered form	Work problems when you review
Review every 5-10 minutes	Write an explanatory paragraph
Flash cards for difficult material	Identify problem solution steps
Identify problem solution steps	Practice solving varied problems
Practice solving varied problems	Personalize / personify content
NT	**NF**
Read abstract / chapter summary	Read abstract / chapter summary
Review charts / tables / photos	Review charts / tables / photos
Then read chapter for details	Then read chapter for details
Take notes showing patterns	Note important vocabulary, patterns
See variables and systems change	Rewrite details with colors to fix mental image of diagrams / graphs
Study alone first, then in groups	Find metaphors for concepts
Find logical patterns in problems	Visualize equations and behavior
Emphasize understanding "why?"	Emphasize "what if?"
Vary approach to problems	Try new, even strange, approaches

Contributions to Engineering and Science from Each of the Four Profiles

This section lists typical contributions to engineering and science made by workers in each of the four quadrants. Engineers and scientists, like other professionals, need to develop skills in all four quadrants. You may find that you are most strongly drawn to work that is in your dominant quadrant.

Team Contributions and Careers

ST

Memorization is good for legal and medical careers – good partners to engineering / science

Skill at complex, repetitive tasks; make good experimentalists

Construction is sequential

SF

People skills lend themselves to careers in consulting, teaching, and technical sales

These interpersonal skills are also powerful assets in leadership and teamwork

Ability to personalize knowledge; relate technical subjects to the general public

NT

Seeing all sides of an issue is a management skill; seeing all sides of a problem is a design and development skill

Pattern recognition is a powerful research skill

Visualization skills are an asset to 3 dimensional design work

NF

Emotional awareness results in a heightened social consciousness – particularly useful in work that affects politics and the public

Creativity and passion is well suited to entrepreneurs and inventors

Famous People from Each of the Four Profiles

FAMOUS STs

Amelia Earhart
Henry Ford
Colin Powell
George Washington
Chuck Yeager

FAMOUS NTs

Marie Curie
Thomas Edison
Bill Gates
Thomas Jefferson
Sir Isaac Newton

FAMOUS SFs

Bill Clinton
Charles Dickens
Kyle Petty
Elvis Presley
Mother Teresa

FAMOUS NFs

Jimmy Carter
Theodore "Dr." Seuss Geisel
Martin Luther King, Jr.
Ronald Reagan
Oprah Winfrey

HOW MBTI HELPS INDIVIDUAL AND TEAM INTERACTIONS

First, a short anecdote about two of the authors.

> *We—Dr. Stephan and Dr. Park—are complete MBTI opposites. I, Dr. Stephan, am ESTJ and he, Dr. Park, is INFP—notice no letters in common. Our first few years together at Clemson University, we very nearly came to blows. We each thought the other was "so smart they were stupid," mentally ill, or probably both. However, we have learned to work together quite effectively, and we each value what the other brings to the team. Admittedly, we frequently have difficulty understanding what the other is saying, occasionally even having to ask someone else to translate English to English for us. (And no, this is not an exaggeration!) At this point, we laugh about it a fair amount.*

The moral of our little tale is this: just because a colleague makes little sense to you does not necessarily means that he or she is not firing on all thrusters—it may simply be a difference in personality types. You should try to understand how your fellow students or coworkers think and act, not simply immediately condemn them. As always, there are exceptions to every rule and you will indeed sometimes stumble on others that really are not an asset to the team in any way, but you will have to deal with that by other methods. In general, everyone on a team brings their own unique skills to the table. The best teams, in fact, have a variety of personality types.

As a point of interest, each of the four two-letter profiles mentioned earlier is represented among the authors of this text. Hopefully, our melding of the various ways the five of us perceive the world and communicate our knowledge into one unified presentation will make sense to most of our readers. Each author is described below as their four-letter code, followed by the official MBTI description from the Myers-Briggs Foundation website.

Dr. Stephan is an ESTJ:

> *Practical, realistic, matter-of-fact. Decisive, quickly move to implement decisions. Organize projects and people to get things done, focus on getting results in the most efficient way possible. Take care of routine details. Have a clear set of logical standards, systematically follow them and want others to also. Forceful in implementing their plans.*

Dr. Park is the exact opposite of Dr. Stephan, an INFP:

> *Idealistic, loyal to their values and to people who are important to them. Want an external life that is congruent with their values. Curious, quick to see possibilities, can be catalysts for implementing ideas. Seek to understand people and to help them fulfill their potential. Adaptable, flexible, and accepting unless a value is threatened.*

Dr. Sill is an ESFP, a blend of codes from Dr. Stephan (ES) and Dr. Park (FP):

> *Outgoing, friendly, and accepting. Exuberant lovers of life, people, and material comforts. Enjoy working with others to make things happen. Bring common sense and a realistic approach to their work, and make work fun. Flexible and spontaneous, adapt readily to new people and environments. Learn best by trying a new skill with other people.*

Prof. Bowman is an INTJ. Notice he is the opposite of Dr. Sill:

Have original minds and great drive for implementing their ideas and achieving their goals. Quickly see patterns in external events and develop long-range explanatory perspectives. When committed, organize a job and carry it through. Skeptical and independent, have high standards of competence and performance—for themselves and others.

Dr. Ohland is an ENFP, almost, but not quite, the same as Dr. Park and Dr. Sill:

Warmly enthusiastic and imaginative. See life as full of possibilities. Make connections between events and information very quickly, and confidently proceed based on the patterns they see. Want a lot of affirmation from others, and readily give appreciation and support. Spontaneous and flexible, often rely on their ability to improvise and their verbal fluency.

Elsewhere in this book you will find some hints on studying. There, we will revisit the ideas of the MBTI to illustrate how each of us uniquely developed different strategies for success in college.

How I Study—An ESTJ Reflection

by Dr. Beth Stephan

Two weeks before the exam:

- **Review all book and lecture materials.** Take notes on important formulas and concepts as you reread the text (assuming you have already read it once). This will become the foundation of a note sheet. Work the example questions on your own, and compare the answers. Determine how the lectures were different from the book (different examples, different equation derivations). Determine if particular topics were emphasized more than others.
- **Reexamine assignments.** Determine if there was a recurring type of question. Does this match the emphasis areas given in the lecture materials? Decide if there are concepts you did not understand. Work these questions again, and determine if you now understand the concept or need to ask for more help.
- **Divide the studying over several days.** Create a list of materials to review each day. Study in small chunks, not hours at a time on one subject. If you become tired—stop! Get something to eat, take a walk, or work on a different subject. You won't absorb the material well if you can't focus.

One week before the exam:

- **Figure out the exam structure.** If you don't already know what will be covered on the exam, ask the professor. Determine the material covered on the exam, the type and number of questions, and review problems if available. Note any special things to bring (calculator, extra batteries, laptop, note sheet, textbook).
- **Attend any review session.** This is a good place to cover the text examples or assignment questions you still do not understand. Be specific—come prepared with a list of things to ask. Do not worry about your class image. If it is "uncool" to ask questions, imagine how "uncool" it will be when you fail an exam because you failed to ask.

Three days before:

- **Create a "brain dump" sheet.** The basic idea is you, your memory, alone, pencil, paper; fill it up with everything you can remember. Compare with book review notes sheet. Review any concepts you incorrectly listed or omitted on the brain dump sheet.

- **Make a notes sheet (see below for more information).**
- **Take a sample exam.** Create a sample exam and put it away, and then take it a few days later under actual exam conditions (time, materials allowed—not in front of the television). This is a great place to use a study group to generate questions and check your answers.

The night before and day of:

- **Pack all materials needed for exam before going to sleep.**
- **Get a good night's sleep.**
- **Get up early.** You don't want to be in a rush or panic to get to class.
- **Take a shower and get dressed up.** Confidence comes from knowing the material and feeling good about yourself. Don't wear a suit, but wear something you would wear to dinner with friends. Jeans and a tucked-in shirt improve your mood compared to sleep pants and shower shoes.
- **Eat some good food.** It is impossible to ignore a growling stomach during an exam.

Notice here it does not say "Study" or "Cram" or "Pull an All Nighter." At this point, you will do better if you get a good night's sleep rather than studying all night. A rested brain can pull up information far better and faster than a tired one.

On the exam:

- **Read the directions and questions carefully.** Look over the whole exam before beginning to answer any questions. Be sure you are answering the right question. Watch for details of how the answer should be presented and what to include in your solution. But don't overanalyze—go with your instincts.
- **Answer the easy questions first.** Skip what you don't know—maximize your points! Exams are not always ordered easy to hard, so there may be some simple questions on page 5 buried behind several pages of derivations.
- **Try to answer all questions, even if you are unsure of what you are doing.** Sometimes in trying to work out a problem, the solution will become clear to you. Start by writing down what you know or equations you think may apply. If you get stuck, move on. You may answer a few more questions and the solution to the "stuck" question will come to you subconsciously.
- **Ask the instructor to explain unclear points.**
- **Use all of the time allotted for the test.** Go back and review that your answers match the question asked. Be sure you have used consistent unit systems. Double-check your calculations. Take the exam twice if time permits. It is difficult to find a mistake when checking over your own work, especially under stress conditions. Think of it as proof-reading: it is always better to have someone else check it for you or to read it out loud. By working the problem all over again, you may write something differently and be able to compare it to find your mistake.
- **Relax!** Have confidence to know you are prepared, and do your best. At this point, that is all you can do, so don't fret about the outcome!

What goes on a note sheet?

- **Important formulas and definitions.** Determine if any will be given to you during the exam. If they will be provided (such as a conversion table), don't waste the time and effort to put them on your note sheet and risk writing them incorrectly. Check your note sheet carefully against the text and lecture materials. Make sure the formulas are written correctly.

- **Procedures.** Write general steps that can be followed for any problem. Do *not* include example problems. You will not have time on an exam to figure them out if you haven't already done so.
- **Reminders.** Include tips to yourself on things you have made mistakes on during the assignments, such as "Don't forget to convert 1 atm = 101,325 Pa" or "Check to make sure input power is always greater than output power." Be sure to write down important facts, such as "The density of water = 1 g/cc" even if you have this memorized. Under pressure, it becomes harder to recall small details.

Tips and Techniques

- **Be sure you can read it!!** Often, symbols have two or three different meanings. Be sure you can decipher what you write. Write neatly. If you can't read it on the exam, it won't help! Don't copy old exams. Most professors are not lazy enough to give you the exact same question from the text all over again on the exam, but will alter it to stretch your mind.
- **Be sure it is your own work.** Each student understands the material in a unique way. Things you struggle with may be easy for your neighbor, and chances are your neighbor organizes material differently than you do. Do *not* copy anyone else's note sheet. Sometimes, students find it helpful to write encouraging meditative thoughts, such as "Remain CALM" or "You can do this!" or an inspirational quote. Write something that will ground you, and if you begin to panic, you can read your notes to yourself and try to relax.
- **Create a note sheet even if you can't use it on the exam.** A note sheet is a great organizational tool to make sure you have the important stuff all in one place. It makes a great last-minute review paper to look over those last 10 minutes before the exam. Since you composed it, you should (in theory) understand everything on it. Flipping through the text at the last minute often causes panic if you come across a topic you are unsure of.

How I Study—an INFP Reflection

by Dr. Bill Park

Engineering is about solving problems. I did not learn to solve the problems in my field (electrical engineering, EE) by reading books, or highlighting stuff, or listening to professors rattle on interminably; *I slogged through lots and lots of problems*! If you look at my old textbooks, you will find no handwritten notes, no highlighting, no dog-eared corners. By the way, I kept *all* of my texts, and I am very glad I did. Even today, decades later, I not infrequently look something up in them.

This is not to say that I did not go to class every day (unless I was deathly ill) and take detailed notes—I absolutely did. The professor's explanations or examples frequently proved invaluable in helping me solve the problems later. My first rule for success in college was *never* cut class unless absolutely necessary. You never know when the professor will say the one thing that will make everything fall into place in your brain. This happened to me many times during my formal education.

As an EE, I never had to do much unit conversion to a large extent, EEs just have to move the decimal place around efficiently and correctly using appropriate metric prefixes; because everything is in SI units. However, I would have studied the units material in this book the same way I studied electronics, communication theory, or digital signal processing.

When learning new material, I would look at the first problem at the end of the chapter being studied. Note that I did *not* say the first problem the professor assigned. It would be obvious that I knew how to work the problem, probably because I had worked a few similar exercise problems, or I thought I could work the problem but was not sure about all of the details, or I was totally clueless.

If the solution and all of the steps necessary was obvious, I'd move on to the next problem. There is little to be gained from working problems with which you are already proficient, except possibly self-esteem. However, I promise you that there is more self-esteem to be had by conquering a new type of problem than by being bored to tears working the same old problems a zillion times.

As I read through the problems, eventually I would find one I was not certain about, so I would have a go at it. If I figured it out, I'd move on to the next. If I got stuck, was uncertain, or came upon a problem that I had no idea how to approach, I would try to find an example problem that was similar, whether in the text or in my class notes. Working through a solved example or two (or ten) would usually give me the insight I needed to solve the problem at hand.

Some of you are thinking at this point, "This guy is insane. I do not have time to work that many problems!" Well, sorry folks. Engineering school is *far* more than a full-time job, at least if you wish to live up to your potential. Unless you have an IQ of about 175 or higher, you will need to spend far more than 40 hours per week to *excel* in a curriculum as difficult as engineering. There is nothing anyone can do to change that.

If working through the available examples failed, then, *and only then*, would I resort to reading the book, and even then I would try to isolate the sections that had bearing on the problem I was trying to work. Many engineering books are deadly dull (in my opinion) and are excellent cures for insomnia!

Sometimes, particularly if I had to resort to the book, things were still not clear. Time to go see the professor!

You will find that if you ask professors for help, most of them will quickly know whether or not you have put an appropriate amount of effort into learning on your own, or whether you tried for a few minutes and gave up. Professors' time is a limited commodity, and if you have obviously done your part, they are generally far more willing to help you. I *never* went to a professor about a problem unless I had beat my head against the wall to the point where a few more blows would gelatinize my brain. All of this effort almost always paid off, because invariably I could articulate exactly where my comprehension failed and more often than not the professor could clear up my misunderstanding within one to five minutes—my brain was primed for the answer.

If you say to a professor, "I don't understand this problem about power," then the professor will have to start asking questions to try to find exactly where your problem is. If, on the other hand, you say "I understand how to get the total amount of energy required to lift the load, and I understand how to get the power since I know how long it takes, but I am confused about the relationship between input power, output power, and efficiency," then the professor immediately knows which part of the problem to explain, using less professorial time, and keeping her or him happy. Do you want a happy professor or an irritable professor? (No-brainer!)

Another issue: note sheets. I do not believe I ever made out a note sheet for a test, even when it was allowed. I recall in first semester calculus, we were allowed to have a note sheet on tests. For the first test, I showed up with no notes. The student next to me leaned over, almost aghast, just before the test was handed out: "Where is your note sheet?!?" My reply was, "What in the world would I put on the note sheet? That the derivative of x with respect to x is 1? If you had done enough problems, you'd know all this stuff!"

Another time, in second semester electronics, we had open-book tests. One of the problems required an equation I did not recall. Rather than waste my time flipping frantically through the book, I derived the necessary formula from fundamental principles—much quicker than finding it in the book. This is the level of understanding for which I strove.

More germane to the topic of units, when I teach this material, it is sometimes necessary to express a complex unit, like Pascal, in terms of base SI units. I *do not* have this stuff memorized, but I can easily derive it from my understanding of what pressure and force are. Pressure is force per area, and by Newton's second law, force is mass times acceleration. Stuffing in the base units gives

$$(\text{kg m/s}^2)/\text{m}^2 = \text{kg}/(\text{m s}^2)$$

I do not use this frequently enough to have it memorized, but I can derive it in just a few seconds. Things I use frequently I memorize without trying; things I only use occasionally are not worth cluttering up my brain with.

To summarize my approach to studying engineering:

- The focus is on solving problems, not on reading books and notes. I work problems until the solutions to similar problems are almost blindingly obvious.
- I strive to understand concepts and the overall picture, not memorize formulae and details—the details I can fill in later. To me, memorization (at least in engineering) is double-plus unclever.
- I do my work before seeking help. Not only is this more effective for me, it uses less time on the part of the person from whom I seek help.

With all of this said, you need to tailor your study methods to the subject. I have taken courses (e.g., plant taxonomy and Chinese) that require rather prodigious feats of memorization. However, for me, engineering is not something to be memorized, but mastered by understanding concepts.

LEARNING OBJECTIVES

The overall learning objectives for this unit include:

- Apply basic principles from mathematical and physical sciences, such as trigonometry, Hooke's law, and the ideal gas law, to solve engineering problems.
- Convert units for physical and chemical parameters such as density, energy, pressure and power as required for different systems of units.
- Identify basic and derived dimensions and units; accurately express relevant observations and parameters such as distance, force, and temperature in appropriate units.
- Use "practical" skills, such as visualizing common units and conducting simple measurements, calculations, and comparisons to make estimations.

CHAPTER 10

FUNDAMENTAL DIMENSIONS AND BASE UNITS

As aspiring engineers you must learn to distinguish among many terms that laymen tend to use interchangeably. You must also understand the technical meaning of terms that are misunderstood by those untrained in science or engineering. One term that is often misunderstood is "dimension." To most people, a dimension refers to a straight line (length, one dimension), a flat surface (area, two dimensions), or a solid object (volume, three dimensions). Some slightly more educated folks might include time as a fourth dimension. The number of concepts classified as dimensions is far broader.

A **dimension** is a measurable physical idea; it generally consists solely of a word description with no numbers. A **unit** allows us to quantify a dimension—to state a number describing how much of that dimension exists in a specific situation. Units are defined by convention and related to an accepted standard.

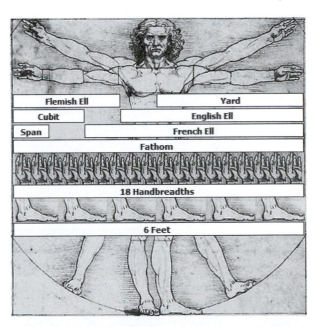

- Length is a dimension. There are many units for length, such as mile, foot, meter, light-year, and fathom.
- Time is a dimension. There are many units for time, such as second, minute, hour, day, fortnight, year, and century.
- Temperature is a dimension. There are many units for temperature, such as Celsius, Fahrenheit, and kelvin.

One of Leonardo da Vinci's most famous drawings is the Vitruvian Man. He drew this figure to depict the proportions of the human body, based upon the work of the Roman architect Vitruvian. Shown on the drawing are nine historical units in their proper ratios.

The dimensions of length, time, and temperature are familiar to us, but in reality, we don't often use these words since they are fairly vague.

We don't say . . .

It's really a long **length** to Lumberton.

Bake the cake for a **time**.

Set the oven to a high **temperature**.

We do say . . .

Lumberton is about 175 **miles** away.

Bake the cake for 35 **minutes**.

Set the oven to 450 **degrees Fahrenheit**.

The difference between the left and right columns is that the statements on the left refer to dimensions and those on the right refer to established standards or units.

10.1   THE METRIC SYSTEM

Within this text, dimensions are shown in braces { } and units in brackets [ ].

The SI system (Le Système International d'Unités), commonly known as the metric system, is the standard set of units for most of the world. Originally developed by French scientists under King Louis XVI, the SI system was finalized by the international scientific community as the standard unit system in 1971. This system defines seven base units, from which all others are derived. Table 10-1 shows the seven base units and their corresponding fundamental dimensions.

Table 10-1 Fundamental dimensions and base units

Dimension	Symbol	Unit	Symbol
Length	L	meter	m
Mass	M	kilogram	kg
Time	T	second	s
Temperature	Θ	kelvin	K
Amount of substance	N	mole	mol
Light intensity	J	candela	cd
Electric current	I	ampere	A

SI Prefixes

The SI system is based upon multiples of 10. By using an **SI prefix** when reporting numbers, we avoid scientific notation or long strings of zeros. For example, instead of saying, "The distance to Atlanta is 198,000 meters," we would say, "The distance to Atlanta is 198 kilometers."

For a list of SI prefixes, refer to the inside cover of this book or to Table 10-2. Note that the abbreviations for all SI prefixes from kilo- down to yocto- are lower case, whereas from Mega- up to Yotta- are upper case.

Table 10-2 SI prefixes (Example: 1 millimeter [mm] = 1×10^{-3} meters [m])

Numbers Less Than One			Numbers Greater Than One		
Power of 10	Prefix	Abbreviation	Power of 10	Prefix	Abbreviation
10^{-1}	deci-	d	10^{1}	deca-	da
10^{-2}	centi-	c	10^{2}	hecto-	h
10^{-3}	milli-	m	10^{3}	kilo-	k
10^{-6}	micro-	μ	10^{6}	Mega-	M
10^{-9}	nano-	n	10^{9}	Giga-	G
10^{-12}	pico-	p	10^{12}	Tera-	T
10^{-15}	femto-	f	10^{15}	Peta-	P
10^{-18}	atto-	a	10^{18}	Exa-	E
10^{-21}	zepto-	z	10^{21}	Zetta-	Z
10^{-24}	yocto-	y	10^{24}	Yotta-	Y

EXAMPLE 10-1

Express the following values using the correct SI prefix.

(a) 3,100 joules   *3.1 kilojoules*   *3.1 kJ*
(b) 2.51×10^6 watts   *2.51 megawatts*   *2.51 MW*
(c) 4×10^{-18} seconds   *4 attoseconds*   *4 as*
(d) 3×10^{-6} grams   *3 micrograms*   *3 μg*

> In engineering contexts using SI prefixes, it is common practice to adjust the decimal point such that there are one, two, or three digits to the left of the decimal point coupled with the appropriate SI prefix.

Determining the appropriate SI prefix to use becomes simple when the number is placed in engineering notation: just examine the exponent.

As a reminder, scientific and engineering notation, discussed previously in Chapter 5, are defined as follows:

Scientific notation is typically expressed in the form $\#.\#\#\# \times 10^N$, where the digit to the left of the decimal point is the most significant non-zero digit of the value being represented. Sometimes, the digit to the right of the decimal point is the most significant digit instead. The number of decimal places can vary, but is usually two to four. N is an integer, and multiplying by 10^N serves to locate the true position of the decimal point.

Engineering notation is expressed in the form $\#\#\#.\#\#\# \times 10^M$, where M is an integer multiple of 3, and the number of digits to the left of the decimal point is 1, 2, or 3 as needed to yield a power of 10 that is indeed a multiple of 3. The number of digits to the right of the decimal point is typically between two and four.

EXAMPLE 10-2

Express the following values using scientific notation, engineering notation, or with the correct SI prefix.

Standard	Scientific	Engineering	With Prefix
43,480,000 m	4.348×10^7 m	43.48×10^6 m	43.48 Mm
0.0000003060 V	3.060×10^{-7} V	306.0×10^{-9} V	306.0 pV
9,860,000,000 J	9.86×10^9 J	9.86×10^9 J	9.86 GJ
0.0351 s	3.51×10^{-2} s	35.1×10^{-3} s	35.1 ms
0.0000000522 μA	5.22×10^{-14} A	52.2×10^{-15} A	52.2 fA
456200 kW	4.562×10^8 W	456.2×10^6 W	456.2 MW

Note that the numeric values of the mantissa are the same in the last two columns, and the exponent in engineering notation specifies the metric prefix.

Comprehension Check 10-1

In January 2008, *Scientific American* reported that physicists Peter Sutter and Eli Sutter of Brookhaven National Laboratories made a pipette to measure droplets in units of a zeptoliter. Previously, the smallest unit of measure in a pipette was an attoliter. Express the answers to the following questions in scientific notation.

(a) Convert the measurement of 5 zeptoliters into units of picoliters.
(b) Convert the measurement of 80 zeptoliters into units of microliters.
(c) Convert the measurement of 300 zeptoliters into units of liters.

Official SI Rules

When reporting units using the SI system, follow these official rules.

- **If a unit abbreviation appears as a capital letter, it has been named after a person; all other abbreviations appear as lowercase letters.** For example, the abbreviation "N" stands for "newton," the SI unit of force named after Isaac Newton.

    Correct: The book weighs 5 N.   Incorrect: The book weighs 5 n.
    Correct: The rod is 5 m long.   Incorrect: The rod is 5 M long.

    The one exception to this rule is the volumetric unit of liter. The abbreviation is shown as L, since a lowercase l can be confused with both the number 1 and the uppercase letter I.

- **Symbols of units are not shown as plural.**

    Correct: 10 centimeters = 10 cm   Incorrect: 10 centimeters ≠ 10 cms

- **Symbols are not shown with a period unless they appear at the end of a sentence.**

    Correct: The rod length is 5 mm   Incorrect: The rod is 5 mm. long

- **Symbols are written in upright Roman type (m, k, L) to distinguish them from mathematical variables (m, k, l), which are indicated by italics.**

- **One space separates the number and symbol, except with the degree symbol referring to an angle.**

    Correct: 5 mm   or   5°   Incorrect: 5mm or   5 °

- **Spaces or commas may be used to group digits by threes.**

    Correct: 1 000 000   or   1,000,000

- **Symbols for derived units formed by multiple units are joined by a space or center dot. Care must be taken to avoid confusing SI prefixes with units.**

    Correct: kg m   or   kg·m   Incorrect: kgm   or   mkg

This is particularly important when confusion might arise. For example "ms" stands for millisecond, but "m s" stands for meter second. In cases like this, using a center dot is preferable since it is less likely to be misunderstood.

- **Symbols for derived units formed by dividing units are joined by a virgule (the "slash" /) or shown with a negative exponent. Care must be taken to appropriately display the entire denominator.**

    Correct: $N / (m\,s^2)$ or $N\,m^{-1}\,s^{-2}$ Incorrect: $N / m\,s^2$

- **Do not combine prefixes to form compound prefixes. Use the single correct prefix.**

    Correct: picojoules (pJ)   Incorrect: millinanojoules (mnJ)
    Correct: Gigaseconds (Gs)   Incorrect: kiloMegaseconds (kMs)

Comprehension Check 10-2

Indicate if the following units are correctly expressed according to the official SI rules. If the unit is incorrectly displayed, show the correction.

(a) Reading this sentence took 5 Sec.
(b) The average person's pupils are 60 mms. apart.
(c) One gallon is the same as 3,800 microkiloliters.

10.2   OTHER UNIT SYSTEMS

Prior to the adoption of the SI unit system by the scientific community, several other systems of units were used and are still used today, particularly in the United States. The other countries that use non-SI units are Liberia and Myanmar. Great Britain officially converted to metric in 1965, but it is still common there to see nonmetric units used in communications for the general public.

It is important to know how to convert between all unit systems. Table 10-3 compares several systems. The system listed as AES (American Engineering System) is in common use by the general public in the United States. The USCS (United States Customary System) is commonly called "English" units.

Table 10-3 Comparison of unit system, with corresponding abbreviations

Dimension	SI (MKS)	AES	USCS
Length {L}	meter [m]	foot [ft]	foot [ft]
Mass {M}	kilogram [kg]	pound-mass [lb$_m$]	slug
Time {T}	second [s]	second [s]	second [s]
Relative Temperature {Θ} Absolute Temperature {Θ}	Celsius [°C] Kelvin [K]	Fahrenheit [°F] Rankine [°R]	Fahrenheit [°F] Rankine [°R]

Accepted Non-SI Units

The units in Table 10-4 are not technically in the SI system, but due to their common usage, are acceptable for use in combination with the base SI units.

Table 10-4 Acceptable non-SI units

Unit	Equivalent SI	Unit	Equivalent SI
astronomical unit [AU]	1 AU = 1.4959787×10^{11} m	day [d]	1 d = 86400 s
atomic mass unit [amu]	1 amu = $1.6605402 \times 10^{-24}$ g	hour [h]	1 h = 3600 s
electronvolt [eV]	1 eV = $1.6021773 \times 10^{-19}$ J	minute [min]	1 min = 60 s
liter [L]	1 L = 0.001 m^3	year [yr]	1 yr = 3.16×10^7 s
		degree [°]	1° = 0.0175 rad or 1 rad = 57.3°

NOTE

1 liter does not equal 1 cubic meter!

10.3   CONVERSION PROCEDURE FOR UNITS

We use conversion factors to translate from one set of units to another. This must be done correctly and consistently to obtain the right answers. Some common conversion factors can be found inside the cover of this book, categorized by dimension. Although many more conversions are available, all the work for a typical engineering class can be accomplished using the conversions found in this table.

LENGTH

1 m = 3.28 ft

1 km = 0.621 mi

1 in = 2.54 cm

1 mi = 5280 ft

1 yd = 3 ft

Let's examine the conversions found for the dimension of length, as shown in the box, beginning with the conversion: 1 meter [m] = 3.28 feet [ft]. By dividing both sides of this equation by 3.28 feet, we obtain

$$\frac{1 \text{ m}}{3.28 \text{ ft}} = 1$$

or in other words, "There is 1 meter per 3.28 feet." If we divide both sides of the original expression by 1 meter, we obtain

$$1 = \frac{3.28 \text{ ft}}{1 \text{ m}}$$

or in other words, "In every 3.28 feet there is 1 meter."

The number 1 is dimensionless, a pure number. *We can multiply any expression by 1 without changing the expression.* We do this so as to change the units to the standard we desire.

For example, on a trip we note that the distance to Atlanta is 123 miles [mi]. How many kilometers [km] is it to Atlanta? From the conversion table, we can find that 1 kilometer [km] = 0.621 miles [mi], or

$$1 = \frac{1 \text{ km}}{0.621 \text{ mi}}$$

Place the units you are converting TO in the NUMERATOR; place the units you are converting FROM in the DENOMINATOR.

By multiplying the original quantity of 123 miles by 1, we can say

$$(123 \text{ mi})(1) = (123 \text{ mi})\left(\frac{1 \text{ km}}{0.621 \text{ mi}}\right) = 198 \text{ km}$$

Note that we could have multiplied by the following relationship:

$$1 = \frac{0.621 \text{ mi}}{1 \text{ km}}$$

We would still have multiplied the original answer by 1, but the units would not cancel and we would be left with an awkward, meaningless answer.

$$(123 \text{ mi})(1) = (123 \text{ mi})\left(\frac{0.621 \text{ mi}}{1 \text{ km}}\right) = 76\frac{\text{mi}^2}{\text{km}}$$

As a second example, we are designing a reactor system using 2-inch [in] diameter plastic pipe. The design office in Germany would like the pipe specifications in units of centimeters [cm]. From the conversion table, we find that 1 inch [in] = 2.54 centimeters [cm], or

$$1 = \frac{1 \text{ in}}{2.54 \text{ cm}}$$

By multiplying the original quantity of 2 inches by 1, we can say

$$(2 \text{ in}) (1) = (2 \text{ in}) \left(\frac{2.54 \text{ cm}}{1 \text{ in}}\right) = 5 \text{ cm}$$

In a final example, suppose a car travels at 40 miles per hour (abbreviated mph). Stated in words, "a car traveling at a rate of 40 mph will take one hour to travel 40 miles if the velocity remains constant." By simple arithmetic this means that the car will travel 80 miles in two hours or 120 miles in three hours. In general,

$$\text{distance} = (\text{velocity})(\text{time elapsed at that velocity})$$

Suppose the car is traveling at 40 mph for 6 minutes. How far does the car travel? Simple calculation shows

$$\text{distance} = (40)(6) = 240$$

Without considering units, the preceding example implies that if we drive our car at 40 mph, we can cover the distance from Charlotte, North Carolina, to Atlanta, Georgia, in 6 minutes! What is wrong? Note that the velocity is given in miles per hour, and the time is given in minutes. If the equation is written with consistent units attached, we get

$$\text{distance} = \left(\frac{40 \text{ mi}}{\text{h}}\right)\left(\frac{6 \text{ min}}{1} \left| \frac{1 \text{ h}}{60 \text{ min}}\right.\right) = 4 \text{ mi}$$

IMPORTANT CONCEPT

Be sure to *always* include units in your calculations *and* your final answer!

It seems more reasonable to say "traveling at a rate of 40 miles per hour for a time period of 6 minutes will allow us to go 4 miles."

To convert between any set of units, the following method demonstrated in Examples 3–6 is very helpful. This procedure is easy to use, but take care to avoid mistakes. If you use one of the conversion factors incorrectly, say, with 3 in the numerator instead of the denominator, your answer will be in error by a factor of 9.

EXAMPLE 10-3

Convert the length 40 yards [yd] into units of feet [ft].

Method	Steps
(1) Term to be converted	40 yd
(2) Conversion formula	1 yd = 3 ft
(3) Make a fraction (equal to one)	$\frac{3 \text{ ft}}{1 \text{ yd}}$
(4) Multiply	$\frac{40 \text{ yd}}{1}\left|\frac{3 \text{ ft}}{1 \text{ yd}}\right.$
(5) Cancel, calculate, be reasonable	120 ft

Comprehension Check 10-3

(a) The highest mountain in the world is Mount Everest in Nepal. The peak of Mount Everest is 29,029 feet above sea level. Convert the height from feet to miles.
(b) The Mariana Trench is the deepest known part of the ocean, at a depth of 10.91 kilometers. Convert the depth from kilometers to miles.
(c) The shortest race in the Olympics is the 100-meter dash. Convert the length from meters to feet.

10.4 CONVERSIONS INVOLVING MULTIPLE STEPS

Sometimes, more than one conversion factor is needed. We can multiply by several conversion factors, each one of which is the same as multiplying by 1, as many times as needed to reach the desired result. For example, suppose we determined that the distance to Atlanta is 123 miles [mi]. How many yards [yd] is it to Atlanta? From the conversion table, we do not have a direct conversion between miles and yards, but we see that both can be related to feet. We can find that 1 mile [mi] = 5280 feet [ft], or

$$1 = \frac{5280 \text{ ft}}{1 \text{ mi}}$$

We can also find that 1 yard [yd] = 3 feet [ft], or

$$1 = \frac{1 \text{ yd}}{3 \text{ ft}}$$

By multiplying the original quantity of 123 miles by 1 using the first set of conversion factors, we can say:

$$(123 \text{ mi}) \, (1) = (123 \text{ mi}) \left(\frac{5280 \text{ ft}}{1 \text{ mi}} \right) = 649{,}440 \text{ ft}$$

If we multiply by 1 again, using the second set of conversion factors and applying reasonableness:

$$(649{,}440 \text{ ft}) \, (1) = (649{,}440 \text{ ft}) \left(\frac{1 \text{ yd}}{3 \text{ ft}} \right) = 216{,}000 \text{ yd}$$

This is usually shown as a single step:

$$(123 \text{ mi}) \left(\frac{5280 \text{ ft}}{1 \text{ mi}} \right) \left(\frac{1 \text{ yd}}{3 \text{ ft}} \right) = 216{,}000 \text{ yd}$$

UNIT CONVERSION PROCEDURE

1. Write the value and unit to be converted.
2. Write the conversion formula between the given unit and the desired unit.
3. Make a fraction, equal to 1, of the conversion formula in Step 2, such that the original unit in Step 1 is located either in the denominator or the numerator, depending on where it must reside so that the original unit will cancel.
4. Multiply the term from Step 1 by the fractions developed in Step 3.
5. Cancel units, perform mathematical calculations, and express the answer in "reasonable" terms (i.e., not too many decimal places).

EXAMPLE 10-4

Convert the length 40 yards [yd] into units of millimeters [mm].

Method	Steps
(1) Term to be converted	40 yd
(2) Conversion formula	1 yd = 3 ft 1 ft = 12 in 1 in = 2.54 cm 1 cm = 10 mm
(3) Make fractions (equal to one)	$\dfrac{3\ ft}{1\ yd}\ \dfrac{12\ in}{1\ ft}\ \dfrac{2.54\ cm}{1\ in}\ \dfrac{10\ mm}{1\ cm}$
(4) Multiply	$\dfrac{40\ yd}{}\left\|\dfrac{3\ ft}{1\ yd}\right\|\dfrac{12\ in}{1\ ft}\left\|\dfrac{2.54\ cm}{1\ in}\right\|\dfrac{10\ mm}{1\ cm}$
(5) Cancel, calculate, be reasonable	37,000 mm

EXAMPLE 10-5

Convert 55 miles per hour [mph] to units of meters per second [m/s].

Note that we have two units to convert here, miles to meters, and hours to seconds.

TIME
1 d = 24 h
1 h = 60 min
1 min = 60 s
1 yr = 365 d

Method	Steps
(1) Term to be converted	55 mph
(2) Conversion formula	1 km = 0.621 mi 1 km = 1000 m 1 h = 60 min 1 min = 60 s
(3) Make fractions (equal to one) (4) Multiply	$\dfrac{55\ mi}{h}\left\|\dfrac{1\ km}{0.621\ mi}\right\|\dfrac{1000\ m}{1\ km}\left\|\dfrac{1\ h}{60\ min}\right\|\dfrac{1\ min}{60\ s}$
(5) Cancel, calculate, be reasonable	24.6 m/s

EXAMPLE 10-6

VOLUME
1 L = 0.264 gal
1 L = 0.0353 ft^3
1 L = 33.8 fl oz
1 mL = 1 cm^3

Convert the volume of 40 gallons [gal] into units of cubic feet [ft^3].

By examining the "Volume" box in the conversion table, we see that the following facts are available for use:

$$1\ L = 0.264\ gal \quad and \quad 1\ L = 0.035\ ft^3$$

By the transitive property, if a = b and a = c, then b = c. Therefore, we can directly write

$$0.264\ gal = 0.0353\ ft^3$$

Method	Steps
(1) Term to be converted	40 gal
(2) Conversion formula	$0.264 \text{ gal} = 0.0353 \text{ ft}^3$
(3) Make a fraction (equal to one)	$\dfrac{0.0353 \text{ ft}^3}{0.264 \text{ gal}}$
(4) Multiply	$\dfrac{40 \text{ gal}}{} \left\vert \dfrac{0.0353 \text{ ft}^3}{0.264 \text{ gal}}\right.$
(5) Cancel, calculate, be reasonable	5.3 ft^3

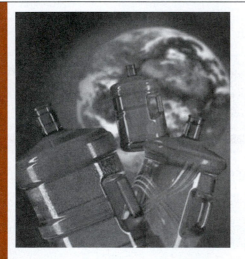

This picture shows 5-gallon water bottles made from polycarbonate. Millions of these bottles are made each year around the world to transport clean water to remote locations.

The use of polycarbonate to contain products for consumption has raised safety concerns because bisphenol A is leached from the plastic into the stored liquid. As of March 2009, six U.S. manufacturers have stopped using polycarbonate in bottles and cups used by infants and small children.

Photo courtesy of Kautex Machines

One frequently needs to convert a value that has some unit or units raised to a power, for example, converting a volume given in cubic feet to cubic meters. It is critical in this case that the power involved be applied to the *entire* conversion factor, both the numerical values and the units.

EXAMPLE 10-7

Convert 35 cubic inches [in^3] to cubic centimeters [cm^3 or cc].

When raising a quantity to a power, be sure to apply the power to both the value and the units.

Method	Steps
(1) Term to be converted	35 in^3
(2) Conversion formula	$1 \text{ in} = 2.54 \text{ cm}$
(3) Make fractions (equal to one) (4) Multiply	$\dfrac{35 \text{ in}^3}{} \left\vert \dfrac{(2.54 \text{ cm})^3}{(1 \text{ in})^3}\right.$ $\dfrac{35 \text{ in}^3}{} \left\vert \dfrac{16.387 \text{ cm}^3}{1 \text{ in}^3}\right.$
(5) Cancel, calculate, be reasonable	574 cm^3

Note that in some cases, a unit that is raised to a power is being converted to another unit that has been defined to have the same dimension as the one raised to a power. This is difficult to say in words, but a couple of examples should clarify it.

If one is converting square meters [m^2] to acres, the conversion factor is *not* squared, since the conversion provided is already in terms of length squared: 1 acre = 4047 m^2.

If one is converting cubic feet [ft^3] to liters [L], the conversion factor is *not* cubed, since the conversion provided is already in terms of length cubed: 1 L = 0.0353 ft^3.

EXAMPLE 10-8

Convert a velocity of 250 kilometers per second [km/s] to units of millimeters per picosecond [mm/ps].

Method	Steps
(1) Term to be converted	250 km/s
(2) Conversion formula	1 km = 10^3 m 1 mm = 10^{-3} m 1 ps = 10^{-12} s
(3) Make fractions (equal to one) (4) Multiply	$\dfrac{250 \text{ km}}{\text{s}} \left\| \dfrac{10^3 \text{ m}}{1 \text{ km}} \right\| \dfrac{1 \text{ mm}}{10^{-3} \text{ m}} \left\| \dfrac{10^{-12} \text{ s}}{1 \text{ ps}} \right.$
(5) Cancel, calculate, be reasonable	250 × 10^{-6} mm/ps

Specific instructions usually override the general rules.

- *Comment: Following the rules for use of prefixes given earlier would indicate that the term in the denominator should not have a prefix at all, and it should be transferred to the numerator, giving 250 × 10^6 mm/s. Beyond that, it was earlier stated that in general, the prefix should be adjusted to give one, two, or three digits to the left of the decimal place (with no power of 10). This would yield 250 km/s, right back where we started.*
- *Moral: specific instructions (such as "convert to mm/ps") usually override the general rules. In this case, perhaps we are studying the velocity of protons in a particle accelerator, and for comparison with other experiments, we need to know how many millimeters the particles go in one picosecond, rather than ending up with the units the general rules would dictate.*

Comprehension Check 10-4

Officially, a hurricane is a tropical storm with sustained winds of at least 74 miles per hour. Convert this speed into units of kilometers per minute.

Comprehension Check 10-5

A common unit of volume in the United States is one gallon. Convert one gallon to units of cubic inches.

REASONABLENESS RULES	
One quart ≈ 1 liter	One cubic foot ≈ 7.5 gallons
One cubic meter ≈ 250 gallons	1 cubic meter ≈ 5, 55-gallon drums
One cup ≈ 250 milliliters	1 golf ball ≈ 1 cubic inch

LESSONS OF THE MARS CLIMATE ORBITER

Some of you may have heard that the loss of the Mars Climate Orbiter (MCO) spacecraft in 1999 was due to a unit conversion error. The complete story is rather more complicated and illustrates a valuable lesson in engineering design. Most engineering failures are not due to a single mistake, since built-in redundancies and anticipation of failure modes make this unlikely. Three primary factors (plus bad luck) conspired to send the MCO off course.

First, the spacecraft was asymmetrical, with the body of the spacecraft on one side and a large solar panel on the other. You might think shape is not an issue in the vacuum of space, but in fact it is, and the NASA engineers were well aware of the potential problems. The panel acted like a sail, causing the craft to slowly change its orientation and requiring the MCO to make occasional small corrections by firing thrusters onboard the craft. This was a perfectly manageable "problem."

Second, the software on the spacecraft expected thruster data in SI units, requiring the force expressed in newtons. On Earth, a separate system calculated and sent instructions to the MCO concerning when and how long to fire its thrusters. The Earth-based system relied on software from an earlier Mars mission, and the thruster equations had to be modified to correct for the thrusters used on the new spacecraft. The original software had been written correctly, with the conversion factor from pound-force to newtons included. However, this conversion was neither documented nor obvious from the code, being buried in the equations. When the equations were rewritten, the programmers were unaware of the conversion factor and it was left out of the new code. This sent incorrect thruster-firing data to the MCO, specifically being too small by a factor of 4.45. This problem alone was manageable by comparing the calculated trajectory with tracking data.

Finally, after the third trajectory correction, the MCO entered "safe mode" while adjusting the solar panel, indicating a fault on the craft. At about the same time, the first preliminary indications that the spacecraft trajectory was flawed began to come in. Unfortunately, the engineers spent the next several weeks trying to determine what caused the craft to enter safe mode, falsely assuming the preliminary trajectory data was in error and waiting for longer-term tracking to give a better estimate. In the end, the spacecraft arrived at Mars about 100 kilometers off course.

Here is where the bad luck comes in. Other configurations of the craft or trajectory might have caused the 100 kilometer error to be away from Mars or parallel to the surface, in which case the trajectory could have been corrected later. Unfortunately, the trajectory was 100 kilometers lower than expected, and the MCO was probably destroyed by heating and stresses as it plunged through the Martian atmosphere. Cost: well over $100 million.

10.5 | CONVERSIONS INVOLVING "NEW" UNITS

**"NEW" UNITS
YOU ALREADY KNOW . . .**

1 tablespoon [tbsp] = 15 milliliters

1 teaspoon [tsp] = 5 milliliters

20 drops = 1 milliliter

In the past, many units were derived from common physical objects. The "inch" was the width of man's thumb, and the "foot" was the heel-to-toe length of a king's shoe. Obviously, when one king died or was deposed and another took over, the unit of "foot" changed, too. Over time, these units were standardized and have become common terminology.

New units are added as technology evolves; for example, in 1999 the unit of katal was added as an SI derived unit of catalytic activity used in biochemistry. As you proceed in your engineering field, you will be introduced to many "new" units. The procedures discussed here apply to *any* unit in *any* engineering field.

EXAMPLE 10-9

According to the U.S. Food and Drug Administration (21CFR101.9), the following definition applies for nutritional labeling:

1 fluid ounce means 30 milliliters

Using this definition, how many fluid ounces [fl oz] are in a "U.S. standard" beverage can of 355 milliliters [mL]?

Method	Steps
(1) Term to be converted	355 mL
(2) Conversion formula	1 fl oz = 30 mL
(3) Make a fraction (equal to one) (4) Multiply	$\dfrac{355 \text{ mL}}{} \left\| \dfrac{1 \text{ fl oz}}{30 \text{ mL}} \right.$
(5) Cancel, calculate, be reasonable	11.8 fl oz

EXAMPLE 10-10

The volume of water in a reservoir or aquifer is often expressed using the unit of acre-foot. A volume of 1 acre-foot is the amount of water covering an area of 1 acre to a depth of 1 foot.

Lake Mead, located 30 miles southeast of Las Vegas, Nevada, is the largest man-made lake in the United States. It holds approximately 28.5 million acre-feet of water behind the Hoover Dam. Convert this volume to units of gallons.

Method	Steps
(1) Term to be converted	28.5×10^6 acre feet
(2) Conversion formula	1 acre = 4047 m^2 1 m = 3.28 ft 1 m^3 = 100^3 cm^3 1000 cm^3 = 0.264 gal
(3) Make a fraction (4) Multiply	28.5×10^6 acre ft $\left\| \dfrac{4047 \text{ m}^2}{1 \text{ acre}} \right\| \dfrac{1 \text{ m}}{3.28 \text{ ft}} \left\| \dfrac{100^3 \text{ cm}^3}{1 \text{ m}^3} \right\| \dfrac{0.264 \text{ gal}}{1000 \text{ cm}^3}$
(5) Cancel, calculate, be reasonable	9.3×10^{12} gal

Comprehension Check 10-6

(a) A hogshead is a unit of volume describing a large barrel of liquid. Convert 10 hogsheads into units of cubic feet. Conversion factor: 1 hogshead = 63 gallons.

(b) A barrel in petroleum engineering is a unit of volume describing a measure of crude oil. Convert 1 barrel into units of cubic meters. Conversion factor: 1 barrel = 42 gallons.

(c) A drum is a shipping container used to transport liquids. Convert 5 drums into units of liters. Conversion factor: 1 drum = 55 gallons.

10.6 DERIVED DIMENSIONS AND UNITS

With only the seven base dimensions in the metric system, all measurable things in the known universe can be expressed by various combinations of these concepts. These are called **derived dimensions**. As simple examples, area is length squared, volume is length cubed, and velocity is length divided by time.

As we explore more complex parameters, the dimensions become more complex. For example, the concept of force is derived from Newton's second law, which states that force is equal to mass times acceleration. Force is then used to define more complex dimensions such as pressure, which is force acting over an area, or work, which is force acting over a distance. As we introduce new concepts, we introduce the dimensions and units for each parameter.

Sometimes, the derived dimensions become quite complicated. For example, electrical resistance is mass times length squared divided by both time cubed and current squared. Particularly in the more complicated cases like this, a **derived unit** is defined to avoid having to say things like "The resistance is 15 kilogram-meters squared divided by second cubed ampere squared." It is much easier to say "The resistance is 15 ohms," where the derived unit "ohm" equals one $(kg\ m^2)/(s^3\ A^2)$.

Within this text, dimensions are presented in exponential notation rather than fractional notation. If a dimension is not present in the quantity, it is noted by a zero exponent.

Quantity	Fractional Notation	Exponential Notation
Velocity	$\dfrac{L}{T}$	$M^0\ L^1\ T^{-1}\ \Theta^0$
Acceleration	$\dfrac{L}{T^2}$	$M^0\ L^1\ T^{-2}\ \Theta^0$

Currently, there are officially 22 named derived units in the SI system. All are named after famous scientists or engineers who are deceased. Five of the most common derived units can be found in Table 10-5. It is worth noting that numerous common derived dimensions do not have a corresponding named derived SI unit. For example, there is no named derived SI unit for the derived dimension velocity as there is for force (newton) or electrical resistance (ohm).

Table 10-5 Common derived units in the SI system

Dimension	SI Unit	Base SI Units	Derived from
Force (F)	newton [N]	$1\,N = 1\dfrac{kg\,m}{s^2}$	$F = m\,a$ Force = mass times acceleration
Energy (E)	joule [J]	$1\,J = 1\,N\,m = 1\dfrac{kg\,m^2}{s^2}$	$E = F\,d$ Energy = force times distance
Power (P)	watt [W]	$1\,W = 1\dfrac{J}{s} = 1\dfrac{kg\,m^2}{s^3}$	$P = E/t$ Power = energy per time
Pressure (P)	pascal [Pa]	$1\,Pa = 1\dfrac{N}{m^2} = 1\dfrac{kg}{m\,s^2}$	$P = F/A$ Pressure = force per area
Voltage (V)	volt [V]	$1\,V = 1\dfrac{W}{A} = 1\dfrac{kg\,m^2}{s^3\,A}$	$V = P/I$ Voltage = power per current

- A note of caution: One letter can represent several quantities in various engineering disciplines. For example, the letter "P" can indicate pressure, power, or vertical load on a beam. It is important to examine and determine the nomenclature in terms of the context of the problem presented.
- Always remember to use the units and the symbol in calculations.

Comprehension Check 10-7

Complete Table 10-6 by finding the exponents to express the parameters in the form $\{=\}\,M^a\,L^b\,T^c\,\Theta^d$.

Table 10-6 Dimensions of some common parameters

Quantity	Common Units	Exponents M	L	T	Θ
Fundamental quantities						
mass	kg	1	0	0	0
length	m	0	1	0	0
time	s	0	0	1	0
temperature	K	0	0	0	1
Geometric quantities						
area	ft^2				
	acre	0	2	0	0
volume	gal				
	L				
Rate quantities						
velocity	mi/h	0	1	−1	0
acceleration	ft/s^2	0	1	−2	0
flowrate	gal/min				
evaporation	kg/h				

SPECIAL UNIT: RADIAN

The derived unit of **radian** is defined as the angle at the center of a circle formed by an arc (S) equal in length to the radius (r) of that circle. In a complete circle there are 2π radians. Since by definition a radian is a length (S) divided by a length (r), it is a dimensionless ratio.

$$1 \text{ radian [rad]} = S/r$$

Thus, an angle has units, but is dimensionless! In addition to radians, another common unit used for angle is the degree [°]. There are 360° in a complete circle.

$$360° = 2\pi \text{ radians}$$

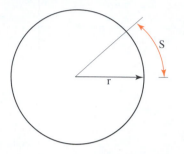

10.7 EQUATION LAWS

Equations are mathematical "sentences" composed of "words" (terms) that are strung together with "punctuation marks" (mathematical symbols, such as $+$, $-$, $\times$, $\div$, and $=$). Just as there are rules in the English language that govern how a sentence is structured, there exists a set of "rules" for equations.

Addition and Subtraction

Suppose we are interested in the manufacture and use of sandpaper for furniture construction. We think for a while and then develop a list of the important quantities that affect the final product, along with their respective units and dimensions:

W	Wood removed	[inches]	L
R	Roughness diameter	[millimeters]	L
D	Density of grains	[kilograms per cubic meter]	$\dfrac{M}{L^3}$
A	Adhesive thickness	[millimeters]	L
H	How heavy the paper is	[newtons]	$\dfrac{M\,L}{T^2}$
O	Operation stroke length	[centimeters]	L
K	Kernel (grain) spacing	[millimeters]	L

Let us propose a simple equation with only plus and minus signs that could possibly relate several of these parameters. If we are interested in how heavy the product would be, we might assume this would depend on the thickness of the adhesive, the diameter of the roughness, and the grain density. We will try

$$H = A + R + D$$

Each of these terms represents something "real," and consequently we expect that each term can be expressed in terms of fundamental dimensions. Writing the equation in terms of dimensions given:

$$\frac{M\,L}{T^2} = L + L + \frac{M}{L^3}$$

It is obvious that this is just terrible! We cannot add length and mass or time; as the adage goes, "You can't add apples and oranges!" The same holds true for dimensions. As a result of this observation, we see that this cannot possibly be a valid equation. This gives one important "law" governing equations, the **Plus law**.

Let's try this again with another equation to see if we can determine how effective the sandpaper will be, or how much wood will be removed after each stroke. We might assume this depends on the operation stroke length, the roughness diameter, and the spacing of the grains.

$$W = O + R + K$$

Substituting dimensions,

$$L = L + L + L$$

We see that at least dimensionally, this can be a valid equation, based on the Plus law. Next, units can be inserted to give

$$\text{inches} = \text{centimeters} + \text{millimeters} + \text{millimeters}$$

Dimensionally, this equation is fine, but from the perspective of units, we cannot carry out the arithmetic above without first converting all the length dimensions into the same units, such as millimeters. We can state an important result from this observation as well, forming the **Unit law**.

It is important to state a corollary to this observation. If two parameters have the same dimensions and units, it is not always meaningful to add or subtract them.

Two examples show this.

1. If Student A has a mass m_A [kilograms] and Student B a mass m_B [kilograms], then the total mass of both students [kilograms] is the sum of the two masses. This is correct and meaningful in both dimensions and units.
2. Suppose we assume that an equation to predict the mass of a car is this: mass of the car in kilograms = mass of an oak tree in kilograms + mass of an opossum in kilograms. This equation has three terms; all with the dimension of mass and units of kilograms; thus the terms can be added, although the equation itself is nonsense.

Consequently the requirement that each term must have the same dimensions and units is a necessary, but not a sufficient, condition for a satisfactory equation.

Multiplication and Division

There are many ways to express the rate at which things are done. Much of our daily life is conducted on a "*per*" or rate basis. We eat three meals *per* day, have five fingers *per* hand, there are 11 players *per* team in football, 3 feet *per* yard, four tires *per* car, 12 fluid ounces *per* canned drink, and four people *per* quartet.

Although it is incorrect to add or subtract parameters with different dimensions, it is perfectly permissible to divide or multiply two or more parameters with different dimensions. This is another law of dimensions, the **Per law**.

When we say 65 miles *per* hour, we mean that we travel 65 miles in one hour. We could say we travel at 130 miles per 2 hours, and it would mean the same thing. Either way, this rate is expressed by the "*per*" ratio, distance per time.

One of the most useful applications of your knowledge of dimensions is in helping to determine if an equation is dimensionally correct. This is easy to do and only involves the substitution of the dimensions of every parameter into the equation and simplifying the resulting expressions. A simple application will demonstrate this process.

EXAMPLE 10-11

Is the following equation dimensionally correct?

$$t = \sqrt{\frac{d_{final} - d_{initial}}{0.5\,a}}$$

where: t is time
         d is distance
         a is acceleration
         0.5 is unitless

Determine the dimensions of each parameter:

acceleration	(a)	$\{=\}\, L^1 T^{-2}$
distance	(d)	$\{=\}\, L$
time	(t)	$\{=\}\, T$

Substitute into the equation: $T = \sqrt{(L - L)\left|\dfrac{T^2}{L}\right.}$

Simplifying $T = \sqrt{L\left|\dfrac{T^2}{L}\right.} = \sqrt{T^2} = T$

Yes, the equation is dimensionally correct. Both sides of the equation have the same dimensions.

EXAMPLE 10-12

We can use dimensional arguments to help remember formulas. We are interested in the acceleration of a body swung in a circle of radius (r), at a constant velocity (v). We remember that acceleration depends on r and v, and one is divided by the other, but can't quite remember how. Is the acceleration (a), given by one of the following?

$$a = \frac{v}{r} \quad \text{or} \quad a = \frac{v}{r^2} \quad \text{or} \quad a = \frac{v^2}{r} \quad \text{or} \quad a = \frac{r}{v} \quad \text{or} \quad a = \frac{r}{v^2} \quad \text{or} \quad a = \frac{r^2}{v}$$

Determine the dimensions of each parameter:

acceleration	(a)	$\{=\}\, L^1 T^{-2}$
radius	(r)	$\{=\}\, L$
velocity	(v)	$\{=\}\, L^1 T^{-1}$

Original Equation	Substituting into the Equation	Simplify	Correct?
$a = v/r$	$LT^{-2} = (LT^{-1})\,L^{-1}$	$LT^{-2} = T^{-1}$	No
$a = v/r^2$	$LT^{-2} = (LT^{-1})\,L^{-2}$	$LT^{-2} = L^{-1}\,T^{-1}$	No
$a = v^2/r$	$LT^{-2} = (LT^{-1})^2\,L^{-1}$	$LT^{-2} = LT^{-2}$	YES
$a = r/v$	$LT^{-2} = L\,(LT^{-1})^{-1}$	$LT^{-2} = T$	No
$a = r/v^2$	$LT^{-2} = L\,(LT^{-1})^{-2}$	$LT^{-2} = L^{-1}\,T^2$	No
$a = r^2/v$	$LT^{-2} = L^2\,(LT^{-1})^{-1}$	$LT^{-2} = LT$	No

Comprehension Check 10-8

Weirs are typically installed in open channels such as streams to determine discharge (flow rate). A Cipoletti weir is described by the following equation, where Q is flow rate [gallons per minute, or gpm], L is length [cm], and H is height [cm]. For the following equation to be dimensionally correct, what are the units on the constant value of 3.37?

$$Q = 3.37\ LH^{\frac{3}{2}}$$

10.8 CONVERSION INVOLVING EQUATIONS

Engineering problems are rarely as simple as converting from one set of units to another. Normally, an equation is involved in the problem solution. To minimize the likelihood of mistakes, we adopt the following procedure for all problems. While this procedure may seem to overanalyze simple problems, it is relatively foolproof and will become more and more useful as the material progresses in difficulty.

EQUATION PROCEDURE

1. Given a problem, first convert all parameters into base SI units, combinations of these units, or accepted non-SI units. Use the five-step conversion procedure previously described.

2. Perform all necessary calculations, as follows:

   (a) Determine the appropriate equation.

   (b) Insert the known quantities and units. Be sure to carry the units through until the end!

   (c) Calculate the desired quantity and express the answer in "reasonable" terms.

   This gives the answer in SI units.

3. Convert the final answer to the required units and express the answer in "reasonable" terms.

EXAMPLE 10-13

On a trip from Alphaville to Betaville, you can take two main routes. Route #1, which goes through Gammatown, is 50 kilometers [km] long; however, you can only drive an average speed of 36 miles per hour [mph]. Route #2 travels along the freeway, at an average speed of 50 mph, but it is 65 km long. How long does it take to complete each route? State the time for each route in minutes [min].

STEP ONE: Convert to base SI units		
Method	Route #1	Route #2
(1) Term to be converted	36 mph	50 mph
(2) Conversion formula (3) Make a fraction (equal to one) (4) Multiply	$\dfrac{36\ mi}{h}\left\|\dfrac{1\ km}{0.621\ mi}\right.$	$\dfrac{50\ mi}{h}\left\|\dfrac{1\ km}{0.621\ mi}\right.$
(5) Cancel, calculate, be reasonable	58 km/h	81 km/h
STEP TWO: Calculate		
Method	Route #1	Route #2
(1) Determine appropriate equation	distance = (velocity) (time) which can be rewritten as . . . time = distance/velocity	
(2) Insert known quantities	time = $\dfrac{50\ km}{58\frac{km}{h}}$	time = $\dfrac{65\ km}{81\frac{km}{h}}$
(3) Calculate, be reasonable	time = 0.86 h	time = 0.8 h
STEP THREE: Convert from base SI units to desired units		
Method	Route #1	Route #2
(1) Term to be converted	0.86 h	0.8 h
(2) Conversion formula (3) Make a fraction (equal to one) (4) Multiply	$\dfrac{0.86\ h}{}\left\|\dfrac{60\ min}{1\ h}\right.$	$\dfrac{0.8\ h}{}\left\|\dfrac{60\ min}{1\ h}\right.$
(5) Cancel, calculate, be reasonable	52 min	48 min

WATCH

$$\frac{\frac{km}{km}}{h} = \frac{km\ h}{km}$$

Comprehension Check 10-9

The dimensions of a window are 81 centimeters by 7 feet. What is the surface area of the window in units of square millimeters?

Comprehension Check 10-10

A basketball has a diameter of approximately 27 centimeters. Find the volume of the basketball in units of gallons.

In-Class Activities

ICA 10-1

Complete the following table. Place all answers in scientific notation in the form $Y \times 10^Z$.

	meters	centimeters	millimeters	micrometers	nanometers
abbreviation	[m]				
(a)	9E-08	9E-06	9E-05	0.09	90
(b)		50			
(c)				5	
(d)			300		

ICA 10-2

Complete the following table. Place all answers in scientific notation in the form $Y \times 10^Z$.

	grams	milligrams	micrograms	nanograms	kilograms
abbreviation	[g]				
(a)					0.4
(b)		20			
(c)				800	
(d)			3		

ICA 10-3

Complete the following table.

	inches	feet	yards	meters	miles
abbreviation	[in]				
(a)		90			
(b)					2
(c)				52	
(d)			53.5		

ICA 10-4

Complete the following table.

	kilograms	grams	pound-mass	slugs	ton
abbreviation	[kg]				
(a)		700			
(b)					1.5
(c)				75	
(d)			150		

ICA 10-5

Complete the following table.

	cubic inch	fluid ounces	gallon	liter	cubic foot
abbreviation	[in^3]				
(a)	716	400	3.12	11.8	0.414
(b)					3
(c)				5	
(d)			42		

ICA 10-6

Complete the following table.

	square feet	square meter	acre	square inch	square mile
abbreviation	[ft^2]				
(a)		0.06			
(b)					3
(c)				500	
(d)			1.5		

ICA 10-7

Complete the following table.

	miles per hour	kilometers per hour	yards per minute	feet per second
abbreviation	[mph] or [mi/h]			
(a)		100		
(b)	55			
(c)				110
(d)			2.25	

ICA 10-8

Complete the following table.

	gallons per minute	cubic feet per hour	liters per second	fluid ounces per day
abbreviation	[gpm] or [gal/min]			
(a)		15		
(b)	20			
(c)				64
(d)			8	

ICA 10-9

In China, one "fen" is defined as 3.3 millimeters, one "cun" is 3.5 centimeters, and one "bu" is 1.66 meters.

(a) A "league" is defined as 3 miles. Convert 2 leagues to units of bu.
(b) The average height of a human is 5 feet, 7 inches. Convert this height to units of bu.
(c) A "cubit" is defined as 18 inches. Convert 50 cubits to units of cun.
(d) Ten nanometers is the thickness of a cell membrane. Convert 10 nanometers to units of fen.

ICA 10-10

One "shake" is equal to 10 nanoseconds.

(a) A blink of a human eye takes approximately 300 to 400 milliseconds. Convert 350 milliseconds to units of shake.
(b) Andy Warhol stated that "Everyone will be famous for fifteen minutes." Convert your 15 minutes of fame to units of shake.
(c) If the SI prefix system was expanded to other units, there would be such definitions as a "millihour," meaning 1/1000 of an hour. Convert 1 millihour to units of shake.
(d) A "jiffy" is defined as 1/60 of a second. Convert 20 jiffys to units of shake.

ICA 10-11

Rather than using the traditional MKS system (meter - kilogram - second) unit system, an unusual unit system is the FFF system: furlong - firkin - fortnight. One furlong is equal to 201 meters, one firkin is equal to 40 kilograms, and one fortnight is 14 days.

(a) A football field is 120 yards long by 53.5 yards wide. Calculate the area of the football field, and convert this area into units of furlong squared.
(b) A "knot" is a unit of speed in marine travel. One knot is 1.852 kilometers per hour. Convert the speed of 20 knots to units of furlong per fortnight.
(c) The Earth's escape velocity is 7 miles per second. Convert this to units of furlong per fortnight.
(d) A manufacturing process uses 10 pound-mass of plastic resin per hour. Convert this rate to units of firkin per fortnight.

ICA 10-12

One "metric drop" is equal to 5 microliters [μL].

(a) A "pinch" is defined as ⅛ of a teaspoon, or 0.616 milliliters [mL]. Convert 1 pinch to units of drops.
(b) One metric teaspoon is defined as 5 milliliters [mL]. Convert 2 teaspoons to units of drops.
(c) One cup is 250 milliliters [mL]. Convert 1/2 cup to units of drops.
(d) A standard can of soda is 355 milliliters [mL]. Convert 355 milliliters to units of drops.

ICA 10-13

(a) A category F5 tornado can have wind speeds of 300 miles per hour [mph]. What is this velocity in units of meters per second?
(b) A new hybrid automobile with regenerative braking has a fuel economy of 55 miles per gallon [mpg] in city driving. What is this fuel economy expressed in units of feet per milliliter?
(c) The AbioCor™ artificial heart pumps at a rate of 10 liters per minute. Express this rate in units of gallons per second.
(d) If a pump moves water at 2 cubic feet per hour, what is this rate in units of cubic centimeters per second?

ICA 10-14

(a) A circular window has a 10-inch radius. What is the surface area of one side of the window in units of square centimeters?
(b) When shipping freight around the world, most companies use a standardized set of containers to make transportation and handling easier. The 40-foot container is the most popular container worldwide. If the container is 2.4 meters wide and has an enclosed volume of 2,385 cubic feet, what is the height of the container in units of inches?
(c) We know a speed boat can travel at 30 knots. How long (in minutes) will it take to cross the Chesapeake Bay at a place where the bay is 24 miles across? 1 knot = 1 nautical mile per hour; 1 nautical mile = 6,076 feet.
(d) We have constructed a model car that can accelerate (a) over time (t) and travel a distance (d). A friend who is knowledgeable in physics tells us that we can compute the distance traveled with the equation $d = \frac{1}{2}at^2$. Given the following values, compute the distance traveled in units of miles: a = 6.5 feet per second squared, and t = 0.25 hour.

ICA 10-15

Complete the following table, expressing each unit in terms of fundamental dimensions.

	Quantity and/or Unit	Symbol	M	L	T	Θ
			\multicolumn{4}{c}{Exponents}
(a)	acre	acre	0	2	0	0
(b)	atmospheres	atm	1	−1	−2	0
(c)	calorie	cal				
(d)	calories per minute	cal/min				
(e)	cubic feet	ft^3				
(f)	day	d				
(g)	degrees Fahrenheit	°F				
(h)	foot pound-force	ft lb$_f$				
(i)	gallon	gal				
(j)	horsepower	hp				
(k)	inches of mercury	in Hg				
(l)	joule	J				
(m)	kelvin	K				
(n)	kilogram	kg				
(o)	liter	L				
(p)	mile	mi				
(q)	newton	N				
(r)	pascal	Pa				
(s)	pound-force	lb$_f$				
(t)	pound-force per square inch	psi				
(u)	pound-mass	lbm				
(v)	slug	slug				
(w)	square feet	ft^2				
(x)	watt	W				

CHAPTER 11
UNIVERSAL UNITS

In Chapter 10, the concepts of derived dimensions and units were introduced. Five of the most common named units were given in Table 10-5, which is so critical it is repeated here as Table 11-1. Recall that numerous common derived dimensions do not have a corresponding derived SI unit. For example, there is no named SI unit for the derived dimension velocity as there is for force (newton) or electrical resistance (ohm).

The conversion table inside the front cover of this book groups unit conversion factors by dimension. Several fundamental dimensions, such as length and time, were previously discussed in Chapter 10. The remaining derived dimensions shown in the table are discussed below, such as force, along with several named derived dimensions that do not have corresponding simple unit conversions, such as density.

Table 11-1 Common derived units in the SI system

Dimension	SI Unit	Base SI Units	Derived from
Force (F)	newton [N]	$1\,N = 1\dfrac{kg\,m}{s^2}$	$F = m\,a$ Force = mass times acceleration
Energy (E)	joule [J]	$1\,J = 1\,N\,m = 1\dfrac{kg\,m^2}{s^2}$	$E = F\,d$ Energy = force times distance
Power (P)	watt [W]	$1\,W = 1\dfrac{J}{s} = 1\dfrac{kg\,m^2}{s^3}$	$P = E/t$ Power = energy per time
Pressure (P)	pascal [Pa]	$1\,Pa = 1\dfrac{N}{m^2} = 1\dfrac{kg}{m\,s^2}$	$P = F/A$ Pressure = force per area
Voltage (V)	volt [V]	$1\,V = 1\dfrac{W}{A} = 1\dfrac{kg\,m^2}{s^3\,A}$	$V = P/I$ Voltage = power per current

11.1   FORCE

When you push a grocery cart, it moves. If you keep pushing, it keeps moving. The longer you push, the faster it goes; the velocity increases over time, meaning that it accelerates. If you push a full grocery cart that has a high mass, it doesn't speed up as much, meaning it accelerates less than a cart with low mass. Simply put, the acceleration (a) of a body depends on the force (F) exerted on it and its mass (m). This is a simple form of "Newton's second law of motion" and is usually written as $F = ma$.

The SI unit of force, the **newton** [N], is defined as the force required to accelerate a mass of one kilogram at a rate of 1 meter per second squared (see Table 11-2). It is named for Sir Isaac Newton (1643–1727). Newton's *Principia* is considered one of the world's greatest scientific writings, explaining the law of universal gravitation and the three laws of motion. Newton also developed the law of conservation of momentum, the law of cooling, and the reflecting telescope. He shares credit for the development of calculus with Gottfried Leibniz.

Table 11-2 Dimensions of force

Quantity	Common Units	Exponents			
		M	L	T	Θ
force	N	1	1	−2	0

In general, a "pound" can be used as a unit of mass or force.

For distinction, the following convention is used:

– pound-mass [lb_m]

– pound-force [lb_f]

In the SI system, mass, length and time are base units and force is a derived unit; force is found from combining mass, length, and time using Newton's second law. The SI system is called "coherent," because the derived unit is set at one by combing base units. The AES system is considered non-coherent as it uses units that do not work together in the same fashion as the SI units do. There are two uses of the term "pound" in the AES system, which occurred in common usage long before Newton discovered gravity. To distinguish mass in pounds and force in pounds, the unit of mass is given as pound-mass (lb_m) and the unit of force is given as pound-force (lb_f). One pound-force is the amount of force needed to accelerate 1 pound-mass at a rate of 32.2 feet per second squared. Since this relationship is not easy to remember or use in conversions, we will stick with SI units for problem solving, following the procedure discussed in Chapter 10.

EXAMPLE 11-1

A professional archer is designing a new longbow with a full draw weight of 63 pounds-force [lb_f]. The draw weight is the amount of force needed to hold the bowstring at a given amount of draw, or the distance the string has been pulled back from the rest position. What is the full draw weight of this bow in units of newtons [N]?

Method	Steps
(1) Convert term	63 lb_f
(2) Apply conversion formula	1 N = 0.225 lb_f
(3) Make a fraction (4) Multiply	$\dfrac{63\ lb_f}{} \bigg| \dfrac{1\ N}{0.225\ lb_f}$
(5) Cancel, calculate, be reasonable	280 N

11.2 WEIGHT

Objects in space are weightless, not massless.

The **mass** of an object is a fundamental dimension. Mass is a quantitative measure of how much of an object there is, or in other words, how much matter it contains. The **weight** (w) of an object is a force equal to the mass of the object (m) times the acceleration of **gravity** (g).

While mass is independent of location in the universe, weight is dependent upon both mass and gravity.

On Earth, the pull of gravity is approximately 9.8 meters per second squared [m/s²]. On the moon, gravity is approximately 1/6 this value, or 1.6 m/s². A 1 kilogram [kg] object acted on by Earth's gravity would have a weight of 9.8 N, but on the moon it would have a weight of 1.6 N. Unless otherwise stated, assume all examples take place on Earth.

Table 11-3 Dimensions of weight

Quantity	Common Units	Exponents			
		M	L	T	Θ
weight	N	1	1	−2	0

DEVILISH DERIVATION

Mass of an object: A quantitative measure of how much of an object there is.

Weight of an object: A quantitative measure of the force exerted on the object due to gravity; we are concerned about the force between Earth (subscript e) and a body on Earth (or another planet).

Newton's Law of Universal Gravitation

$$F = G \frac{mm_e}{R^2}$$

■ G is universal gravitational constant.

$$G = 6.673 \times 10^{-11}(N\ m^2)/kg^2$$

■ m is the mass.
■ R is the distance between the centers of mass of the two bodies.

On Earth, the distance between the center of a body and the center of the earth is approximately the radius of Earth, R_e.

The mass of one of the bodies (Earth) is the mass of Earth (with an "e" subscript). Rewrite the equation:

$$F = m \left[G \frac{m_e}{R_e^2} \right]$$

The quantity in square brackets is a constant (call it "g"). We call the force "the weight (w) of the body." So,

$$w = mg$$

This is the common equation that relates weight and mass. The value for g is calculated to be 9.8 meters per second squared [m/s²], or 32.2 feet per second squared [ft/s²]. Note that g has the units of acceleration.

REASONABLENESS RULES

1 gallon of water
  ~ 8 pound-mass

1 liter of water ~ 1 kilogram

One penny ~ 2.5 grams

One nickel ~ 5 grams

EXAMPLE 11-2

What is the weight of a 225 kilogram [kg] bag of birdseed in units of newtons [N]?

Step One: Convert to Base SI Units	
No Conversion Necessary	
Step Two: Calculate	
Method	Steps
(1) Determine appropriate equation	$w = mg$
(2) Insert known quantities	$w = \dfrac{225 \text{ kg} \mid 9.8 \text{ m}}{s^2}$
(3) Calculate, be reasonable	$w = 2205 \dfrac{\text{kg m}}{s^2}$

This is apparently our final answer, but the units are puzzling. If the units of force are newtons, and if this is a valid equation, then our final result for force should be newtons. If we consider the dimensions of force:

	Common	Exponents			
Quantity	Units	M	L	T	Θ
force	N	1	1	−2	0

A unit of force has dimensions $F \{=\} \dfrac{ML}{T^2}$, *which in terms of base SI units would be* $F [=] \dfrac{\text{kg m}}{s^2}$. *This term occurs so frequently it is given the special name "newton" (see Table 11-1). Any time we see the term* $\left(\dfrac{\text{kg m}}{s^2}\right)$, *we know we are dealing with a force equal to a newton.*

(3) Calculate, be reasonable	$w = 2205 \dfrac{\text{kg m}}{s^2} \left| \dfrac{1 \text{ N}}{1 \frac{\text{kg m}}{s^2}} \right. = 2205 \text{ N}$
Step Three: Convert from Base SI Units to Desired Units	
No Conversion Necessary	

Comprehension Check 11-1

The mass of the human brain is 1360 grams. State the weight of the human brain in units of newtons.

Comprehension Check 11-2

What is the weight of a 5-kilogram bowling ball in units of pounds-force?

11.3   DENSITY

Density (ρ, Greek letter rho) is the mass of an object (m) divided by the volume the object occupies (V). Density should not be confused with weight—think of the old riddle: which weighs more, a pound of feathers or a pound of bricks? The answer is they both weigh the same amount, one pound, but the density of each is different. The bricks have a higher density than the feathers, since the same mass takes up less space.

Specific weight (γ, Greek letter gamma) is the weight of an object (w) divided by the volume the object occupies (V).

Table 11-4 Dimensions of density and specific weight

Quantity	Common Units	Exponents			
		M	L	T	Θ
density	kg/m^3	1	−3	0	0
specific weight	N/m^3	1	−2	−2	0

EXAMPLE 11-3

The density of sugar is 1.61 grams per cubic centimeter [g/cm^3]. What is the density of sugar in units of pound-mass per cubic foot [lb_m/ft^3]?

Upon conversion from units of grams per cubic centimeter to pound-mass per cubic foot, the answer should be $\approx$ 60 times larger.

Method	Steps
(1) Term to be converted	1.61 g/cm^3
(2) Conversion formula	
(3) Make fractions	$1.61\dfrac{g}{cm^3}\left\|\dfrac{2.205\ lb_m}{1000\ g}\right\|\dfrac{1000\ cm^3}{0.0353\ ft^3}$
(4) Multiply	
(5) Cancel, calculate, be reasonable	101 lb_m/ft^3

EXAMPLE 11-4

The density of a biofuel blend is 0.72 grams per cubic centimeter [g/cm^3]. What is the density of the biofuel in units of kilograms per cubic meter [kg/m^3]?

Upon conversion from units of grams per cubic centimeter to kilograms per cubic meter, the answer should be 1,000 times larger.

Method	Steps
(1) Term to be converted	0.72 g/cm^3
(2) Conversion formula	
(3) Make fractions	$0.72\dfrac{g}{cm^3}\left\|\dfrac{1\ kg}{1000\ g}\right\|\dfrac{(100\ cm)^3}{1\ m^3}$
(4) Multiply	
(5) Cancel, calculate, be reasonable	720 kg/m^3

A vast array of valuable compounds can be formed by microbial cultures. Oil produced by the fungi *Pythium irregulare* can be extracted and used for biofuels or pharmaceutical compounds. Biosystems engineers culture the microorganism, design the bioreactor, and extract the valuable compounds using sustainable, ecoprocessing techniques.

Photo courtesy of C. Drapcho

EXAMPLE 11-5

What is the weight of water, in units of pounds-force [lb$_f$], in a 55-gallon drum completely full? Assume the density of water to be 1 gram per cubic centimeter. Ignore the weight of the drum.

Step One: Convert to Base SI Units		
Method	**Steps**	
(1) Term to be converted	55 gal	1 g/cm^3
(2) Conversion formula (3) Make fractions (4) Multiply	$\dfrac{55 \text{ gal}}{} \left\| \dfrac{1000 \text{ cm}^3}{0.264 \text{ gal}} \right\| \dfrac{1 \text{ m}^3}{100^3 \text{ cm}^3}$	$\dfrac{1 \text{ g}}{\text{cm}^3} \left\| \dfrac{1 \text{ kg}}{1000 \text{ g}} \right\| \dfrac{100^3 \text{cm}^3}{1 \text{ m}^3}$
(5) Cancel, calculate, be reasonable	0.208 m^3	1000 kg/m^3
Step Two: Calculate		
Method	**Steps**	
(1) Determine appropriate equation	$w = mg$	
(2) Insert known quantities	$w = \dfrac{m}{} \left\| \dfrac{9.8 \text{ m}}{\text{s}^2} \right.$	
For unknown quantities, repeat the process		
Method	**Steps**	
(1) Determine appropriate equation	$m = \rho V$	
(2) Insert known quantities	$m = \dfrac{1000 \text{ kg}}{\text{m}^3} \left\| 0.208 \text{ m}^3 \right.$	
(3) Calculate, be reasonable	$m = 208$ kg	
(2) Insert known quantities	$w = \dfrac{208 \text{ kg}}{} \left\| \dfrac{9.8 \text{ m}}{\text{s}^2} \right.$	
(3) Calculate, be reasonable	$w = 2038 \dfrac{\text{kg m}}{\text{s}^2} \left\| \dfrac{1 \text{ N}}{1 \frac{\text{kg m}}{\text{s}^2}} \right. = 2038$ N	
Step Three: Convert from Base SI Units to Desired Units		
Method	**Steps**	
(1) Term to be converted	2038 N	
(2) Conversion formula (3) Make a fraction (4) Multiply	$\dfrac{2038 \text{ N}}{} \left\| \dfrac{0.225 \text{ lb}_f}{1 \text{ N}} \right.$	
(5) Cancel, calculate, be reasonable	460 lb$_f$	

Specific Gravity

In technical literature, density is rarely given; instead, the **specific gravity** is reported. The specific gravity (SG) of an object is a dimensionless ratio of the density of the object to the density of water (see Table 11-5). It is convenient to list density in this

IMPORTANT CONCEPT

$$SG = \frac{\rho_{object}}{\rho_{water}}$$

fashion so *any* unit system may be applied by our choice of the units of the density of water. The specific gravities of several common substances are listed in Table 11-6.

Table 11-5 Dimensions of specific gravity

Quantity	Common Units	Exponents			
		M	L	T	Θ
specific gravity	–	0	0	0	0

IMPORTANT CONCEPT

Density of water
= 1 g/cm^3
= 1 kg/L
= 1000 kg/m^3
= 62.4 lb$_m$/ft^3
= 1.94 slug/ft^3

Table 11-6 Specific gravity values for common substances

Liquids	SG
Acetone	0.785
Benzene	0.876
Citric Acid	1.67
Gasoline	0.739
Glycerin	1.26
Iodine	4.93
Mercury	13.6
Mineral Oil	0.900
Olive Oil	0.703
Propane	0.806
Sea Water	1.03
Toluene	0.865
Water	1.00

Solids	SG
Aluminum	2.70
Baking Soda	0.689
Brass	8.40–8.75
Concrete	2.30
Copper	8.96
Gallium	5.91
Gold	19.3
Graphite	2.20
Iron	7.87
Lead	11.4
Polyvinyl Chloride (PVC)	1.38
Silicon	2.33
Zinc Oxide	5.60

REASONABLENESS RULES

When calculating or considering specific gravities, it is helpful to keep in mind the range of values that you are likely to have.

The densest naturally occurring elements at normal temperature and pressure are osmium and iridium, both with a specific gravity close to 22.6. The *densest substances that a normal person is likely to encounter are platinum (SG = 21.5) and gold (SG = 19.3)*. Thus, if you calculate a specific gravity to be higher than about 23, you have almost certainly made an error.

Most liquids are similar to water, with a specific gravity around 1. One notable exception is mercury, with a specific gravity of 13.

On the lower end of the scale, the *specific gravity of air is about 0.001*, whereas hydrogen has a specific gravity of slightly less than 0.0001. Therefore, if you get a specific gravity value less than about 10^{-4}, you need to check your work very carefully.

EXAMPLE 11-6

The specific gravity of butane is 0.599. What is the density of butane in units of kilograms per cubic meter?

Step One: Convert to Base SI Units	
No Conversion Needed	
Step Two: Calculate	
Method	**Steps**
(1) Determine appropriate equation	$\rho_{object} = (SG)(\rho_{water})$
(2) Insert known quantities	$\rho_{object} = (0.599)\left(1000\,\dfrac{kg}{m^3}\right)$
(3) Calculate, be reasonable	$\rho_{object} = 599\,\dfrac{kg}{m^3}$
Step Three: Convert from Base SI Units to Desired Units	
No Conversion Needed	

EXAMPLE 11-7

Mercury has a specific gravity of 13.6. What is the density of mercury in units of slugs per cubic foot?

Step One: Convert to Base SI Units	
No Conversion Needed	
Step Two: Calculate	
Method	**Steps**
(1) Determine appropriate equation	$\rho_{object} = (SG)(\rho_{water})$
(2) Insert known quantities	$\rho_{object} = (13.6)\left(1\,\dfrac{g}{cm^3}\right)$
(3) Calculate, be reasonable	$\rho_{object} = 13.6\,\dfrac{g}{cm^3}$
Step Three: Convert from Base SI Units to Desired Units	
Method	**Steps**
(1) Term to be converted	$13.6\ g/cm^3$
(2) Conversion formula	$\dfrac{13.6\ g}{cm^3}\left\|\dfrac{0.0685\ slug}{1000\ g}\right.\left\|\dfrac{1000\ cm^3}{0.0353\ ft^3}\right.$
(3) Make a fraction	
(4) Multiply	
(5) Cancel, calculate, be reasonable	$26.6\ slug/ft^3$

Comprehension Check 11-3

Convert 50 grams per cubic centimeter into units of pounds-mass per cubic foot.

Comprehension Check 11-4

A 75-gram cylindrical rod is measured to be 10 centimeters long and 2.5 centimeters in diameter. What is the specific gravity of the material?

11.4 AMOUNT

Some things are really, really large and some are very, very small. Stellar distances are so large that it becomes inconvenient to report values such as 235 trillion miles, or 6.4×10^{21} feet when we are interested in the distance between two stars or two galaxies. To make things better, we use a new unit of length that itself is large—the distance that light goes in a year; this is a very long way, 3.1×10^{16} feet. As a result, we don't have to say that the distance between two stars is 620,000,000,000,000,000 feet, we can just say that they are 2 light-years apart.

This same logic holds when we want to discuss very small things such as molecules or atoms. Most often we use a constant that has been named after Amedeo Avagadro, an Italian scientist (1777–1856) who first proposed the idea of a fixed ratio between the amount of substance and the number of elementary particles. The Avogadro constant has a value of 6.022×10^{23} particles per mole. If we have 12 of something, we call it a dozen. If we have 20, it is a score. If we have 6.022×10^{23} of anything, we have a mole. If we have 6.022×10^{23} baseballs, we have a mole of baseballs. If we have 6.022×10^{23} elephants, we have a mole of elephants, and if we have 6.022×10^{23} molecules, we have a mole of molecules. Of course, the mole is never used to define amounts of macroscopic things like elephants or baseballs, being relegated to the realm of the extremely tiny. In the paragraphs below we will see how this rather odd value originated and how this concept simplifies our calculations.

The mass of a nucleon (neutron or proton) is about 1.673×10^{-24} grams. To avoid having to use such tiny numeric values when dealing with nucleons, physicists defined the **atomic mass unit** [amu] to be approximately the mass of one nucleon. Technically, it is defined as one-twelfth of the mass of a carbon twelve atom. In other words, 1 amu = 1.673×10^{-24} g. The symbol "u" is often used for amu, which is also known as a **Dalton** [Da].

If there are $(1.673 \times 10^{-24} \text{ g})/(1 \text{ amu})$, then there is $(1 \text{ amu})/(1.673 \times 10^{-24} \text{ g})$. Dividing this out gives 6.022×10^{23} amu/g. This numeric value is used to define the **mole** [mol]. One mole of a substance (usually an element or compound) contains exactly 6.022×10^{23} fundamental units (atoms or molecules) of that substance. In other words, there are 6.022×10^{23} fundamental units per mole. This is often written as

$$N_A = 6.022 \times 10^{23} \, \text{mol}^{-1}$$

As mentioned above, this is called Avogadro's constant or **Avogadro's number**, symbolized by N_A. So why is this important? Consider combining hydrogen and oxygen to get water (H_2O). We need twice as many atoms of hydrogen as atoms of oxygen for this reaction; thus, for every mole of oxygen, we need two moles of hydrogen, since one mole of anything contains the same number of fundamental units, atoms in this case.

The problem is that it is difficult to measure a substance directly in moles, but it is easy to measure its mass. *Avogadro's number affords a conversion path between moles and mass.* Consider hydrogen and oxygen in the above. The atomic mass of an atom in **atomic mass units** [amu] is approximately equal to the number of nucleons it contains. Hydrogen contains one proton, and thus has an atomic mass of 1 amu. We can also say that there is one amu per hydrogen atom. Oxygen has an atomic mass of 16; thus, there are 16 amu per oxygen atom. Since atomic mass refers to an individual specific atom, the term **atomic weight** is used, representing the average value of all isotopes of the element. This is the value commonly listed on periodic tables.

Let's use this information, along with Avogadro's number, to determine the mass of one mole of each of these two elements.

$$\text{Hydrogen:} \quad \frac{1\,\text{amu}}{\text{H atom}} \left| \frac{1\,\text{g}}{6.022 \times 10^{23}\,\text{amu}} \right| \frac{6.022 \times 10^{23}\,\text{atom}}{1\,\text{mol}} = \frac{1\,\text{g}}{1\,\text{mol H}}$$

$$\text{Oxygen:} \quad \frac{16\,\text{amu}}{\text{O atom}} \left| \frac{1\,\text{g}}{6.022 \times 10^{23}\,\text{amu}} \right| \frac{6.022 \times 10^{23}\,\text{atom}}{1\,\text{mol}} = \frac{16\,\text{g}}{1\,\text{mol O}}$$

> If Element Z has an atomic mass of X amu, there are X grams per mole of Element Z.

The numerical value for the atomic mass of a substance is the same as the number of grams in one mole of that substance, often called the **molar mass**.

$$\text{Atomic Weight} = \text{Molar Mass}$$

> If Molecule AB has a molecular weight of X amu, there are X grams per mole of Molecule AB.

Avogadro's number is the link between the two. Hydrogen has a molar mass of 1 gram per mole; oxygen has a molar mass of 16 grams per mole.

When groups of atoms react together, they form molecules. Consider combining hydrogen and oxygen to get water (H_2O). Two atoms of hydrogen combine with one atom of oxygen, so 2 * 1 amu H + 16 amu O = 18 amu H_2O. The **molecular mass** of water is 18 amu. By an extension of the example above, we can also state that one mole of water has a mass of 18 grams, called the **formula weight**.

$$\text{Molecular Weight} = \text{Formula Weight}$$

The difference between these ideas is summarized in Table 11-7.

This text assumes that you have been exposed to these ideas in an introductory chemistry class and so does not cover them in any detail. In all problems presented, you will be given the atomic weight of the elements or the formula weight of the molecule,

Table 11-7 Definitions of "amount" of substance

	measures the . . .	in units of . . .	and is found by . . .
Atomic Mass	Mass of one atom of an individual isotope of an element	[amu]	Direct laboratory measurement
Atomic Weight	Average mass of all isotopes of an element	[amu]	Listed on Periodic Table
Molar Mass	Mass of one mole of the atom	[g/mol]	Listed on Periodic Table
Molecular Mass or Molecular Weight	Sum of average weight of isotopes in molecule	[amu]	Combining atomic weights of individual atoms represented in the molecule
Formula Weight	Mass of one mole of the molecule	[g/mol]	Combining molar mass of individual atoms represented in the molecule

depending on the question asked. This topic is briefly introduced because Avogadro's number (N_A) is important in the relationship between several constants, including the following:

- The gas constant (R [=] J/(mol K)) and the Boltzmann constant (k [=] J/K), which relates energy to temperature: $R = k N_A$.
- The elementary charge (e [=] C) and the Faraday constant (F [=] C/mol), which is the electric charge contained in 1 mole of electrons: $F = e N_A$.
- An electron volt [eV] is a unit of energy describing the amount of energy gained by one electron accelerating through an electrostatic potential difference of 1 volt: $1 \text{ eV} = 1.602 \times 10^{-19}$ J.

EXAMPLE 11-8

Let's return to the problem of combining hydrogen and oxygen to get water. Assume you have 50 grams of oxygen with which you want to combine the proper mass of hydrogen to convert it completely to water. The atomic weight of hydrogen is 1 and the atomic weight of oxygen is 16.

First determine how many moles of oxygen are present.

$$\frac{50\,g\,O}{} \left| \frac{1\,mol\,O}{16\,g\,O} = 3.125\,mol\,O$$

We need twice as many moles of hydrogen as oxygen (H_2O), so we need 6.25 moles of hydrogen. Converting to mass gives

$$\frac{6.25\,mol\,H}{} \left| \frac{1\,g\,H}{1\,mol\,H} = 6.25\,g\,H$$

EXAMPLE 11-9

Acetylsalicylic acid (aspirin) has the chemical formula $C_9H_8O_4$. How many moles of aspirin are in a 1-gram dose? Use the following facts:

- Atomic weight of carbon = 12
- Atomic weight of hydrogen = 1
- Atomic weight of oxygen = 16

First, determine how many grams are in one mole of aspirin (determine formula weight).

$$\text{FW of aspirin} = \left[\frac{12\frac{grams}{mole}}{1\,molecule\,C} \middle| 9\,C\,molecules\right] + \left[\frac{1\frac{grams}{mole}}{1\,molecule\,H} \middle| 8\,H\,molecules\right]$$

$$+ \left[\frac{16\frac{grams}{mole}}{1\,molecule\,O} \middle| 4\,O\,molecules\right] = 180\frac{g}{mole}$$

Finally, convert to moles per dose.

$$\frac{1\,g\,aspirin}{dose} \left| \frac{1\,mol\,aspirin}{180\,g\,aspirin} = 5.56 \times 10^{-3}\frac{mol\,aspirin}{dose}$$

EXAMPLE 11-10

Many gasses exist as diatomic compounds in nature, meaning two of the atoms are attached to form a molecule. Hydrogen, oxygen, and nitrogen all exist in a gaseous diatomic state under standard conditions.

Assume there are 100 grams of nitrogen gas in a container. How many moles of nitrogen (N_2) are in the container? Use the fact that the atomic weight of nitrogen is 14.

First, determine how many grams are in one mole of nitrogen (determine the formula weight).

$$\text{FW of } N_2 = \frac{14 \frac{\text{grams}}{\text{mols}}}{1 \text{ molecule of N}} \left| \frac{2 \text{ molecules of N}}{} = 28 \frac{\text{grams}}{\text{mole}} \right.$$

Next, convert mass to moles.

$$\frac{100 \text{ grams of } N_2}{} \left| \frac{\text{mole}}{28 \text{ gram}} = 3.57 \text{ moles } N_2 \right.$$

11.5   TEMPERATURE

Temperature was originally conceived as a description of energy: heat (thermal energy) flows spontaneously from "hot" to "cold." But how hot is "hot"? The thermometer was devised as a way to measure the "hotness" of an object. As an object gets warmer, it usually expands. In a thermometer, a temperature is a level of hotness that corresponds to the length of the liquid in the tube. As the liquid gets warmer, it expands and moves up the tube. To give temperature a quantitative meaning, numerous temperature scales have been developed.

Many scientists, including Isaac Newton, have proposed temperature scales. Two scales were originally developed about the same time—**Fahrenheit** [°F] and **Celsius** [°C]—and have become widely accepted in laymen use. These are the most frequently used temperature scales by the general public. Gabriel Fahrenheit (1686–1736), a German physicist and engineer, developed the Fahrenheit scale in 1708. Anders Celsius (1701–1744), a Swedish astronomer, developed the Celsius scale in 1742. The properties of each scale are in Table 11-8.

You may wonder why the Celsius scale seems so reasonable, and the Fahrenheit scale so random. Actually Mr. Fahrenheit was just as reasonable as Mr. Celsius. Mr. Celsius set the freezing point of water to be 0 and the boiling point to be 100. Mr. Fahrenheit took as 0 a freezing mixture of salt and ice, and as 100 body temperature. With this scale, it just so happens that the freezing and boiling points of water work out to be odd numbers.

Some units can cause confusion in conversion. One of those is temperature. One reason for this is that we use temperature in two different ways: 1) reporting an actual temperature value and 2) discussing the way a change in temperature affects a material property. To clarify, we resort to examples.

Human body
  ~ 37 °C = 98 °F

Room temperature
  ~ 21 °C = 70 °F

Melting point of
  mercury ~ −39 °C
  = −38 °F

Melting temperature
  of lead ~ 330 °C
  = 620 °F

Is 180 really an odd value? No, not in Fahrenheit's day. The number 180 was familiar to mathematicians and scientists as the number of degrees in half a circle (360/2 = 180).

Table 11-8 Properties of water

Scale	Freezing Point	Boiling Point	Divisions between Freezing and Boiling
Fahrenheit [°F]	32	212	180
Celsius [°C]	0	100	100
Kelvin [K]	273	373	100
Rankine [°R]	492	672	180

Conversion Between Temperature Values

Fahrenheit

Celsius

Boiling point of water

212 °F   100 °C

Temperature of interest

T [°F]   T [°C]

Freezing point of water

32 °F   0 °C

When an actual temperature reading is reported, such as "the temperature in this room is 70 °F," how do we convert from one temperature unit to another? The scales have different zero points, so they cannot be converted with a single conversion factor as done previously but require a conversion formula. Most of you are familiar with the conversion between Fahrenheit and Celsius, but this equation is cumbersome to remember.

$$T[°F] = \frac{9}{5}T[°C] + 32$$

Let us imagine we have two thermometers, one with the Fahrenheit scale and one with the Celsius scale. We set two thermometers side by side so that the freezing point and the boiling point of water are at the same location on both thermometers. We are interested in the relationship between these two scales. From this figure we see that the fraction of the distance from the freezing point to the boiling point in both scales is the same. This means that we can write

$$\frac{T[°F] - 32}{212 - 32} = \frac{T[°C] - 0}{100 - 0}$$

This relationship is really all we need to know to relate a temperature in degrees Fahrenheit to one in degrees Celsius. You can easily do the algebra to convert from Fahrenheit to Celsius, or vice versa. By remembering this form, you don't have to remember if the conversion is 9/5 or 5/9, or to add or subtract 32. This formula is determined by the method of interpolation, previously discussed in Chapter 8.

There are numerous other temperature scales, but two are worth mentioning: **Kelvin** [K] and **Rankine** [°R]. The Kelvin scale is named for First Baron William Thomson Kelvin (1824–1907), an English mathematician and physicist. Kelvin first proposed the idea of "infinite cold," or absolute zero, in 1848, using the Celsius scale for comparison. The Rankine scale is named for William J. M. Rankine (1820–1872), a Scottish engineer and physicist, who proposed an analogy to the Kelvin scale, using the Fahrenheit scale. Both men made significant contributions to the field of thermodynamics.

The Kelvin and Rankine scales are "absolute," which means that at absolute zero, the temperature at which molecules have minimum possible motion, the temperature is zero. Absolute temperature scales therefore have no negative values. In the Kelvin scale, the degree sign is not used; it is simply referred to as "kelvin," not "degrees kelvin." It is the base SI unit for temperature and the most frequently used temperature unit in the scientific community.

IMPORTANT CONCEPT

When actual temperature readings are converted:

$$\frac{T[°F] - 32}{180} = \frac{T[°C] - 0}{100}$$

$$T[K] = T[°C] + 273$$

$$T[°R] = T[°F] + 460$$

IMPORTANT CONCEPT

Absolute temperature scales are never negative.

EXAMPLE 11-11

The hottest temperature in the United States ever recorded by the National Weather Service, 56.7 degrees Celsius [°C], occurred in Death Valley, California, on July 10, 1913. State this value in units of degrees Fahrenheit [°F].

Step One: Convert to Base SI Units	
No Conversion Needed	
Method	**Steps**
(1) Determine appropriate equation	$\dfrac{T[°F] - 32}{212 - 32} = \dfrac{T[°C] - 0}{100 - 0}$
(2) Insert known quantities	$\dfrac{T[°F] - 32}{180} = \dfrac{56.7}{100}$
(3) Calculate, be reasonable	$T = 134\ °F$
Step Three: Convert from Base SI Units to Desired Units	
No Conversion Needed	

IMPORTANT CONCEPT

When properties that contain temperature are converted:

$$\frac{1°C}{1.8°F} \quad \frac{1\,K}{1°C} \quad \frac{1°R}{1°F}$$

When this type of conversion is needed, we read the units under consideration as "*per* degree F" or "*per* degree C," with the clue being the word "*per.*"

Conversions Involving Temperature Within a Material Property

When considering how a change in temperature affects a material property, we use a scalar conversion factor. In general, we encounter this in sets of units relating to the property of the material; for example, the units of the thermal conductivity are given by $\frac{W}{m\,K}$, which is read as "watts per meter kelvin." When this is the case, we are referring to the size of the degree, not the actual temperature.

To find this relationship, remember that between the freezing point and the boiling point of pure water, the Celsius scale contains 100 divisions whereas the Fahrenheit scale contains 180 divisions. The conversion factor between Celsius and Fahrenheit is $100\ °C \equiv 180\ °F$, or $1\ °C \equiv 1.8\ °F$.

EXAMPLE 11-12

Specific Heat [J/(g K)]
water = 4.184
aluminum = 0.897
copper = 0.385
lead = 0.127
air = 1.012
helium = 5.1932

The specific heat (C_p) is the ability of an object to store heat. Specific heat is a material property, and values are available in technical literature. The specific heat of copper is 0.385 $\frac{J}{g\,°C}$, which is read as "joules per gram degrees Celsius." Convert this to units of $\frac{J}{lb_m\,°F}$, which reads "joules per pound-mass degree Fahrenheit."

Method	Steps
(1) Term to be converted	$0.385\ \dfrac{J}{g\,°C}$
(2) Conversion formula (3) Make a fraction (4) Multiply	$\dfrac{0.385\,J}{g\,°C} \left\vert \dfrac{1000\,g}{2.205\,lb_m} \right\vert \dfrac{1°C}{1.8°F}$
(5) Cancel, calculate, be reasonable	$97\ \dfrac{J}{lb_m\,°F}$

A note of clarification about the term "PER": When reading the sentence: "Gravity on earth is commonly assumed to be 9.8 meters per second squared," there is often little confusion in translating the words to symbols: $g = 9.8\ m/s^2$. For a more complex unit however, this can present a challenge. For example, the sentence "The thermal conductivity of aluminum is 237 calories per hour meter degrees Celsius," can be confusing because it can be interpreted as:

$$k = 237 \frac{cal}{h\,m\,°C} \quad or \quad k = 237 \frac{cal}{h}(m\,°C) \quad or \quad k = 237 \frac{cal}{h\,m}°C$$

Officially, according to SI rules, when writing out unit names anything following the word "per" appears in the denominator of the expression. This implies the first example above is correct.

The thermal conductivity of aluminum is 237 calories per hour meter degrees Celsius: $k = 237 \frac{cal}{h\,m\,°C}$

Comprehension Check 11-5

The temperature of dry ice is −109.3 degrees Fahrenheit [°F]. Convert this temperature into units of kelvin [K].

Comprehension Check 11-6

The specific heat capacity of copper is 0.09 British thermal units per pound-mass degrees Fahrenheit $\left[\frac{BTU}{lb_m\,°F}\right]$. Convert into units of British thermal units per gram kelvin $\left[\frac{BTU}{g\,K}\right]$.

11.6 PRESSURE

IMPORTANT CONCEPT

Pressure = Force/Area

SI unit of pressure = Pascal

$$Pa = \frac{N}{m^2}$$

Pressure is defined as force acting over an area, where the force is perpendicular to the area. In SI units, a **pascal** [Pa] is the unit of pressure, defined as 1 newton of force acting on an area of 1 square meter. The unit pascal is named for the French mathematician and physicist Blaise Pascal (1623–1662) who made great contributions to the study of fluids, pressure, and vacuums. His contributions with Pierre de Fermat on the theory of probability were the groundwork for calculus.

Table 11-9 Dimensions of pressure

Quantity	Common Units	Exponents			
		M	L	T	Θ
pressure	Pa	1	−1	−2	0

EXAMPLE 11-13

An automobile tire is pressurized to a 40 pound-force per square inch [psi or lb_f/in^2]. State this pressure in units of atmospheres [atm].

PRESSURE

1 atm = 1.01325 bar
= 33.9 ft H_2O
= 29.92 in Hg
= 760 mm Hg
= 101325 Pa
= 14.7 psi

By examining the "Pressure" box in the conversion table on the inside front cover, we see that the following facts are available for use: 1 atm = 14.7 psi.

Method	Steps
(1) Term to be converted	40 psi
(2) Conversion formula	1 atm = 14.7 psi
(3) Make a fraction (4) Multiply	$\frac{40\,psi}{} \left\| \frac{1\,atm}{14.7\,psi}\right.$
(5) Cancel, calculate, be reasonable	2.7 atm

Comprehension Check 11-7

If the pressure is 250 feet of water [ft H_2O], what is the pressure in units of inches of mercury [in Hg]?

In this chapter, we consider four forms of pressure, all involving fluids. The general term **fluid** applies to a gas, such as helium or air, or a liquid, such as water or honey.

- **Atmospheric pressure**—the pressure created by the weight of air above us.
- **Hydrostatic pressure**—the pressure exerted on a submerged object by the fluid in which it is immersed.
- **Total pressure**—the combination of atmospheric and hydrostatic pressure.
- **Gas pressure**—the pressure created by a gas inside a closed container.

Atmospheric Pressure

1 atm ~ 14.7 psi ~ 101 kPa

Atmospheric pressure results from the weight of the air above us, which varies both with altitude and weather patterns. Standard atmospheric pressure is an average air pressure at sea level, defined as 1 atmosphere [atm], and is approximately equal to a 14.7 pound-force per square inch [psi].

Pressure Measurement

When referring to the measurement of pressure, two types of reference points are commonly used.

REASONABLENESS RULES

Car tires are inflated with between 30 and 40 psi (gauge pressure).

"Normal" blood pressure is 120 mm Hg/80 mm Hg (gauge pressure).

    Absolute pressure uses a perfect vacuum as a reference point. Most meteorological readings are given as absolute pressure, using units of atmospheres or bars.

    Gauge pressure uses the local atmospheric pressure as a reference point. Note that local atmospheric pressure is generally *not* standard atmospheric pressure at sea level (1 atm). Measurements such as tire pressure and blood pressure are given as gauge pressure.

    Absolute pressures are distinguished by an "a" after the pressure unit, such as "psia" to signify "pound-force per square inch absolute." Gauge pressure readings are distinguished by a "g" after the pressure unit, such as "psig" to signify "pound-force per square inch gauge." When using instrumentation to determine the pressure, be sure to note whether the device reads absolute or gauge pressure.

    Gauge pressure, absolute pressure, and atmospheric pressure are related by

$$P_{absolute} = P_{gauge} + P_{atmospheric}$$

A few notes on absolute and gauge pressure. Except as otherwise noted, assume that local atmospheric pressure is 14.7 psia.

- 35 psig = 49.7 psia (35 psi + 14.7 psi).
- Using gauge pressure, local atmospheric pressure would be 0 psig, although this would seldom be used.
- If a gauge pressure being measured is less than the local atmospheric pressure, this is usually referred to as **vacuum pressure**, and the negative sign is dropped. Thus, a perfect vacuum created at sea level on Earth would read about 14.7 psig vacuum pressure. (A perfect vacuum is 0 psia, and thus is about 14.7 psi less than atmospheric pressure.)
- A vacuum pressure of 10 psig is an absolute pressure of 4.7 psia (14.7 psi − 10 psi).

■ To illustrate the effect of local atmospheric pressure, consider the following scenario. You fill your automobile's tires to 35 psig on the shore of the Pacific Ocean in Peru, and then drive to Lake Titicaca on the Bolivian border at about 12,500 feet above sea level. Your tire pressure now reads about 40 psig due to the decreased atmospheric pressure (about 9.5 psia at 12,500 feet altitude versus about 14.7 psia at sea level).

Occasionally in industry, it may be helpful to use a point of reference other than atmospheric pressure. For these specific applications, pressure may be discussed in terms of **differential pressure**, distinguished by a "d" after the pressure unit, such as "psid."

IMPORTANT CONCEPT
Pascal's Law
Phydro = ρ g H

Hydrostatic Pressure

Hydrostatic pressure (P_{hydro}) results from the weight of a liquid or gas pushing on an object. *Remember, weight is a force!* A simple way to determine this is to consider a cylinder with a cross-sectional area (A) filled with a liquid (of density ρ).

The pressure (P) at the bottom of the container can be found by **Pascal's law**, named after (once again) Blaise Pascal. Pascal's law states the hydrostatic pressure of a fluid is equal to the force of the fluid acting over an area.

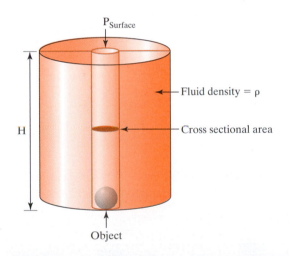

$P_{Surface}$

Fluid density = ρ

H

Cross sectional area

Object

DEVILISH DERIVATION

How does the term ρ g H equal a pressure?

$$P_{hydro} = \frac{F}{A}$$

weight is a force
$$= \frac{w}{A}$$

$$w = mg \qquad = \frac{mg}{A}$$

$$\rho = m/V \qquad = \frac{\rho V g}{A}$$

$$V = AH \qquad = \frac{\rho (AH) g}{A}$$

$$P_{hydro} = \rho\, g H$$

Recreational scuba diving takes place at depths between 0–20 meters. At deeper depths, additional training is usually required because of the increased risk of narcosis, a state similar to alcohol intoxication. The relationship between depth and the level of narcosis is called the "Martini Effect," as it is said divers feel like they have drunk one martini for every 20 meters they descend.

Photo courtesy of E. Stephan

EXAMPLE 11-14

We want to know the hydrostatic pressure in a lake at a depth of 20 feet in units of pascals.

For hydrostatic pressure, we need to know the density of the fluid in the lake. Since a density is not specified, we assume the density to be the standard density of water. We want all quantities in units of kilograms, meters, and seconds, so we use a density of 1,000 kilograms per cubic meter for water.

Step One: Convert to Base SI Units	
Method	**Steps**
(1) Term to be converted	20 ft
(2) Conversion formula (3) Make a fraction (4) Multiply	$\dfrac{20\text{ ft}}{}\Bigg|\dfrac{1\text{ m}}{3.28\text{ ft}}$
(5) Cancel, calculate, reasonable	6.1 m
Step Two: Calculate	
Method	**Steps**
(1) Determine appropriate equation	$P_{hydro} = \rho g H$
(2) Insert known quantities	$P_{hydro} = \dfrac{1000\text{ kg}}{m^3}\Bigg|\dfrac{9.8\text{ m}}{s^2}\Bigg|\dfrac{6.1\text{ m}}{}$
(3) Calculate, be reasonable	$P_{hydro} = 59{,}760\,\dfrac{kg}{m\,s^2}$

This is apparently our final answer, but the units are puzzling. If the units of pressure are pascals and if this is a valid equation, then our final result for pressure should be pascals. If we consider the dimensions of pressure:

Quantity	Common Units	Exponents			
		M	**L**	**T**	**Θ**
pressure	Pa	1	−1	−2	0

A unit of pressure has dimensions, $P\{=\}\dfrac{M}{LT^2}$, which in terms of base SI units would be $P[=]\frac{kg}{m\,s^2}$. This term occurs so frequently it is given the special name "pascal." When we see this term, we know we are dealing with a pressure equal to a pascal.

(3) Calculate, be reasonable	$P_{hydro} = 59{,}760\dfrac{kg}{m\,s^2}\Bigg|\dfrac{1\text{ Pa}}{1\frac{kg}{m\,s^2}} = 59{,}760\text{ Pa}$
Step Three: Convert from Base SI Units to Desired Units	
No Conversion Needed	

IMPORTANT CONCEPT

$P_{total} = P_{surface} + P_{hydro}$
$P_{total} = P_{surface} + \rho gH$

Total Pressure

We need to realize that Pascal's law is only a part of the story. Suppose we dive to a depth of 5 feet in a swimming pool and measure the pressure. Now we construct an enclosure over the pool and pressurize the air above the water surface to 3 atmospheres. When we dive back to the 5-foot depth, the pressure will have increased by 2 atmospheres.

Consequently, we conclude that total pressure at any depth in a fluid is the sum of hydrostatic pressure and pressure in the air above the fluid.

EXAMPLE 11-15

REASONABLENESS RULES

A diver at 15 feet will feel a ½ atmosphere pressure increase.

A diver at 10 meters will feel a 1 atmosphere pressure increase.

When you dive to the bottom of a pool, at 12 feet under water, how much total pressure do you feel in units of atmospheres?

Step One: Convert to Base SI Units	
Method	**Steps**
(1) Term to be converted	12 ft
(2) Conversion formula	
(3) Make a fraction	$\dfrac{12\ ft}{}\bigg|\dfrac{1\ m}{3.28\ ft}$
(4) Multiply	
(5) Cancel, calculate, reasonable	3.66 m

For hydrostatic pressure, we need to know the density of the fluid in the pool. Since a density is not specified, we assume the density to be the standard density of water. We want all quantities in units of kilograms, meters, and seconds, so we use a density of 1,000 kilograms per cubic meter for water.

For total pressure, we need to know the surface pressure on top of the pool. Since a surface pressure is not specified, we assume the pressure to be 1 atmosphere. We want all quantities in units of kilograms, meters, and seconds, so we use a pressure of 101,325 pascals, or 101,325 kilograms per meter second squared.

Step Two: Calculate	
Method	**Steps**
(1) Determine appropriate equation	$P_{total} = P_{surface} + \rho gH$
(2) Insert known quantities	$P_{total} = \dfrac{101,325\ kg}{m\ s^2} + \dfrac{1,000\ kg}{m^3}\bigg|\dfrac{9.8\ m}{s^2}\bigg|3.66\ m$
(3) Calculate, reasonable	$P_{total} = 137,193\dfrac{kg}{m\ s^2}\bigg|\dfrac{Pa}{\frac{kg}{m\ s^2}} = 137,193\ Pa$
Step Three: Convert from Base SI Units to Desired Units	
Method	**Steps**
(1) Term to be converted	137,193 Pa
(2) Conversion formula	
(3) Make a fraction	$\dfrac{137,193\ Pa}{}\bigg|\dfrac{1\ atm}{101,325\ Pa}$
(4) Multiply	
(5) Cancel, calculate, be reasonable	1.35 atm

Comprehension Check 11-8

An object is completely submerged in a liquid of density 0.75 grams per cubic centimeter at a depth of 3 meters. What is the total pressure on the object? State your answer in atmospheres.

At an elevation of 3 miles, the pressure will be about half that at sea level. At an elevation of 10 miles, the pressure is about 1/10 that at sea level.

Cakes and meats may require as much as 1/4 more cooking time at elevations of 5,000 feet because of reduced pressure and lower boiling points.

11.7   GAS PRESSURE

Gas pressure results when gas molecules impact the inner walls of a sealed container. The **ideal gas law** relates the quantities of pressure (P), volume (V), temperature (T), and amount (n) of gas in a closed container:

$$PV = nRT$$

IMPORTANT CONCEPT

Ideal Gas Law

$$PV = nRT$$

Only absolute temperature units (K or °R) can be used in the ideal gas equation.

In this equation, R is a fundamental constant called the **gas constant.** It can have many different numerical values, depending on the units chosen for pressure, volume, temperature, and amount, just as a length has different numerical values, depending on whether feet or meters or miles is the unit being used. Scientists have defined an "ideal" gas as one where 1 mole [mol] of gas at a temperature of 273 kelvin [K] and a pressure of 1 atmosphere [atm] will occupy a volume of 22.4 liters [L]. Using these values to solve for the constant R yields

$$R = \frac{8314\ Pa\ L}{mol\ K}$$
$$= 0.08206\ \frac{atm\ L}{mol\ K}$$

$$R = \frac{PV}{nT} = \frac{1\ [atm]\ 22.4\ [L]}{1\ [mol]\ 273\ [K]} = 0.08206\ \frac{atm\ L}{mol\ K}$$

Note that we must *use* absolute *temperature units in the ideal gas equation.* We cannot begin with relative temperature units and then convert the final answer. Also, all pressure readings must be in absolute, not gauge, units.

In previous chapters, we have suggested a procedure for solving problems involving equations and unit conversions. For ideal gas law problems, we suggest a slightly different procedure.

IDEAL GAS LAW PROCEDURE

1. Examine the units given in the problem statement. Choose a gas constant (R) that contains as many of the units given in the problem as possible.
2. If necessary, convert all parameters into units found in the gas constant (R) that you choose.
3. Solve the ideal gas law for the variable of interest.
4. Substitute values and perform all necessary calculations.
5. If necessary, convert your final answer to the required units and apply reasonableness.

EXAMPLE 11-16

A container holds 1.43 moles of nitrogen (formula: N_2) at a pressure of 3.4 atmospheres and a temperature of 500 degrees Fahrenheit. What is the volume of the container in liters?

Method	Steps
(1) Choose ideal gas constant	Given units: mol, atm, °F, L Select R: $0.08206 \dfrac{atm\ L}{mol\ K}$
(2) Convert to units of chosen R	500 °F = 533 K
(3) Solve for variable of interest	$V = \dfrac{nRT}{P}$
(4) Calculate	$V = \dfrac{1.43\ mol}{} \left\vert \dfrac{0.08206\ atm\ L}{mol\ K} \right\vert \dfrac{533\ K}{3.4\ atm}$
(5) Convert, be reasonable	V = 18.4 L

EXAMPLE 11-17

A container holds 1.25 moles of nitrogen (formula: N_2) at a pressure of 350 kilopascals and a temperature of 160 degrees Celsius. What is the volume of the container in liters?

Method	Steps
(1) Choose ideal gas constant	Given units: mol, Pa, °C, L Select R: $8314 \dfrac{Pa\ L}{mol\ K}$
(2) Convert to units of chosen R	160 °C = 433 K
(3) Solve for variable of interest	$V = \dfrac{nRT}{P}$
(4) Calculate	$V = \dfrac{1.25\ mol}{} \left\vert \dfrac{8314\ Pa\ L}{mol\ K} \right\vert \dfrac{433\ K}{350,000\ Pa}$
(5) Convert, be reasonable	V = 13 L

EXAMPLE 11-18

A gas originally at a temperature of 300 kelvin and 3 atmospheres pressure in a 3.9 liter flask is cooled until the temperature reaches 284 kelvin. What is the new pressure of gas in atmospheres?

Since the volume and the mass of the gas remain constant, we can examine the ratio between the initial condition (1) and the final condition (2) for pressure and temperature. The volume of the container (V) and the amount of gas (n) are constant, so $V_1 = V_2$ and $n_1 = n_2$.

Method	Steps
(1) Choose ideal gas constant	Given units: mol, K, L Select R: $0.08206 \dfrac{\text{atm L}}{\text{mol K}}$
(2) Convert to units of chosen R	None needed
(3) Solve for variable of interest, eliminating any variables that remain constant between the initial and final state	$\dfrac{P_1 V_1}{P_2 V_2} = \dfrac{n_1 R T_1}{n_2 R T_2}$ $\dfrac{P_1}{P_2} = \dfrac{T_1}{T_2}$
(4) Calculate	$\dfrac{3 \text{ atm}}{P_2} = \dfrac{300 \text{ K}}{284 \text{ K}}$
(5) Convert, be reasonable	$P = 2.8$ atm

Comprehension Check 11-9

A 5 gallon container holds 35 grams of nitrogen (formula: N_2, molecular weight = 28 grams per mole) at a temperature of 400 kelvin. What is the container pressure in units of kilopascals?

Comprehension Check 11-10

An 8-liter container holds nitrogen (formula: N_2, molecular weight = 28 grams per mole) at a pressure of 1.5 atmospheres and a temperature of 310 kelvin. If the gas is compressed by reduction of the volume of the container until the gas pressure increases to 5 atmospheres while the temperature is held constant, what is the new volume of the container in units of liters?

11.8   ENERGY

IMPORTANT CONCEPTS

Work
  $W = F \, \Delta x$

Potential Energy
  $PE = m \, g \, \Delta H$

Kinetic Energy, translational
  $KE_T = \frac{1}{2} m \, (v_f^2 - v_t^2)$

Kinetic Energy, rotational
  $KE_R = \frac{1}{2} I \, (\omega_f^2 - \omega_t^2)$

Kinetic Energy, total
  $KE = KE_T + KE_R$

Thermal Energy
  $Q = m \, C_p \Delta T$

Energy is an abstract quantity with several definitions, depending on the form of energy being discussed. You may be familiar with some of the following types of energy.

Types of Energy

- **Work** (W) is energy expended by exertion of a force (F) over a distance (d). As an example, if you exert a force on (push) a heavy desk so that it slides across the floor, which will make you more tired: pushing it 5 feet or pushing it 50 feet? The farther you push it, the more work you do.
- **Potential energy** (PE) is a form of work done by moving a weight (w)—which is a force—a vertical distance (H). Recall that weight is mass (m) times gravity (g). Note that this is a special case of the work equation, where force is weight and distance is height.
- **Kinetic energy** (KE) is a form of energy possessed by an object in motion. If a constant force is exerted on a body, then by $F = ma$, we see that the body experiences a constant acceleration, meaning the velocity increases linearly with time. Since the velocity increases as long as the force is maintained, work is being done on the

object. Another way of saying this is that the object upon which the force is applied acquires kinetic energy, also called **energy of translational motion**. For a nonrotating body moving with some velocity (v) the kinetic energy can be calculated by $KE_T = (1/2) mv^2$.

This, however, is not the entire story. A rotating object has energy whether it is translating (moving along a path) or not. If you have ever turned a bicycle upside down, spun one of the wheels fairly fast, then tried to stop it with your hand, you understand that it has energy. This is **rotational kinetic energy**, and for an object spinning in place (but not going anywhere), it is calculated by $KE_R = (1/2) I\omega^2$. The Greek letter omega (ω) symbolizes angular velocity or the object's rotational speed, typically given in units of radians per second. The moment of inertia (I) depends on the mass and the geometry of the spinning object. The table shown lists the moments of inertia for a few common objects.

MOMENTS OF INERTIA (I) m = mass, r = radius	
Object	I for KE_R
Cylinder: thin shell	mr^2
Cylinder: solid	$mr^2/2$
Sphere: thin shell	$2\,mr^2/3$
Sphere: solid	$2\,mr^2/5$

For an object that is rotating and translating, such as a bowling ball rolling down the lane toward the pins, the total kinetic energy is simply the sum of the two:

$$KE = KE_T + KE_R = (1/2)mv^2 + (1/2)I\omega^2 = (1/2)(mv^2 + I\omega^2)$$

- **Thermal Energy** or heat (Q) is energy associated with a change in temperature (ΔT). It is a function of the mass of the object (m) and the specific heat (Cp), which is a property of the material being heated:

$$Q = mCp\Delta T$$

DEVILISH DERIVATION

Where does KE = $1/2\ mv^2$ come from?

If a constant force is applied to a body,

- That body will have a constant acceleration (remember F = ma);
- Its velocity will increase linearly with time;
- Its average velocity is the average of its initial and final values.

This is	$v = (v_f + v_i)/2$
Distance travelled is	$d = v\,t$
The work done is	$W = F\,d$
	$= (m\,a)\,d$
	$= m\,a\,v\,t$

Acceleration is the change in velocity over time, or $\qquad a = (v_f - v_i)/t$

Substituting for a and v in the work equation

$$W = m\,a\,v\,t = m\,[(v_f - v_i)/t]\,[(v_f + v_i)/2][t]$$
$$W = (1/2)\,m\,(v_f^2 - v_i^2)$$

This is given the name *kinetic energy*, or the energy of motion. Remember, work and energy are equivalent.

This expression is for the *translation* of a body only.

Calories and BTUs and Joules — Oh My!

The SI unit of work is **joule**, defined as 1 newton of force acting over a distance of 1 meter. The unit is named for James Joule (1818–1889), an English physicist responsible for several theories involving energy, including the definition of the mechanical equivalent of heat and Joule's law, which describes the amount of electrical energy converted to heat by a resistor (an electrical component) when an electric current flows through it. In some mechanical systems, work is described in units of foot pound-force [ft lb$_f$].

For energy in the form of heat, units are typically reported as British thermal units and calories instead of joules. A **British thermal unit** [BTU] is the amount of heat required to raise the temperature of 1 pound-mass of water by 1 degree Fahrenheit. A **calorie** [cal] is amount of heat required to raise the temperature of 1 gram of water by 1 degree Celsius.

Table 11-10 Dimensions of energy

Quantity	Common Units	Exponents			
		M	L	T	Θ
work	J	1	2	−2	0
thermal energy	BTU	1	2	−2	0
	cal	1	2	−2	0

EXAMPLE 11-19

A 50-kilogram load is raised vertically a distance of 5 meters by an electric motor. How much work in units of joules was done on the load?

First, we must determine the type of energy. The parameters we are discussing include mass (kilograms) and height (meters). Examining the energy formulas given above, the equation for potential energy fits. Also, the words "load is raised vertically a distance" fits with our understanding of potential energy.

Step One: Convert to Base SI Units	
No Conversion Needed	
Step Two: Calculate	
Method	**Steps**
(1) Determine appropriate equation	$PE = m\,g\,\Delta H$
(2) Insert known quantities	$PE = \dfrac{50\text{ kg}}{}\left\|\dfrac{9.8\text{ m}}{s^2}\right\|\dfrac{5\text{ m}}{}$
(3) Calculate, be reasonable	$PE = 2450\,\dfrac{\text{kg m}^2}{s^2}$

This is apparently our final answer, but the units are puzzling. If the units of energy are joules and if this is a valid equation, then our final result for energy should be joules. If we consider the dimensions of energy:

Quantity	Common Units	Exponents			
		M	L	T	Θ
energy	J	1	2	−2	0

A unit of energy has dimensions $E \{=\} \frac{M\,L^2}{T^2}$, which in terms of base SI units would be $E [=] \frac{kg\,m^2}{s^2}$. This term occurs so frequently it is given the special name "joule." Anytime we see this term $\left(\frac{kg\,m^2}{s^2}\right)$, we know we are dealing with an energy, equal to a joule.

(3) Calculate, be reasonable	$PE = 2450\dfrac{kg\,m^2}{s^2}\left|\dfrac{1\,J}{1\frac{kg\,m^2}{s^2}}\right. = 2450\,J$
Step Three: Convert from Base SI Units to Desired Units	
No Conversion Needed	

EXAMPLE 11-20

In the morning, you like to drink your coffee at a temperature of exactly 70 degrees Celsius [°C]. The mass of the coffee in your mug is 470 grams. To make your coffee, you had to raise the temperature of the water by 30 °C. How much energy in units of British thermal units [BTU] did it take to heat your coffee? The specific heat of water is 4.18 joules per gram degree Celsius [J/(g °C)].

First, you must determine the type of energy we are using. The parameters discussed include mass, temperature, and specific heat. Examining the energy formulas given above, the equation for thermal energy fits. Also, the words "How much energy . . . did it take to heat your coffee" fits with an understanding of thermal energy.

Step One: Convert to Base SI Units	
No Conversion Needed	
Step Two: Calculate	
Method	**Steps**
(1) Determine appropriate equation	$Q = m\,C_p\Delta T$
(2) Insert known quantities	$Q = \dfrac{470\,g}{}\left|\dfrac{4.18\,J}{g\,°C}\right|\dfrac{30\,°C}{}$
(3) Calculate, be reasonable	$Q = 59{,}370\,J$
Step Three: Convert from Base SI Units to Desired Units	
Method	**Steps**
(1) Term to be converted	$59{,}370\,J$
(2) Conversion formula (3) Make a fraction (4) Multiply	$\dfrac{59370\,J}{}\left|\dfrac{9.48\times10^{-4}\,BTU}{1\,J}\right.$
(5) Cancel, calculate, be reasonable	$56\,BTU$

Comprehension Check 11-11

You push an automobile with a constant force of 20 pounds-force until 1500 joules of energy has been added to the car. How far did the car travel in units of meters during this time? You may assume that frictional losses are negligible.

Comprehension Check 11-12

One gram of Material A is heated until the temperature rises by 10 kelvin. If the same amount of heat is applied to one gram of Material B, what is the temperature rise of Material B in units of kelvin?

The specific heat (C_p) of Material A = 4 joules per gram kelvin [J/(g K)]
The specific heat (C_p) of Material B = 2 joules per gram kelvin [J/(g K)]

11.9 POWER

IMPORTANT CONCEPTS

Power = energy/time

SI unit of power = Watt

W = J/s

Power is defined as energy per time. The SI unit of power is **watt**, named for James Watt (1736–1819), a Scottish mathematician and engineer whose improvements to the steam engine were important to the Industrial Revolution. He is responsible for the definition of **horsepower** [hp], a unit of power originally used to quantify how the steam engine could replace the work done by a horse.

Table 11-11 Dimensions of power

Quantity	Common Units	Exponents			
		M	L	T	Θ
power	W	1	2	−3	0

To help understand the relationship between energy and power, imagine the following. Your 1000-kilogram car has run out of gas on a level road. There is a gas station not far ahead, so you decide to push the car to the gas station. Assume that you intend to accelerate the car up to a speed of 1 meter per second (about 2.2 miles per hour), and then continue pushing at that speed until you reach the station. Ask yourself the following questions:

- Can I accelerate the car to 1 meter per second in one minute?
- On the other hand, can I accelerate it to 1 meter per second in one second?

Power is the RATE at which energy is delivered over time.

Most of you would probably answer "yes" to the first and "no" to the second, but why? Well, personal experience! But that is not really an explanation. Since the change in kinetic energy is the same in each case, to accelerate the car in 1 second, your body would have to generate energy at a rate 60 times greater than the rate required if you accelerated it in one minute. The key word is rate, or how much energy your body can produce per second. If you do the calculations, you will find that for the one-minute scenario, your body would have to produce about $1/90$ horsepower, which seems quite reasonable. On the other hand, if you try to accomplish the same acceleration in one second, you would need to generate $2/3$ horsepower. Are you two-thirds as powerful as a horse?

As another example, assume that you attend a class on the third floor of the engineering building. When you are on time, you take 2 minutes to climb to the third floor. On the other hand when you are late for class, you run up the three flights in 30 seconds.

(a) In which case do you do the most work (expend the most energy)?
(b) In which case do you generate the most power?

EXAMPLE 11-21

A 50-kilogram load is raised vertically a distance of 5 meters by an electric motor in 60 seconds. How much power in units of watts does the motor use, assuming no energy is lost in the process?

This problem was started in Example 11-19 above. The energy used by the system was found to be 2450 J, the analysis of which is not repeated here.

Step One: Convert to Base SI Units	
No Conversion Needed	
Step Two: Calculate	
Method	Steps
(1) Determine appropriate equation	$\text{Power} = \dfrac{\text{Energy}}{\text{Time}}$
(2) Insert known quantities	$\text{Power} = \dfrac{2450 \text{ J}}{60 \text{ s}}$
(3) Calculate, be reasonable	$\text{Power} = 41 \dfrac{\text{J}}{\text{s}} \left\| \dfrac{1 \text{ W}}{1 \frac{\text{J}}{\text{s}}} \right. = 41 \text{ W}$
Step Three: Convert from Base SI Units to Desired Units	
No Conversion Needed	

Note that since power = energy/time, energy = power * time. We pay the electric company for energy calculated this way as kilowatt-hours. If power is constant, we can obtain the total energy involved simply by multiplying the power by the length of time that power is applied. If power is *not* constant, we would usually use calculus to determine the total energy, but that solution is beyond the scope of this book.

11.10 EFFICIENCY

IMPORTANT CONCEPT

Efficiency is always less than 100%.

Efficiency (η, Greek letter eta) is a measure of how much of a quantity, typically energy or power, is lost in a process. In a perfect world, efficiency would always be 100%. All energy put into a process would be recovered and used to accomplish the desired task. We know that this can never happen, so *efficiency is always less than 100%*. If a machine operates at 75% efficiency, 25% of the energy is lost. This means you have to put in "extra" energy to complete the work.

The use of the terms "input" and "output" require some explanation. The **input** is the quantity of energy or power or whatever required by the mechanism from some source to operate and accomplish its task. The **output** is the amount of energy or power or whatever is actually applied to the task itself by the mechanism. Note that the rated power of a device, whether a light bulb, a motor, or an electric heater, refers to the input power—the power needed to operate the device—not the output power. In an ideal, 100% efficient system, the input and output would be equivalent. In an inefficient system (the real world), the input is equivalent to the sum of the output and the power or energy lost. This is perhaps best explained by way of examples.

REASONABLENESS RULES

The table below gives you an idea of orders of magnitude of power and energy as related to real-world objects and phenomena. All values are approximate and, in most cases, have been rounded to only one significant figure. Thus, if you actually do the calculations from power to energy, you will find discrepancies.

Unless you are already familiar with the standard metric prefixes, refer to the list on the inside back cover of the book.

A few things of possible interest:

- U.S. power consumption (all types) is one fifth of the total world power consumption.
- The Tsar Bomba generated 1.5% of the power of the sun, but only lasted 40 nanoseconds.
- Total human power consumption on the planet is about 0.01% (one ten-thousandth) of the total power received from the sun.

Power	"Device"	Energy per hour	Energy per year
10 fW	Minimum reception power for cell phone	40 pJ	300nJ
1 pW	Single human cell	4 nJ	30 mJ
10 mW	DVD laser	40 J	300 kJ
500 mW	Cell phone microprocessor	2 kJ	15 MJ
50 W	Modern 2GHz microprocessor	200 kJ	1.5 GJ
100 W	Human at rest	400 kJ	3 GJ
500 W	Human doing strenuous work	2 MJ	15 GJ
750 W	Power per meter bright sunshine	3 MJ	25 GJ
2 kW	Maximum human power for short period	(NA)	
20 kW	Average U.S. home	80 MJ	600 GJ
100 kW	Typical automobile	400 MJ	3 TJ
150 MW	Boeing 747 jet	500 GJ	5 PJ
1 GW	Large commercial nuclear reactor	4 TJ	30 PJ
20 GW	Three Gorges Hydroelectric Dam (China)	80 TJ	600 PJ
4 TW	US total power consumption	15 PJ	100 EJ
20 TW	Total human power consumption	80 PJ	600 EJ
100 TW	Average hurricane	400 PJ	(NA)
200 PW	Total power received on Earth from sun	1 ZJ	6 YJ
5 YW	Largest fusion bomb ever built (Russian Tsar Bomba)	(NA) Total yield 200 PJ	
400 YW	Total power of sun	10^6 YJ	10^9 YJ

IMPORTANT CONCEPTS

Efficiency (η) = Output/Input

Efficiency (η) = Output/(Output + Loss)

Input = Quantity required by mechanism to operate

Output = Quantity actually applied to task

Loss = Quantity wasted during the application

EXAMPLE 11-22

A standard incandescent light bulb has an efficiency of about 5%; thus, $\eta = 0.05$. An incandescent bulb works by heating a wire (the filament) inside the bulb to such a high temperature that it glows white. About 95% of the power delivered to an incandescent bulb is discarded as heat. Only 5% results in "light" energy.

Light energy output 5%

Energy input (100%)

Heat energy lost (95%)

If a 100 watt bulb is turned on for 15 minutes, how much energy is "lost" as heat during the 15-minute period?

Step One: Convert to Base SI Units	
No Conversion Needed	
Step Two: Calculate	
Method	**Steps**
(1) Determine appropriate equation	power = energy/time
(2) Insert known quantities	power = energy/15 min
For unknown quantities, repeat the process	
Method	**Steps**
(1) Determine appropriate equation	input power = output power/efficiency "lost" power = input power − output power
(2) Insert known quantities	100 W = output power/(0.05)
(3) Calculate, be reasonable	output power = 5 W "lost" power = 100 W − 5 W = 95 W
(2) Insert known quantities	95 W = energy/15 min
(3) Calculate, reasonable	$E \ = \ 95 \dfrac{J}{s} \left\vert \dfrac{15 \text{ min}}{} \right\vert \dfrac{60 \text{ s}}{1 \text{ min}}$ Energy lost = 85,500 J
Step Three: Convert from Base SI Units to Desired Units	
No Conversion Needed	

EXAMPLE 11-23

Over the past few decades, the efficiency of solar cells has risen from about 10% to the most recent technologies achieving about 40% conversion of solar energy to electricity. The losses are due to several factors, including reflectance and resistive losses, among others.

Assume you have an array of solar cells mounted on your roof with an efficiency of 28%. If the array is delivering 750 watts of electricity to your home, how much solar power is falling on the photoelectric cells?

NOTE

How much power is wasted by the array of solar cells? If the array received 2680 W and delivered 750 W, then the difference is 2680 − 750 = 1930 W; thus, 1930 watts are wasted.

Step One: Convert to Base SI Units	
No Conversion Needed	
Step Two: Calculate	
Method	Steps
(1) Determine appropriate equation	input power = output power/efficiency
(2) Insert known quantities	input power = 750 W/0.28
(3) Calculate, reasonable	input power = 2680 W
Step Three: Convert from Base SI Units to Desired Units	
No Conversion Needed	

EXAMPLE 11-24

If your microwave takes 2 minutes to heat your coffee in Example 11-20, how many watts of power does your microwave require, assuming it is 80% efficient? Remember that our answer was 59,370 joules, before we converted the final answer to units of BTU.

Step One: Convert to Base SI Units	
No Conversion Needed	
Step Two: Calculate	
Method	**Steps**
(1) Determine appropriate equation	input power = output power/efficiency
(2) Insert known quantities	input power = output power/0.8
For unknown quantities, repeat the process	
Method	**Steps**
(1) Determine appropriate equation	output power = energy/time
(2) Insert known quantities	output power = $\dfrac{59370 \text{ J}}{2 \text{ min}}\left\|\dfrac{1 \text{ min}}{60 \text{ s}}\right\|\dfrac{1 \text{ W}}{1 \frac{\text{J}}{\text{s}}}$
(3) Calculate, be reasonable	output power = 493 W
(2) Insert known quantities	input power = 493 W/0.8
(3) Calculate, be reasonable	input power = 615 W
Step Three: Convert from Base SI Units to Desired Units	
No Conversion Needed	

Comprehension Check 11-13

A motor with an output power of 100 watts is connected to a flywheel. How long, in units of hours, must the motor operate to transfer 300,000 joules to the flywheel, assuming the process is 100% efficient?

Comprehension Check 11-14

If a 50-kilogram load was raised 5 meters in 50 seconds, determine the minimum rated wattage of the motor needed to accomplish this, assuming the motor is 80% efficient.

EXAMPLE 11-25

A simple two-stage machine is shown in the diagram on the next page. Initially, an electric motor receives power from the power grid, accessed by being plugged into a standard electrical wall socket. The power received by the motor from the wall socket is the "input" power or the power the motor uses or requires (Point A).

The spinning drive shaft on the motor can then be used to power other devices, such as a hoist or a vacuum cleaner or a DVD drive; the power available from the spinning shaft is the "output" power of the motor (Point B). In the process of making the drive shaft spin, however, some of the input power is lost (Point C) because of both frictional and ohmic heating as well as other wasted forms such as sound.

Since some energy is being lost as heat or other unusable forms every second of operation, (remember, power is energy per time!) the output power *must* be less than the input power—the efficiency of the motor is less than 100%. The power available from the spinning shaft of the motor (Point B) is then used to operate some device: the hoist or vacuum cleaner or DVD drive (Point E). In other words, the output power of the motor is the input power to the device it drives. This device will have its own efficiency, thus wasting some of the power supplied to it by the motor (Point D).

Assume that a motor with an efficiency of 80% is used to power a vacuum cleaner that has an efficiency of 70%. If the input power to the motor (Point A) is one-half horsepower, what is the output power of the vacuum (Point E)? The output from the motor (Point B) is $P_B = \eta\,P_A = 0.8\,(0.5\text{ hp}) = 0.4$ hp.

This is the input to the vacuum, so the output from the vacuum (Point E) is

$$P_E = \eta P_B = 0.7\,(0.4\text{ hp}) = 0.28 \text{ hp}$$

What is the overall efficiency of this machine? The input power (Point A) is 0.5 hp, and the output power (Point E) is 0.28 hp; thus, the efficiency of this linked system is

$$\eta \;=\; P_A/P_E = 0.28\text{ hp}/0.5\text{ hp} = 0.56,\text{ or an efficiency of } 56\%.$$

> **NOTE**
>
> The overall efficiency of two linked devices is the product of the two efficiencies.

SO WHAT WAS THE ANSWER TO THE BOAT AND THE STONE QUESTION FROM THE INTRODUCTION?

This is a problem that can be analyzed without the need for any equations or formal mathematics. We will start with Archimedes' principle that states simply: "When a body is submerged in a fluid, the buoyancy force (which pushes the body upward) is equal to the weight of the fluid which the body displaces."

If an object is floating at equilibrium in a fluid, the buoyancy force is equal to the weight of the object. The upward buoyancy force and downward weight of the object (remember, weight is a force!) are the same so the net force on the object is zero. For example, when a 500 pound-mass boat is placed in a pool, its weight acts downward (with a value of 500 pounds-force if at sea level on Earth), and it sinks deeper and deeper, displacing more and more fluid, until it has displaced 500 pounds-mass of fluid. At that point, the buoyancy force is 500 pounds-force and this opposes the weight of the boat so that it rests at equilibrium.

If the body we place in a fluid weighs more than the fluid it displaces when completely submerged, then the body sinks, but it still experiences a buoyancy force equal to the weight of the fluid displaced. This is one reason astronauts train under water—the net downward force of themselves and of objects they are manipulating is greatly reduced although their mass is the same.

If we place the same 500 pound-mass boat in a pool, and put a 200 pound-mass person in it, the boat will sink in the fluid more deeply than before, until it displaces 700 pound-mass of fluid. At this new level of submergence, the upward and downward forces are equal and the new "system" of boat and person will rest at equilibrium. If we add a stone in the boat, the new system of the boat, the person and the stone will sink even deeper until it displaces a mass of fluid equal to the combined masses of the boat, the person, and the stone. In each of these situations the displaced fluid has to go somewhere, so the level of the liquid in the pool rises.

If we assume that the stone will sink when tossed overboard, it will displace a volume of liquid equal to the volume of the stone. Since the stone is denser than the liquid (has more mass per volume), it now displaces less liquid than it did when it was in the boat. Thus: *When the stone is placed in the pool, the water level will drop.*

In-Class Activities

ICA 11-1

Complete the following table, using the equation weight = (mass) (gravity) as needed.

	pound-mass	kilogram	newton	pound-force
abbreviation		[kg]		
(a)	*0.33*	0.15	*1.47*	*0.33*
(b)	10			
(c)			700	
(d)				25

ICA 11-2

Complete the following table, using the equation weight = (mass) (gravity) as needed.

	slug	grams	newton	pound-force
abbreviation		[g]		
(a)		130		
(b)	50			
(c)			50	
(d)				30

ICA 11-3

Complete the following table.

	Compound	SG	Density [lb_m/ft^3]	Density [g/cm^3]	Density [kg/m^3]
(a)	Ethyl Alcohol	1.025	*64.0*	*1.025*	*1,025*
(b)	Acetic Acid			1.049	
(c)	Octane				703
(d)	Tetrachloroethane		100		
(e)	Chloroform	1.489			

ICA 11-4

Complete the following table.

	Compound	SG	Density [lb_m/ft³]	Density [g/cm³]	Density [kg/m³]
(a)	Gallium	5.91			
(b)	Aluminum		168.5		
(c)	Zinc Oxide				5,600
(d)	Sugar, powdered			0.801	
(e)	Lead	11.4			

ICA 11-5

Complete the following table. Assume you have a cube composed of each material, with "Length" indicating the length of one side of the cube.

	Material	Mass [g]	Length [cm]	Volume [cm³]	Density [g/cm³]
(a)	Tungsten	302	2.5	*15.6*	*19.3*
(b)	Zinc			25	7.14
(c)	Copper	107			8.92
(d)	Lead	1413			11.35

ICA 11-6

Complete the following table. Assume you have a cylinder, composed of each material, with "Radius" and "Height" indicating the dimensions of the cylinder.

	Material	Mass [kg]	Radius [m]	Height [m]	Density [kg/m³]
(a)	Silicon	2,750	0.50	1.5	
(b)	Aluminum		1.25	0.75	2,700
(c)	Tin	10,000	0.25		7,290
(d)	Titanium	8,000		1.0	4,540

ICA 11-7

Complete the following table.

	Compound	Boiling Temperature [°F]	Boiling Temperature [°C]	Boiling Temperature [K]	Boiling Temperature [°R]
(a)	Acetic Acid	180			
(b)	Ethyl Alcohol				815
(c)	Octane		126		
(d)	Tetrachlorethane			419	

ICA 11-8

Complete the following table.

	Compound	Eutectic Temperature			
		[°F]	[°C]	[K]	[°R]
(a)	48% Sn, 52% In	244			
(b)	42% Sn, 58% Bi		138		
(c)	91% Sn, 9% Zn			472	
(d)	96.5% Sn, 3.5% Ag				890

ICA 11-9

(a) In the Spring of 2004, NASA discovered a new planet beyond Pluto named Sedna. In the news report about the discovery, the temperature of Sedna was reported as "never rising above −400 degrees." What are the units of the reported temperature?

(b) Is there a physical condition at which a Fahrenheit thermometer and Celsius thermometer will read the same numerical value? If so, what is this value?

(c) Is there a physical condition at which a Fahrenheit thermometer and a Kelvin thermometer will read the same numerical value? If so, what is this value?

ICA 11-10

(a) Determine the mass in units of grams of 0.025 moles of caffeine (formula: $C_8H_{10}N_4O_3$). The components are hydrogen (formula: H, amu = 1); carbon (formula: C, amu = 12); nitrogen (formula: N, amu = 14); and oxygen (formula: O, amu = 16).

(b) Determine the mass in units of grams of 0.15 moles of the common analgesic naproxen sodium (formula: $C_{14}H_{13}NaO_3$). The components are hydrogen (formula: H, amu = 1); carbon (formula: C, amu = 12); sodium (formula: Na, amu = 23); and oxygen (formula: O, amu = 16).

(c) Determine the amount in units of moles of 5 grams of another common analgesic acetaminophen (formula: $C_8H_9NO_2$). The components are hydrogen (formula: H, amu = 1); carbon (formula: C, amu = 12); nitrogen (formula: N, amu = 14); and oxygen (formula: O, amu = 16).

ICA 11-11

Complete the table.

	Atmosphere	Pascal	Inches of Mercury	Pound-force per Square Inch
abbreviation	[atm]			
(a)	0.030	3000	0.886	0.435
(b)			30	
(c)				50
(d)	2.5			

ICA 11-12

Complete the table.

	Atmosphere	Pascal	Inches of Mercury	Pound-force per Square Inch
abbreviation	[atm]			
(a)		500,000		
(b)				5
(c)	10			
(d)			25	

ICA 11-13

Complete the table, using the equation for total pressure.

	Fluid Density [kg/m^3]	Height of Fluid [ft]	Surface pressure [atm]	Total pressure [atm]
(a)	787	10	1	
(b)		50	1	2
(c)	1,263		3	5.4
(d)	1,665	35		3.25

ICA 11-14

Complete the table, using the equation for total pressure.

	Fluid Specific Gravity	Height of Fluid [m]	Surface Pressure [atm]	Total Pressure [psi]
(a)	0.787	15	1	
(b)		75	1	30
(c)	1.263		3	80
(d)	1.665	20		48

ICA 11-15

Complete the table, using the ideal gas law.

	Compound	Mass [g]	MW [g/mol]	Amount [mol]	Pressure [atm]	Volume [L]	Temperature [K]
(a)	Oxygen (O_2)	8	32		2.0		400
(b)	Nitrogen (N_2)	5.6	28			5	373
(c)	Chlorine (Cl_2)		71	0.5	2.5	5	
(d)	Xylene	265		2.5	2.5		273

ICA 11-16

Complete the table, using the ideal gas law.

	Compound	Mass [lb$_m$]	MW [g/mol]	Amount [mol]	Pressure [Pa]	Volume [gal]	Temperature [°C]
(a)	Acetylene (C$_2$H$_2$)	0.1	26		303,975		−23
(b)	Benzene (C$_6$H$_6$)	0.04	78			1.32	100
(c)	Napthalene	0.07		0.25	131,723	1.32	
(d)	Toluene	0.425		2.1	354,638		27

ICA 11-17

Complete the table for specific heat conversions.

	Compound	[cal/(g °C)]	[BTU/(lb$_m$ °F)]	[J/(kg K)]
(a)	Neon			1,030
(b)	Benzene		0.0406	
(c)	Mercury	0.03325		
(d)	Aluminum			0.897
(e)	Ethanol			2,440
(f)	Copper		0.0912	

ICA 11-18

Complete the table for thermal conductivity conversions.

	Compound	[W/(m °C)]	[BTU/(ft h °F)]	[cal/(cm min K)]
(a)	Zinc		122	
(b)	Silver	420		
(c)	Lead	35		
(d)	Sandstone		1.387	
(e)	Copper		232	
(f)	Glass	1.1		

ICA 11-19

Complete the table. This problem involves the power required to raise a mass a given distance in a given amount of time.

	Mass [kg]	Distance [m]	Energy [J]	Time [s]	Power [W]
(a)	250	11		200	
(b)	440	52			100
(c)	60		2940	60	
(d)		10		50	100

ICA 11-20

Complete the table. This problem involves the power required to raise a mass a given distance in a given amount of time.

	Mass [lb$_m$]	Distance [ft]	Energy [J]	Time [min]	Power [hp]
(a)	220	15		0.5	
(b)	1875	145			0.268
(c)	200		4,000	1	
(d)		35		2	0.134

ICA 11-21

(a) The specific heat of copper is 0.09 British thermal units per pound-mass degree Fahrenheit. Convert this value into units of joules per gram kelvin.
(b) The specific heat of helium is 5.24 joules per gram kelvin. Convert this value into units of British thermal units per pound-mass degree Fahrenheit.
(c) The thermal conductivity of a plastic is 0.325 British thermal units per foot hour degree Fahrenheit. Convert this value in units of watts per meter kelvin.
(d) The heat transfer coefficient of steel is 25 watts per square meter degree Celsius. Convert this value into units of calories per square centimeter second kelvin.

ICA 11-22

One problem with solar energy is that any given point on the planet is illuminated by the sun for only half of the time at best. It would be helpful, therefore, if there were a simple, affordable, and efficient means for storing any excess energy generated on sunny days for use during the night or on cloudy days.

    You are investigating the electrodes used in electrolysis cells as part of a three-stage process for solar energy collection and storage.

1. Convert sunlight to electricity with photovoltaic cells.
2. Use the electricity generated in an electrolysis cell to split water into its component elements, hydrogen and oxygen. The hydrogen can be stored indefinitely. The oxygen can simply be released into the atmosphere.
3. Use a fuel cell to recombine the stored hydrogen with oxygen from the atmosphere to generate electricity.

You have obtained an array of new high-efficiency, thin-film photovoltaic cells with an efficiency of 41%. The efficiency of fuel cells varies with the current demands placed on them, but the cells you have obtained yield an overall efficiency of 37% at the anticipated load.

Assume the total solar power on the solar cells is 2000 watts. You conduct four experiments, each with a different alloy of palladium, platinum, gold, copper, and/or silver for the electrodes in the electrolysis cell. The final output power from the fuel cell is measured for each case, and the results are tabulated below. Determine the efficiency of each electrolysis cell and complete the table.

Alloy	Output power (P_0) [W]	Electrolysis Cell Efficiency (η)
Alloy A	137	
Alloy B	201	
Alloy C	67	
Alloy D	177	

ICA 11-23

Materials

Bag of Cylinders	Scale	Calipers	Ruler

Procedure

For each cylinder, record the mass, length, and diameter in the table provided.

Analysis

- Calculate the volume and density for each cylinder, recording the results in the table provided.
- Rank the rods in order of increasing density, with the least dense rod first on the list.
- Using the density, determine the material of each rod.

Data Worksheet

Description	Measured Values			Calculated Values	
	Mass	Length	Diameter	Volume	Density
Units					
Rod #1					
Rod #2					
Rod #3					
Rod #4					
Rod #5					
Rod #6					
Rod #7					
Rod #8					
Rod #9					
Rod #10					

Rank Rods Increasing Density	Density	Material

ICA 11-24

Materials

25 mL graduated cylinder	Scale	Paper towels
Unknown liquids	Basket labeled "Wash"	Water bottle
Wastewater bucket		

Procedure

Record the following data in the table provided.

1. Weigh the empty graduated cylinder and record the value.
2. Pour 15 milliliters [mL] of water into the cylinder.
3. Weigh the cylinder with water and record the value.
4. Pour the water into the wastewater bucket.
5. Pour 15 milliliters [mL] of unknown liquid (UL) #1 into a cylinder.
6. Weigh the cylinder with UL #1 and record the value.
7. Pour the UL back into the original container.
8. Place the graduated cylinder in the "Wash" basket if more than one cylinder is available. If only one cylinder is being used, wash the cylinder with water so no trace of the UL is left in the cylinder.
9. Repeat Steps 5–9 with each unknown liquid provided.

Analysis

- Calculate the density and specific gravity for each fluid, recording the results in the table provided.
- Rank the fluids in order of increasing density, with the least dense fluid first on the list.
- Using the specific gravity, determine the type of liquid in each container.

Data Worksheet

Description	Measured			Calculated	
	Total Mass	Volume	Liquid Mass	Density	Specific Gravity
Units					
Empty Cylinder					
Water					
UL #1					
UL #2					
UL #3					
UL #4					
UL #5					

Rank Liquids Increasing Density	Density	Liquid

Recall that in Chapter 10, we discussed the concept of dimensions. A **dimension** is a measurable physical idea; it generally consists solely of a word description with no numbers. A **unit** allows us to quantify a dimension, to state a number describing how much of that dimension exists in a specific situation. Units are defined by convention and related to an accepted standard. Table 12-1 (which is the same as Table 10-1) shows the seven base units and their corresponding fundamental dimensions.

Table 12-1 Fundamental dimensions and base units

Dimension	Symbol	Unit	Symbol
Length	L	meter	m
Mass	M	kilogram	kg
Time	T	second	s
Temperature	Θ	kelvin	K
Amount of substance	N	mole	mol
Light intensity	J	candela	cd
Electric current	I	ampere	A

12.1 COMMON DIMENSIONLESS NUMBERS

A dimensionless number is not able to be described by any dimension, but rather by the lack of dimenions, shown as $M^0 L^0 T^0 Theta^0$

Sometimes, we form the ratio of two parameters, where each parameter has the same dimensions. Sometimes, we form a ratio with two groups of parameters, where each group has the same dimensions. The final result in both cases is dimensionless.

Pi (π): One example is the parameter π, used in the calculation of a circumference or area of a circle. The reason π is dimensionless is that it is actually defined as the ratio of the circumference of a circle to its diameter:

$$\pi \equiv \frac{C}{D} = \frac{\text{circumference}}{\text{diameter}} \{=\} \frac{\text{length}}{\text{length}} = \frac{L^1}{L^1} = L^0$$

The ratio of one length to another length yields a dimensionless ratio. We can see this in another way through reversing the process. For the circumference of a circle:

$$C = \pi D$$

and if dimensions are inserted,

$$[L^1] = \pi [L^1]$$

This equation is dimensionally correct only if π has no dimensions. The same result is obtained for the equation of the area of a circle.

$$A = \pi r^2$$

Inserting dimensions:

$$[L^2] = \pi \, [L^1] \, [L^1] = \pi \, [L^2]$$

Again, this equation is dimensionally correct only if π is dimensionless.

Specific Gravity (SG): The specific gravity is the ratio of the density of an object to the density of water.

$$\text{Specific Gravity} = \frac{\text{Density of the Object}}{\text{Density of Water}} = \frac{\text{mass/volume}}{\text{mass/volume}} \{=\} \frac{(M/L^3)}{(M/L^3)} = M^0 L^0$$

Mach Number (Ma): We often describe the speed at which an airplane or rocket travels in terms of the Mach number, named after Ernst Mach, an Austrian physicist. This number is the ratio of the speed of the plane compared with the speed of sound in air.

$$\text{Mach Number} = \frac{\text{Speed of the Object}}{\text{Speed of Sound in Air}} \{=\} \frac{(L/T)}{(L/T)} = L^0 T^0$$

Radian [rad]: The derived unit of a radian is defined as the angle at the center of a circle formed by an arc (S) equal in length to the radius (r) of that circle. In a complete circle there are two π radians. Since by definition a radian is a length (S) divided by a length (r), it is a dimensionless ratio. Note, however, that angles are often measured using the radian as a unit, even though it is dimensionless.

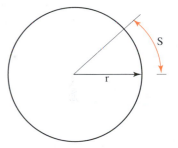

$$1 \text{ radian [rad]} = \frac{S}{r} \{=\} \frac{L}{L} = L^0$$

Often, when a ratio with two groups of parameters with the same dimensions occurs frequently in calculations, the ratio is given a special name. A dimensionless number is often named for a famous scientist followed by the word "Number," like the Mach Number.

When describing the speed, "Mach 2" indicates the object is traveling twice the speed of sound.

Table 12-2 Some common dimensionless parameters

Name	Phenomena Ratio	Symbol	Expression
Coefficient of friction	sideways force (F)/weight of object (w) [object static or kinetic (object sliding)]	μ_{st} & μ_k	F/w
Drag coefficient	drag force (F_d)/inertia force (ρ, density; v, speed; A body area)	C_d	$F_d/[1/2 \, \rho \, v^2 \, A]$
Mach number	object speed (v)/speed of sound (v_{sound})	Ma	v/v_{sound}
pi	circle circumference (C)/circle diameter (D)	π	C/D
Poisson's ratio	transverse contraction (ε_{trans})/longitudinal extension (ε_{long})	v	$\varepsilon_{trans}/\varepsilon_{long}$
Radian	arc length (S)/circle radius (r)	rad	S/r
Specific gravity	object density/density of water	SG	ρ/ρ_{H_2O}

We must remind ourselves that it is always essential to use the appropriate dimensions and units for every parameter. Suppose that we are interested in computing the sine of an angle. This can be expressed as a dimensionless number by forming the ratio of the length of the opposite side divided by the length of the hypotenuse of a right triangle:

$$\sin(x) = \frac{\text{Length Opposite Side}}{\text{Length Hypotenuse}} \; \{=\} \; \frac{L}{L} = L^0$$

In addition to the ratio of two lengths, you will know from one of your math classes that the sine can be also be expressed as an infinite series given by

$$\sin(x) = x - \frac{x^3}{3!} + \frac{x^5}{5!} - \frac{x^7}{7!} + \cdots$$

> ### ARGUMENTS LAW
>
> Any function that can be computed using a series must employ a *dimensionless* argument.
>
> This includes all the trigonometric functions, logarithms and e^x, where e is the base of natural logarithms.

Let us suppose that the argument x had the units of length, say, feet. The units in this series would then read as

$$\text{ft} - \frac{\text{ft}^3}{3!} + \frac{\text{ft}^5}{5!} - \frac{\text{ft}^7}{7!} + \cdots$$

We already know that we cannot add two terms unless they have the same units; recall the Plus law from Chapter 10. This is clearly not the case in the example above. The only way we can add these terms, all with different exponents, is if each term is dimensionless. Consequently, when we calculate $\sin(x)$, we see that the x must be dimensionless, which is why we use the unit of radians. This conclusion is true for any function that can be computed using a series form, leading to the Law of Arguments.

EXAMPLE 12-1

What are the dimensions of k in the following equation, where d is distance and t is time?

$$d = Be^{kt}$$

Since exponents must be dimensionless, the product of k and t must not contain any dimensions. The dimensions of time are {T}

$$k \, T^1 \; \{=\} \; M^0 L^0 T^0 \Theta^0$$

Solving for k yields:

$$k \; \{=\} \; T^{-1}$$

k is expressed in dimensions of inverse time or "per time."

12.2   DIMENSIONAL ANALYSIS

Dimensionless quantities are generated as a result of a process called **dimensional analysis**. As an example, suppose we want to study rectangles, assuming that we know nothing about rectangles. We are interested in the relationship between the area of a rectangle, the width of the rectangle, and the perimeter of the rectangle. We cut out a lot of paper rectangles and ask students in the class to measure the area, the perimeter, and the width.

Table 12-3 Rectangle measurements

Perimeter (P) [cm]	Area (A) [cm^2]	Width (W) [cm]
4.02	1.0	1.1
8.75	4.7	1.9
6	2.3	1.55
13.1	6.0	1.1
17.75	19	5.25
10.25	1.2	0.25
12.1	3.0	5.5
6	0.3	2.9
16.25	15.4	5.1
17	7.8	1.05

If we graph the area against the perimeter, we obtain the following plot.

From this, we see that the data are scattered. We would not have a great deal of confidence in drawing conclusions about how the area depended on the perimeter of the rectangle (or in trying to draw a line through the data). About the best we could do is make a statement such as "It seems that the larger the perimeter, the larger the area." However, close examination of the data table shows that as the perimeter increases from 8.75 to 10.25 centimeters and from 16.25 to 17 centimeters, the area actually decreases in each case. One reason for this problem is that our plot has omitted one important parameter: the width.

Analysis shows that one way in which to generalize plots of this type is to create dimensionless parameters from the problem variables. In this case, we have perimeter with dimension of length, width with the dimension of length, and area with the dimension of length squared. A little thought shows that we could use the ratio of P/W (or W/P) instead of just P on the abscissa. The ratio W/P has the dimensions of length/ length so it is dimensionless. It doesn't matter whether this is miles/miles, or centimeters/ centimeters, the ratio is dimensionless. Similary, we could write $A/(W^2)$, and this would also be dimensionless.

These ratios are plotted and shown below. The scatter of the first plot disappears and all the data appear along a single line.

Question: If a rectangle has a perimeter of 20 feet and a width of 2 foot, what is the area?

Step 1: P/W = (20 ft)/(2 ft) = 10 (with no units)

Step 2: From the chart, at a P/W value of 10 we read a value from the line of $A/(W^2)$ = 3.5

Step 3: Calculate A from this as A = 3.5 * (2 ft * 2 ft) = 14 ft^2

Some of you may be thinking that we made this problem unnecessarily difficult. After all, anyone who manages to get to college knows that the "sensible" measurements to make are length, width, and area. However, many phenomena are far more complicated than simple rectangles, and it is often not at all obvious what parameters should be measured to characterize the behavior of the system we are studying. In situations of this type, dimensionless analysis can become a powerful tool to help us understand *which* parameters affect the behavior of the system and *how* they affect it. With this in mind, let's look at a slightly more complicated example.

EXAMPLE 12-2

A not-so-famous scientist, Dr. Triticale, decided to apply his scientific skills to cooking. He had always been fascinated with the process of cooking pancakes, so it seemed reasonable that he start there. He wanted to learn how to flip the flapjacks in a graceful arc in the air and then catch them. Well, he spent long summer days pondering this process until he finally was able to produce a list of the parameters that he felt were important. He kept asking himself, "If I change this parameter, will the trajectory of the pancake change?" If he could answer "Yes!" or even "Probably," he then considered the parameter as important enough to include on his list. As he saw it, these parameters were:

Speed of the frying pan, U	Mass of the flapjack, m
Height of the flip, H	Gravity (it pulls the flapjack back down), g

He then wrote this dependency in equation form as H = f:(U, m, g).

Dr. Triticale realized that while he felt that gravity was important, it would not be easy to change the value of gravity in his tests (he could have gone to a high mountain or the moon, but this was too hard). His plan was to do many tests (and consequently eat many pancakes). He would make many measurements for many different flipping speeds and pancake masses, and try to fit a curve to the data.

Based on his work with conversion factors and his knowledge of the Per law, he reasoned that it is acceptable to multiply parameters with different dimensions. It is also fine to raise a parameter (and its associated units) to a power. Based on his understanding of the Plus law, he knew it is not acceptable to add parameters with different dimensions. If you need to refresh yourself on the Per and Plus laws of dimensions, refer to Chapter 10.

Using this information, he decided that it would be permissible to try to "fit" the dependence of pancake flipping to the important parameters raised to different powers and multiplied together. This would create a single term like $k_1 U^{a_1} m^{b_1} g^{c_1}$.

He also knew that if this term were made to have the same dimensions as H, it just might be a legitimate expression. In fact, if this were the case, he could use many terms, each of which had the dimensions of H and add them all together. While this might not be a valid equation, at least it would satisfy the Per and Plus laws, and with many terms he would have a good chance of his equation fitting the data. So, he boldly decided to try the following series:

$$H = k_1 U^{a_1} m^{b_1} g^{c_1} + k_2 U^{a_2} m^{b_2} g^{c_2} + k_3 U^{a_3} m^{b_3} g^{c_3} + \cdots$$

He needed to determine the values of the dimensionless k constants as well as all of the exponents. He knew that all the terms on the right-hand side must have the same dimensions, or they couldn't be added together. He also knew that the dimensions on the left and right sides must match.

With this, he then realized that he could examine the dimensions of any term on the right-hand side since each had to be the same dimensionally. He did this by comparing a typical right-hand term with the left-hand side of the equation, or

$$H = k U^a m^b g^c$$

The next step was to select the proper values of a, b, and c so that the dimensions of the right-hand side would match those on the left side. To do this, he substituted the dimensions of each parameter:

$$L^1 M^0 T^0 = (LT^{-1})^a (M)^b (LT^{-2})^c = (L)^{a+c} (M)^b (T)^{-a-2c}$$

For this to be dimensionally correct, the exponents for L, M, and T on the right and left would have to match, or

$$L: 1 = a + c \qquad M: 0 = b \qquad T: 0 = -a - 2c$$

This yields

$$a = 2 \qquad b = 0 \qquad c = -1$$

From this, Dr. Triticale settled on a typical term as

$$k\, U^2\, m^0\, g^{-1}$$

Finally, he wrote the "curve fitting" equation (with a whole series of terms) as

$$H = \sum_{i=1}^{\infty} k_1\, U^2\, g^{-1} + k_2\, U^2\, g^{-1} + k_3\, U^2\, g^{-1} + \cdots = \frac{U^2}{g} \sum_{i=1}^{\infty} k_i = (K)\left(\frac{U^2}{g}\right)$$

Now, armed with this expression, he was sure that he could flip flapjacks with the best, although he knew that he would have to conduct many experiments to make sure the equation was valid (and to determine the value of K). What he didn't realize was that he had just performed a procedure called "dimensional analysis."

12.3   RAYLEIGH'S METHOD

In this section we follow a method devised by Lord Rayleigh, John William Strutt, the third Baron Rayleigh (1842–1919). Three detailed examples illustrate his approach to dimensionless analysis:

- Example 12-3, in which we analyze factors affecting the distance traveled by an accelerating object.
- Example 12-4, in which we determine the most famous named dimensionless number, Reynolds number.
- Example 12-5, in which we simplify one use of Rayleigh's method.

No matter the problem, the way we solve it stays the same:

Step 1: Write each variable and raise each to an unknown exponent (use *all* the variables, even the dependent variable). Order and choice of exponent do not matter.

Step 2: Substitute dimensions of the variables into Step 1. Be sure to raise each dimension to the proper exponent groups from Step 3 from Step 1.

Step 3: Group by dimension.

Step 4: Exponents on each dimension must equal zero for dimensionless numbers, so form a set of equations by setting the exponent groups from Step 3 for each dimension equal to zero.

Step 5: Solve the simultaneous equations (as best as you can).
     Hint: # of unknowns − # of equations = # of groups you will have!

Step 6: Substitute results of Step 5 back into Step 1 exponents

Step 7: Group variables by exponent. These resulting groups are your dimensionless numbers.

Step 8: Be sure to *check* it out!! Are *all* of the ratios really dimensionless?
     Hint: If the resulting groups are *not* dimensionless, you most likely goofed in either Step 2 or Step 5!

Rayleigh's analysis is quite similar to the Buckingham Pi method, another method to determine dimensionless groups. Rayleigh's method is however, a bit more direct and often seems less "mysterious" to those who are new to dimensional analysis. Both methods use a general form with multiplied and exponentiated variables. Any inspection of physics, engineering and mathematical texts reveal many examples of this from of equation governing a myriad of behaviors."

EXAMPLE 12-3

To develop an understanding of how initial velocity, acceleration, and time all affect the distance traveled by an accelerating object, we conduct some experiments and then analyze the resulting data. We asked a student to conduct a series of tests for us. She observed 25 different moving bodies with a wide range of initial speeds and different accelerations. For each, she measured the distance the bodies traveled for some prescribed time interval. Results are given in Table 12-4.

Table 12-4 Position of a body as a function of initial velocity, acceleration, and time

Test #	Initial Velocity (v_o) [m/s]	Acceleration (a) [m/s^2]	Time (t) [s]	Distance Traveled (d) [m]
1	3	1	6	36
2	3	2	6	54
3	1.5	5	6	99
4	5	4	6	44
5	5	3	8	136
6	5	5	2	20
7	10	1	9	131
8	14	2	11	275
9	20	3	4	104
10	10	2	4	56
11	10	4	3	48
12	10	6	2	32
13	5	2	2	14
14	8	2	10	180
15	12	2	4	64
16	6	1	4	32
17	2	2	9	99
18	3	3	12	252
19	6	4	6	108
20	15	5	2.4	50
21	4	7	7.2	210
22	2	2	8	80
23	9	8	6.2	210
24	6.7	2	1.7	14
25	3.1	2	10	131

In addition, we would like to use this data set to help make predictions of the distance traveled by other bodies under different conditions. For example, we might want to answer the following question:

- What is the acceleration needed to travel 4800 meters in 200 seconds, if the initial velocity is 8 meters per second?

There are several independent variables (initial velocity, acceleration, and time), so it is not obvious what to plot. We can write the dependency as

$$d = f:(v_o, a, t)$$

We anticipate that it is difficult to draw conclusions regarding the interdependence of all of these variables. Realizing this, we plot distance against time without worrying about the initial velocity and the acceleration.

It seems that, in general, the longer one travels, the farther one goes. Upon closer inspection, however, it is obvious that this is not always the case. For example, for a travel time of 6 seconds, the distance traveled varies from about 25 meters to 210 meters. Since not enough tests were conducted with systematic variation of the initial velocity or acceleration, it is not possible to do much better than this. We certainly have no hope of answering the questions above with any confidence. In addition, since the values are so scattered, we realize that we have no good way to determine if any of our measurements were "bad."

With this disheartening conclusion, we perform a dimensional analysis in an attempt to place the parameters into fewer groups. The immediate problem we face is how to combine the parameters to give dimensionless groups. For the rectangles, this was relatively easy to do by inspection. The parameters in this new problem are slightly more complicated, and although we might find suitable groups by inspection, as the problems become a lot harder (and they will), we need some sort of standard method, or algorithm, to define suitable dimensionless groups.

This technique is relatively simple and comprises the eight steps listed below.

Step 1: Raise each variable to a different unknown power, using symbols for the power variables that do not already appear in the problem, and then multiply all of these individual terms together. For example, since the current problem has both an **a** and **d** used as variables, we should not use a and d for the powers; hence we choose the letters p through s for the exponents. The order in which we list the variables and assign exponents is completely random.

This gives us the term: $d^p v_o^q a^r t^s$

Step 2: Substitute the correct dimensions for each variable.

$$\{L\}^p \left\{\frac{L}{T}\right\}^q \left\{\frac{L}{T^2}\right\}^r \{T\}^s$$

Step 3: Expand the expression to have each dimension as a base raised to some power.

$$(L)^{p+q+r} (T)^{-q-2r+s}$$

Step 4: For the expression in Step 3 to be dimensionless, each exponent must equal zero, or

$$p + q + r = 0 \quad \text{and} \quad -q - 2r + s = 0$$

Step 5: Solve for the exponents. In this case, we have two equations and four unknowns, so it is not possible to solve for *all* the unknowns in terms of an actual number. We must be satisfied with finding two of the exponents in terms of the other two. This might seem problematic, but we will find that this not only is not a difficulty, but also is quite common in this type of analysis. *Note that there are many ways to do this and all will lead to two dimensionless ratios. If you don't like the plot you get from doing it one way, try solving for different exponents and see if that provides a better plot. All will be correct, but some are easier to use than others.* Although several procedures will lead to solutions, in general, we will solve for one of the variables, and then substitute into the other equations to reduce the number of variables. If a variable appears in all or most of the equations, that may be a good one to begin with.

In our example, we solve the second equation for q.

$$q = -2r + s$$

Substituting for q into the other equation gives

$$p - 2r + s + r = 0$$

thus,

$$p = r - s$$

At this point, we have defined p and q in terms of the other two variables, r and s.

Step 6: Substitute into the original expression.

$$d^{r-s} v_o^{-2r+s} a^r t^s$$

Note that all of the exponents are now expressed in terms of only two variables, r and s.

Step 7: Simplify by collecting all terms associated with the remaining exponential variables (r and s in this case).

$$\left(\frac{d\,a}{v_o^2}\right)^r \left(\frac{t\,v_o}{d}\right)^s$$

Step 8: The simplification in Step 7 gives the dimensionless ratios we are looking for. Dropping the exponents assumed in Step 1 gives the following groups:

$$\left(\frac{d\,a}{v_o^2}\right) \text{ and } \left(\frac{t\,v_o}{d}\right)$$

We need to double-check to make sure that both of these groups are dimensionless. Before plotting them, we make one additional observation. If the denominator of a ratio goes to zero, the fraction will then be undefined (and we can't plot it). In our example, a body may easily have a zero initial speed, so v_o would be zero. Since these ratios are still dimensionless if we invert them, we may want to use the inverse of the first ratio to eliminate this potential problem.

Compute these two parameters from the original data and plot to obtain the figure below.

NOTE

There seems to be one "bad" data point. Would you have been able to pick out this point from the original data *or* from the dimensional plot above?

Now the scatter from the original dimensional plot is gone, and all but one data point seem to lie on a smooth curve. We can use this plot to determine the relationship between distance traveled, acceleration, time, and initial velocity. Let's see how to do this for the question we posed earlier.

What acceleration is required to go 4800 meters in 200 seconds if the initial car velocity is 8 meters per second?

On the x-axis: [(8 m/s) (200 s)]/(4800 m) = 0.33
Reading from the graph: $v_0^2/(d\,a) = 0.075$
Solving for acceleration: a = (8 m/s)^2/(0.075) (4800 m) = 0.18 m/s^2

Several important conclusions can be drawn from this exercise.

- Dimensionless parameters often allow us to present data in an easily interpretable fashion when the "raw" data have no recognizable pattern.
- When we use the exponent approach to find dimensionless parameters, we need to remember that the exponents could be either positive or negative. Thus, in the case of the rectangle, P/W is just as good as W/P. You can always try both to see which gives the best-looking results. The choice is yours, and sometimes depends on whether one of the variables goes to zero; since you can't divide by zero, that variable should not be in the denominator, if possible.
- If we "collapse" the data by using dimensionless parameters so that a single curve can fit through the resulting points, *bad data points will usually become obvious.*

■ Finally, this approach can reduce literally thousands of different measurements into one simple curve. In the case of the car, the single line we obtained will work for all possible combination of times, initial velocities, accelerations, and distance. If we used dimensional plots, we would need many, many plots, with many lines on each plot. This would require an entire book of plots rather than the single plot with a single line that we obtained above by dimensionless ratios.

EXAMPLE 12-4

To classify the smoothness of a flowing fluid, Osborne Reynolds (1842–1912, British, ME) developed the now famous dimensionless quantity of **Reynolds number**. His theory stated that the smoothness or roughness (a lot of eddies or swirling) of a fluid depended upon:

How fast the fluid was moving (velocity)	$v \, [=] \, m/s$
The density of the fluid	$\rho \, [=] \, kg/m^3$
The diameter of the pipe	$D \, [=] \, m$
How hard it was to move the fluid (viscosity)	$\mu \, [=] \, g/(cm \, s)$

Reynolds knew the smoothness depended upon these quantities:

$$\text{Smoothness of the flow} = f:(v, \rho, D, \mu)$$

But how did they depend on one another? We could write the four variables above as

$$v^a \, \rho^b \, D^c \, \mu^d$$

and if this was dimensionless, it would appear as $M^0 \, L^0 \, T^0 \, \Theta^0$.

To make this grouping dimensionless, first we substitute in the dimensions of the four variables to obtain:

$$\left(\frac{L}{T}\right)^a \left(\frac{M}{L^3}\right)^b L^c \left(\frac{M}{LT}\right)^d = M^{b+d} \, L^{a-3b+c-d} \, T^{-a-d}$$

Note that we generally have dimensions of mass, length, time, and temperature, or M, L, T, and Θ; in this problem there is no temperature dimension. If this is to be dimensionless, then the exponents on all of the dimensions must equal zero, therefore:

M:	$b + d = 0$
L:	$a - 3b + c - d = 0$
T:	$-a - d = 0$

This gives three equations in four unknowns, so we will have to solve for three of the variables in terms of the fourth. In this example, we solve for the three unknowns a, b, c in terms of d:

M:	$b = -d$
T:	$a = -d$
L:	$c = -a + 3b + d = d - 3d + d = -d$

Substituting these back into the original parameters gives:

$$v^{-d} \quad \rho^{-d} \quad D^{-d} \quad \mu^d$$

We see that there is one dimensionless group since all the parameters have an exponent of d. We can write

$$\frac{\mu}{v\,\rho D} \{=\} \frac{\frac{M}{LT}}{\frac{L}{T}\frac{M}{L^3}L}$$

Since the variables of diameter and velocity can approach zero, the Reynolds number is commonly written

$$Re = \frac{\rho\,D\,v}{\mu}$$

> The Reynolds number is used to describe fluid flow.
>
> Re < 2,000 = laminar
>
> 2,000 < Re < 10,000 = transitional
>
> Re > 10,000 = turbulent

If the Reynolds number has a value less than 2,000, the flow is described as **laminar**, meaning it moves slowly and gently with no mixing or churning. If the Reynolds number has a value greater than 10,000, the flow is described as **turbulent**, meaning it moves quickly with much mixing and churning (lots of eddies) occurring. The region in between (2,000–10,000) is called the **transition region**.

EXAMPLE 12-5

Suppose we conduct an experiment with a ball that we throw from the top of a tall tower of height H. We throw it directly downward with some initial velocity v, and then measure the elapsed time t until it hits the ground. We vary the initial height and the initial velocity. The variables of interest in this problem are H, v, and t. A little thought leads us to include g since it is the force of gravity that causes the ball to fall in the first place.

Using Rayleigh's method, find a set of dimensionless ratios that can be used to correlate our data.

Step 1: Write each variable and raise each to an unknown exponent (use *all* the variables, even the dependent variable).

$$t^a \quad H^b \quad v^c \quad g^d$$

Step 2: Substitute dimensions of the variables into Step 1. Be sure to raise each dimension to the proper exponent from Step 1.

$$t^a \{=\} T^a \quad H^b \{=\} L^b \quad v^c \{=\} L^c T^{-c} \quad g^d \{=\} L^d T^{-2d}$$

Step 3: Group by dimension.

$$L^{b+c+d} \quad T^{a-c-2d}$$

Step 4: Exponents on each dimension must equal zero for dimensionless numbers! Form a set of equations by setting the exponents for each dimension equal to zero.

$$b + c + d = 0 \qquad a - c - 2d = 0$$

Step 5: Solve the simultaneous equations (as best as you can).

$$b = -c - d \qquad a = c + 2d$$

Step 6: Substitute results of Step 5 back into Step 1 exponents.

$$t^{c+2d} \quad H^{-c-d} \quad v^c \quad g^d$$

Step 7: Group variables by exponent. These resulting groups are your dimensionless numbers.

$$\left[\frac{v\,t}{H}\right]^{c} \qquad \left[\frac{g\,t^{2}}{H}\right]^{d}$$

Step 8: Be sure to *check* it out!! Are *all* of the ratios really dimensionless?

$$\left[\frac{v\,t}{H}\right]\{=\}\frac{L}{T}\left|\frac{T}{}\right|\frac{}{L} = \text{YES!} \qquad \left[\frac{g\,t^{2}}{H}\right]\{=\}\frac{L}{T^{2}}\left|\frac{(T)^{2}}{}\right|\frac{}{L} = \text{YES!}$$

> We are not sure that the results of this technique are *physically correct*, only that they are *dimensionally correct*.

Always remember that we initiate this procedure simply by providing a list of parameters we think are important to the situation at hand. If we use the wrong parameters, our final result will likely be incorrect.

From the preceding statement, we should understand that unless we consider a parameter important and place it in the initial equation (infinite series), then it cannot possibly appear in the final result. Thus, if we omit an important parameter, our final result will not be physically correct, even if it is dimensionally correct. Consequently, if we select an improper parameter, then when tests are conducted, we will discover that it was not important to the problem and we can drop it from further consideration. *We cannot decide whether any variable is important until we conduct some experiments.* This leads us to one final rule:

> At the beginning of the analysis, when in doubt about the importance of a parameter, put it in the list of important parameters.

> If we "know" a parameter is in fact important "for sure" and *must* appear in the final result, then it can appear only if the other important parameters among themselves contain all the dimensions contained in the parameter.

Consequently, if we are *sure* that a parameter is important, then we know that it should *not* drop from the analysis. If it drops out, then the only way it can be retained is if at least one other parameter contains the missing dimension. In this case, we need to ask ourselves what other parameters might be important, add them to our list, and rework the analysis. This practice can be of enormous help in analyzing an unfamiliar problem.

Dimensional analysis helps us organize data by allowing us to plot one dimensionless parameter against an other (which in this case gives only one line on a single plot). This is a powerful result, and reduces a problem of initial parameters to one containing only two. This discussion leads to the **Problem Simplification rule**.

PROBLEM SIMPLIFICATION RULE

By performing a dimensional analysis of the parameters in a problem, we can generally find dimensionless groupings that will allow us to effectively reduce the number of parameters, thus facilitating the presentation of interdependencies and often greatly simplifying the problem.

When we complete our analysis, we must always check the results against experiments. This is the only way to ensure that our analysis is correct.

In-Class Activities

ICA 12-1

Complete the following table.

	Quantity	SI Units	Dimensions				
			M	L	T	Θ	N
(a)	Acoustic impedance	(Pa s)/m	*1*	*−2*	*−1*	*0*	*0*
(b)	Fuel consumption	kg/(kW h)					
(c)	Heat transfer coefficient	W/(m² K)					
(d)	Ideal gas constant	(Pa L)/(mol K)					
(e)	Latent heat	BTU/lb$_m$					
(f)	Mass density	kg/m³					
(g)	Molar heat capacity	cal/(mol °C)					
(h)	Rate of drying	lb$_m$/(ft² h)					
(i)	Specific heat	J/(g °C)					
(j)	Specific weight	N/m³					
(k)	Surface tension	J/m²					
(l)	Thermal conductivity	cal/(cm s °C)					
(m)	Thermal resistance	(K m²)/W					

ICA 12-2

Calculate the numerical value of each of the dimensionless parameters listed in the table. Be sure to check that the ratio is actually dimensionless after you insert the values.

	Situation	Name	Expression	Value
(a)	Water in a pipe	Reynolds number, Re	$\dfrac{\rho\, v_W D_{pipe}}{\mu}$	
(b)	Hot water	Prandtl number, Pr	$\dfrac{\mu\, C_p}{k}$	
(c)	Air over a flat plate	Nusselt number, Nu	$\dfrac{h\, L}{k}$	
(d)	Sphere in air	Drag coefficient, Cd	$\dfrac{F}{\frac{1}{2}\rho\, v_a^2\, A}$	
(e)	Water: Effect of surface tension	Weber number, We	$\dfrac{\rho\, v_w^2\, d}{\sigma}$	
(f)	Water in a river	Froude number, Fr	$\dfrac{v_W}{\sqrt{gH}}$	
(g)	Wind making a wire "sing"	Strouhal number, St	$\dfrac{\omega\, D_{wire}}{v_a}$	

Properties and definitions for this problem:

Property	Symbol	Units	Air	Water
Density	ρ	slugs/ft^3	0.002378	1.94
Dynamic viscosity	μ	kg/(m s)	2×10^{-5}	4×10^{-4}
Heat transfer coefficient	h	W/(m^2 °C)	20	—
Thermal conductivity	k	W/(m K)	0.025	0.7
Surface tension	σ	dynes/cm	—	70
Specific heat	C_p	cal/(g °C)	0.24	1

Property	Symbol	Units	Value
Plate length	L	ft	2
Silhouette area of object	A	in^2	120
Water depth	H	m	3
Water speed	v_w	cm/s	210
Air speed	v_a	mph	60
Pipe diameter	D_{pipe}	in	6
Drag force on sphere	F	N	30
Depth of water film	d	cm	3
Oscillation frequency	ω	Hz (or cycles/s)	140
Wire diameter	D_{wire}	mm	20

ICA 12-3

A fluid with a specific gravity of 0.91 and a viscosity of 0.38 pascal-seconds is pumped through a 25-millimeter diameter smooth pipe at an average velocity of 2.6 meters per second.

(a) Determine the Reynolds number in the pipe for the system.
(b) Is the flow laminar, transitional, or turbulent?

ICA 12-4

Brine, with a density of 1.25 grams per cubic centimeter and a viscosity of 0.015 grams per centimeter second is pumped through a 5-centimeter radius steel pipe at an average velocity of 15 centimeters per second.

(a) Determine the Reynolds number in the pipe for the system.
(b) Is the flow laminar, transitional, or turbulent?

ICA 12-5

The Euler number is a function of the pressure drop, velocity, and density. Determine the form of the Euler number with Rayleigh's method.

Pressure drop	ΔP	pascal
Density	ρ	grams per cubic centimeter
Velocity	v	meters per second

ICA 12-6

When a simple turbine is used for mixing, the following variables are involved:

Power requirement	P	watt
Shaft speed	N	hertz
Blade diameter	D	meters
Blade width	W	meters
Liquid density	ρ	kilograms per meter cubed

Determine a set of dimensionless groups for the turbine, using Rayleigh's method.

ICA 12-7

We assume that the total storm water runoff (R, given in volume units) from a plot of land depends on the length of time that it rains (t), the area of the land (A), and the rainfall rate (r, given in inches per hour).

Rainfall Rate (r) [in/h]	Land Area (A) [acres]	Rainfall Duration (t) [h]	Measured Runoff (R) [ft³]
0.50	2	3.0	49
0.30	14	2.5	172
0.78	87	4.1	4563
0.15	100	2.2	541
0.90	265	0.4	1408
0.83	32	7.6	3310
1.40	18	1.8	744
0.22	6	4.7	102
0.67	26	3.1	886
0.48	62	4.9	2392

(a) Using the data, construct a plot of the runoff versus the time that it rains. You should see that this plot is of little help in understanding the relationships between the various parameters.
(b) Using this plot, estimate the total run off from 200 acres if rain falls for 3 hours at a rate of 1.2 inches per hour.
(c) Using the variables in the table, complete a dimensional analysis to help you plot the data. You should obtain two dimensionless ratios. Plot these, with the ratio containing the total runoff on the ordinate. This plot should collapse the values to a single line. Draw a smooth curve through the values.
(d) Use this line to answer question (b) again.

ICA 12-8

We are interested in analyzing the velocity of a wave in water. By drawing a sketch of the wave, and labeling it, we decide that the velocity depends on the wave length (λ), the depth of the water (H), the density of the water (ρ), and the effect of gravity (g). We have measured wave speeds in many situations; the data are given below.

(a) Construct a plot of wave velocity versus either wave length or water depth. You will see substantial scatter.

(b) Using your plot, estimate the wave velocity for a wave length of 25 meters in water that is 25 meters deep.

Water Depth (H) [m]	Wave Length (λ) [m]	Velocity (v) [m/s]
1.0	10.0	7.4
1.0	20.0	7.7
2.0	30.0	10.8
9.0	40.0	18.7
0.2	13.0	3.5
4.0	24.0	13.6
18.0	20.0	14.0
0.3	2.6	4.0
33.0	30.0	17.1
5.0	15.0	11.9

(c) Perform a dimensional analysis on the parameters (v, ρ, H, g, and λ). After calculating the new dimensionless ratio values, make a dimensionless plot.

(d) Recalculate the answer to question (b).

1. Insert the appropriate mathematical symbol in each comparison below ($>, <, =$).

(a) 10 centimeter	1 meter
(b) 1 mile	1000 meters
(c) 1 mile	1000 feet
(d) 50 °C	100 °F
(e) 75 pounds-mass	75 kilograms
(f) 1 gallon	1 cubic foot
(g) 1 joule	1 foot pounds-force
(h) 10 horsepower	10 kilowatts
(i) 700 millimeters of mercury	15 pounds-force per square inch
(j) 1000 kilograms per cubic meter	1 gram per cubic centimeter
(k) 50 miles per hour	50 kilometers per hour

2. Convert the following:

   (a) 2 feet to millimeters
   (b) 4000 millimeters to feet
   (c) 600 decimeters to feet
   (d) 27 kilometers to feet
   (e) 27 kilometers to miles
   (f) 30 miles per hour to miles per day
   (g) 4 feet per second to miles per day

3. A basketball is approximately 40 centimeters in diameter. Find the following:

   (a) Surface area in units of square inches
   (b) Surface area in units of square centimeters
   (c) Surface area in units of square feet
   (d) Volume in units of cubic inches
   (e) Volume in units of cubic centimeters
   (f) Volume in units of cubic feet

4. How large a surface area in units of square feet will 1 gallon of paint cover if we apply a coat that is:

   (a) 0.1 millimeter thick
   (b) 0.1 centimeter thick
   (c) 0.1 inches thick

5. An unmanned X-43A scramjet test vehicle has achieved a maximum speed of Mach number $= 9.68$ in a test flight over the Pacific Ocean. Mach number is defined as the speed of an object divided by the speed of sound. Assuming the speed of sound is 343 meters per second, what is this record speed expressed in units of miles per hour?

6. If hair grows at a rate of 1.5 inches per month, how fast does hair grow in units of millimeters per day? Assume a month has 30 days.

7. A common unit of volume for large water bodies is acre-foot, which is the product of the area of 1 acre multiplied by a depth of 1 foot. If a lake contains 43 million cubic feet of water, what is the volume of the lake expressed in units of acre-foot?

8. The top speed of a Boeing-747 is 290 meters per second. What is the speed of the airplane in units of:

   (a) miles per hour
   (b) feet per hour
   (c) centimeters per minute

9. If a liquid evaporates at a rate of 50 kilograms per minute, what is this evaporation rate in units of:

   (a) pounds-mass per second
   (b) slugs per hour
   (c) grams per second

10. If a pump moves water at 70 gallons per minute, what is the volumetric flow rate in units of:

    (a) cubic inches per second
    (b) cubic feet per hour
    (c) liters per day

11. The distance travelled in a car depends on several parameters:

    - The initial position from the starting point: x_0
    - The initial velocity: v_i
    - The rate of acceleration: a
    - The time of travel: t

    The equation that governs the location of the car is $x = x_0 + v_i t + \frac{1}{2} a t^2$.

    If a car has a head start of 0.3 miles from the starting point, has an initial speed of 50 miles per hour, and accelerates at a constant rate of 10 feet per second squared, how far is it from the starting point after 1.5 hours? Express your answer in units of:

    (a) feet
    (b) meters
    (c) yards
    (d) miles

12. In an attempt to convert the United States to the metric system, a proposal is introduced to force all car manufacturers to report engine power in units of kilowatts. Convert a 250 horsepower motor rating into the metric unit kilowatts.

13. The oxgang is unit of area equal to 20 acres. Express an area of 12 oxgangs in units of square meters.

14. We know a speed boat can travel at 30 knots. How long (in minutes) will it take to cross the Chesapeake Bay at a place where the bay is 24 miles across? 1 knot = 1 nautical mile per hour; 1 nautical mile = 6076 feet.

15. In an effort to modernize the United States interstate system, the Department of Transportation proposes to change speed limits from miles per hour to "flashes." A flash is equal to 10 feet per second. On a car speedometer, what will the new range be in units of "flashes" if the old scale was set from 0 to 120 miles per hour?

16. In a thunderstorm we see the lightning flash before we hear the thunder. By measuring the time between seeing the flash and hearing the thunder, we can determine how far away the lightning was. Determine a number N so that if the number of seconds (t) between flash and thunder is divided by N, the result is the distance (d) in miles. Be sure to specify the units of N. The speed of sound in air is 343 m/s. Assume the travel time of light is negligible.

$$d = t/N$$

17. If a body of mass (m) is moving at a given velocity (v), which has the largest kinetic energy? Complete the following table. Assume the bodies are not rotating.

Body	Mass (m)	Velocity (v)	Kinetic Energy (KE) [J]
1	100 grams	30 feet per second	
2	0.5 pounds-mass	10 meters per second	
3	2 slugs	3 centimeters per second	

18. Old Mississippi River paddle wheelers routinely measured the river depths to avoid running aground. They used the unit "fathoms." The pilot would sing out "mark three" when the river was 3 fathoms deep and "mark twain" at 2 fathoms. The American writer Samuel Clemens took this as his pen name, Mark Twain. If we take 2 fathoms as a new unit, "twain," express the following values in units including the twain as distance. One fathom is equal to 6 feet.
   (a) 60 miles per hour = __?__ twains per second
   (b) 11,000 meters per second = __?__ twains per second
   (c) 40 meters per second squared = __?__ twains per second squared
   (d) 32.2 feet per second squared = __?__ twains per second squared

19. Here are some new units:

Length:	1 stride = 1.5 meters	1 car = 20 feet
Time:	1 blink = 0.3 second	1 class = 50 minutes
Mass:	1 heavy = 5 kilograms	1 light = 2 pound-mass

Determine X in the following expressions.
   (a) Speed limit, 60 miles per hour = X cars per minute
   (b) Boeing 747 cruising speed, 550 miles per hour = X strides per blink
   (c) Force, 1 newton = X heavy stride per blink squared
   (d) Force, 1 pound-force = X light car/class squared

20. A box has a volume of 10 gallons. If two sides of the box measure 2.4 meters × 2.4 feet, what is the length of the third side of the box in units of inches?

21. We turn on our garden hose and point it straight up. It seems reasonable to assume that the height (H) to which the jet of water rises depends on the initial velocity of the water (v_0) and the acceleration due to gravity (g) as expressed by the relationship

$$H = K \frac{v_0^2}{g}$$

The constant (K) is unitless. If the value of K is 25, what initial velocity, in units of meters per second, will give a water height of 0.5 meters?

22. If a person weighs 700 newtons, what is the mass of the person in units of:
   (a) kilograms
   (b) pounds-mass
   (c) slugs

23. A football lineman weighs 300 pounds-force. What is his mass in units of:
   (a) kilograms
   (b) pounds-mass
   (c) slugs

24. The weight of a can of soda on the moon (where the acceleration of gravity is 1.6 meters per second squared) is 0.6 newtons. What is the mass of the can of soda on Earth in units of:

(a) kilograms   (c) pounds-mass
(b) grams   (d) slugs

25. If the density of silicon is 10.5 grams per cubic centimeter, what is this in units of:

(a) pounds-mass per cubic foot
(b) kilograms per cubic meter
(c) slugs per gallon

26. If the density of sulphur is 131 pounds-mass per cubic foot, what is this in units of:

(a) grams per cubic centimeter
(b) kilograms per cubic meter
(c) slugs per gallon

27. Convert 200 grams per cubic centimeter into units of:

(a) pounds-mass per cubic foot
(b) kilograms per cubic meter
(c) slugs per liter

28. The specific gravity of acetic acid (vinegar) is 1.049. State the density of acetic acid in units of:

(a) pounds-mass per cubic foot
(b) grams per cubic centimeter
(c) kilograms per cubic meter
(d) slugs per liter

29. The specific gravity of octane is 0.703. State the density of octane in units of:

(a) pounds-mass per cubic foot
(b) grams per cubic centimeter
(c) kilograms per cubic meter
(d) slugs per liter

30. The specific gravity of gold is 19.3. What is the length of one side of a 0.4 kilogram cube of solid gold, in units of inches?

31. The density of gasoline is 0.72 grams per cubic centimeter. What is the mass in units of kilograms of a 5-gallon container filled completely with gasoline? Ignore the mass of the container.

32. A lab reports that the density of a new element is X kilograms per cubic foot and Y grams per cubic meter. Which of the following statements are true?

(a) X > Y
(b) X < Y
(c) X = Y
(d) Cannot be determined

33. A new headache medicine has a volume of 3 milliliters and weighs 0.010 kilograms. What is the specific gravity of the medicine?

34. A golden bar of metal (5 centimeters by 18 centimeters by 4 centimeters) being transported by armored car is suspected of being fake, made from a less valuable metal with a thin coating of pure gold. The bar is found to have a mass of 2.7 kilograms. If the specific gravity of gold is 19.3, is the bar fake? Justify your answer.

35. I want to bake a cake at 350 degrees Fahrenheit. What is the temperature in units of:

(a) degrees Celsius
(b) kelvin
(c) degrees Rankine

36. The "normal" temperature of a chicken is 40 degrees Celsius. State this temperature in units of:

(a) degrees Fahrenheit
(b) kelvin
(c) degrees Rankine

37. If we increase the temperature in a reactor by 90 degrees Fahrenheit, how many degrees Celsius will the temperature increase?

38. We are making a cup of coffee and want the temperature to be just right, so we measure the temperature with both Fahrenheit and Celsius thermometers. The Fahrenheit meter registers 110 degrees Fahrenheit, but you prefer to it to be slightly hotter at 119 degrees Fahrenheit, so we heat it up a little. How much will the Celsius thermometer increase when we make this change?

39. Ice cream hardens (freezes) at −10 degrees Celsius. What is this temperature in:

    (a) degrees Fahrenheit
    (b) kelvin
    (c) degrees Rankine

40. The tiles on the space shuttle are constructed to withstand a temperature of 950 kelvin. What is the temperature in units of:

    (a) degrees Celsius
    (b) degrees Fahrenheit
    (c) degrees Rankine

41. The boiling point of propane is −43 degrees Celsius. What is the temperature in units of:

    (a) degrees Fahrenheit
    (b) kelvin
    (c) degrees Rankine

42. Two new temperature scales have been recently developed, the Quick scale and the Mega scale. Here are the calibration points:

	Water freezes	Water boils
Quick	50 °Q	60 °Q
Mega	−100 °M	200 °M

Answer the following, giving reasons for your answers:

    (a) Which scale is better for reporting weather phenomena?
    (b) Which scale is better for cooking?
    (c) Which scale is better for metallurgy?
    (d) Which scale is more accurate?
    (e) Which scale is better for medical use?
    (f) Find x: 100 °Q = x °M
    (g) Find x: 150 °M = x °Q

43. We want to construct a thermometer using mercury (Hg). As the mercury in the bulb is heated, it expands and moves up the thin capillary tube connected to the bulb. The symbol used for the coefficient of volume expansion of a substance due to a temperature increase is β. It is used in the following equation:

$$\Delta V = \beta V(\Delta T)$$

Here, ΔV is the increase in volume, V is the original volume, and ΔT is the temperature increase. The value of β for mercury is 1.8×10^{-4} [1/degrees Celsius]. If the bulb contains 0.2 milliliters and the tube has a diameter of 0.2 millimeters, how much will the mercury rise in the tube in units of centimeters if we increase the temperature from 30 degrees Fahrenheit to 70 degrees Fahrenheit?

44. If the pressure is 250 feet of water, what is the pressure in units of:

    (a) inches of mercury
    (b) millimeters of mercury
    (c) pounds-force per square inch
    (d) pascals
    (e) atmospheres

45. If the pressure is 100 millimeters of mercury, what is the pressure in units of:

   (a) inches of mercury
   (b) feet of water
   (c) pounds-force per square inch
   (d) pascals
   (e) atmospheres

46. A car tire is inflated to 30 pounds-force per square inch. If the tire has an area of 0.25 square feet in contact with the road, how much force is exerted by all four tires?

47. A "normal" blood pressure is 120 millimeters of mercury (systolic reading) over 80 millimeters of mercury (diastolic reading). Convert 120 millimeters of mercury into units of:

   (a) inches of mercury
   (b) pounds-force per square inch
   (c) pascals
   (d) atmospheres

48. The force on the inside of a cork in a champagne bottle is 10 pound-force. If the cork has a diameter of 0.5 inches, what is the pressure inside the bottle in units of feet of water?

49. If a force of 15 newtons is applied to a surface and the pressure is measured as 4,000 pascals, what is the area of the surface in units of square meters?

50. A sensor is submerged in a silo to detect any bacterial growth in the stored fluid. The stored fluid has a density of 2.2 grams per cubic centimeters. What is the hydrostatic pressure felt by the sensor at a depth of 30 meters in units of atmospheres?

51. If the hydrostatic pressure at the bottom of a 10 foot deep ketchup vat is 34,360 pascals, what is the density of the ketchup in kilograms per cubic meter?

52. A cylindrical tank filled to a height of 25 feet with tribromoethylene has been pressurized to 2 atmospheres ($P_{surface}$ = 2 atmospheres). The total pressure at the bottom of the tank is 4 atmospheres. Determine the density of tribromoethylene in units of kilograms per cubic meter.

53. A submersible vehicle is being designed to operate in the Atlantic Ocean. Density of ocean water is 1.025 grams per cubic centimeter. For a maximum depth of 300 feet, what is the total pressure the hull of the submersible must be designed to withstand? Give your answer in units of pounds per square inch.

54. If the total pressure on the bottom of an open container is 2.5 atmospheres, what would the atmospheric pressure have to be to double the total pressure on the bottom?

55. An object completely submerged in a pressurized tank feels 3 atmospheres of hydrostatic pressure. If the density of the fluid is 1.025 grams per cubic centimeter, how deeply is the object submerged in the fluid in units of meters?

56. Airspeed (v), is determined from dynamic pressure using the following formula: $P_{DYNAMIC} = \frac{1}{2}\rho v^2$. Determine the dynamic pressure, in units of pascals, for an aircraft moving at an airspeed of 600 miles per hour. Air density is 1.20 kilograms per cubic meter.

57. A 10-liter flask contains 1.3 moles of an ideal gas at a temperature of 20 degrees Celsius. What is the pressure in the flask in units of atmospheres?

58. An ideal gas, kept in a 5-liter container at 300 kelvin, exhibits a pressure of 2 atmospheres. If the volume of the container is decreased to 2.9 liters, but the temperature remains the same, what is pressure in the new container in units of atmospheres?

59. An ideal gas is kept in a 10-liter container at a pressure of 1.5 atmospheres and a temperature of 310 kelvins. If the gas is compressed until its pressure is raised to 3 atmospheres while holding the temperature constant, what is the new volume in units of liters?

60. A 5-liter container holds nitrogen (formula: N_2, molecular weight = 28 grams per mole) at a pressure of 1.1 atmospheres and a temperature of 400 kelvins. What is the amount of nitrogen in the container, in units of grams?

61. An ideal gas in a 1-gallon container is at a temperature of 35 degrees Celsius and pressure of 1.5 atmospheres. If the gas is oxygen (formula: O_2, molecular weight = 32 grams per mole), what is the amount of gas in the container in units of grams?

62. A 10-liter flask contains 5 moles of gas at a pressure of 15 atmospheres. What is the temperature in the flask in units of kelvin?

63. A container holding 1.5 moles of oxygen (formula: O_2, molecular weight = 32 grams per mole) at a pressure of 1.5 atmospheres and a temperature of 310 kelvins is heated to 420 kelvins, while maintaining constant volume. What is the new pressure inside the container in units of pascals?

64. You push an automobile 100 meters with a constant force, and 1500 joules of energy is added to the car. With what constant force, in units of pounds-force, were you pushing on the car? Assume frictional losses are negligible.

65. The specific heats of aluminum and iron are 0.214 calories per gram degrees Celsius [cal/(g °C)] and 0.107 calories per gram degrees Celsius [cal/(g °C)] respectively. If we add the same amount of energy to a cube of each material (of the same mass) and find that the temperature of the aluminum increases by 30 degrees Fahrenheit, how much will the iron temperature increase in degrees Fahrenheit?

66. We go out to sunbathe on a warm summer day. If we soak up 100 British thermal units per hour [BTU / h] of energy, how much will the temperature of 60,000-gram person increase in 2 hours in units of degrees Celsius? We assume that since our bodies are mostly water they have the same specific heat as water (4.18 joules per gram degree Celsius [J/(g °C)]).

67. We are furniture movers and have to slide a large sofa across the room. It takes a push of 100 pound-force to do this. If the room is 20 feet across and we accomplish the move in 5 seconds, how much power in units of horsepower did it take?

68. Which object—A, B, or C—has the most potential energy when held a distance H above the surface of the ground?

Object A:	mass = 1 kilogram	H = 3 meters
Object B:	mass = 1 slug	H = 3 feet
Object C:	mass = 1 gram	H = 1 centimeter

69. Measurements indicate that boat A has twice the kinetic energy of boat B of the same mass. How fast is boat A traveling if boat B is moving at 30 knots? 1 knot = 1 nautical mile per hour; 1 nautical mile = 6076 feet.

70. If a ball is dropped from a height (H) its velocity will increase until it hits the ground (assuming that aerodynamic drag due to the air is negligible). During its fall, its initial potential energy is converted into kinetic energy. Find an expression for the impact velocity.

    (a) If the mass of the ball is doubled, how will the impact velocity change?
    (b) If the ball is dropped from a height of 800 centimeters, and the impact velocity is 41 feet per second, determine the value of gravity in units of meters per second.

71. A ball is thrown vertically into the air with an initial kinetic energy of 2500 joules. As the ball rises, it gradually loses kinetic energy as its potential energy increases. At the top of its flight, when its vertical speed goes to zero, all of the kinetic energy has been converted into potential energy. Assume that no energy is lost to frictional drag, etc. How high does the ball rise in units of meters if it has a mass of 5 kilograms?

72. If a 10-kilogram rotating solid cylinder moves at a velocity (v), it has a kinetic energy of 36 joules. Determine the velocity the object is moving in units of meters per second if the kinetic energy is given by $KE = \frac{1}{2}mv^2 + \frac{1}{4}mv^2$

73. If a rotating solid cylinder moves at a velocity (v) of 5 meters per second, it has a kinetic energy of 94 joules. Determine the mass of the object in units of kilograms if the kinetic energy is given by $KE = \frac{1}{2}mv^2 + \frac{1}{4}mv^2$

74. A horse pulling a plow has to overcome 50 pounds-force of drag on the plow as he pulls. He can plow an acre in an hour. If the plot of land is square and each furrow occupies a strip that is 4 feet wide, how much power (in horsepower) will the horse need to expend to plow the acre?

75. How much power (in units of horsepower) is required by a 200 pounds-force person to run up three flights of stairs (10 feet each) in 20 seconds?

76. When we drive our car at 100 feet per second, we measure an aerodynamic force (called drag) of 66 pounds-force that opposes the motion of the car. How much horsepower is required to overcome this drag?

77. A 100-watt electric motor is tested and found to be able to raise a 10-kilogram mass by 12 meters in 15 seconds. What is the efficiency of this motor?

78. A 100-watt motor (60% efficient) is used to raise a 100-kilogram load 5 meters into the air. How long, in units of seconds, will it take the motor to accomplish this task?

79. A 100-watt motor (60% efficient) is available to raise a load 5 meters into the air. If the task takes 65 seconds to complete, how heavy was the load in units of kilograms?

80. You purchase a 50-watt motor (60% efficient) and use it to raise a 100-kilogram load 5 meters into the air. How long in units of seconds will it take you to achieve this task, assuming the motor is capable of lifting this load?

81. We have learned that only 5% of the energy that goes into an incandescent light bulb is emitted as light while the remainder is emitted as heat. What is the thermal efficiency of an incandescent light bulb?

82. We place 1 gallon of water, at a temperature of 70 degrees Fahrenheit, in a pan on the stove, turn on the burner, and wait for it to boil (don't watch it to closely or it will never boil). If it boils in 20 minutes and the burner is 1400 watts, how efficient is the burner?

83. A 3-kilogram projectile traveling at 100 meters per second is stopped by being shot into an insulated tank containing 100 kilograms of water. If the kinetic energy of the projectile is completely converted into thermal energy with no energy lost, how much will the water increase in temperature in units of degrees Celsius? The specific heat of water is 1 calorie per gram per degree Celsius [cal/(g °C)].

84. A 10-gram rubber ball is released from a height of 6 meters above a flat surface on the moon. Gravitational acceleration on the moon is 1.62 meters per second squared. Assume that no energy is lost from frictional drag. What is the velocity, in units of meters per second, of the rubber ball the instant before it strikes the flat surface?

85. You need to purchase a motor to supply 400 joules in 10 seconds. All of the motors you can choose from are 80% efficient. What is the minimum wattage on the motor you need to choose?

86. The specific heat of neon is 1.03 $\frac{\text{J}}{\text{g K}}$. Convert this value into units of:

   (a) $\dfrac{\text{cal}}{\text{kg °C}}$

   (b) $\dfrac{\text{BTU}}{\text{lb}_\text{m}\text{ °F}}$

87. The specific heat of octane is 0.51 $\frac{\text{cal}}{\text{g °C}}$. Convert this value into units of:

   (a) $\dfrac{\text{cal}}{\text{kg °C}}$

   (b) $\dfrac{\text{BTU}}{\text{lb}_\text{m}\text{ °F}}$

88. The thermal conductivity of ice is 1.6 $\frac{\text{w}}{\text{m K}}$. Convert this into units of:

   (a) $\dfrac{\text{cal}}{\text{cm s °C}}$

   (b) $\dfrac{\text{Btu}}{\text{ft h °F}}$

89. The heat transfer coefficient of water is $300 \dfrac{\text{Btu}}{\text{ft}^2 \text{ h } ^\circ\text{F}}$. Convert this into units of

    (a) $\dfrac{\text{W}}{\text{m}^2 \text{ K}}$

    (b) $\dfrac{\text{cal}}{\text{cm}^2 \text{ s } ^\circ\text{C}}$

90. A rod on the surface of Jupiter's moon Callisto has a volume of 0.3 cubic meters. Determine the weight of the rod in units of pounds-force. The density is 4700 kilograms per cubic meter. Gravitational acceleration on Callisto is 1.25 meters per second squared.

91. The maximum radius a falling liquid drop can have without breaking apart is given by the equation $R = \sqrt{\dfrac{\sigma}{g\rho}}$, where σ is the liquid surface tension, g is the acceleration due to gravity, and ρ is the density of the liquid. For acetone at 20 degrees Celsius, determine the surface tension (σ) in units of joules per meter squared if the maximum radius of a drop is 1 centimeter and the specific gravity of the liquid is 0.79.

92. When a flowing fluid is stopped, its pressure increases. This is called stagnation pressure. The stagnation pressure is determined by: $P_{\text{stagnation}} = \frac{1}{2}\rho v^2 + P_{\text{surface}}$, where ρ is the fluid density, v the fluid speed, and P_{surface} the atmospheric pressure. Calculate the stagnation pressure in units of atmospheres for acetone flowing at 15 feet per second. Assume the density of acetone to be 790 kilograms per cubic meter.

93. When a flowing fluid is stopped, its pressure increases. This is called stagnation pressure. The stagnation pressure is determined by $P_{\text{stagnation}} = \frac{1}{2}\rho v^2 + P_{\text{surface}}$, where ρ is the fluid density, v the fluid speed, and P_{surface} the atmospheric pressure. If the stagnation pressure is 18 pounds-force per square inch, what is the fluid speed in units of feet per minute? Assume the fluid is methyl ethyl ketone (MEK) and the density of is 805 kilograms per cubic meter.

94. Complete the following table.

Quantity	Symbol	"Derived" from	Typical SI Units	Dimensions			
				M	L	T	Θ
Dynamic viscosity	μ	Newton's law of viscosity	g/(cm s)				
Mass density	ρ	$\rho = m/V$	kg/m^3				
Specific heat	C_p	$Q = m\ C_p\ \Delta T$	J/(g $^\circ$C)				
Surface tension	σ	from theory or experiment	J/m^2				
Thermal conductivity	k	Fourier's law of heat transfer	W/(m $^\circ$C)				
Young's modulus	E	stress/strain	Pa				

95. The Darcy-Weisbach friction formula for pipes is used to calculate the frictional energy loss per unit mass of fluid flowing through the pipe. Determine the correct formula for this equation. In these equations,

h_L	= energy loss per unit mass	[=] J/kg
f	= friction factor	[=] dimensionless
L	= length of pipe	[=] ft
D	= diameter of pipe	[=] cm
v	= average velocity	[=] cm/min
g	= acceleration of gravity	[=] m/s^2

I. $h_L = f\dfrac{L}{D}\dfrac{v}{2g}$

II. $h_L = fL\dfrac{v^2}{2g}$

III. $h_L = f\dfrac{L}{D}\dfrac{v^2}{2}$

IV. $h_L = f\dfrac{L}{v}\dfrac{2D}{g}$

96. We wish to analyze the velocity (v) of a fluid exiting an orifice in the side of a pressurized tank. The tank contains a fluid to a depth (H) above the orifice. The air above the fluid in the tank is pressurized to a value of (P_i). We realize the greater the pressure inside, the greater the velocity. We also believe the greater the depth of fluid, the greater the velocity. With this information, examine the equations below and determine which one(s) could be the correct equation(s), and which one(s) could not possibly be correct; justify your answer for each case. In these expressions, g is the acceleration of gravity and ρ is the fluid density.

(a) $v = \dfrac{P}{\rho} + \sqrt{2\,g\,H}$

(b) $v = \sqrt{\dfrac{P}{\rho} + 2\,H}$

(c) $v = \sqrt{2\,g\,H}$

(d) $v = \sqrt{\dfrac{2\,P}{\rho} + 2\,g\,H}$

(e) $v = \sqrt{2\,P + 2\,g\,H}$

97. Wind energy uses large fans to extract energy from the wind and turn it into electric power. Which of the following equations is valid for this process? In these expressions, P is the power, η is efficiency, ρ is the density of the air, A the area swept out by the fan blades, and the velocity of the wind. Show clearly why you select your answer.

I. $P = \eta \rho A^2 v^2$

II. $P = \eta\, \rho\, A v^2$

III. $P = \eta\, \rho\, A v^3$

IV. $P = \eta\, \rho^2\, A v$

V. $P = \eta \sqrt{\rho\, A v^3}$

98. One of the most famous experiments in physics is the Millikan oil-drop experiment. In it, very tiny drops are sprayed into a closed container by an atomizer. The drops tend to fall, but they are so small that the air slows them down. Sir George Stokes studied this part of the oil-drop experiment to see how fast the drops would fall in the still air in the container. He found that the fall speed (v) depended on the mass of the drop (m), the acceleration of gravity (g), the radius of the drop (r), and the viscosity of the air (μ). The equation governing the motion of the drop was determined to be

$$mg - 6\,\pi\mu rv = 0$$

(a) The constant (the value 6) in the second term of the equation is dimensionless. What are the dimensions of viscosity (μ)?

(b) What kind of quantity is the second term ($6\pi\mu rv$)? Choose from the following and prove your answer in terms of dimensions.

- Pressure
- Force
- Velocity
- Energy

99. The friction factor (f) is a dimensionless number used in fluid flow calculations. It can be related to velocity in the pipe (v) and the fluid density (ρ) by the following equation:

$$f = \frac{2\varphi}{\rho v^2}$$

   (a) Using the laws of dimensions in equations, determine the dimensions of the variable φ.
   (b) What kind of quantity is the variable φ? Choose from the following and prove your answer in terms of dimensions.

   - Pressure
   - Force
   - Velocity
   - Energy

100. In previous examples, we discussed quantities with fundamental dimensions of mass {M}, length {L}, time {T}, and temperature {Θ}. As we study other fields, such as electricity and magnetism, we must introduce new fundamental dimensions, such as {Q}, representing charge, which cannot be expressed in terms of mass, length, time, or temperature. Electric power can be calculated as $P = I^2 R$, where R is the resistance [ohms, Ω] and I is the electrical current [amperes, A or coulombs per second]. The coulomb is a unit with a fundamental dimension of {Q}, as described above. Assuming that electrical power must have the same dimension as mechanical power, derive the fundamental dimension of the ohm in terms of mass {M}, length {L}, time {T}, temperature {Θ}, and charge {Q}.

101. A biodegradable fuel having a specific gravity of 0.95 and a viscosity of 0.04 grams per centimeter second is draining by gravity from the bottom of a tank. The drain line is a plastic 3-inch diameter pipe. The velocity is 5.02 meters per second.

   (a) Determine the Reynolds number in the pipe for the system.
   (b) Is the flow laminar, in transition, or turbulent?

102. A sludge mixture having a specific gravity of 2.93 and a viscosity of 0.09 grams centimeter per second is pumped from a reactor to a holding tank. The pipe is a 2½ inch diameter pipe. The velocity is 1.8 meters per second.

   (a) Determine the Reynolds number in the pipe for the system.
   (b) Is the flow laminar, in transition, or turbulent?

103. Water (specific gravity = 1.02; viscosity = 0.0102 grams per centimeter second) is pumped through a 0.5 meter diameter pipe. If the Reynolds number is 1800 for the system, determine the velocity of the water in units of meters per second.

104. Water (specific gravity = 1.02; viscosity = 0.0102 grams per centimeter second) is pumped through 0.5 meter diameter pipe. If the Reynolds number is 5800 for the system, determine the velocity of the water in units of meters per second.

105. The Peclet number is used in heat transfer in general and forced convection calculations in particular. It is a function of the two other dimensionless groups, the Reynolds number and the Prandlt number. Determine the functional form of these dimensionless groups, using Rayleigh's method. The problem depends on the following variables:

   - Liquid density, ρ [=] kg/m^3
   - Specific heat of liquid, C_p [=] J/(g °C)
   - Liquid viscosity, μ [=] kg/(m s)
   - Thermal diffusivity, α [=] m^2/s
   - Thermal conductivity of the plate, k [=] W/(m °C)
   - Distance from edge of the plate, x [=] m
   - Liquid velocity, v [=] m/s

106. When a fluid flows slowly across a flat plate and transfers heat to the plate, the following variables are important. Analyze this system using Rayleigh's method.

- Liquid density, ρ [=] kg/m^3
- Specific heat of liquid, C_p [=] J/(g °C)
- Liquid viscosity, μ [=] kg/(m s)
- Thermal conductivity of the plate, k [=] W/(m °C)
- Heat transfer coefficient, h [=] W/(m^2 °C)
- Distance from edge of the plate, x [=] m
- Liquid velocity, v [=] m/s

107. A projectile is fired with an initial velocity (v_0) at an angle (θ) with the horizontal plane. Find an expression for the range (R). The data are given in the table below. Use the data in the table to create one or more *dimensional* plots (for example, launch speed on the abscissa and range on the ordinate). From these plots, answer the following question as best you can:

Launch Angle (θ) [°]	Launch Speed (v_0) [m/s]	Measured Range (R) [m]
4	70	73
50	50	230
3	50	30
45	18	32
37	27	75
35	60	325
22	8	4.4
10	30	34
88	100	77
45	45	210

(a) If the launch speed is 83 meters per second and the launch angle is 64 degrees, what is the range? You will likely find it difficult to provide a good estimate of the range, but do the best you can.

(b) Complete a dimensional analysis of this situation. In this case, you would assume that the important parameters are θ, v_0, and R. Upon closer examination, however, it would seem that the range on Earth and on the moon would be different. This suggests that gravity is important, and that you should include g in the list of parameters. Finally, since it's not clear how to include θ, you could omit it and replace the velocity by v_x and v_z, where x is the distance downrange and z the height. You should use this information to determine dimensionless parameters. Also, you must decide how the lengths in R and g should appear. When you complete the analysis, you should find that these four parameters will be grouped into a single dimensionless ratio.

(c) Use the data from the table to calculate the numerical value of the ratio for each test. Note that $v_x = v \cos(\theta)$ and that you can find a similar expression for v_z. Insert these expressions into your dimensionless ratio.

(d) Assuming that you performed the dimensional analysis correctly, you should find that the ratio you obtained will always give the same value (at least nearly, within test-to-test error). Calculate the average value of the tests, and if it is nearly an integer, use the integer value.

(e) Finally, set this ratio equal to this integer, and then solve for the range R. Write your final equation for the range (i.e., R = xxxxx). Now using this equation, answer question (a) again.

108. The drag on a body moving in a fluid depends on the properties of the fluid, the size and shape of the body, and probably most importantly, the velocity of the body. We find that for high velocities, the fluid density is important but the "stickiness" (or viscosity) of the fluid is not. The frontal area of the object is important. You might expect that there is more drag on a double-decker bus moving at 60 miles per hour than there is on a sports car moving at 60 miles per hour.

The table below gives some data for tests of several spheres placed in air and in water. The terminal velocity, the point at which the velocity becomes constant when the weight is balanced by the drag, is shown. Plot the drag on the ordinate and the velocity of the object on the abscissa. Is this plot useful?

Object	Drag (F) [lb$_f$]	Velocity (v) [ft/s]	Diameter (D) [in]	Fluid
Table tennis ball	0.005	12	1.6	air
Bowling ball	6	60	11	air
Baseball	0.18	41	3	air
Cannon ball	33	174	9	air
Table tennis ball	0.0028	0.33	1.6	water
Bowling ball	12.4	3.1	11	water
Baseball	0.31	1.7	3	water
Cannon ball	31	6.2	9	water

(a) Use the data in this table for different spheres to answer the following question: What is the drag on a baseball in gasoline (specific gravity = 0.72) at a speed of 30 feet per second? You may struggle with this, but do the best you can.

(b) Now complete a dimensional analysis of this situation and replot the data. First, recognize that the important parameters are the ball diameter (use the silhouette area of a circle), the density of the fluid, the drag, and the velocity. You will find a single dimensionless ratio that combines these parameters.

(c) Compute the value of this ratio for the eight tests. Be sure in your analysis (calculations) that you use consistent units so that the final ratio is truly unitless. Summarize your results.

(d) Use this result to help you answer question (a) again.

PART 4

SCRUPULOUS SPREADSHEETS

Chapters in Part 4

Everything should be made as simple as possible, but not simpler.

ALBERT EINSTEIN

Nothing is so simple it cannot be misunderstood.

TEAGUE'S PARADOX

Scrupulous: **scroop·yə·ləs** ~ *adjective*;
  definition - _____

Microsoft Excel is a spreadsheet computer program used internationally for an incalculable number of different applications. A **spreadsheet** is a document that contains data separated by rows and columns. The idea of using a spreadsheet to solve different types of problems originated before the advent of computers in the form of bookkeeping ledgers. The first graphical spreadsheet computer program for personal computers, VisiCalc, was released in 1979 for the Apple II® computer.

Modern spreadsheet computer programs, like Excel, are significantly more powerful than earlier versions like VisiCalc; a comparison of the interface is shown in Figure P4-1. Excel contains text-formatting controls, built-in functions to perform common calculations, and a number of different plotting capabilities that make it an extremely powerful data analysis tool for engineers. Part 4 introduces the Microsoft Excel interface, the formatting controls used to create organized worksheets, and many built-in functions to assist in analyzing data or performing calculations on data contained in the worksheet.

Figure P4-1 Comparison of VisiCalc and Excel interfaces.

A successful engineer must rely on knowledge of the way things work in order to develop solutions to problems, whether ameliorating climate change or trapping cockroaches. In many cases, the behavior of systems or phenomena can be described mathematically. These mathematical descriptions are often called mathematical models. The variables in the model vary with respect to one another in the same way that the corresponding parameters of the real physical system change.

As a very simple example, imagine you are driving your car on a country road at a constant speed of 30 miles per hour. You know that at this speed, you travel one-half mile every minute. If you drive at this speed for 44 minutes, you cover a distance of 22 miles.

A mathematical model for this is $d = 0.5\,t$, where d is distance in miles, t is time in minutes, and the value 0.5 has units of miles per minute. If you substitute *any* number of minutes in this equation for time (including 44), the distance (in miles) will be exactly half of the time numerical value. This allows you to predict what would happen in the "real world" of cars and roads without having to actually go out and drive down the road to determine what would happen if you drove 30 miles per hour for 44 minutes.

Needless to say, the mathematical descriptions for some physical systems can be extremely complicated, such as models for the weather, global economic fluctuations, or the behavior of plasma in an experimental fusion reactor.

As it turns out, a significant number of phenomena important in engineering applications can be described mathematically with only three simple types of models. Also in Part 4, we introduce these three models and their characteristics, as well as discuss the use of Excel to determine a mathematical model from a set of data determined by experimentation.

A few notes about this section of the book:

- Within the examples given in this portion of the text, note that any information you are asked to type directly into Excel will be found in quotations. Do not type the quotation marks, only the information found between the quotation marks.
- In hardcopy, the data needed to create a chart will be shown in columns or rows, depending on the size of the data, to efficiently use space and save a few trees by using less textbook paper. In the worksheets containing the starting data online, the data will be shown in columns.
- Files available online are indicated by the symbol
- ✆ This symbol indicates directions for an important process to follow. Step-by-step instructions are given once for each procedure.
- ⌘ This symbol indicates special instructions for Mac OS users.

TIME MANAGEMENT

The bad news is time flies. The good news is you are the pilot.

Michael Altshuler

If you are using this book sequentially, by this point you are probably starting to feel a bit overwhelmed with all you need to do. While many introductory textbooks cover time management during the first few weeks of the semester, the authors have found it more useful to cover it a little later. In week 2 of your first semester of college, you are probably feeling like you still have things under control and don't need help. By week 10, however, you may be struggling to keep everything together and are more open to try some time management suggestions. Please note these are just that—suggestions—and each person must develop a time management system that works best for him or her. It may take you a few attempts to find a process you can actually use, so keep making adjustments until you find your own personal solution.

There are 24 hours in each day, and 7 days in a week. Each week, you have 168 hours, or an estimated 170 hours, to spend doing something—sleeping, going to class, doing homework, or attending a football game. How are you using all this time?

> | Total available |
> | hours = 170 |
> | Sleep = 50 |
> | Meals = 20 |
> | Personal |
> | hygiene = 10 |
> | Classes = 20 |
> | Studying = 40 |
> | "Free" hours = 30 |

- To get enough rest, you should be sleeping at least 7 hours each night, or about 50 hours each week.
- If you spend 1 hour for each meal during the day, about 20 hours of your week will be spent eating.
- If you allow 1 hour per day for personal hygiene and a few hours for laundry (your classmates will thank you for showering and having clean clothes), this takes about 10 hours per week.
- Attending class is critical, and with lectures and labs you are probably in the classroom for 20 hours.
- If you spend the maximum recommended study time on each course, this will take another 30–45 hours each week.

So what's left? Actually, quite a bit of time remains: 30 hours. While that may not seem like much, remember we assumed the maximum limits in our analysis.

- It may only take you 30 minutes each day to get showered and dressed, saving you 3.5 hours per week.
- Your lab may be canceled, freeing up an additional 3 hours.
- While there are weeks when it will be necessary to study the maximum amount, this will also be balanced by weeks when you can study the minimum amount.

How, exactly, can you balance this "free" time with the "required" time? To be successful at time management, you must plan. If you carve out 1 hour each week to determine your plan for the upcoming days, you will be able to find time to work in any activities you want to do and still find time to study, eat, and sleep. Here, we present a PLAN with four steps: Prioritize, Leave time for fun, Anticipate delays, and No—learn to say it.

PRIORITIZE

Ask:

- What must be completed this week (required assignments)?
- What can I begin to work on for next week (upcoming project, exams)?
- What would be nice to do if I have the time (recommended problems, reading)?

Rules:

- Schedule all courses in your plan. Attend every class. Be sure to include travel time, especially if you are commuting.
- Select a study time for each class and stick to it. As a general rule, plan for 2–3 hours of studying for each hour in class. For a 15-credit-hour course load, this is 30–45 hours.
- Determine when you study best. Are you an early riser or a night owl?
- Be specific in your plan. Listing "Read Chapter 2, pages 84–97" is much better than "Read chemistry." Break down large projects into smaller tasks, each with a deadline.
- Don't study more than two hours at a time without a break. Pay attention to how long it takes you to become distracted easily.
- Schedule time daily to read course email and check any online course management system.
- If you are working during college, don't forget to schedule in this time. As a general rule, you should not plan to work more than 10 hours per week while taking a 15-credit- hour course load. If you are working more, you may want to consult your financial aid office for advice.

LEAVE TIME FOR FUN (AND CHORES)

Ask:

- What has to get done this week (chores)?
- What activities do I want to take part in (fun stuff you really want to do)?
- What would be nice to do if I have time (fun stuff you want to do if you have time)?

Rules:

- Schedule time for planning each week. Adopt your weekly schedule to meet the upcoming week deadlines and assignments.
- Schedule time for meals. Relax and talk with friends, read an engrossing book. Do not study during meals!
- Schedule time for sleep. Stick to this schedule—you will feel better if you go to sleep and awake each day at the same time . . . yes, even on weekends!
- Schedule time for physical activity. This can be hitting the gym, playing intramurals, or talking a walk. Staying healthy will help you stay on track.
- Schedule time for chores, such as laundry, paying bills, and keeping your room clean.
- Allow time for technology on a limited basis. If you have a favorite TV show, schedule time to watch. If you want to surf on a social network, do so for a limited time each day.
- Plan outings. Colleges are wonderful resources for arts, music, theater, and athletics. Explore and find activities to enjoy, but don't compromise study time to attend.
- Leave some open time. It is not necessary to schedule every minute of every day. Free time is a wonderful stress reliever!

ANTICIPATE DELAYS

Ask:

- What can go wrong this week?
- What activities will alter my plans?

Rules:

- Plan time for "Murphy's Law:" broken computers, running out of paper, getting sick, or helping a friend. If none of these occur, you will have extra hours in your plan.
- Leave time to proof-read your work, or better yet, have someone else help you. Utilize your course teaching assistants, professor, or college academic facilities to assist you in polishing your final product.
- Plan to finish large projects one week before they are due to allow for any unexpected delays.

NO — LEARN TO SAY IT!

Ask:

- Will this activity help me reach my goal?
- If I do this activity, what will alter in my plan?

Rules:

- Schedule social activities around academics. Say "no" if you are not finished with your coursework.
- Remember, you are here to get an education. Employers will not care that you attended every basketball game or that you have 10,000 online friends if you have poor grades.

In absence of clearly defined goals, we become strangely loyal to performing daily acts of trivia.

Unknown

GOAL SETTING, BE REALISTIC

By Ms. Solange Dao, P.E., President DAO Consultants, Inc, Orlando, Florida and Vice President of Tau Beta Pi National Engineering Honor Society

Love to have fun? Many of us go out of our way to have fun or explore something intriguing. I will wait in a long line for an hour to ride a roller coaster, or detour 5 miles to drive down a new Interstate off ramp. Fun, exploration, and intrigue naturally soak our minds with positive stimulation.

If fun could be channeled into traditionally boring things, productivity would skyrocket. What if all Americans went to the gym regularly because it was truly fun to sweat off calories? How about smiling all day at work because the job was a pleasurable challenge? Many New Year's resolvers give up by January 3. Why is it so hard to keep a promise to yourself just 72-hrs after you've sworn to change? As an adult, you can choose to have fun in your type-A goal oriented world. The key to mental stimulation and success is to trick yourself into having fun.

There are Three Main Reasons Why Goals Fail: 1) They are Too Big to Achieve Quickly, 2) Pitfalls were Not Foreseen, and 3) No Back-Up Plan was Devised.

Too big to achieve quickly

In our on-demand society, we tend to equate movement with progress. When you set a large goal (e.g. lose 50-lbs, or save $1,000), it takes time to get there. The long-term goals seem more unreachable since daily movement toward them is miniscule. Since big goals are our vision of a greater life for ourselves, how can we better handle our big goals?

Breaking the big goal into small chunks is how you win. Break the "lose weight" goal down to its essentials. For example, to lose weight you have to eat less calories than you use, or burn more calories than you consume. Try to conquer each meal one at a time. This month, every breakfast will be healthy oatmeal with fruit. Then next month, add healthy choices for lunch. By focusing on the immediate task, you can win and be proud daily.

If you plan to exercise your way to a slimmer waist, think of small increments. Saying you'll run 10k races every month is a big bite to chew. How about just parking in the farthest spot from the building door? Commit to taking stairs instead of the elevator, even when it's ten flights. Stop using the TV remote and get up to change the channel. Small victories add up to a big win.

What if it's an academic goal such as getting into an honor society? Break it down to conquering one test at a time. Plan small study periods well in advance of each test instead of attempting a marathon cram session 24 hours before the tests.. Micromovements are more mentally manageable.

Pitfalls were not foreseen

Minimize your disappointments in goal setting by attempting to foresee possible problems. Thinking it will be smooth sailing to your destination will only lead to bailing out. Challenges are needed to test us so we can build character. However, there is some good news: we can cheat at this obstacle course. We can cheat fate by predicting the obstacles to come. Like a football coach who studies his opponents' play tapes, he sees patterns for which he devises a counter play. You've just been assigned as Coach of your goals.

In early 2009, I decided to quit sodas. I knew this was going to be a big task. I usually had a Diet Coke everyday around 3:00 p.m. to give me a little pep. I foresaw myself dragging at 3:00 p.m. without a Diet Coke, so I planned ahead: if I craved a soda, I would replace it with a non-carbonated iced tea. I still got the caffeine pep, yet the acidity was greatly reduced, not to mention the artificial additives.

What pitfall are you not foreseeing in your grand plan to achieve a goal? If it is financial, will you let a tire blow-out tumble you into despair and give up the goal of saving $1,000? As much as possible, plan your reactions to detours and missteps along the path. If the unexpected new tire costs $100, then accept it and keep going. You just extend your goal deadline by a month.

Many people look at pitfalls or setbacks as a sign from the Divine that they won't ever make their goal, or that they are too weak to change. The challenges are a part of the game. In the child's game Chutes and Ladders, sometimes you climb up, sometimes you slide down. Who quits Chutes and Ladders upon their first chute down? You roll the dice again and keep going. You may land on another chute, but you can't predict that. Don't be the child who gets mad, smashes the game for being non-fun, and pouts for the rest of the day.

Challenge yourself to foresee the obstacles you will face. Brainstorm pitfalls with your friends. Make it a game by adding really goofy ones like participating in the county fair pie-eating contest. Thus, when problems arise, you can actually laugh at them by saying to yourself, "Ha! I knew this molasses and rhubarb pie was going to show up one of these days."

What if a friend invites you to a sold-out concert-of-the-decade that interferes with your planned study time for your engineering midterm exams? If you plan some extra time into your schedule, you will be able to take advantage of unforeseen opportunities. If the study plan is too tight, any deviation will seem insurmountable.

Back-up plan not ready

It is impossible to predict everything that Life will throw at you. Despite your best efforts to predict the unexpected often a misstep occurs. You can also plan for failure. If you set several micro-goals which will lead to your bigger vision, a total failure of one small goal is much more manageable.

You ate the cake. OK, fine. Stop with the guilt and get to action. What is your back-up plan? For example, the 500-calorie cake slice means salads all day tomorrow, or it means walk a half-mile during each of your 15-min work breaks. Make the consequences funny by putting up a sign at your desk that says "Vegan in Training."

Spent $300 on stereo equipment in a moment of insanity? No problem, that means making coffee at home. What is the back-up plan to keep you going forward?

What if your grades were not good enough to enter the honor society? There are other ways to gain recognition on campus such as joining service organizations and helping with charity events. Remember, there is always more than one path to any goal.

By pre-planning micro movement to our large goals, we can have fun predicting the inevitable, skirting around obstacles, and pulling ourselves up if we stumble.

LEARNING OBJECTIVES

The overall learning objectives for this part include:

- Use Microsoft Excel to enhance problem solution techniques, including:
  - Entering, sorting, and formatting data;
  - Analyzing functions, including mathematical, statistical, and trigonometric;
  - Creating and formatting various types of graphs, including scatter, line, and column charts; rectilinear, semi-logarithmic, and logarithmic graphs;
  - Utilizing conditional statements, conditional formatting, data validation, built-in functions, and iteration.
- Select and interpret mathematical models to describe physical data.
- Apply basic principles from mathematical and physical sciences, such as Hooke's law, Ohm's law, and Newton's law of viscosity, to solve engineering problems.

CHAPTER 13
EXCEL WORKBOOKS

The following is an example of the level of knowledge of Excel need to proceed into Chapter 13. *If you are not able to quickly recreate an Excel Worksheet similar to the one shown, including equations and formatting, please review worksheet basics in Appendix B before proceeding into Chapter 13.*

The wind chill temperature is calculated using the following empirical formula:

$$WC = 35.74 + 0.6215\,T - 35.75\,v^{0.16} + 0.4275\,Tv^{0.16}$$

where WC = Wind Chill Temperature [degrees Fahrenheit, °F], T = Actual Air Temperature [degrees Fahrenheit, °F], and v = Wind Speed [miles per hour, mph]. Create a single cell to hold the wind speed, set at 5 miles per hour, in Column A. Create a vertical series of actual air temperatures: T = 30, 10, 0, −10. Enter the equation for wind chill, referencing the cells and not the actual numerical values. A sample worksheet is shown below.

	A	B	C	D	E
1					
2					
3					
4					
5					
6					
7	Wind Speed	Actual Temperature (T) [deg F]			
8	(v) [mph]	30	10	0	−10
9	5	24.7	1.2	-10.5	-22.3

13.1 CELL REFERENCES

EXAMPLE 13-1

Suppose we are given a list of XY coordinates in a worksheet. We want to calculate the distance between each point. We can find the distance between two XY coordinates by using Pythagoras' theorem:

$$d = \sqrt{(x_2 - x_1)^2 + (y_2 - y_1)^2}$$

Distance Between XY Coordinates
This example demonstrates how to handle Excel's order of operations and cell references.

Point 1		Point 2	
X	Y	X	Y
27	20	25	10
25	4	7	8
4	6	24	3
25	26	13	24
19	24	26	1
29	10	0	5
7	29	13	13
3	20	19	16
20	7	5	17
20	26	19	3
13	15	13	14
23	22	17	25
3	27	10	22
30	16	30	17

Distance Between XY Coordinates
This example demonstrates how to handle Excel's order of operations and cell references.

Point 1		Point 2			Distance
X	Y	X	Y		
27	20	25	10		10.20
25	4	7	8		18.44
4	6	24	3		20.22
25	26	13	24		12.17
19	24	26	1		24.04
29	10	0	5		29.43
7	29	13	13		17.09
3	20	19	16		16.49
20	7	5	17		18.03
20	26	19	3		23.02
13	15	13	14		1.00
23	22	17	25		6.71
3	27	10	22		8.60
30	16	30	17		1.00

Distance Between XY Coordinates
This example demonstrates how to handle Excel's order of operations and cell references.

Point 1		Point 2			Distance
X	Y	X	Y		
27	20	25	10		10.20
25	4				6.00
4	6				21.38
25	26				16.00
19	24				15.23
29	10				4.00
7	29				26.17
3	20				24.17
20	7				5.83
20	26				16.76
13	15				13.00
23	22				12.17
3	27				27.80
30	16				7.81

To solve this problem, we must adhere to the default behavior of Excel in order to properly calculate the distance between the coordinates. First, we must observe the order of operations that Excel follows to determine how we need to write our equations. Second, we must determine how to use **cell references** to translate the $x_2, x_1, y_2,$ and y_1 values in the equation shown above into locations in our worksheet.

Before moving on, let's rewrite Pythagoras' theorem in the notation shown above using what we know about order of operations in Excel:

$$d = ((x_2 - x_1) \wedge 2 + (y_2 - y_1) \wedge 2) \wedge (1/2)$$

Let's calculate the distance between Point 1 and Point 2 in column G. In cell G7, we need to translate the equation into an equation that replaces the x_1, y_1 and x_2, y_2 variables with addresses to cells in the worksheet. Since each row represents a single calculation, we know that for the first data pair, x_1 is located in cell A7, y_1 is in B7, x_2 is in D7, and y_2 is in E7. The equation we need to type into cell G7 becomes

$$= ((D7 - A7) \wedge 2 + (E7 - B7) \wedge 2) \wedge (1/2)$$

If we copy that equation down for the other pairs of XY coordinates, our sheet should now contain a column of all the distance calculations.

Suppose we start off with a slightly modified worksheet that requires us to calculate the distance between all the points in the first column of XY values to a single point in the second column.

We can calculate the distance between all the points in the first column to the single point through the use of absolute addressing. An **absolute address** allows an equation to reference a single cell that will remain constant regardless of where the equation is copied in the worksheet. An absolute reference is indicated by a dollar sign ($) in front of the row and column designators. In this example, we want to use an absolute reference on cells D7 and E7 in all distance calculations. The equation we need to type in cell G7 becomes:

$$= ((\$D\$7 - A7) \wedge 2 + (\$E\$7 - B7) \wedge 2) \wedge (1/2)$$

Relative Addressing

- A **relative cell address** used in a formula will always *refer to the cell in the same relative position* to the cell containing the formula no matter where formula is copied in the worksheet. For example, if "=B2" is typed into cell C4 and then copied to cell C7, the formula in cell C7 would read "=B5". In this case, the cell reference is to call the cell two rows up and one cell to the left.

- When we insert or change cells, the formulas automatically update. This is one of a worksheet's major advantages: easily applying the same calculation to many different sets of data.

Absolute Addressing

- Absolute addressing is indicated by the presence of a dollar sign ($) immediately before both the column and row designators in the formula (e.g., $C$5; $AB$10).
- An **absolute cell address** will *always refer to the same cell* if the formula is copied to another location. For example, if "=$B$2" is typed into cell C4 and then copied to cell C7, the formula in cell C7 would read "=$B$2".

Mixed Addressing

- In **mixed addressing**, *either the row or the column designator is fixed* (by the $), but the other is relative (e.g., $C5; AB$10; $AB10).
- It may not be immediately obvious why this capability is desirable, but many problems are dramatically simplified with this approach. We will study this in more detail later.

Comprehension Check 13-1

Launch a new worksheet.
    Type "5" in cell E22 and "9" in cell E23; type "=E22 + 4" in cell F22.
    Copy cell F22 to cell F23.
Is this an example of absolute, mixed, or relative addressing?
What is displayed in cell F23?

Type "20" into cell G22 and "=$G$22 + 10" in cell H22.
    Copy cell H22 down to row 26 using the fill handle.
Is this an example of absolute, mixed, or relative addressing?
What is displayed in cell H26?

Type "25" into cell A28 and "=A$28 + 5" in cell D28.
    Copy cell D28 down to row 30 using the fill handle.
    Copy cell D28 across to column F using the fill handle.
Is this an example of absolute, mixed, or relative addressing?
What is displayed in cell D30?
What is displayed in cell F28?

Type "=$A28 + 5" in cell G28.
    Copy cell G28 down to row 30 using the fill handle.
    Copy cell G28 across to column J using the fill handle.
Is this an example of absolute, mixed, or relative addressing?
What is displayed in cell G30?
What is displayed in cell J28?

13.2 FUNCTIONS IN EXCEL

Hundreds of functions are built into Excel. Tables 13-1 through 13-4 list a few functions commonly used in engineering applications. There are several things you should note when using these functions.

- You must make certain to *use the correct name of the function*. For example, the average function is written as AVERAGE and cannot be abbreviated AVE or AVG.
- *All functions must be followed by parentheses.* For example, the value of π is given as PI(), with nothing inside the parentheses.
- The *argument* of the function (the stuff in the parentheses) can include numbers, text, expressions, or cell references, as long as they are appropriate for the function.
- Many functions can *accept a list or range of cells as the argument*. These can be expressed as a list separated by commas [e.g., A6, D7, R2, F9], as a rectangular block designated by the top-left cell and bottom-right cell separated by a colon [e.g., D3:F9], or as a mixed group [e.g., A6, R2, D3:F9]. To insert cells into a formula, type the formula up to the open parenthesis and select the desired cells. You can also type in the references directly into the formula.
- Most functions will also *accept another function as the argument*. These can be fairly simple [e.g., SIN (RADIANS (90))] or more complicated [e.g., AVERAGE (SQRT(R2), COS(S4 + C4), MIN (D3:F9) + 2)].
- Some functions, such as trigonometric functions, require specific arguments. *Trigonometric functions must have an argument in units of radians, not units of degrees.* Be sure you are aware of any limitations of the functions you are using. Look up an unfamiliar function in the **HELP** menu.
- Note that *some functions can be expressed in several different ways*. For example, raising the number 2 to the fifth power can be written as = 2 ^ 5 or as POWER(2,5).

Table 13-1 Trigonometric functions in Excel

Function as written in Excel	Definition
ACOS (cell)	Calculates the inverse cosine of a number (also ASIN)
COS (angle in radians)	Calculates the cosine of an angle (also SIN)
DEGREES (angle in radians)	Converts radians to degrees
PI()	Calculates pi (π) to about 15 significant figures
RADIANS (angle in degrees)	Converts degrees to radians

Table 13-2 Mathematical functions in Excel

Function as written in Excel	Definition
EXP (cell)	Raises e (base of the natural log) to the power "cell"
POWER (cell, power)	Raises the cell to "power"
PRODUCT (cells)	Finds the product of a list of cells
SQRT (cell)	Finds the square root of cell
SUM (cells)	Finds the sum of a list of cells

Table 13-3 Statistical functions in Excel

Function as written in Excel	Definition
AVERAGE (cells)	Finds the mean or average value of a list of cells
MAX (cells)	Finds the maximum value in a list of cells
MEDIAN (cells)	Finds the median value of a list of cells
MIN (cells)	Finds the minimum value in a list of cells
STDEV (cells)	Finds the standard deviation value of a list of cells
VAR (cells)	Finds the variance value of a list of cells

Table 13-4 Miscellaneous functions in Excel

Function as written in Excel	Definition
COUNT (cells)	Counts number of cells that are not blank and that do not contain an error
COUNTIF (cells, criteria)	Counts number of cells that meet the stated criteria, such as a numerical value, text, or a cell reference
INTERCEPT (y values, x values)	Calculates linear line for range of (x, y) pairs and returns the intercept value of y (where x = 0)
ROUND (cell, number of decimal places)	Rounds a number to a specific number of decimal places
SLOPE (y values, x values)	Calculates linear line for range of (x, y) pairs and returns the slope value
TRUNC (cell, number of digits)	Truncates a number to a specific number of digits

EXAMPLE 13-2

Assume we are studying the number of fatal accidents that occur during different times of the day. Given the data shown, we want to use Excel to analyze our data to determine the average, minimum, or maximum number of accidents, as well as a few other items that might be of significance.

Total Accidents: = SUM (B5:B18)
Total Samples: = COUNT (B5:B18)
Mean: = AVERAGE (B5:B18)
Median: = MEDIAN (B5:B18)
Variance: = VAR (B5:B18)
Standard deviation: = STDEV (B5:B18)

	A	B	
1	Fatal Vehicular Accidents		
2	This worksheet demonstrates the proper use of Excel functions		
3			
4	Week	Number Fatal Accidents	
5	A	190	
6	B	202	
7	C	179	
8	D	211	
9	E	160	
10	F	185	
11	G	172	
12	H	205	
13	I	177	
14	J	120	
15	K	235	
16	L	183	
17	M	177	
18	N	193	
19			

Note that decimal values appear when we calculate the mean, variance, and standard deviation of the accident data. Since it makes sense to round these values up to the nearest whole number, we need to type those functions as the argument to a rounding function. Start by modifying the equation for the mean by typing the ROUND function. Notice that as you start typing the ROUND function in the cell, a drop-down menu with a list of all

2589	
14	
=round	
fx ROUND	
703	*fx* ROUNDDOWN	
26.5	*fx* ROUNDUP	Rounds a number up, away from zero

Total Accidents	2589
Total Samples	14
Me	=ROUNDUP(
	ROUNDUP(**number**, num_digits)
Median	
Variance	703.1483516
Standard Deviation	26.51694461

of the functions that start with the letters ROUND appears below the cell. Note that Excel contains a function called ROUNDUP that will round a number up to the nearest whole value away from zero.

After we select the ROUNDUP function, a new box below the cell documents the arguments the function requires. Note that we need to provide the value we want to round as the first argument and the number of decimal places to which we want to round the number (in this case, 0).

The new function we need to type ultimately becomes

$$= ROUNDUP (AVERAGE (B5:B18), 0)$$

Repeat this with the equations for calculating the median and the standard deviation.

Suppose we want to determine how many of the samples reported accidents greater than the calculated average number of accidents. Note that the COUNTIF function requires a "criteria" argument, which can take on a number of different values. For example, if we want to count the number of values greater than 200 in the range B5:B18, we need to type the criteria ">200" (in double quotes) as the 2nd argument to the COUNTIF function.

$$= COUNTIF (B5:B18,">200")$$

In this example, we want to compare our COUNTIF result to a value calculated in a different cell. Since we can't type cell references inside of double quotes (">E21"), we need to use the **ampersand** operator (&) to **concatenate** the logical operator to the cell reference (">"&E21).

NOTE

The ROUND function refers to number of decimal places, although the Excel help menu calls this "num_digits". Be sure to always read ALL the help menu file when using a new function.

	A	B
1	Fatal Vehicular Accidents	
2	This worksheet demonstrates the proper use of Excel functions	
3		
4	**Week**	**Number of Fatal Accidents**
5	A	190
6	B	202
7	C	179
8	D	211
9	E	160
10	F	185
11	G	172
12	H	205
13	I	177
14	J	120
15	K	235
16	L	183
17	M	177
18	N	193
19		
20	Total Accidents	2589
21	Total Samples	14
22	Mean	185
23	Median	184
24	Variance	704
25	Standard Deviation	27
26		

Samples Greater than Mean: = COUNTIF (B5:B18,">"&B22).

Similarly, we could use the COUNTIFS function to calculate the number of samples that have a number of accidents between (and including) 180 and 200. COUNTIFS is a special function that contains a variable number of arguments, with a minimum of two arguments required (range1, criteria1) to use the function. Since we have two criteria that must be met (>180 and <200), we must pass in four arguments to the COUNTIFS function (range1, criteria1, range2, criteria2). In this example, range1 and range2 must be the same range of cells since we are enforcing the criteria on the same set of data. We will place the bounds in the worksheet as follows:

Lower Bound in D23: 180   Upper Bound in E23: 200

Samples Between: = COUNTIFS (B5:B18,">="&D23, B5:B18, "<="&E23). Your final worksheet should appear as shown.

	A	B	C	D	E
1	Fatal Vehicular Accidents				
2	This worksheet demonstrates the proper use of Excel functions				
3					
4	**Week**	**Number of Fatal Accidents**			
5	A	190			
6	B	202			
7	C	179			
8	D	211			
9	E	160			
10	F	185			
11	G	172			
12	H	205			
13	I	177			
14	J	120			
15	K	235			
16	L	183			
17	M	177			
18	N	193			
19					
20	Total Accidents	2589	Samples Greater than Mean		
21	Total Samples	14	6		
22	Mean	185			
23	Median	184	Samples Between	180	200
24	Variance	704	4		
25	Standard Deviation	27			

Comprehension Check 13-2

Launch a new worksheet. Type the following Excel expressions into the specified cells. Be certain you understand *why* each of the following yields the specific result. If the formula returns an error, how can the formula be changed to correctly display the desired result?

In cell . . .	Enter the formula . . .	The cell will display . . .
A1	= SQRT (144)	
A2	= MAX (5, 8, 20/2, 5 + 6)	
A3	= AVERAGE (5, SQRT(100), 15)	
A4	= POWER (2, 5)	
A5	= PI()	
A6	= PI	
A7	= PRODUCT (2, 5, A2)	
A8	= SUM (2 + 7, 3 * 2, A1:A3)	
A9	= RADIANS (90)	
A10	= SIN (RADIANS (90))	
A11	= SIN (90)	
A12	= ACOS (0.7071)	
A13	= DEGREES(ACOS(0.7071))	
A14	= CUBRT(27)	

EXAMPLE 13-3

The maximum height (H) an object can achieve when thrown can be determined from the velocity (v) and the launch angle with respect to the horizontal (θ):

$$H = \frac{v^2 \sin(\theta)}{2g}$$

Note the use of a cell (E7) to hold the value of the acceleration due to gravity. This cell will be referenced in the formulae instead of our inserting the actual value into the formulae.

This will allow us to easily work the problem in a different gravitational environment (e.g., Mars) simply by changing the one cell containing the gravitational constant.

For the following, assume that the angle 50° is in cell B10. After setting up the column of velocities and the row of angles, we type the following into cell B11 (immediately below 50°)

$$= \$A11{\wedge}2 * SIN (RADIANS (B\$10)) / (2*\$E\$7)$$

Note the use of absolute addressing (for gravity) and mixed addressing (for angle and velocity). For the angle, we allow the column to change (since the angles are in different columns) but not the row (since all angles are in row 10). For the velocity, we allow the row to change (since the velocities are in different rows) but the column is fixed (since all velocities are in column A). This allows us to write a *single* formula and replicate it in both directions.

The sine function requires an argument in units of radians, and the angle is given in units of degrees in the problem statement. In this example, we used the RADIANS function to convert from degrees into radians. Another method is to use the relationship 2π radians is equal to 360 degrees, or

$$= \$A11{\wedge}2 * SIN ((2 * PI() / 360) * B\$10) / (2*\$E\$7)$$

We replicate the formula in cell B11 across the row to cell E11, selecting all four formulae in row 11 and replicating to row 16. If done correctly, the values should appear as shown.

Velocity (v) [m/s]	Angle (θ) [°]			
	50	60	70	80
20	3.91	4.42	4.79	5.02
12	5.63	6.36	6.90	7.24
14	7.66	8.66	9.40	9.85
16	10.01	11.31	12.27	12.86
18	12.66	14.32	15.53	16.28
20	15.63	17.67	19.18	20.10

Here, we change the planet to be Mars and the value of gravity to be 3.7 meters per second squared in cell E7. The worksheet should automatically update, and the values should appear as shown.

Velocity (v) [m/s]	Angle (θ) [°]			
	50	60	70	80
10	10.35	11.70	12.70	13.31
12	14.91	16.85	18.29	19.16
14	20.29	22.94	24.89	26.08
16	26.50	29.96	32.51	34.07
18	33.54	37.92	41.14	43.12
20	41.41	46.81	50.79	53.23

Now, we change the planet to be Moon and the value of gravity to be 1.6 meters per second squared in cell E7. The worksheet should automatically update, and the values should appear as shown.

Velocity (v) [m/s]	Angle (θ) [°]			
	50	60	70	80
10	23.94	27.06	29.37	30.78
12	34.47	38.97	42.29	44.32
14	46.92	53.04	57.56	60.32
16	61.28	69.28	75.18	78.78
18	77.56	87.69	95.14	99.71
20	95.76	108.25	117.46	123.10

13.3   LOGIC AND CONDITIONALS

Outside of the realm of computing, logic exists as a driving force for decision-making. Logic transforms a list of arguments into outcomes based on a decision.

Some examples of every day decision-making:

- If the traffic light is red, stop. If the traffic light is yellow, slow down. If the traffic light is green, go.

    Argument: 3 traffic bulbs   Decision: is bulb lit?   Outcomes: stop, go, slow

- If the milk has passed the expiration date, throw it out; otherwise, keep the milk

    Argument: expiration date   Decision: before or after?   Outcomes: garbage, keep

To bring decision-making into our perspective on problem solving, we need to first understand how computers make decisions. **Boolean logic** exists to assist in the decision-making process, where each argument has a binary result and our overall outcome exhibits binary behavior. **Binary behavior**, depending on the application, is any sort of behavior that results in two possible outcomes. That is:

In computing, we often refer to the outcome of Boolean calculations as 'yes' and 'no.' Alternatively, we may refer to the outcomes as 'true' and 'false,' or '1' and '0.' To connect all the Boolean arguments to make a logical decision, we have a few operators that allow us to relate our arguments in order to determine a final outcome.

> AND is true if and only if all arguments are true.
>
> OR is true if at least one of the arguments is true.

- **AND:** The AND logical operator enables us to connect two Boolean arguments and return the result as TRUE if and only if *both* Boolean arguments have the value of TRUE. In Excel, the AND function accepts more than two arguments and is TRUE if all the arguments are TRUE.
- **OR:** The OR logical operator enables us to connect two Boolean arguments and return the result as TRUE if *only one* of the Boolean arguments has the value of TRUE. In Excel, the OR function accepts two or more arguments and is TRUE if at least one of the arguments is TRUE.
- **NOT:** The NOT logical operator enables us to invert the result of a Boolean operation. In Excel, the NOT function accepts one argument. If the value of that argument is TRUE, the NOT function returns FALSE. Likewise, if the argument of the function is FALSE, the NOT function returns TRUE.

To determine the relationship between two cells (containing numbers or text), we have a few operators, listed in Table 13-5, that allow us to compare two cells to determine whether or not the comparison is true or false.

Table 13-5 Relational operators in Excel

Operator	Meaning
>	Greater than
<	Less than
>=	Greater than or equal to
<=	Less than or equal to
=	Equal to
<>	Not equal to

These relational operators are usually placed between two different cells in order to determine the relationship between the two values. This expression of cell-operator-cell is typically called a **relational expression**. Relational expressions can be combined by means of logical operators to create a **logical expression**. If no logical operator is required in a particular decision, then the single relational expression can be the logical expression.

Conditional statements are commands that give some decision-making authority to the computer. Specifically, the user asks the computer a question using conditionals statements, and then the computer selects a path forward based on the answer to the question. Sample statements are given below:

- If the water velocity is fast enough, switch to an equation for turbulent flow!
- If the temperature is high enough, reduce the allowable stress on this steel beam!
- If the RPM level is above red line, issue a warning!
- If your grade is high enough on the test, state: You Passed!

In these examples, the comma indicates the separation of the condition and the action that is to be taken if the condition is true. The exclamation point marks the end of the statement. Just as in language, more complex conditional statements can be crafted with the use of "else" and "otherwise" and similar words. In these statements, the use of a semicolon introduces a new conditional clause, known as a nested conditional statement. For example:

- If the collected data indicate the process is in control, continue taking data; otherwise, alert the operator.
- If the water temperature is at or less than 10 degrees Celsius, turn on the heater; or else if the water temperature is at or greater than 80 degrees Celsius, turn on the chiller; otherwise, take no action.

Single Conditional Statements

In Excel, conditional statements can be used to return a value within a cell based upon specified criteria. The IF conditional statement within Excel takes the form

=IF (logical test, value if true, value if false)

Every statement must contain three and only three parts:

1. A logical test, or the question to be answered
The answer to the logical test must be TRUE or FALSE.
Is the flow rate in Reactor #1 higher than Reactor #5?

2. **A TRUE response,** if the answer to the question is yes
   Show the number 1 to indicate Reactor #1.

3. **A FALSE response,** if the answer to the question is no
   Show the number 5 to indicate Reactor #5.

The whole statement for the above example would read:

$$= IF (B3 > B4, 1, 5)$$

	A	B	C
1			
2			
3	Reactor #1 Flowrate	10	[gpm]
4	Reactor #5 Flowrate	25	[gpm]
5	Maximum Flowrate in Reactor #	5	

Special Things to Note

- **To leave a cell blank, type a set of quotations with nothing in between ("").** For example, the statement =IF (C3>10, 5, " ") is blank if C3 is less than 10.
- For display of a text statement, the *text must be stated within quotes ("text goes in here")*. For example, the statement = IF (E5 > 10, 5, "WARNING") would display the word WARNING if E5 is less than 10.

EXAMPLE 13-4

For the following scenarios, write a conditional statement to be placed in cell B5 to satisfy the conditions given. Below each statement are sample outcomes of the worksheet in different scenarios.

(a) Display the pressure difference between upstream station 1 (displayed in cell B3) and downstream station 2 (displayed in cell B4) if the pressure difference is positive; otherwise, display the number 1.

	A	B	C
1			
2			
3	Station #1 Pressure	2.4	[atm]
4	Station #2 Pressure	2.8	[atm]
5	Pressure Difference	1	[atm]

	A	B	C
1			
2			
3	Station #1 Pressure	3.2	[atm]
4	Station #2 Pressure	2.8	[atm]
5	Pressure Difference	0.4	[atm]

Answer: =IF ((B3 − B4) > 0, B3 − B4, 1)

(b) Display the value of the current tank pressure if the current pressure is less than the maximum tank pressure; otherwise, display the word "MAX".

	A	B	C
1			
2			
3	Maximum Tank Pressure	5	[atm]
4	Current Tank Pressure	2	[atm]
5	Pressure Status	2	[atm]

	A	B	C
1			
2			
3	Maximum Tank Pressure	5	[atm]
4	Current Tank Pressure	10	[atm]
5	Pressure Status	MAX	[atm]

Answer: =IF (B3 > B4, B4, "MAX")

(c) If the sum of the temperature values shown in cell B2, B3, and B4 is greater than or equal to 100, leave the cell blank; otherwise, display a warning to the operator that the temperature is too low.

	A	B	C
1			
2	Temperature Reading #1	25	[°C]
3	Temperature Reading #2	50	[°C]
4	Temperature Reading #3	45	[°C]
5	Cumulative Temperature		

	A	B	C
1			
2	Temperature Reading #1	25	[°C]
3	Temperature Reading #2	10	[°C]
4	Temperature Reading #3	45	[°C]
5	Cumulative Temperature	Too Low	

Answer: =IF (SUM (B2:B4) >= 100, "", "Too Low")

Comprehension Check 13-3

Are the logical expressions displayed in Example 13-4 the only logical expressions that could be used to display the required outcomes? Write out a logically equivalent expression that could be typed into cell B5 for each situation.

Nested Conditional Statements

If more than two outcomes exist, the conditional statements in Excel can be nested. The nested IF conditional statement within Excel can take the form

=IF (logical test #1, value if #1 true, IF (logical test #2, value if #2 true, value if both false))

Note that the number of parenthesis must match (open and closed) and must be placed in the proper location. Recall that every statement must contain three and only three parts. For the first IF statement, they are

1. **The first logical test, or the first question to be answered**
   The answer to the logical test must be TRUE or FALSE.
   Is the score for Quiz #1 less than the score for Quiz #2?

2. **A true response**, or what to do if the answer to the first question is yes
   Show the score for Quiz #1.

3. **A false response**, or what to do if the answer to the first question is no
   Proceed to the logical question for the second IF statement.

For the second IF statement, the three parts are

1. **The second logical test, or the second question to be answered**
   The answer to the logical test must be TRUE or FALSE.
   Is the score for Quiz #2 less than the score for Quiz #1?

2. **A true response**, or what to do if the answer to the second question is yes
   Show the score for Quiz #2.

3. **A false response**, or what to do if the answer to the second question, and by default both questions, is no
   Show the text "equal".

The whole statement typed in cell B5 for the above example would read

$$= IF\,(B3 < B4,\,B3,\,IF\,(B3 > B4,\,B4,\,"equal"))$$

	A	B	C
1			
2			
3	Quiz Grade #1	70	
4	Quiz Grade #2	70	
5	Lowest Quiz Score	Equal	

	A	B	C
1			
2			
3	Quiz Grade #1	90	
4	Quiz Grade #2	70	
5	Lowest Quiz Score	70	

	A	B	C
1			
2			
3	Quiz Grade #1	90	
4	Quiz Grade #2	70	
5	Lowest Quiz Score	70	

There can be a maximum of 64 nested IF statements within a single cell. The nested IF can appear as either the true or false response to the first IF logical test. In the above example, only the false response option is shown.

EXAMPLE 13-5

Write the conditional statement to display the state of water (ice, liquid, or steam) based upon temperature displayed in cell B4, given in degrees Celsius. Below are sample outcomes of the worksheet in different scenarios.

	A	B	C
1			
2			
3			
4	Temperature of Mixture	75	[°C]
5	State of Mixture	Liquid	

	A	B	C
1			
2			
3			
4	Temperature of Mixture	110	[°C]
5	State of Mixture	Steam	

	A	B	C
1			
2			
3			
4	Temperature of Mixture	10	[°C]
5	State of Mixture	Ice	

Here, there must be *two* conditional statements because there are three responses:

- *If the temperature is less than or equal to zero, display "ice";*
- *If the temperature is greater than or equal to 100, display "steam";*
- *Otherwise, display "liquid".*

   Answer: $=IF\,(B4 <= 0,\,"ice",\,IF\,(B4 >= 100,\,"steam",\,"liquid"))$

> **Comprehension Check 13-4**
>
> Is the logical expression displayed in Example 13-5 the only logical expression that could be used to display the phase of the water? Write out a logically equivalent expression that could be typed into cell B5.

Compound Conditional Statements

If more than two logic tests exist for a single condition, conditional statements can be linked together by AND, OR, and NOT functions. Up to 255 logical tests can be compared in a single IF statement (only two are shown in the box below). The compound IF conditional statement takes the form

---

=IF (AND (logical test #1, logical test #2), value if both tests are true, value if either test is false)

=IF (OR (logical test #1, logical test #2), value if either test is true, value if both tests are false)

EXAMPLE 13-6

Write the conditional statement that meets the following criteria.

(a) If the product has cleared all three quality checks (given in cells B2, B3, and B4) with a score of 80 or more on each check, mark the product as "OK" to ship; otherwise, mark the product as "Recycle".

	A	B	C
1			
2	Quality Check #1 Rating	90	
3	Quality Check #2 Rating	80	
4	Quality Check #3 Rating	85	
5	Mark Product	OK	

	A	B	C
1			
2	Quality Check #1 Rating	60	
3	Quality Check #2 Rating	80	
4	Quality Check #3 Rating	85	
5	Mark Product	Recycle	

Answer: =IF (AND (B2 >= 80, B3 >= 80, B4 >= 80),"OK", "Recycle")

(b) If the product has cleared all three quality checks (given in cells B2, B3, and B4) with a minimum score of 80 on each check, mark the product as "OK" to ship; otherwise, if the product scored a 50 or below on any check, mark the product as "Rejected"; otherwise, mark the product as "Rework".

	A	B	C
1			
2	Quality Check #1 Rating	90	
3	Quality Check #2 Rating	80	
4	Quality Check #3 Rating	85	
5	Mark Product	OK	

	A	B	C
1			
2	Quality Check #1 Rating	40	
3	Quality Check #2 Rating	80	
4	Quality Check #3 Rating	85	
5	Mark Product	Rejected	

	A	B	C
1			
2	Quality Check #1 Rating	60	
3	Quality Check #2 Rating	80	
4	Quality Check #3 Rating	85	
5	Mark Product	Rework	

Answer: =IF (AND (B2 >= 80, B3 >= 80, B4 >= 80),"OK", IF (OR (B2 <= 50, B3 <= 50, B4 <= 50), "Rejected", "Rework"))

Comprehension Check 13-5

Is the logical expression displayed in Example 13-6(a) the only logical expression that could be used to display the desired outcome? Write out a logically equivalent expression that could be typed into cell B5.

13.4 LOOKUP AND DATA VALIDATION

The LOOKUP function enables Excel to locate information from a table of data in a worksheet. There are two LOOKUP functions: VLOOKUP, which searches vertically, and HLOOKUP, which searches horizontally. In the following example, we focus on VLOOKUP, but the same principles could easily be applied to HLOOKUP. To use the VLOOKUP function, we need to pass in four different arguments:

> VLOOKUP (lookup_value, table_array, col_index_num, [range_lookup])

- The ***lookup_value*** argument is the value we want to look up in the table. Typically, this value is a string, but it can be a numerical value. Note that whatever we use as the *lookup_value*, Excel will perform a case-insensitive search of the data for the value; that means that any special characters used in the string, like punctuation or spaces, must appear the same in the *lookup_value* and the table, and must be a unique identifier in the table.
- The ***table_array*** is the range of cells that encapsulates the entire data table we want to search. Since we're using VLOOKUP, it's important to realize that our *table_array* must have at least two columns of data. Note that the *lookup_value* we're passing in to the VLOOKUP function will only search the first column of the *table_array*, so it might be necessary to move the data around.
- The ***col_index_num*** argument is the column number that contains the data we want as a result of our search. By default, Excel will refer to the first column where the *lookup_value* is located as the number 1, so the *col_index_num* will always be a number greater than 1.
- The last argument, [***range_lookup***], is an optional argument (so it's listed in square brackets.) This argument tells the function what type of search to perform and can only take on two values: TRUE or FALSE.
  - Passing in TRUE tells Excel to conduct an approximate search of the data. That is, Excel will search the data table for the largest value that is less than the *lookup_value* and use that result as the selected value. Note that for an approximate search, the first column of the *table_array* must be sorted in ascending order.
  - Passing in FALSE tells Excel to conduct an exact search of the data. The data need not be sorted for this option. If an exact match is not found, the function returns an error.
  - If we don't specify TRUE or FALSE, Excel attempts to match the data exactly, and if a match is not found, Excel returns an approximate value. This may give undesired results. It's good practice to tell Excel which searching algorithm to use to search the *table_array*.

Assume we're given the following table of data on students. To determine what Sally's eye color is from (column C) in cell A5, we could type

$$= VLOOKUP (\text{"Sally"}, A1{:}D4, 3, FALS)$$

since the data are unsorted and we're looking for an exact match on Sally.

	A	B	C	D
1	Joe	18	Blue	EE
2	John	19	Brown	ME
3	Sally	18	Brown	IE
4	Julie	18	Blue	CE

EXAMPLE 13-7

Digital audio is a relatively new medium for storing and reproducing music. Before albums were sold on CD and other digital media formats, analog recordings were commonly sold as vinyl records, 8-track tapes, and cassette tapes. We want to build to build a worksheet to help us compare these different media formats to observe how information storage has progressed over the past 50 years.

Note the following media equivalencies:

- A 74-minute CD (44.1 KHz, 2 channel, 16-bit digital audio) can hold 650 MB of data.
- A single-sided, single-layer DVD can hold 4.7 GB of data (approx. 4813 MB, 547 minutes of 44.1 KHz, 2 channel, 16-bit digital audio).
- A single-sided, single-layer Blu-ray disc can hold 25 GB of data (approx. 25600 MB, 2914 minutes of 44.1 KHz, 2 channel, 16-bit digital audio).
- A 7″ vinyl record recorded at 45 rpm can hold 9 minutes of music.
- A 7″ vinyl record recorded at 33 $\frac{1}{3}$ rpm can hold 12 minutes of music.
- A 12″ vinyl record recorded at 45 rpm can hold 24 minutes of music.
- A 12″ vinyl record recorded at 33 $\frac{1}{3}$ rpm can hold 36 minutes of music.
- An 8-track tape can hold 46 minutes of music.
- A typical cassette tape can hold 60 minutes of music.

	A	B	C	D	E
1	Digital Audio Media				
2	This worksheet demonstrates the use of VLOOKUP and data validation				
3					
4					
5	Quantity	Format	is equivalent to	Quantity	Format
6	100	Blu-ray Disc		3938	CD
7				533	DVD
8				100	Blu-ray Disc
9				32378	7" @ 45 rpm
10				24284	7" @ 33 1/3 rpm
11				12142	12" @ 45 rpm
12				8095	12" @ 33 1/3 rpm
13				6335	8-track tape
14				4857	Cassette tape

To determine audio equivalencies between these different storage formats, we first create a worksheet. We want to allow the user to input the media type and quantity of the desired format to be converted. To complete the comparison, it would seem like each calculation requires a statement with nine questions to ask (Is it a CD? Is it a DVD? Is it a Blu-ray? . . .).

Rather than requiring the user to type the name of the media each time (CD, DVD, Blu-ray, etc), Excel can do *data validation*, so we can give the user of our worksheet a drop-down menu from which to select the media. We need to add a table that contains the name of each media type along with the length of the audio we can fit on each media. We will place this table below our initial data, in cells A17:B26.

Next, we need to calculate the quantity of each item. Since the name of the media will appear in cell B6, we use that as the lookup value in our VLOOKUP statement. To calculate the quantity for each equivalent media, we look up the length of the format specified in B6, divide that by the length of each media given in column E, and multiply that by the number of the original media provided in A6. Note that we need to round this number up since it doesn't make sense to have a noninteger value in our count.

For CDs, the calculation in Cell D6 should be

= ROUNDUP (VLOOKUP ($B$6, $A$18:$B$26, 2, FALSE)/
VLOOKUP (E6, $A$18:$B$26, 2, FALSE) *$A$6, 0)

A sample calculation using 100 Blu-ray discs is shown.

The next step to finish our worksheet is to include a drop-down menu of the different media formats. To insert data validation on the media format, we click Cell B6 and go to **Data** > **Data Tools** > **Data Validation**.

	A	B
16	Storage Information	
17	Format	Length [min]
18	CD	74
19	DVD	547
20	Blu-ray Disc	2914
21	7" @ 45 rpm	9
22	7" @ 33 1/3 rpm	12
23	12" @ 45 rpm	24
24	12" @ 33 1/3 rpm	36
25	8-track tape	46
26	Cassette tape	60

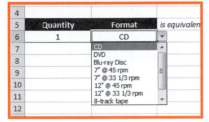

⌘ **Mac OS: Data > Validation**

The **Data Validation** window is displayed. Under the **Settings** tab, the **Allow:** menu lets us specify the type of data that can be provided in the cell we selected. Since we want to restrict the data to a list of values, we select **List**.

Under the **Source:** option, we select the range of all of the media types, $A\$18:\$A\$26$, and click **OK** to close the Data Validation window.

Notice the drop-down handle next to cell B6. When the user of the worksheet clicks B6, a drop-down menu appears that lists all of the possible media types so that the user can quickly select an item from the list. Furthermore, this feature prevents the user from typing items that aren't on the list, making a typo, or entering any other information that will cause an error in calculations that rely on the value in B6.

In addition to controlling the input type to a cell, it's also possible to give feedback to the person using the worksheet using pop up messages. In this example, the quantity can't be a negative number, so we need to bring up the Data Validation window again and restrict the input to only allow whole numbers that are greater than or equal to zero.

Next, we need to click the Input Message tab to type in a message that will appear below the cell when the person using our worksheet clicks on the cell to type in a quantity.

Finally, we need to click the Error Alert tab to provide the message that should pop up when an invalid number is typed into the cell.

Comprehension Check 13-6

Open the worksheet you created in Example 13-3 and modify it to use VLOOKUP and data validation to allow the user of the worksheet to select the planet and automatically fill in the gravity for each planet.

Planet	Gravity (g) [m/s^2]
Earth	9.8
Jupiter	24.8
Mars	3.7
Mercury	3.7
Moon	1.6
Neptune	11.2
Pluto	0.7
Saturn	10.4
Uranus	8.9
Venus	8.9

13.5 CONDITIONAL FORMATTING

You can use conditional formatting to change the font color or background of a cell based upon the values found in that cell. As an example:

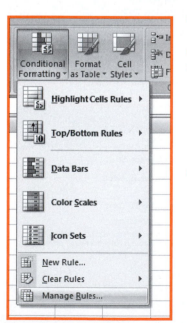

- On a blank worksheet, type the value of 20 in cell A4, a value of 30 in cell B4, and a value of 50 in cell C4.
- Select cells A4 to C4.
- Select the **Conditional Formatting** drop-down menu in **Home > Styles** and click the **Manage Rules** item.

⌘ **Mac OS: Format > Conditional Formatting**

- In the Conditional Formatting Rules Manager window, click the New Rule button to open the New Formatting Rule wizard. In this window, you can now select several options. Under Select a Rule Type, click the option **Format only cells that contain** to bring up the interface for creating custom formats based on certain conditions. In the first drop-down menu, you can choose among **Cell Value** and a few other options. For this activity, use only Cell Value.

- In the second drop-down menu, you can choose various conditional statements such as "between" or "less than." To begin this problem, choose "less than" from this list.

- The choice of "less than" will combine the next two boxes into a single box. You can enter a number or formula, or reference a cell within the worksheet.

  For this example, enter the value "25". Note: If you enter a formula, the same rules apply for absolute and relative referencing. In addition, if you select a cell within the worksheet, the program automatically defaults, to an absolute reference.

- Click the **Format** button. Select the formatting you want to apply when the cell value meets the condition or the formula returns the value TRUE. You can change the font, border, or pattern (background) of the cell. For this example, choose a light grey background on the Pattern tab. When you're finished, click **OK**.

- To add another condition, click **New Rule**. Additionally, in this window, you can edit or delete existing conditional formatting options. For this condition, make it greater than 40, with a font of white, bolded on a dark grey background. Click **OK**. To apply the conditions in the Conditional Formatting Rules Manager window, click the **Apply** button, then click **OK**.

  Your worksheet should now look like the one shown. If none of the specified conditions are TRUE, the cells keep their existing formats.

EXAMPLE 13-8

Let's assume we want to build an interactive worksheet that changes the format of a cell to model the behavior of a traffic light. We want the user to input the number of seconds it takes for a light (which is initially green) to turn red. In addition, the user must also be able to provide the "warning" so that the light can switch from green to yellow to red.

In this text, the colors will not show correctly as red, yellow, and green. Please refer to the online materials to see the correct colors.

- The green light (bottom) will only be lit if the time remaining is greater than the warning time.
- The yellow light (middle) will only be lit if the time remaining is greater than 0 seconds, but less than the warning time.
- The red light (top) will only be lit if the time remaining is 0 seconds.

Before we set up the conditional formatting for each cell, we need to write IF statements in the light cells that will be used as a trigger for conditional formatting.

For Cell E5 (the red light): =IF (A5=0, "R", "")
For Cell E9 (the yellow light): =IF (AND (A5<=B5,A5>0), "Y", "")
For Cell E13 (the green light): =IF (A5 > B5, "G", "")

Next, we add a set of conditional formatting rules for each cell.

For the red light, we click E5 and select the Manage Rules item from the Conditional Formatting menu. We add two rules:

- The first rule turns the fill and font color red when cell has the letter "R" in the text.
- Otherwise, the second rule turns the fill and font color black (off).

For the yellow light, we click E9 and select the Manage Rules item from the Conditional Formatting menu. We add two rules:

- The first rule turns the fill and font color yellow when cell has the letter "Y" in the text.
- Otherwise, the second rule turns the fill and font color black (off).

For the green light, we click E13 and select the Manage Rules item from the Conditional Formatting menu. We add two rules:

- The first rule turns the fill and font color green when cell has the letter "G" in the text.
- Otherwise, the second rule turns the fill and font color black (off).

The final worksheet should appear as shown. Note that cell formats should change when the time remaining changes. In a later section, we discuss how to insert dynamic controls to automatically change cell values, thereby giving this model even more interactive capability.

Comprehension Check 13-7

Open the worksheet you created in Example 13-3 and modify it to highlight all heights greater than 100 meters with a light grey background and all heights less than 25 meters with a dark grey background with a white font.

13.6 SORTING AND FILTERS

Excel provides a number of built-in tools for sorting and filtering data in a worksheet. This section describes how to use these tools effectively without causing unintended side effects.

Each year, the federal government publishes a list of fuel economy values. The complete lists for recent years can be found at http://www.fueleconomy.gov/feg. A partial list of 2009 vehicles is shown below. In the table, MPG = miles per gallon.

Make	Model	MPG City	MPG Highway	Annual Fuel Cost
Jeep	Liberty 4WD	15	21	$1835
BMW	X5 xDrive 30i	15	21	$1927
Honda	Civic Hybrid	40	45	$ 743
Volkswagen	Jetta 2.5L	20	29	$1301
Ford	Mustang 5.4L	14	20	$2166
Bentley	Continental GTC	10	17	$2886
Honda	Fit	27	33	$1039

Given this information, assume you are to present it with some sort of order. What if you want to sort the data on text values (Make or Model) or numerical values (MPG City, MPG Highway, Annual Fuel Cost), or what if you want to view only certain vehicles that meet a certain condition?

Sorting Data in a Worksheet

- Select the cells to be sorted. You can select cells in a single column or row, or in a rectangular group of cells.
- Select **Home > Editing > Sort & Filter.** By default, two commonly used sorting tools (Sort A to Z, Sort Z to A) appear, in addition to a button for Custom Sort. With a group of cells selected, the common sorting tools will sort according to the values in the left most column. If the leftmost column contained numerical values, the options would have read Sort Smallest to Largest/Largest to Smallest. Since it's often desired to involve multiple sorting conditions, click **Custom Sort.**

- The sorting wizard is displayed as shown. If your selected group of cells had a header row (a row that displays the names of the columns and not actual data) the "My data has headers" checkbox should be selected.

By default, Excel automatically detects whether the top row of your selected data is a header or a data row. Since you selected the data including the header rows, the "Sort by" drop-down menu will contain the header names. If you had not included the header row, the "Sort by" drop-down menu would show the column identifiers as options. It is good practice to select the headers in addition to the data to make sorting easier to understand.

 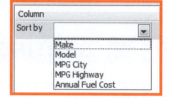

■ Assume you want to sort the list alphabetically (A to Z) by the make, then by smallest-to-largest annual fuel cost. Click the **Add Level** button to add two levels of sorting since there are two conditions. In the sorting wizard, the topmost sorting level will be the sort applied first, then the next level will sort each data group that forms from the first sort. In the example there is more than one Honda vehicle, so the second level will place the Civic Hybrid above the Fit, since the Civic Hybrid has a smaller annual fuel cost.

The resulting sorted data appear as shown.

	A	B	C	D	E
1	**Fuel Economy of Vehicles**				
2	This worksheet demonstrates the use of sorting and filtering in Excel				
3					
4	**Make**	**Model**	**MPG City**	**MPG Highway**	**Annual Fuel Cost**
5	Bentley	Continental GTC	10	17	$2,885
6	BMW	X5 xDrive 30i	15	21	$1,927
7	Ford	Mustang 5.4L	14	20	$2,165
8	Honda	Civic Hybrid	40	45	$743
9	Honda	Fit	27	33	$1,039
10	Jeep	Liberty 4WD	15	21	$1,835
11	Volkswagen	Jetta 2.5L	20	29	$1,301
12					

It's important to be sure to select all of the data when using the sort functions because it is possible to corrupt your data set. To demonstrate, select only the first three columns (Make, Model, MPG City) and sort the data smallest to largest on the MPG City column.

To "undo" a sort, either choose the "Undo" arrow button on the top menu or use CTRL-Z.

Notice after sorting that the last two columns (MPG Highway, Annual Fuel Cost) are not the correct values for the vehicle. There is no way to recover the original association if you were to save the file and open it at a later time, so it's critical that when using the built-in sorting functions, you verify the correctness of your data before saving your workbook. In this case, you can click Excel's Undo button or CTRL-Z to unapply the last sort.

	A	B	C	D	E
1	Fuel Economy of Vehicles				
2	This worksheet demonstrates the use of sorting and filtering in Excel				
3					
4	Make	Model	MPG City	MPG Highway	Annual Fuel Cost
5	Bentley	Continental GTC	10	17	$2,886
6	Ford	Mustang 5.4L	14	21	$1,927
7	BMW	X5 xDrive 30i	15	20	$2,166
8	Jeep	Liberty 4WD	15	45	$743
9	Volkswagen	Jetta 2.5L	20	33	$1,039
10	Honda	Fit	27	21	$1,835
11	Honda	Civic Hybrid	40	29	$1,301
12					

Improving automotive gas mileage while keeping costs under control is a complex puzzle, involving many different types of engineers. Below are some ways to possibly improve fuel efficiency. Some really work, some are false claims, and some are fictitious. Can you tell the difference? What other ways can you think of to improve today's automobiles?

- Minimize wait time at traffic signals
- Place a magnet on the fuel line
- Inflate tires with gas other than air
- Improve car pool regulations

- Put wind turbine on roof
- Add alternators to each wheel
- Make speed limits on weather
- Use solid rubber tires

- Limit acceleration rates
- Change tire pressure
- Add wind sail on roof
- Limit max speed

Comprehension Check 13-8

In 1980, the Environmental Protection Agency (EPA) began the Superfund Program to help clean up highly polluted areas of the environment. There are over 1,300 Superfund sites across the country. Not all Superfund sites are from deliberate pollution. Some sites are old factories where chemicals were dumped on the ground; landfills where garbage was dumped along with other poisonous waste; remote places where people secretly dumped hazardous waste because they did not know what to do with it; or old coal, iron ore, or silver mines.

According to the EPA (http://www.epa.gov/superfund/index.htm), the following groundwater contaminants were found in South Carolina Superfund sites in Greenville, Pickens, Oconee, and Anderson counties.

Sort by city in ascending order. Examine the result: Which city appears first?

Sort again: first by city in descending order, then by site name in descending order. Examine the results: Which site name now appears first?

Sort again by contaminant in ascending order, then by site name in ascending order. Examine the results: Which site name appears last?

Contaminants	Site Name	City
Polycyclic aromatic hydrocarbons	Sangamo Weston	Pickens
Volatile organic compounds	Beaunit Corporation	Fountain Inn
Polycyclic aromatic hydrocarbons	Beaunit Corporation	Fountain Inn
Polycyclic aromatic hydrocarbons	Para-Chem Southern, Inc.	Simpsonville
Volatile organic compounds	Golden Strip Septic Tank Service	Simpsonville
Volatile organic compounds	Para-Chem Southern, Inc.	Simpsonville
Metals	Para-Chem Southern, Inc.	Simpsonville
Polycyclic aromatic hydrocarbons	Rochester Property	Travelers Rest
Volatile organic compounds	Sangamo Weston	Pickens
Polychlorinated biphenyl	Sangamo Weston	Pickens
Metals	Rochester Property	Travelers Rest
Metals	Golden Strip Septic Tank Service	Simpsonville
Metals	Beaunit Corporation	Fountain Inn
Volatile organic compounds	Rochester Property	Travelers Rest

Filtering Data in a Worksheet

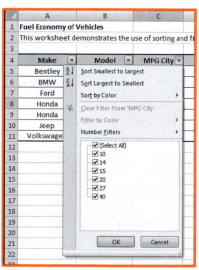

Assume you want to look only at a specific portion of the data set and hide all the other rows of data. For example, you might want to look only at Honda vehicles or all vehicles that have an MPG City rating between 10 and 15 MPG. Excel has a built-in filtering capability by which you can conditionally display rows in a data set.

■ Select the header row for a data set and click the **Sort & Filter** button in the **Home > Editing** ribbon. Click the **Filter** option to enable filtering for each column of data. Each column label contains a drop-down menu with various sorting options, as well as a number of different approaches for filtering.

● For data sets that contain a small number of options, use the checkboxes in the drop-down filter to manually check certain options to display.

● For numerical values, use the Number Filters submenu to filter on certain conditional expressions. The Custom Filter option in the Number Filters submenu lets you combine up to two logical expressions to filter a single column of data.

Assume you want to revisit your fuel economy data set and add in a number of statistical functions to assist in analysis.

	A	B	C	D	E
1	Fuel Economy of Vehicles				
2	This worksheet demonstrates the use of sorting and filtering in Excel				
3					
4	Make	Model	MPG City	MPG Highway	Annual Fuel Cost
5	Bentley	Continental GTC	10	17	$2,886
6	BMW	X5 xDrive 30i	15	21	$1,927
7	Ford	Mustang 5.4L	14	20	$2,166
8	Honda	Civic Hybrid	40	45	$743
9	Honda	Fit	27	33	$1,039
10	Jeep	Liberty 4WD	15	21	$1,835
11	Volkswagen	Jetta 2.5L	20	29	$1,301
12					
13		Average:	20	27	$1,700
14		Min:	10	17	$743
15		Max:	40	45	$2,886
16					

Suppose you filter the data set to look only at the Honda vehicles.

	A	B	C	D	E
1	Fuel Economy of Vehicles				
2	This worksheet demonstrates the use of sorting and filtering in Excel				
3					
4	Make	Model	MPG City	MPG Highway	Annual Fuel Cost
8	Honda	Civic Hybrid	40	45	$743
9	Honda	Fit	27	33	$1,039
12					
13		Average:	20	27	$1,700
14		Min:	10	17	$743
15		Max:	40	45	$2,886
16					

Notice that the statistical calculations at the bottom are still referencing the entire data set, even though, because of the filter, only a subset of the data is displayed. For data comparisons, this will be a valuable side effect; however, if you want the calculations to apply only to the visible data, you will need to use built-in functions other than the traditional functions (AVERAGE, MIN, MAX).

Using the SUBTOTAL Function

The **SUBTOTAL** function allows the worksheet to dynamically recalculate expressions generated with a filtered list. In the example where only Honda vehicles are selected, only the two visible vehicles will be used in the calculations if you modify your worksheet to use the SUBTOTAL function instead of the traditional statistical functions. To use the SUBTOTAL function, pass in two different arguments:

= SUBTOTAL (function_num, range)

- The *function_num* argument is a number associated to various built-in Excel functions. Table 13-6 lists the available functions for use with the SUBTOTAL function.
- The *range* argument is the range of cells to which the function should be applied.

In the example, use the following calculation in cell C13 to calculate the average of

MPG City: = AVERAGE (C5:C11)

Table 13-6 Available functions in SUBTOTAL

function_num	Function	Definition
1	AVERAGE	Computes the average value of the range
2	COUNT	Counts the number of cells in the range that contain numbers
3	COUNTA	Counts the number of nonempty cells in the range
4	MAX	Calculates the maximum value of the range
5	MIN	Calculates the minimum value of the range
6	PRODUCT	Calculates the product of each number in the range
7	STDEV	Calculates the standard deviation of the numbers in the range ("n − 1" method)
8	STDEVP	Calculates the standard deviation of the numbers in the range ("n" method)
9	SUM	Calculates the sum of all of the numbers in the range
10	VAR	Calculates the variance of the numbers in the range ("n" method)
11	VARP	Calculates the variance of the numbers in the range ("n − 1" method)

The AVERAGE function corresponds to function_num 1, so the resulting calculation in cell C13 using the SUBTOTAL function would appear as follows:

$$= \textbf{SUBTOTAL (1, C5:C11)}$$

After you modified all of the statistical calculations in the worksheet to use the SUBTOTAL function, the sheet should appear as shown in the examples below. Note that the values recalculate automatically according to the filtered data.

Filter on Make: Honda Only

	A	B	C	D	E
1	Fuel Economy of Vehicles				
2	This worksheet demonstrates the use of sorting and filtering in Excel				
3					
4	Make	Model	MPG City	MPG Highway	Annual Fuel Cost
8	Honda	Civic Hybrid	40	45	$743
9	Honda	Fit	27	33	$1,039
12					
13		Average:	34	39	$891
14		Min:	27	33	$743
15		Max:	40	45	$1,039
16					

Filter on Annual Fuel Cost: Less than $1,500

	A	B	C	D	E
1	Fuel Economy of Vehicles				
2	This worksheet demonstrates the use of sorting and filtering in Excel				
3					
4	Make	Model	MPG City	MPG Highway	Annual Fuel Cost
8	Honda	Civic Hybrid	40	45	$743
9	Honda	Fit	27	33	$1,039
11	Volkswagen	Jetta 2.5L	20	29	$1,301
12					
13		Average:	29	36	$1,028
14		Min:	20	29	$743
15		Max:	40	45	$1,301
16					

In-Class Activities

ICA 13-1

	A	B	C
4			
5			
6			
7	**Neper Frequency (α_0)**	25	[rad / s]
8	**Resonant Frequency (ω_0)**	400	[rad / s]
9	**Initial Voltage (V_0)**	15	[V]
10			
11	**Damped Frequency (ω_d)**		[rad / s]
12			
13			
14	**Time (t) [s]**	**Voltage (V) [V]**	
15			
16			
17			

You want to set up a worksheet to investigate the oscillatory response of an electrical circuit. Create a worksheet similar to the one shown, including the proper header information.

First, calculate another constant, the damped frequency ω_d, which is a function of the neper frequency (α_0) and the resonant frequency (ω_0). This can be calculated with the formula

$$\omega_d = \sqrt{\omega_0^2 - \alpha_0^2}$$

Next, create a column of times (beginning in A15) used to calculate the voltage response, ranging from 0 to 0.002 seconds at an increment of 0.0002 seconds.

In column B, calculate the voltage response with the following equation, formatted to one decimal place:

$$V(t) = V_0 e^{-\alpha_0 t} \cos(\omega_d t)$$

(a) Change neper frequency to 200 radians per second, resonant frequency to 800 radians per second, and initial voltage to 100 volts. At a time of 0.0008 seconds, what is the voltage?
(b) Change neper frequency to 100 radians per second, resonant frequency to 600 radians per second, and initial voltage to 100 volts. At a time of 0.0008 seconds, what is the voltage?
(c) Change neper frequency to 200 radians per second, resonant frequency to 400 radians per second, and initial voltage to 75 volts. At a time value of 0.0008 seconds, what is the voltage?

ICA 13-2

When liquid and vapor coexist in a container at equilibrium, the pressure is called vapor pressure. Several models predict vapor pressure. One, called the *Antoine equation*, first introduced by Ch. Antoine in 1888, yields vapor pressure in units of millimeters of mercury [mm Hg].

$$P = 10^{\left(A - \frac{B}{T+C}\right)}$$

	A	B	C	D	E	F
1						
2						
3						
4						
5						
6						
7			Constants			Validity Range	
8	Compound	A	B [°C]	C [°C]	T_{min} [°C]	T_{max} [°C]
9	Ethanol	8.204	1642.89	230.300	-57	80
10						
11						
12						
13	Temperature (T) [°C]	Pressure (P) [mm Hg]				
14	-100					
15	-95					
16	-90					
17	-85					
18	-80					
19	-75					
20	-70					
21	-65					
22	-60					
23	-55	0.0679				
24	-50	0.1236				
25	-45	0.2177				
26	-40	0.3723				

The constants A, B, and C are called the *Antoine constants;* they depend on both fluid type and temperature. Note that "B" and "C" must be in the same units as "T," and A" is a dimensionless number, all determined by experiment.

Create a worksheet similar to the one shown, including the proper header information. The Antoine constants should automatically fill in after the user selects one from a drop-down menu in Cell A9 of the compounds shown below. (Hint: use data validation and LOOKUP expressions.) The data in the table below has been provided online.

Compound	Constants			Validity Range	
	A	B [°C]	C [°C]	T_{min} [°C]	T_{max} [°C]
Acetic acid	7.5596	1644.05	233.524	17	118
Acetone	7.1327	1219.97	230.653	−64	70
Cyclohexane	6.85146	1206.47	223.136	7	81
Ethanol	8.20417	1642.89	230.3	−57	80
Hexadecane	7.0287	1830.51	154.45	150	321
Methanol	8.08097	1582.27	239.7	15	100
m-Xylene	7.22319	1585.83	226.5	−45	140
Tetrahydrofuran	6.99515	1202.29	226.254	23	100

Next, create a column of temperature (T) beginning at −100 degrees Celsius and increasing in increments of 5 degrees Celsius until a temperature of 400 degrees Celsius.

In column B, calculate the vapor pressure (P, [millimeters of mercury, mm Hg]) using the Antoine equation, formatted to four decimal places. If the equation is outside the valid temperature range for the compound, the pressure column should be blank.

ICA 13-3

The ideal gas law assumes that molecules bounce around and have negligible volume themselves. This is not always true. To compensate for the simplifying assumptions of the ideal gas law, the Dutch scientist Johannes van der Waals developed a "real" gas law that uses several factors to account for molecular volume and intermolecular attraction. He was awarded the Nobel Prize in 1910 for his work. The *van der Waals equation* is as follows:

$$\left(P + \frac{an^2}{V^2} \right)(V - bn) = n\,R\,T$$

P, V, n, R, and T are the same quantities as found in the ideal gas law. The constant "a" is a correction for intermolecular forces [atm L^2/mol^2], and the constant "b" accounts for molecular volume [L/mol]. Each of these factors must be determined by experiment.

	A	B	C
7	Type of Gas	Argon ▾	
8	Quantity [g]	4	
9	Temperature (T) [°C]	15	
10			
11			
12	Molecular Weight (MW) [g / mol]	40	
13	vdW Constant "a" [atm L² / mol²]	1.345	
14	vdW Constant "b" [L / mol]	0.0322	
15			
16	Ideal Gas Constant (R) [(atm L) / (mol K)]	0.08206	
17			
18			
19		Ideal Gas	van der Waals
20	Volume (V) [L]	Pressure (P) [atm]	
21	0.5	4.73	4.70
22	0.6	3.94	3.92
23	0.7	3.38	3.36
24	0.8	2.95	2.95
25	0.9	2.63	2.62
26	1.0	2.36	2.36

Create a worksheet similar to the one shown, except be sure to include the proper header information. The molecular weight, "a," and "b" should automatically fill in after the user selects the type of gas in Cell B7. (Hint: use data validation and LOOKUP expressions.) The user will also set the quantity of gas and the temperature of the system. The data in the table below has been provided online.

Next, create a column of volume beginning in A21 at 0.5 liters and increasing in increments of 0.1 liters to a volume of 5 liters.

In column B, calculate the pressure (P, [atmospheres, atm]) using the ideal gas law.

In column C, calculate the pressure (P, [atmospheres, atm]) using the van der Waals equation.

Compound	a [atm L²/mol²]	b [L/mol]	MW [g/mol]
Acetic acid	17.587	0.1068	60
Acetone	13.906	0.0994	58
Ammonia	4.170	0.0371	17
Argon	1.345	0.0322	40
Benzene	18.001	0.1154	78
Chlorobenzene	25.433	0.1453	113
Diethyl ether	17.380	0.1344	74
Ethane	5.489	0.0638	30
Ethanol	12.021	0.0841	46
Hexane	24.387	0.1735	86
Methanol	9.523	0.0670	32
Neon	0.211	0.0171	20
Oxygen	1.360	0.0318	32
Pentane	19.008	0.1460	72
Propane	8.664	0.0845	44
Toluene	24.061	0.1463	92
Water	5.464	0.0305	18
Xenon	4.194	0.0511	131

ICA 13-4

You have a large stock of several values of inductors and capacitors, and are investigating how many possible combinations of a single capacitor and a single inductor chosen from the ones you have in stock will give a resonant frequency between specified limits.

Create two cells to hold a minimum and maximum frequency the user can enter. If the value entered for the maximum frequency (f_{MAX}) is less than the minimum frequency (f_{MIN}), the maximum frequency cell should be shaded dark grey with white, bold text.

Incorrect data:

Allowable Frequency Range	
f_{min}[Hz]	f_{max}[Hz]
2500	**1000**

Correct data:

Allowable Frequency Range	
f_{min}[Hz]	f_{max}[Hz]
2500	7777

Calculate the resonant frequency (f_R) for all possible combinations of one inductor and one capacitor, rounded to the nearest integer. For a resonant inductor/capacitor circuit, the resonant frequency in hertz [Hz] is calculated by

$$f_R = \frac{1}{2\pi\sqrt{LC}}$$

Here, L is the inductance in units of henry [H] and C is the capacitance in units of microfarads [μF]. Automatically format each result to indicate its relation to the minimum and maximum frequency values as listed below.

- $f_R > f_{MAX}$: The cell should be shaded white with light grey text and no border.
- $f_R < f_{MIN}$: The cell should be shaded light grey with dark grey text and no border.
- $f_{MIN} < f_R < f_{MAX}$: The cell should be shaded white with bold black text and a black border.

If done properly, the table should appear similar to the table below for $f_{MIN} = 2500$ and $f_{MAX} = 7777$.

Resonant Frequency (f_R) [Hz]	Capacitance (C) [μF]							
Inductance (L) [H]	0.0022	0.0082	0.05	0.47	0.82	1.5	3.3	10
0.0005	151748	78601	31831	10382	7860	**5812**	**3918**	2251
0.002	75874	39301	15915	**5191**	**3930**	**2906**	1959	1125
0.01	33932	17576	**7118**	2322	1758	1299	876	503
0.05	15175	7860	**3183**	1038	786	581	392	225
0.068	13012	**6740**	**2729**	890	674	498	336	193
0.22	**7234**	**3747**	1517	495	375	277	187	107
0.75	**3918**	2029	822	268	203	150	101	58

After you have this working properly, modify the frequency input cells to use data validation instead of conditional formatting to warn the user of an invalid value entry.

ICA 13-5

A bioengineer conducts clinical trials on stressed-out college students to see if a sleep aid will help them fall asleep faster. She begins the study by having 20 students take a sleep aid for seven days and records through biofeedback the time when they fall asleep. To analyze the data, she sets up the following worksheet. Evaluate the expressions below; state what will appear in the cell when the command is executed. Column I contains the average time each student took to fall asleep during the seven-day trial. Column J contains any adverse reactions the students experienced (H = headache; N = nausea).

(a) Column K will contain the rating of the time it took the student to fall asleep compared with the control group, who did not take the medication. The statement as it appears in cell K14 is given below. What will appear in cell K14 when this statement is executed?

$$= \text{IF (I14} > \$\text{I}\$2 + \$\text{I}\$3, \text{"MORE"}, \text{IF (I14} < \$\text{I}\$2 - \$\text{I}\$3, \text{"LESS"}, \text{""}))$$

(b) Column L groups the participants into three groups according to their reaction to the drug and the time it took them to fall asleep. Assume the statement for Part (a) is executed in Column K. The statement as it appears in cell L7 is given below. What will appear in cell L7 when this statement is executed?

$$= \text{IF (AND (K7} = \text{"MORE"}, \text{J7} = \text{"H"}), \text{"MH"},$$
$$\text{IF (AND (K7} = \text{"MORE"}, \text{J7} = \text{"N"}), \text{"MN"}, \text{""}))$$

(c) Suppose the formula in Column L was changed to regroup the participants. The statement as it appears in cell L9 is given below. In Excel, this statement would appear as a continuous line, but here it is shown on two lines for space. What will appear in cell L9 when this statement is executed?

$$= \text{IF (AND (K9} = \text{"MORE"}, \text{OR (J9} = \text{"H"}, \text{J9} = \text{"N"})), \text{"Severe"},$$
$$\text{IF (OR (J9} = \text{"H"}, \text{J9} = \text{"N"}), \text{"Mild"}, \text{IF (K9} = \text{"LESS"}, \text{"Helpful"}, \text{""})))$$

(d) Suppose the formula in part (c) was copied into cell L16. What would appear in cell L16 when this statement is executed?

(e) Suppose the formula in part (c) was copied into cell L18. What would appear in cell L18 when this statement is executed?

(f) Does the expression given in part (c) accurately identify all severe reaction conditions? If not, how would you change the expression to include all severe reactions?

	A	B	C	D	E	F	G	H	I	J	K	L
1						Control Group Data						
2						Overall Average			35	[min]		
3						Standard Deviation			4	[min]		
4												
5				Number of Minutes to Fall Asleep								
6	Patient	Day 1	Day 2	Day 3	Day 4	Day 5	Day 6	Day 7	Average	Reaction	Time	Group
7	A	45	39	83	47	39	25	42	46	H		
8	B	35	75	15	36	42	12	29	35			
9	C	42	32	63	45	37	34	31	41	N		
10	D	14	25	65	38	53	33	32	37	H		
11	E	14	71	48	18	29	14	24	31			
12	F	14	25	29	24	18	24	15	21	H N		
13	G	31	14	42	19	28	17	21	25			
14	H	12	24	32	42	51	12	16	27	H N		
15	I	28	29	44	15	43	15	22	28	N		
16	J	21	19	35	41	34	25	18	28	H		
17	K	44	36	51	39	30	26	25	36			
18	L	38	43	36	59	14	34	18	35	N		
19	M	19	15	63	50	55	27	31	37	H		

ICA 13-6

We accidentally drop a tomato from the balcony of a high-rise apartment building. As it falls, the tomato has time to ponder some physics and says, "You know, the distance I have fallen equals $\frac{1}{2}$ gravity times the time I have fallen squared." Create a worksheet to solve the question of when the tomato goes splat.

- The user will input the initial balcony height in units of feet. Use data validation to set a limit for the height of 200 feet.
- Place the acceleration due to gravity in a cell under the balcony height and *not* within the formulas themselves. *Be sure to watch the units for this problem!*
- Column A will be the distance the tomato falls, starting at a distance of zero up to a distance of 200 feet, in 5 foot increments.
- Column B will show the calculated time elapsed at each distance fallen.
- Column C will display the status of the tomato as it falls.

  - If the tomato is still falling, the cell should display the distance the tomato still has to fall.
  - If the tomato hits the ground, the cell should display "SPLAT" on a red background.
  - SPLAT should appear once; the cells below are blank.

Changing the Problem Parameters

One of the most useful things about a worksheet is that it can dynamically change problem parameters. Use the worksheet just created to test the following parameters.

	Balcony Height [ft]
(a)	200
(b)	120
(c)	50
(d)	25

ICA 13-7

You enjoy drinking coffee but are particular about the temperature (T) of your coffee. If its temperature is greater than or equal to 70 degrees Celsius [°C], the coffee is too hot to drink; less than or equal to 45 °C is too cold by your standards. Your coffee pot produces coffee at the initial temperature (T_0). The cooling of your coffee can be modeled by the equation below, where time (t) and the cooling factor (k) are in units per second:

$$T = T_0 e^{-kt}$$

First, the worksheet user should be able to modify four properties of the coffee:

- Initial temperature (T_0); for the initial problem, set to 80 °C
- Cooling factor (k); set to 0.001 s^{-1}
- Temperature above which coffee is "Too Hot" to drink (T_{hot}); set to 70 °C
- Temperature below which coffee is "Too Cold" to drink (T_{cold}); set to 45 °C.

Place these values in cells located near the top of the worksheet.

Second, create a temperature profile for the coffee:

- In column A, generate a time range of 0–300 seconds, in 15 second intervals
- In column B, generate the temperature of the coffee, using the equation given and the input parameters set by the user (T_0 and k).

Third, make the worksheet easy for the user to understand:

- In column B, the temperature values should appear on a red background if the coffee is too hot to drink, and a blue background if it is too cold. Use conditional formatting to make this happen.
- In column C, create a warning next to each temperature that says "Don't Drink" if the calculated temperature in column B is too hot or too cold in comparison with the temperature values the user enters. Use an IF statement to make this happen.

Finally, test the worksheet. Use the conditions at the bottom of the page to see how the values and formatting will change if you enter different values of T_0 and k.

Changing the Problem Parameters

Use the worksheet just created to test the following parameters.

	Initial Temperature (T_0) [°C]	Cooling Factor (k) [1/s]	Don't Drink Range T_{hot} [°C]	Don't Drink Range T_{cold} [°C]
(a)	80	0.001	70	45
(b)	70	0.001	70	45
(c)	90	0.002	75	50
(d)	95	0.002	80	50
(e)	82	0.004	80	55

ICA 13-8

You are interested in calculating the best place to stand to look at a statue. Where should you stand so that the angle subtended by the statue is the largest?

At the top of the worksheet, input the pedestal height (P) and the statue height (S).

In Column A, create a series of distances (d) from the foot of the statue, from 2 feet to 40 feet by 2-foot increments.

In Column B, calculate the subtended angle in radians using the following equation:

$$\theta = \tan^{-1}\left(\frac{P + S}{d}\right) - \tan^{-1}\left(\frac{P}{d}\right)$$

In Column C, write a function to change the angles in Column B from radians to degrees. At the bottom of Column C, insert a function to display the maximum value of all the angles.

In Column D, use a conditional statement whose output is blank except at the single distance where the angle is a maximum; at the maximum, print "Stand Here."

Changing the Problem Parameters

Use the worksheet just created to test the following parameters.

	Pedestal Height (P) [ft]	Statue Height (S) [ft]
(a)	20	10
(b)	10	10
(c)	15	20
(d)	30	20

ICA 13-9

Many college students have compact refrigerator-freezers in their dorm room. The data set below is a partial list of energy efficient models less than 3.6 cubic feet [cft], according to the American Council for an Energy Efficient Economy (www.aceee.org). Starting with the data provided, complete the analysis below.

ENERGY COST

(a) We would like to compute the cost to run each model for a year. Assume that it costs $0.086 per kilowatt-hour [kWh]. Create a new column, "Annual Energy Cost [$/yr]," that calculates the annual energy cost for each refrigerator.

SORTING

(b) First, we will sort the first table by energy usage, with the model with the highest kilowatt-hour rating listed first. Which model appears first?

(c) Next, we will sort by the volume in ascending order and the annual energy cost in ascending order. Which model appears first?

FILTERING

(d) Assume we want to restrict our selection to refrigerators that can contain more than 2.5 cubic feet. Which models appear in the list?

(e) Assume we want to restrict our selection to refrigerators that can contain more than 2.5 cubic feet and only require between 0 and 300 kilowatt-hours per year. Which models appear in the list?

Brand	Model	Defrost Type	Volume [ft^3]	[kWh/yr]
Black & Decker	BCE27	Manual	2.7	261
Danby Millennium	DAR194BL	Automatic	1.8	302
Danby Millennium	DAR194W	Automatic	1.8	302
General Electric	SFR03BAPBB	Automatic	3.2	316
General Electric	SFR03BAPWW	Automatic	3.2	316
Haier	ESRB03B	Manual	2.7	261
MicroFridge	MHR-2.7E	Manual	2.7	260
Silhouette	DAR195BL	Automatic	1.8	302
Silhouette	DAR259BL	Automatic	2.5	309
Silhouette	DAR340BL	Automatic	3.3	316
Sunbeam	SBCR039WE	Manual	1.7	252
U-Line	15R	Automatic	3.5	312
U-Line	2015R	Automatic	3.1	307
Whirlpool	WAR349BSL	Automatic	3.3	316

ICA 13-10

The complexity of video gaming consoles has evolved over the years. The data set below is a list of energy usage data on recent video gaming consoles, according to the Sust-It.Net consumer energy report data (www.sust-it.net). Using the data provided, complete the analysis below.

ENERGY ANALYSIS

(a) Compute the cost to run each gaming console for a year, including the purchase price. Assume that it costs $0.086 per kilowatt-hour [kWh]. Create a new column, "Cost + Energy [$ / yr]," that calculates the total (base + energy) cost for each gaming console.

(b) On average, a consumer will own and operate a video gaming console for four years. Calculate the total carbon emission [kilograms of carbon dioxide, or kg CO_2] for each gaming console over the average lifespan; put the result in a column labeled "Average Life Carbon Emission [kg CO_2]."

SORTING

(c) First, sort the table by total cost, with the console with the highest total cost listed first. Which console appears first?

(d) Next, sort by the original cost in ascending order and the average life carbon emission in ascending order. Which console appears last?

FILTERING

(e) Restrict your selection to video game consoles that originally cost $300. Which models appear in the list?

(f) Restrict your selection to video game consoles that originally cost less than or equal to $300 and have an average life carbon emission less than or equal to 25 kg CO_2. Which models appear in the list?

Brand	Model	Cost [$]	[kWh/yr]	[kg CO_2/yr]
Sega	Dreamcast	$200	16.06	6.91
Nintendo	GameCube	$200	15.33	6.59
Sony	PlayStation	$300	7.30	3.14
Sony	PlayStation 2	$300	36.50	15.70
Sony	PlayStation 3 (40 GB)	$500	204.40	87.89
Sony	PlayStation 3 (60 GB)	$600	277.40	119.28
Nintendo	Wii	$250	12.41	5.34
Microsoft	Xbox	$300	51.10	21.97
Microsoft	Xbox 360	$400	120.45	51.97

CHAPTER 14
EXCEL GRAPHS

The following is an example of the level of knowledge of Excel needed to proceed with Chapter 14. *If you are not able to quickly recreate Excel Graphs similar to the ones shown, including formatting, please review graphing basics in Appendix C before proceeding into Chapter 14.*

Two graphs are given here; they describe the draining of tanks through an orifice in the bottom. When the tank contains a lot of liquid, the pressure on the bottom is large and the tank empties at a higher rate than when there is less liquid. The first graph shows actual data obtained from two different tanks. These data are given in the table below. The second plot shows curves (developed from theoretical equations) for two tanks. The equations for these curves are also given.

Experimental data for first graph:

Time (t) [min]	5	10	15	20	40	55	90
Volume Tank #1 (V1) [gal]	19.0	17.0	15.5	11.5	10.5	9.5	8.0

Time (t) [min]	10	15	20	30	35	40	90
Volume Tank #2 (V2) [gal]	25	21	20	19	17	16	14

Theoretical equations for second graph (with t in minutes):

Tank #1: Volume remaining in tank #1 [gal] $V = 33\,t^{-0.31}$

Tank #2: Volume remaining in tank #2 [gal] $V = 44\,t^{-0.26}$

14.1 AVAILABLE GRAPH TYPES

You can create many different types of charts in Excel. Usually, you will only be concerned with a few main types, shown in Table 14-1.

Table 14-1 Common chart types available in Excel

A **scatter plot** is a graph that numerically represents two-dimensional theoretical or experimental data along the abscissa and ordinate of the graph. It is most commonly used with scientific data. To create a scatter plot, you specify each pair in the graph by selecting two identically sized columns or rows of data that represent the (x, y) values of each experimental symbol or point on a theoretical expression.

A scatter plot can be shown as discrete data point (used to show experimental data) or lines only (used to show theoretical expressions). Excel will also show discrete data points connected by lines; the authors of this text do not find this type of chart useful.

The step size of both axes is evenly spaced as determined by the user and can be customized to show all or part of a data set plotted on a graph.

Use a scatter plot to visualize your data when you want to:

- Observe mathematical phenomena and relationships among different data sets
- Interpolate or extrapolate information from data sets
- Determine a mathematical model for a data set, using trendlines.

Table 14-1 **Common chart types available in Excel (*continued*)**

A **line plot** is a graph that visualizes a one-dimensional set of theoretical or experimental data along the abscissa and ordinate of a graph.

A line plot can be shown as points connected by lines, lines only, or in 3-D.

The y-axis values of a line plot are evenly spaced as determined by the user; however, the x-axis of a line plot is not. As shown in the graph to the right, a line plot places each discrete element evenly along the x-axis regardless of the actual step-spacing of the data.

Use a line plot to visualize your data when you want to:

- Display any evenly spaced or sampled data
- Visualize time-series data taken at even intervals
- Display categorical data (e.g., years, profession, colors).

A **column graph** is used for displaying various types of categorical data.

The y-axis increments are spaced evenly, but the x-axis spacing has no meaning since the items are discrete categories. As a rule of thumb, a column graph can be used to represent the same information shown on a line plot.

A column plot can be shown as bars, cylinders, or cones; as a clustered group or stacked; in 1-D or 3-D.

Use a column graph to visualize your data when you want to:

- Display any categorical data
- Observe differences between categories.

Table 14-1 Common chart types available in Excel (*continued*)

A **bar graph** is identical to a column graph, with the x- and y-categories reversed; the x-category appears on the ordinate and the y-category appears on the abscissa. Because of the similarity, only column graphs are covered in this text.	
A **pie graph** is used on a single column or row of non-negative numbers, graphed as a percentage of the whole. It is typically used for categorical data, with a maximum of seven categories possible. A pie graph can be shown in 1-D or 3-D, with either the percentages or the raw data displayed with the category names. Use a pie graph to visualize your data when you want to: ■ Display categorical data as part of a whole ■ Observe differences between categories Because pie charts are so simple, they are not covered in this text.	

<table>
<tr><td>14.2</td><td colspan="2">

STATISTICS IN EXCEL

</td></tr>
</table>

<table>
<tr>
<td>

For a review of histograms and CDFs, please see Chapter 9.

</td>
<td>

To create **histograms** and **Cumulative Distribution Functions (CDFs)** with Excel, you need to first activate the Analysis ToolPak in Microsoft Excel.

■ In Excel, go to the Office button and click **Excel Options**.

■ Choose the **Add-Ins** tab on the left menu of the Excel Options window to display all the active add-in applications in Excel. Notice in our list that the Analysis ToolPak is listed as inactive.

</td>
</tr>
</table>

- At the bottom of the Excel Options window, select **Excel Add-Ins** in the **Manage** drop-down menu and click **Go**.
- In the Add-Ins window, check the **Analysis ToolPak** option and click **OK**.
- A prompt might pop up telling you to install the add-in—click **Yes** and finish the installation, using the Office Installer.

EXAMPLE 14-1

The outline below gives the steps necessary to use the data analysis tool in Excel for basic statistical analysis of a data set. This is presented with an example of the high and low temperatures during the month of October, 2006.

⌘ MAC OS Users:

Please see instructions on Page 382.

- If necessary, input the data; the data for this example have been provided online. Use Column A to input an identifier for the data point, in this case, the date. Column B and C will contain the actual high and low temperatures for each day, respectively.
- Next, decide on the bin range. This discussion focuses on the high temperatures, but can easily be repeated with the low temperatures.
  - A rule of thumb is that the number of bins is approximately equal to the square root of the number of samples. While it is obvious in this example how many total samples are needed, the COUNT function is often very useful. October has 31 days, and the square root of 31 is 5.57; thus, you should choose either 5 or 6 bins.
    - Examine your data to determine the range of values. Using the MAX and MIN functions, you can determine that the highest high temperature during October was 86 °F and the lowest high temperature was 56 °F. Thus, your range is 86 − 56 = 30 °F.
    - Since 5.57 is closer to 6 than to 5, choose 6 bins. Remember, however, that you might want to try a different number of bins to see if that would result in a clearer representation of the data. With a range of 30 °F, 6 bins gives 30 °F/6 bins = 5 °F per bin.
- Type the range of values that will appear in each bin. For example, the first bin will contain temperatures 55, 56, 57, 58, and 59; the second bin will contain temperatures 60–64, etc.
- In the adjacent column, type the corresponding Upper Value of temperature for each of the bins listed.

Your worksheet should be similar to the one shown.

Statistical Graphing with the Analysis ToolPak

✍ **To create histograms and CDF charts:**

- Go to **Data > Analysis > Data Analysis** and under Analysis Tools choose **Histogram**. Click **OK**.
- In the **Input Range**, click the icon at the right end of the blank box. You can then highlight the range (in this case, $B$6:$B$36). Close the box by clicking the icon at the right-hand end of this small box where the range is shown.
- Repeat this procedure for the **Bin Range**, highlighting the cells that contain the upper values.
- Next, for the **Output Range**, click the circle and identify a single cell to begin the placement of the output data.
- Finally, check the boxes to activate the options of **Cumulative Percentage** calculations and **Chart Output**.
- Click **OK**.

Your worksheet should now look like this:

For a review of "Proper Plots," see Chapter 7.

Helpful Hints:

- **Replace the values in the histogram data table for "Bin" with the "Bin Labels"** you entered earlier. This will change the axis labels to the range, rather than the upper value, for each bin.
- **Move the histogram location** to a new worksheet rather than imbedded in the original worksheet to allow the data to be seen clearly. After selecting the chart, use the **Chart > Location** option to select "As new sheet."
- **Modify the histogram to be a proper plot** just as you would any other chart. The same rules for a "proper plot" apply to a histogram, so make sure the background is white and alter the series colors, etc., as appropriate. The histogram generated with the directions above is shown below, properly formatted.
- **Change the vertical scale on the left axis** to be a multiple of 2, 5, or 10 to allow the cumulative percentages on the right axis to line up with the gridlines. *This is important to do!*
- **Change the vertical scale on the right axis** to be a maximum value of 100%. *This is important to do.* The resulting analysis should appear as follows.

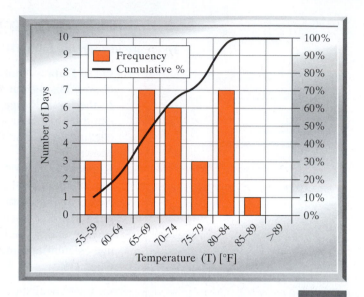

The spreadsheet shows:

	A	B	C	D	E	F	G
1	Weather in Clemson, SC						
2	This example illustrates how to draw a histogram and CDF in Excel using data taken from the						
3	Weather Channel (www.weather.com) on the temperatures in Clemson, SC in October 2006.						
4							
5	Date	High Temperature (HT) [°F]	Low Temperature (LT) [°F]				
6	10/1/06	82	58		HT Analysis		
7	10/2/06	80	57		# of Bins	5.57	
8	10/3/06	81	60		Max	86	
9	10/4/06	84	60		Min	56	
10	10/5/06	86	61		Range	30	
11	10/6/06	82	55		Bin Size	5.39	
12	10/7/06	71	49				
13	10/8/06	68	58		Bin Labels	Upper Value	
14	10/9/06	80	60		55 - 59	59	
15	10/10/06	80	58		60 - 64	64	
16	10/11/06	79	61		65 - 69	69	
17	10/12/06	74	52		70 - 74	74	
18	10/13/06	67	44		75 - 79	79	
19	10/14/06	71	38		80 - 84	84	
20	10/15/06	68	37		85 - 89	89	
21	10/16/06	62	50				
22	10/17/06	65	53				
23	10/18/06	78	62				
24	10/19/06	73	66		Bin	Frequency	Cumulative %
25	10/20/06	72	50		55 - 59	3	9.7%
26	10/21/06	66	45		60 - 64	4	22.6%
27	10/22/06	69	50		65 - 69	7	45.2%
28	10/23/06	56	38		70 - 74	6	64.5%
29	10/24/06	57	32		75 - 79	3	74.2%
30	10/25/06	56	35		80 - 84	7	96.8%
31	10/26/06	64	45		85 - 89	1	100.0%
32	10/27/06	60	50		> 89	0	100.0%
33	10/28/06	68	48				
34	10/29/06	74	43				
35	10/30/06	75	46				
36	10/31/06	64	49				
37							

Comprehension Check 14-1

Repeat this analysis, using the daily low temperatures during October 2006.

⌘ Statistics on the Mac OS

Unfortunately, as this book goes to press, Microsoft has chosen not to include the histogram tool in Excel 2008 for the Mac OS. You have a few options.

- If you have an Intel-based Mac, you can use Excel 2007 for Windows. If you do not know how to activate the Windows option on your machine, ask your friendly local Mac guru at your computer center.
- You can use Excel 2004 for the Mac OS, which did include a histogram tool.
- You can create the histogram manually according to the instructions below.

Create columns for the data, bin ranges, and upper value in each bin as described above.

Next, determine the number of data points in each bin. After doing a few, you will find it easy. Use the advanced Excel function, known as an array function, to accomplish this determination. A detailed explanation of array functions is beyond the scope of this book, but if you follow the instructions *carefully*, you should not have any trouble. The specific function to use is called FREQUENCY.

1. In the cell immediately to the right of the topmost "upper bin value" cell (this would be cell G14 in the example above) enter the formula

   = FREQUENCY (DataRange, UpperBinValueRange)

and press return. In the example above, this would be

   = FREQUENCY (B6:B36, F14:F20)

2. Click-and-hold the cell into which you entered the formula, then drag straight down to the cell in the row *following* the row containing the last "upper bin value." This would be cell G21 in the example above. Release the mouse button. At this point, you will have a vertical group of cells selected (G14:G21 in the example), the top cell will contain the number of data points in the first bin, and the rest of the selected cells *will be blank*. The formula you entered in the topmost of these cells will appear in the formula bar at the top of the window.

3. Click once in the formula in the formula bar. The top cell of the selected group will be highlighted.

4. Hold down the Command (⌘) key and press Return. The selected cells will now contain the number of data points in each bin immediately to the left. The bottommost selected cell will contain the number of data points larger than the upper bin value in the final bin. In the example, this "extra" cell should contain a 0, since no values are larger than those in the final bin. Note the formulae that appear in these cells are all identical—the cell references are exactly the same. This is normal for an array function.

Use these values to create the histogram.

1. Select the cells containing the bin ranges (E14:E20 in our example), then hold down the command (⌘) key while you select the cells containing the number of data points per bin. In our example, since the "extra" cell at the bottom contains a 0, you do not need to include it. If this were non-zero, you might want to add a cell at bottom of the cell ranges that said something like >89. You should now have the two columns for bin ranges and number per bin selected (E14:E20 and G14:G20 in our example).

2. In the toolbar, select **Gallery > Charts > Column**. A row of column chart icons should appear.

3. Click the first icon, which shows pairs of columns. The chart that appears shows the histogram. Be sure to follow all appropriate proper plot rules for completing the histogram.

Finally, generate the cumulative distribution function. If you have survived this far, you should be able to do this with minimal guidance. Create another column of values next to the column containing the number of data points per bin. In the cell next to the topmost bin cell, enter the number of data points in that bin. In the next cell down, enter a formula that will add the cell above to the cell beside it containing the number of data points in that bin. Replicate this formula down to the last bin. Each cell in the new column should now contain the sum of all data points in all bins to that point.

14.3 AUTOMATED CALCULATIONS

Traditional representations of experimental and theoretical data involve the creation of static graphs that visualize some phenomena. This section presents an alternative method that allows you to create rich, interactive graphs in Excel through the use of **iteration.** Iteration is the idea of repeatedly increasing the value of some number until some condition is true. This book addresses two methods for creating iterative graphs in Excel: automatic cell iteration and slider bars. Each method has its own advantages. In this section, we use some of the capabilities of Excel to show in a continuous manner how things change as "time passes" or as some other variable varies. This is best explained through examples.

Iteration—Incrementing a Variable

⌘ **Mac OS**: Excel 2008 does not include the iteration function. You can either run Excel 2004 for Mac or run Windows and use Excel 2007.

✎ **To enable iteration in Excel:**

- Click the **Office** button in the upper left corner of Excel.
- Click **Excel Options**.
- Choose the **Formulas** tab on the left menu of the Excel Options window.
- Click the **Enable iterative calculations** checkbox under the "Calculation Options" section.
- Under the checkbox, set the **Maximum Iterations** to **1**.
- When you're finished, click **OK** to return to the worksheet.

In addition to enabling iterative calculations, make sure your workbook is set to automatic calculations.

- Go to **Formulas > Calculation** and click the **Calculation Options** drop-down menu. Be sure that your workbook is set to manually recalculate by checking the **Manual** option. By default, Excel is set to calculate automatically, but for iteration, it is preferable to control when the calculations are conducted.

This means that the worksheet will now recalculate one time for every time you press the **F9 key**. If you hold down the F9 key, the worksheet will continue to recalculate (over and over very rapidly) until you release the key. By default, iteration is not enabled in a new workbook, so you will need to run through these instructions for each new workbook that requires iteration.

EXAMPLE 14-2

Suppose you want to create an integrative graph in order to visualize the motion of a projectile. Assume you know the launch angle and launch speed of some projectile. From physics, you know how to calculate the horizontal and vertical positions of a projectile in motion, assuming no horizontal forces and gravity as the only vertical force, using the following equations:

$$x(t) = v \sin(\theta)t$$

$$y(t) = -\frac{1}{2}gt^2 + v \cos(\theta)t$$

where

t	time	[s]
x (t)	horizontal position at time t	[m]
y (t)	vertical position at time t	[m]
v	launch speed of the projectile	[m/s]
θ	launch angle of the projectile	[°]
g	acceleration due to gravity	[m/s^2]

To visualize the motion of the projectile, you iterate on time (since it's the value changing in the calculation of horizontal and vertical positions) and graph the horizontal position on the x-axis and the vertical position on the y-axis.

To make this worksheet "iteration ready," you must incorporate at least one cell that will recalculate every time you press the F9 key. Since you are iterating over the time variable, you need to type a special equation into B6, which represents the time in seconds. To take the existing value of time and add the time increment stored in Cell B5, type the following equation into B6: = B5 + B6

In a traditional worksheet, this equation would cause a circular reference error, but since iteration is enabled, this equation is acceptable.

Whenever a horizontal and vertical speed and time is inserted, the horizontal and vertical position of the projectile will appear in cells B10 and B11. This corresponds to the (x, y) coordinates of the projectile.

Create a plot of this single point. For x range, select only the single cell B10. For the y range, select only the single cell B11. Label the ordinate as Elevation (y) [m] and the abscissa as Horizontal Position (x) [m]. For the graph location, select the default choice of "**As object in:**" to insert the graph on the current worksheet. Drag the graph and place it below the cells containing the values. To make sure that you can print it on a single page, go to **File**>**Print Preview** to see if the plot needs to be resized to fit on a single sheet. Now as you change values of the angle, velocity or time, the projectile will move to a different location and you can watch it change location as you select different input values.

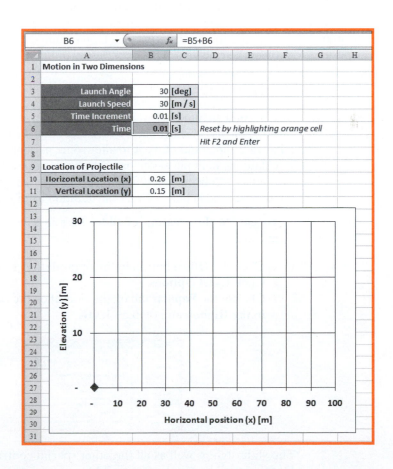

Notice, however, that Excel automatically selects the scale on both axes, so that as you change values, the scales will change. This makes it difficult to track the projectile; it would be best if the scales remained the same.

✥ **To "freeze" the axis scales:**

- Right-click one of the axes and select the **Format Axis** option. The **Format Axis** window will appear.
- On the **Axis Options** tab, select the **Fixed** buttons for the **Minimum**, **Maximum**, and **Major unit** selections. Specify the minimum as zero, the maximum as 30, and the major unit as 10.
- Repeat for the other axis.

Now when you change values the scale will remain the same.

✥ **To change the size of the symbol:**

Since you are only plotting a single symbol, using a larger symbol would allow the visualization to be a little easier to follow. Right-click the symbol and select **Format Data Series**. On the **Marker Options** tab, in the **Size** box increase the value.

✥ **To test your completed worksheet:**

You should now be able to watch the visualization run by holding down the F9 key. To reset the time for a new run, select the cell that calculates the time (Cell B6) and press F2 to edit the equation. Don't change the equation and press the Enter key to reset the calculation.

> To "run" the iteration: press F9.
>
> To reset the iteration: choose the iteration cell and press F2.

Slider Bars

The second method for varying a parameter continuously in an Excel worksheet is through the use of **slider bars**. With slider bars, the user of a worksheet can click and drag a bar in a slider and move it to alter some parameter in the worksheet. Unlike iterative calculations, multiple slider bars can be used in a single worksheet, so more than one "variable" can be included. Example 14-3 revisits Example 14-2 and shows how a slider bar can be used in place of variable iteration in an Excel worksheet.

✥ **To enable the creation of slider bars** in a worksheet, enable the developer tools in Excel:

- Click the **Office** button in the upper-left of Excel.
- Click **Excel Options**.
- Choose the **Popular** tab on the side, click the option that says **Show Developer tab in the Ribbon** and then click **OK**.

The slider bar as well as all the other special controls will now be available under the **Developer** ribbon in the **Controls** box.

EXAMPLE 14-3

Example 14-2 demonstrated that when a projectile is launched from ground level at a set speed and angle with the horizontal, it rises for some distance as it moves downrange, and then gravity gradually pulls it back down to ground level. Rework this example by inserting a slider bar.

✎ **To add a slider bar**

- Click the **Developer** tab.
- If the **Design Mode** icon is active, click it to activate the Design Mode. If it is not active, click "Properties" to engage the Design Mode icon and then click the icon to enter Design Mode.

- Click **Insert** and under ActiveX Controls, click the scroll bar icon. By default, the scroll bar icon is the top-right ActiveX control.
- In the worksheet, click and drag in an empty space on the worksheet to draw the slider bar. After you've drawn the slider bar, you'll be able to resize it and to move it around the worksheet while Design Mode is enabled.
- While still in Design Mode, right-click the scroll bar and click **Properties**. Go to the **LinkedCell** box and type in the cell address of the value that's going to be changed. In this example, you will change the time, so type B5. Go to the **Max** box

LinkedCell	B5
Locked	True
Max	100
Min	0

and type 100 (this sets the maximum value of time) and verify that the **Min** box contains 0. Close the box.

- Click the Design Mode icon to exit this mode.

You should now be able to drag the slider bar and watch the value of cell B5 increase from 0 to 100. Next, modify B7 (Time) to calculate according to the current value of the slider bar and the time increment in B6. By scaling the current value of the slider bar by the size of each time increment B7, you can calculate time by typing

$$= B5 * B6$$

As you move the slider bar, notice that the distances between the calculated data points are pretty far apart. You can increase the resolution of the calculations in your interactive graph by using a smaller time increment. Change the value of the time increment in B6 to 0.01.

Notice that the slider bar doesn't calculate enough values to display each horizontal and vertical position of the flight of the projectile. You can fix this by modifying the properties of the slider bar and changing the Max value to a something much larger than 100 (for this example, 350) to allow the projectile to complete its return to the ground. The maximum value and axis properties require adjustment for different angles and launch speeds to display properly.

Comprehension Check 14-2

Modify the work completed in Example 14-3 to include an additional slider bar to control the launch angle.

Copy the worksheet containing Example 14-3 and insert a new slider bar. Since you now have two slider bars in your current worksheet, label and color-code the controls so that it's apparent to the user of the worksheet which controls will change each variable.

The angle slider bar should be linked to Cell B3 and should vary from 0 to 90 degrees. Using an angle increment of 1 degree is acceptable, so you will not need to calculate a scaled angle as you did with the time. The resulting worksheet should appear as shown below.

In-Class Activities

ICA 14-1

When an attempt is made to stop a car, both the reaction time of the driver and the braking time must be considered. The following data were taken from *www.highwaycode.gov.uk*.

Create a proper plot of these data. Experiment with several different types of graphs, and justify your final choice.

Vehicle Speed (v) [mph]	Distance (d) [m]	
	Reaction (d_r)	Braking (d_b)
20	6	6
30	9	14
40	12	24
50	15	38
60	18	55
70	21	75

ICA 14-2

A biosystems engineer has discovered a relationship between the population density of broomsedge (*Andropogon virginicus*) and levels of sulfur dioxide concentration (SO_2) in the air, which is measured in parts per million [ppm]. The data table gives average kilograms of broomsedge per hectare (ABB) for several values of SO_2 concentration (C).

Draw a proper plot for the data. Assume that the broomsedge biomass is the dependent variable and so will be plotted on the vertical axis. Experiment with several different types of graphs, and justify your final choice.

Sulfur Dioxide Concentration	Average Broomsedge Biomass
75	900
120	2400
190	3500
275	4750
350	5300

ICA 14-3

The number of mosquitoes found in one square meter of the average South Carolina backyard varies according to the total rainfall measured in inches during the summer months of June, July, and August.

Draw a proper plot for the data. Experiment with several different types of graphs, and justify your final choice.

Rainfall	Mosquito Count
6	5
11	8
18	17
24	20
31	15

ICA 14-4

Your team has designed three tennis ball launchers, and you have run tests to determine which launcher best meets the project criteria. Each launcher is set to three different launch angles, and the total distance the ball flies through the air is recorded. These data are summarized in the table. Plot all three sets of data (one for each of the three launchers on a single graph). Launch angle should be plotted on the horizontal axis. Experiment with several different types of graphs, and justify your final choice.

Launcher #1		Launcher #2		Launcher #3	
Launch Angle (θ) [°]	Distance (d) [ft]	Launch Angle (θ) [°]	Distance (d) [ft]	Launch Angle (θ) [°]	Distance (d) [ft]
20	5	10	10	20	10
35	10	45	25	40	20
55	12	55	18	50	15

ICA 14-5

Plastic bottles and containers are often "blow molded" to form the shapes needed in a particular application. This requires energy. Additional energy is also required to operate the robots and other machine movements that complete the process. Consequently, for any particular blow-molding machine, we can measure the amount of energy needed for processing (melting) the plastic as well as the total energy required for the overall process.

The following data were obtained for 10 different machines. Using this information, create an appropriate graph. Experiment with several different types of graphs, and justify your final choice.

Machine Model	Energy for extrusion [W/kg]	Total energy [W/kg]
KLS6	269	494
KLS6D	219	381
KLS8	270	487
KLS8D	221	409
KLS10	236	433
KLS10D	223	395
KLS12	237	444
KLS12D	226	405
KLS14	218	420
KLS14D	227	392

Data provided by Kautex Machines

ICA 14-6

You are asked to relate the power consumption to the mass flow rate of two plastic molding processing machines. The plastic can be melted by the extruder either with an AC or a DC motor (E60 AC water cooled or E60 DC air cooled). The following data were acquired.

E60 AC water-cooled drive		E60 DC air cooled drive	
Power Consumption [kW]	Mass Flow rate [kg/h]	Power Consumption [kW]	Mass Flow rate [kg/h]
3.2	8	12	31
4.3	11	14	42
14	54	17	54
17	65	23	63
22	76	26	74
24	88	31	85
28	99	35	95
33	108	39	104

Data provided by Kautex machines

You need to produce a graphical representation that will allow a user to quickly judge the efficiency between the AC and DC motors. Experiment with several different types of graphs, and justify your final choice.

ICA 14-7

Ten students are asked to participate in a golf-putting contest. Each student is given 5 balls, placed a distance of 60 feet from the hole. They are to try to get each ball in the hole with a single putt. The number of feet that each ball is from the hole when it comes to rest is measured. *For simplicity, assume that none of the students make their putt.*

In Excel, create a histogram of this situation (number of students on the ordinate and distance the putt stops from the hole on the abscissa), using the data given.

Distance from hole [ft]	0	1	2	3	4	5	6	7	8	9	10	11	12	13	14
Number of balls	0	1	3	2	3	5	4	6	9	7	5	3	2	0	0

ICA 14-8

This exercise includes the measurement of a distributed quantity and the graphical presentation of the results. You are to determine how many "flexes" it takes to cause a paper clip to fail.

Test the bending performance of 20 paper clips by doing the following:

- Unfold the paper clip at the center point so that the resulting wire forms an "S" shape.
- Bend the clip back and forth at the center point until it breaks.
- Record the number of flexes required to break the clip.

Use the table below to record the raw data for the paper clips you break. Then, summarize the data for the team by adding up how many clips broke at each number of flexes. Each team member should contribute 20 data points.

Use the data to plot a histogram and cumulative distribution function.

Paper Clip Flexing Data

Paper Clip #	Flexes to Break
1	
2	
3	
4	
5	
6	
7	
8	
9	
10	
11	
12	
13	
14	
15	
16	
17	
18	
19	
20	

Summary of Data

No. of Flexes	No. of Clips

ICA 14-9

An Excel worksheet, titled "Midterm Data," is available online. Use the data provided to determine:

(a) Class mean and median
(b) Class standard deviation
(c) Often in problems such as this, the bin sizes are preset according to certain constraints. Draw a histogram and CDF, based on the letter grade ranges given below. After the histogram is created, change the bin labels to be the letter grade rather than the numerical value.

Grade	F	D	C	B	A
Minimum	0	60	70	80	90
Maximum	59	69	79	89	100

Use the chart to determine:

(d) How many students received a C on the exam?
(e) What percentage of student received an A on the exam?
(f) What percentage of students received a passing grade (C, B, or A) on the exam?

ICA 14-10

The table below lists the number of defective computer chips rejected during random testing on a manufacturing line over the course of a week. Use the following data to generate a histogram and CDF using Excel.

1	1	8	0	2	0
0	2	10	1	3	2
0	1	12	0	2	1
1	6	15	0	0	
3	8	1	2	5	

ICA 14-11

In this exercise and the one that follows, you will develop two worksheets, one for a standing wave and one for a progressive (stationary) wave. In the real world, both of these types of waves occur in lakes, rivers, and oceans. Before starting each of these exercises, first place your worksheet into manual iteration mode.

When you throw a rock in a lake, a series of waves emanates from the impact point. Depending on conditions, these waves may move across the water surface as "progressive waves." The equation of such waves is

$$y = A \sin(x - kt)$$

Be sure to put your worksheet in iteration mode following the instructions on page 384.

In other words, if you "rode the crest of a wave," you would find yourself moving horizontally, much as a surfer. Alternatively, if you positioned yourself at a fixed location (you anchored the boat), as time passed, the water would move up and down at that location.

Create a worksheet similar to the one shown. Create a graph with water height (y) on the ordinate and position (x) on the abscissa imbedded in the worksheet. Set the horizontal scale at a minimum of 0 to a maximum of 25 with a major unit of 5. Set the vertical scale at a minimum of −10 to a maximum of 10 with a major unit of 5 and change the value X axis crosses at to −10. Enable data validation to prevent an amplitude greater than 10 from being selected.

Hold the F9 key down to make the waves move. Experiment with different values of amplitude and k. What happens if you make k negative?

As an alternative, create a slider bar to control the value of A and k.

ICA 14-12

In this exercise and the one preceding, you will develop two worksheets, one for a standing wave and one for a progressive (stationary) wave. In the real world, both of these types of waves occur in lakes, rivers and the ocean. Before starting each of these, first place your worksheet into manual iteration mode.

When a wave is created in a water body with solid boundaries, the waves can reflect and move back in the opposite direction. In such a situation, standing waves may be created where the crest and troughs do not move along the surface but simply move up and down in one location. The equation describing standing waves is:

$$y = A \sin(c\,x)\sin(k\,t)$$

In this equation, A is amplitude, x is position, and t is time. The variable c is the frequency of the standing wave, and controls how far apart the peaks are. The variable k determines how rapidly the standing wave rises and falls.

We can position ourselves at a fixed location (we anchor the boat), and then as time passes, the water moves up and down at that location. If however we position ourselves at a different place, it may be that the water surface remains constant.

Create a worksheet similar to the one shown. Create a graph with water height (y) on the ordinate and position (x) on the abscissa imbedded in the worksheet. Set the horizontal scale at a minimum of 0 to a maximum of 25 with a major unit of 5. Set the vertical scale at a minimum of −10 to a maximum of 10 with a major unit of 5 and change the Value X axis crosses at: to −10. Enable data validation to prevent selecting amplitude greater than 10. Hold the F9 key down to make the waves move.

Experiment with different values of A, c, and k. What happens if you make k negative?

As an alternative, create a slider bar to control the values of A, c, and k.

CHAPTER 15
MODELS AND SYSTEMS

A **model** is an abstract description of the relationship between variables in a system. A model allows the categorization of different types of mathematical phenomena so that general observations about the variables can be made for use in any number of applications.

For example, if we know that t = v + 5 and M = z + 5, any observations we make about v with respect to t also apply to z with respect to M. A specific model describes a *system* or *function* that has the same *trend* or *behavior* as a generalized model. In engineering, many specific models within different subdisciplines behave according to the same generalized model.

This section covers three general models of importance to engineers: **linear**, **power**, and **exponential**. It is worth noting that many applications of models within these three categories contain identical math but apply to significantly different disciplines.

Linear models occur when the dependent variable changes in direct relationship to changes in the independent variable. We discuss such systems, including springs, resistive circuits, fluid flow, and elastic materials, in this chapter by relating each model to Newton's generalized law of motion.

Power law systems occur when the independent variable has an exponent not equal to 1 or 0. We discuss these models by addressing integer and rational real exponents.

Exponential models are used in all engineering disciplines in a variety of applications. We discuss these models by examining the similarities between growth and decay models.

The following is an example of the level of knowledge of Excel need to proceed into Chapter 15. *If you are not able to quickly recreate Excel Graphs similar to the ones shown, including trendlines and formatting, please review graphing basics in Appendix D before proceeding into Chapter 15.*

Energy (E) stored in an **inductor** is related to its inductance (L) and the current (I) passing through it by the following equation.

$$E = \frac{1}{2}LI^2$$

The SI unit of inductance, the **henry** [H], is named for Joseph Henry (1797–1878), credited with the discovery of self-inductance of electromagnets.

Three inductors were tested and the results are given here. Create a proper plot of the data and add a properly formatted power law trendline to each data set.

Current (I) [A]	2	6	10	14	16
Energy of Inductor #1 (E1) [J]	0.002	0.016	0.050	0.095	0.125
Energy of Inductor #2 (E2) [J]	0.010	0.085	0.250	0.510	0.675
Energy of Inductor #3 (E3) [J]	0.005	0.045	0.125	0.250	0.310

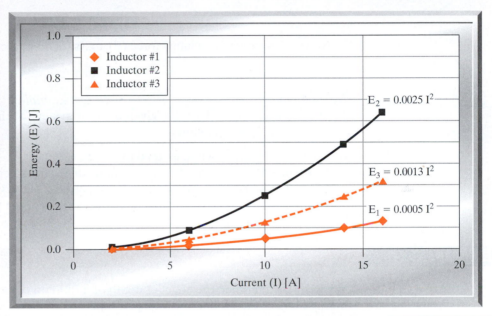

15.1 LINEAR FUNCTIONS

Trend	Equation	Data Form	Graphical Example
Linear	$y = m\,x + c$	Defined value (c) at x = 0 Data appears as a linear (straight) line	Positive value of m Negative value of m

One of the most common models is *linear*, taking the form $y = m\,x + c$, where the ordinate value (y) is a function of the abscissa value (x) and a constant factor called the *slope* (m). At an initial value of the abscissa (x = 0), the ordinate value is equal to the *intercept* (c). Examples include

- Distance (d) traveled at constant velocity (v) over time (t) from initial position (d_0):

$$d = v\,t + d_0$$

- Rate of rotation (ω) as a function of time (t) and angular acceleration (α) from initial rotational rate (ω_i):

$$\omega = \alpha\, t + \omega_i$$

- Total pressure (P_{total}), relating density (ρ), gravity (g), liquid height (H), and the pressure above the surface ($P_{surface}$):

$$P_{total} = \rho\, g\, H + P_{surface}$$

- Newton's second law, relating force (F), mass (m), and acceleration (a):

$$F = m\, a$$

Note that the intercept value (c) is zero in the last example.

   General linear model rules, given a linear system of the form $y = m\, x + c$ and assuming $x \geq 0$, are as follows:

- When m = 1, the function is equal to x + c.
- When m = 0, y = c, regardless of the value of x (y never changes).
- When m > 0, as x increases, y increases, regardless of the value of c.
- When m < 0, as x increases, y decreases, regardless of the value of c.

EXAMPLE 15-1

We want to determine the effect of depth of a fluid on the total pressure felt by a submerged object. Recall that the total pressure is

$$P_{total} = P_{surface} + P_{hydro} = P_{surface} + \rho\, g\, H$$

where P_{total} = total pressure [atm]; $P_{surface}$ = pressure at the surface [atm]; ρ = density [kg/m^3]; g = gravity [m/s^2]; H = depth [m]. We enter the lab, take data, and create the following chart.

$P_{total} = 0.075\ H + 3$

Determine the density of the fluid, in units of kilograms per cubic meter.

*We can determine the parameters by matching the trendline generated in Excel with the theoretical expression. In theory: total pressure = density * gravity * height of fluid + pressure on top of the fluid*
*From graph: total pressure = 0.075 * height + 3*
*By comparison: density * gravity = 0.075 [atm/m]*

$$\frac{0.075 \text{ atm}}{\text{m}} \left| \frac{101{,}325 \text{ Pa}}{1 \text{ atm}} \right| \frac{1 \frac{\text{kg}}{\text{m s}^2}}{1 \text{ Pa}} = \rho \left(9.8 \frac{\text{m}}{\text{s}^2} \right)$$

$$7{,}600 \frac{\text{kg}}{\text{m}^2\text{s}^2} = \rho \left(9.8 \frac{\text{m}}{\text{s}^2} \right)$$

$$\rho = 7{,}600 \frac{\text{kg}}{\text{m}^2\text{s}^2} \left| \frac{\text{s}^2}{9.8 \text{ m}} \right. = 775 \frac{\text{kg}}{\text{m}^3}$$

Determine if the tank is open to the atmosphere or pressurized, and determine the pressure on the top of the fluid in units of atmospheres.

*Once again, we can compare the Excel trendline to the theoretical expression. In theory: total pressure = density * gravity * height of fluid + pressure on top of the fluid*
*From graph: total pressure = 0.075 * height + 3*
By comparison, the top of the tank is pressurized at 3 atm.

Comprehension Check 15-1

The graph shows the ideal gas law relationship ($PV = nRT$) between pressure (P) and temperature (T).

(a) What are the units of the slope (0.0087)?
(b) If the tank has a volume of 12 liters and is filled with nitrogen (formula, N_2; molecular weight, 28 grams per mole), what is the amount of gas in the tank (n) in units of grams?
(c) If the tank is filled with 48 grams of oxygen (formula, O_2; molecular weight, 32 grams per mole), what is the volume of the tank (V) in units of liters?

$$P = 0.0087 \, T$$

Pressure (P) [atm]

Temperature (T) [K]

Increasingly, engineers are working at smaller and smaller scales. Tiny beads made of glass are on the order of 50 micrometers in diameter. They are manufactured so that they become hollow, allowing the wall thickness to be a few nanometers. The compositions of the glass were engineered so when processed correctly, they would sustain a hollow structure and the glass walls would be infiltrated with hundreds of thousands of nanometer-sized pores. These beads can possibly revolutionize the way fluids and gases are stored for use. The pores are small enough that fluids and even gases could be contained under normal conditions. However, if activated properly, the pores would allow a path for a gas to exit the "container" when it is ready to be used.

Photo courtesy of K. Richardson

15.2   LINEAR RELATIONSHIPS

Most physics textbooks begin the study of motion ignoring how that object came to be moving in the first place, since that information is not needed to figure out what will happen to the object in the future. This is appropriate to the way physicists study the world, by observing the world as it is. Engineering is about changing the way things are, such as making things move that are stationary and stopping things that are moving. As a result, engineers are concerned with forces and the changes those forces cause. While physicists study how far a car travels through the air when hit by a truck, engineers focus on stopping the truck before it hits the car or on designing the car so it doesn't fly as far or on designing an air-bag system or crush-proof doors. Engineering has many diverse branches because of the many different kinds of forces and ways to apply them.

Another Way of Looking at Newton's First Law

> A system keeps doing what it is doing unless the forces acting on the system change.

Newton's first law is given as "An object at rest remains at rest and an object in motion will continue in motion with a constant velocity unless it experiences a net external force." As we consider variables other than motion, we want to expand this definition: A system keeps doing what it is doing unless the forces acting on the system change.

Another Way of Looking at Newton's Second Law

> When a force influences a change to a system parameter, the system opposes the change according to its internal resistance.

Newton's second law is given as "The acceleration of an object is directly proportional to the net force acting on it and inversely proportional to its mass." This is summarized by the familiar expression $F = ma$, which can be interpreted as follows: When an external force acts on a system to cause acceleration, the system resists that acceleration according to its mass. Mass, therefore, becomes a resistance to acceleration. Expanding Newton's second law to generalize it for use with variables other than motion: When a force influences a change to a system parameter, the system opposes the change according to its internal resistance. In generalizing these relationships, we can start to establish a pattern observed in a wide variety of phenomena, summarized in Table 15-1.

Table 15-1 Generalized Newton's second law

When a "system" . . .	. . . is acted upon by a "force" . . .	. . . to change a "parameter" . . .	. . . the "system" opposes the change by a "resistance"	Equation
physical object	external push or pull (F)	acceleration (a)	object mass (m)	$F = m\,a$

Springs

When an external force (F), such as a weight, is applied to a spring, it will cause the spring to stretch a distance (x), according to the following expression:

$$F = kx$$

This equation is called **Hooke's law**, named for Robert Hooke (1635–1703), an English scientist. Among other things, he is credited with creating the biological term "cell." The comparison of Hooke's Law and Newton's Second Law is shown in Table 15-2.

Table 15-2 Generalized second law . . . applied to springs

When a "system" . . .	. . . is acted upon by a "force" . . .	. . . to change a "parameter" . . .	. . . the "system" opposes the change by a "resistance"	Equation
physical object	external push or pull (F)	acceleration (a)	object mass (m)	$F = m\,a$
spring	external push or pull (F)	elongation (x)	spring stiffness (k)	$F = k\,x$

The variable k is the **spring constant**, a measure of the stiffness of the spring. Stiff springs are hard to stretch and have high k values; soft springs are easy to stretch and have low k values. The constant k is a material property of the spring, determined by how it is made and what material it is made from.

Property	Symbol	Typical Units
spring constant	k	$\dfrac{N}{m}$

EXAMPLE 15-2

Two springs were tested; a weight was hung on one end and the resulting displacement measured. The results were graphed. Using the graph shown on the following page, give the spring constant of each spring. Which spring is stiffer?

Spring #1 has a linear trendline of F = 66 x. The slope of the line is the spring constant:

$$k_1 = 66 \text{ N/m}$$

Spring #2 has a linear trendline of F = 8 x, which corresponds to:

$$k_2 = 8 \text{ N/m}$$

Spring #1 is stiffer since it has a higher spring constant.

Basics of Electric Circuits

Electric charge (Q) is a property of some subatomic particles, notably the electron and the proton. On a small scale, charge can be measured in terms of the **elementary charge** (e). The magnitude of the elementary charge on the electron or proton equals 1, and by convention, the electron has a negative charge, $e = -1$; the proton has a positive charge $e = +1$.

On larger scales, the elementary charge is far too small to be useful for most practical problems, so a new unit was defined in terms of two base SI units, the ampere and the second. The new unit of charge was called the **coulomb** [C], named for the French physicist Charles-Augustin de Coulomb (1736–1806), who first described and quantified the attractive and repulsive electrostatic force, today known as Coulomb's law. In terms of elementary charges, one coulomb is approximately 6.24×10^{18} e. Thus, 6.24 quintillion electrons is one coulomb of charge.

Electric current (I) is a measure of how many charges (normally electrons) flow through a wire or component in a given amount of time. This is analogous to measuring water flowing through a pipe as amount per time, whether the units are tons per hour, gallons per minute, or molecules per second.

The unit of electric current is the **ampere** [A], one of the seven base SI units. It is named for Andre-Marie Ampere (1775–1836), a French physicist who is credited with discovering electromagnetism. How the ampere is defined as one of the base units is beyond the scope of this book, but can be found in a basic physics book. Of more practical use to the engineer is a definition in terms of charges moving through a conductor per unit time. From this perspective, the ampere is one coulomb per second, or approximately 6.24×10^{18} elementary charges per second. Thus, if 6.24 quintillion electrons move through a wire per second, the wire is carrying one ampere of current.

Voltage (V) is the "force" that pushes the electrons around. Although its effects on charged particles are similar to those of a true force, voltage is quite different dimensionally. The unit of voltage is the **volt** [V], described as the potential difference (voltage) across a conductor when a current of one ampere dissipates one watt of power. (Yes, that probably sounds like gibberish to most readers. We make this a bit more intuitive below.) The volt is named for Italian physicist Alessandro Volta (1745–1827), who perhaps invented the first chemical battery, called the voltaic pile.

We all learned in grade school that "opposite charges attract and like charges repel." This well-known phenomenon underlies the idea of voltage. Imagine an ac-

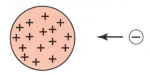

cumulation of electrons (negative charges), all jammed together in a small space and somehow made immovable. Then imagine a similar group of protons (positive charges) a short distance away. Now imagine that a single electron is placed between these two opposite charge accumulations. What happens to the single electron when it is released? It is repelled by the negative charges and attracted by the positive charges; thus, it accelerates toward the positive charges. Recall Newton's second law: a force is required to accelerate a mass (F = ma), even one as tiny as a single electron. Somehow, the accumulated charge is causing a force to be applied to this single electron.

Now imagine a device that could somehow pull electrons out of one piece of material and deposit them on a different piece of material. The material losing the electrons is left with a positive charge since it has more protons than electrons. This is called the cathode. Similarly, the piece of material receiving electrons has an overall negative charge and is called the anode. Now, a charge separation exists, just as in the hypothetical case with the accumulations of charges. This, however, describes a real device, known as a battery, or, more properly, a single Galvanic cell. One can buy these almost anywhere, the supermarket, the drug store, etc., and they power all sorts of devices from singing greeting cards and flashlights to laptop computers and electric automobiles.

In the Galvanic cell, a chemical reaction is responsible for relocating the electrons. When the cell is first constructed, the chemical reaction starts moving electrons from cathode to anode, but as this charge separation increases, the accumulated charges start trying to make the electrons being moved go the other way (they "push back" against the chemical reaction). Very quickly, a stalemate is achieved, with the accumulated charge trying to move the electrons backwards just as hard as the chemical reaction is trying to move them forward. In this condition, the battery can sit in its package on the shelf for months or years without going bad. The charge separation and the relative "push" being exerted by that accumulation against the chemical reaction can be quantified in terms of voltage.

But we still have not really quantified the volt. The question to ask is "how much energy was needed to separate the charges?" If one joule of energy was necessary to separate one coulomb of charge, then that charge separation is quantified as a potential difference of one volt; thus, 1 volt = 1 joule/coulomb: [V = J/C]. Thus, in a standard alkaline D cell, with a voltage of 1.5 volts, for each coulomb of charge that is moved from cathode to anode, the chemical reaction must provide 1.5 joules of energy. Similarly, if such systems were 100% efficient, one joule of energy could be extracted from one coulomb of charge separated by a potential difference of one volt.

Power in an electric circuit is simply the product of voltage and current:

Simplified Galvanic Cell

Insulator

Cathode (+)

Separator

Electrolyte

Anode (−)

$$P = VI$$

A volt times an ampere is a watt.

$V = J/C$

$A = C/s$

$V A = J/s = W$

If one ampere is moving through a potential difference of one volt, the power is one watt. Some electrical devices can use power (converting the energy to heat, light, sound, motion, etc.) and some can provide power (e.g., a battery). Some devices, such as a rechargeable battery can do both; when recharging, it is using or absorbing energy (converting it into chemical energy); when powering a device, it is providing or delivering energy.

Resistance (R) is a measure of how difficult it is to push electrons through a substance or device. When a voltage is applied to a circuit, a current is generated. This current depends on the equivalent resistance of the circuit. Resistance has units of volts per ampere, which is given the special name **ohm** [Ω]. It is named for Georg Ohm (1789–1854), the German physicist who developed the theory, called **Ohm's Law**, to explain the relationship between voltage, current, and resistance. The similarities between Ohm's Law and Newton's Second Law are given in Table 15-3.

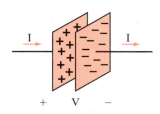

$$V = I R$$

Materials through which electrons can move easily (e.g., most metals) have a low resistance and are called conductors. Materials through which electrons move only with great difficulty have a high resistance and are called insulators (e.g., glass and many other ceramics). Commercially available resistors are available in standard values from less than 1 Ω to over 10 MΩ.

Conductance is the reciprocal of resistance (G = 1/R), and in certain types of problems is much easier to use than resistance. The old unit for conductance was the mho (ohm spelled backwards), but it has been supplanted in more recent times by the **siemens** [S], named after Ernst Werner von Siemens, the German inventor who, among other things, built the first electric elevator and founded the company known today as Siemens AG.

Capacitance (C) is a measure of the ability of a device called a capacitor to store an electric charge. Capacitance is measured in **farads** [F], named for the English scientist Michael Faraday (1791–1867). Among Faraday's numerous contributions to both electromagnetism and chemistry was the development of the principles for the first practical electric motor. We will define the farad after we explain how a capacitor functions.

In its simplest form, a capacitor is made from two flat conducting plates separated by an insulator. When a current enters one plate of the capacitor, the electrons making up the current begin to accumulate on that plate. Since the two plates are separated by an insulator, the electrons cannot cross over to the other plate. With accumulating negative charge on one plate, the electrons in the other plate are repelled, leaving behind an overall positive charge. This superficially gives the appearance that the current is going through the capacitor. However, the electrons entering one plate and leaving the

Table 15-3 Generalized second law . . . applied to circuits

When a "system". . .	. . . is acted upon by a "force" . . .	. . . to change a "parameter". . .	. . . the "system" opposes the change by a "resistance"	Equation
physical object	external push or pull (F)	acceleration (a)	object mass (m)	F = m a
electrical circuit	circuit voltage (V) (electromotive force)	circuit current (I)	circuit resistance (R)	V = I R

other are different electrons, and a charge separation is accumulated on the plates of the capacitor.

Note that the larger the accumulated charge, the greater the voltage. The voltage developed across the plates of a capacitor for a given amount of charge depends primarily on the total area of the plates and the distance between them. If one coulomb of charge is stored on the plates of a capacitor, and the resulting voltage is one volt, then by definition, the capacitance is one farad: a one-farad capacitor requires one coulomb of stored charge to have a potential difference across the plates of one volt [F = C/V].

Capacitors are used in literally hundreds of diverse applications, a few being audio filters, timing circuits, power supplies, and AC motors (efficiency improvement). Commercially available capacitors range from about one picofarad to over one farad. Since the farad is far too large to conveniently describe the majority of practical devices, **microfarads** [μF] are much more commonly used, although picofards or nanofarads, and, rarely, millifarads are also used.

Property	Symbol	Typical Units	Equivalent Units
Charge	Q	coulomb [C]	C = A s
Electric current	I	ampere [A]	A = C/s
Voltage	V	volt [V]	V = J/C
Electric power	P	watt [W]	W = J/s
Resistance	R	ohm [Ω]	Ω = V/A
Conductance	G	siemens [S]	S = A/V
Capacitance	C	microfarad [μF]	F = C/V

Fluid Flow

Velocity (v) increases as distance from wall (y) increases; maximum at center of pipe.

For motion in a fluid to be created, a force must be applied over an area of the fluid. While both liquids and gases can be defined as fluids, we focus on liquids in this section. Imagine a section of fluid-filled pipe sitting on the desk in front of you. While in reality the liquid would flow freely from both sides of the pipe, for this example, imagine that it stays in place. If we apply a force at a single point in the fluid, only the particles at that point will move. To move the entire fluid uniformly, we must apply the force at all points at the pipe entrance simultaneously. Applying a force over the cross-sectional area of the pipe results in the application of a pressure to the fluid. The pressure that results in fluid flow has a special name: **shear stress** (τ, Greek letter tau).

As the fluid moves, we find that the fluid molecules in contact with the wall adhere to the wall and do not move. The motion of the fluid can be visualized as occurring in layers; as the distance from the wall increases, the fluid moves faster. The fluid moves fastest at the farthest point from the wall, which is the center of the pipe. Since the velocity changes depend on the location in the pipe from the wall, the parameter we are changing cannot be expressed as a simple velocity, but rather as a **velocity gradient**, given as ($\Delta v/\Delta y$ or $\dot{\gamma}$). This is sometimes called the *shear rate* or *strain rate.*

Table 15-4 Generalized second law . . . applied to fluid flow

When a "system". . .	. . . is acted upon by a "force" . . .	. . . to change a "parameter" . . .	. . . the "system" opposes the change by a "resistance"	Equation
physical object	external push or pull (F)	acceleration (a)	object mass (m)	$F = m\,a$
fluid	shear stress (τ)	shear rate ($\Delta v / \Delta y$)	dynamic viscosity (μ)	$\tau = \mu \dfrac{\Delta v}{\Delta y}$

Not all fluids respond equally to an applied pressure. The fluid property that represents the resistance of a fluid against flow is called the **dynamic viscosity** (μ, Greek letter mu). The relationship between shear stress and the velocity profile of a fluid is called **Newton's law of viscosity**, named after Isaac Newton. Fluids that behave in this way are called **Newtonian fluids**; some examples are water and oil. The comparison between Newton's Law of Viscosity and Newton's Second Law is given in Table 15-4.

$$\tau = \mu \frac{\Delta v}{\Delta y}$$

Sometimes, a fluid must have a certain amount of stress (called the **yield stress**, τ_0) applied before it will begin to move like a Newtonian fluid. These fluids are called **Bingham plastics**, named after Eugene Bingham, a chemist who made many contributions to the field of **rheology** (the science of deformation and flow of matter, a term he, along with Markus Reiner, is credited in creating). Examples of Bingham plastics include toothpaste and slurries.

$$\tau = \mu \frac{\Delta v}{\Delta y} + \tau_0$$

Common units of dynamic viscosity are **centipoise** [cP], named after the French physician Jean Louis Poiseuille (1799–1869) who studied the flow of blood in tubes. Dynamic viscosity is a function of temperature. In most instances, viscosity decreases with increasing temperature; as the fluid heats up, it becomes easier to move.

Property	Symbol	Typical Units	Equivalent Units
Dynamic viscosity	μ	cP	$P = g/(cm\ s)$
Kinematic viscosity	ν	St	$St = cm^2/s$

Another useful term in describing a fluid is **kinematic viscosity** (ν, Greek letter nu). The kinematic viscosity is the ratio of dynamic viscosity to density and is given the unit of **stoke** [St], named after George Stokes (1819–1903), the Irish mathematician and physicist who made important contributions to science, including Stokes' law, optics, and physics. Several values of dynamic and kinematic viscosity are given on the next page.

Liquid	Specific Gravity	Dynamic Viscosity (μ) [cP]	Kinematic Viscosity (ν) [St]
Acetone	0.791	0.331	0.419
Benzene	0.879	0.647	0.736
Corn syrup	1.36	1380	1015
Ethanol	0.789	1.194	1.513
Glycerin	1.260	1490	1183
Honey	1.36	5000	3676
Mercury	13.600	1.547	0.114
Molasses	1.400	8000	5714
Olive oil	0.703	101	143
SAE 30W oil	0.891	290	325
Toluene	0.865	0.590	0.682
Water	1.000	1.000	1.000

Comprehension Check 15-2

Fluid A has a dynamic viscosity of 0.5 centipoise and a specific gravity of 1.1.

(a) What is the density of Fluid A in units of pound-mass per cubic foot?
(b) What is the dynamic viscosity of Fluid A in units of pound-mass per foot/second?
(c) What is the dynamic viscosity of Fluid A in units of pascal seconds?
(d) What is the kinematic viscosity of Fluid A in units of stokes?

Elastic Materials

Elasticity is the property of an object or material that causes it to be restored to its original shape after distortion. A rubber band is easy to stretch and snaps back to near its original length when released, but it is not as elastic as a piece of piano wire. The piano wire is harder to stretch, but would be said to be more elastic than the rubber band because of the precision of its return to its original length. The term elasticity (*Young's modulus* or *modulus of elasticity*, E) is described as the amount of deformation resulting from an applied force. Young's modulus is named for Thomas Young (1773–1829), a British scientist, who contributed to several fields: material elongation theory; optics, with his "double slit" optical experiment that led to the deduction that light travels in waves; and fluids, with the theory of surface tension and capillary action.

Property	Symbol	Typical Units
Young's modulus	E	Pa

Like fluids, elastic materials accept a force applied over a unit area rather than a point force. **Stress** (σ, Greek letter sigma) is the amount of force applied over a unit area of the material, which has units of pressure [Pa]. The **strain** (ε, Greek letter epsilon) is the ratio of the elongation to the original length, yielding a dimensionless number. Since the modulus values tend to be large, Young's modulus is usually expressed in units of Gigapascals [GPa]. A comparison between Young's Modulus and Newton's Second Law is shown in Table 15-5.

$$\sigma = E\varepsilon$$

Table 15-5 Generalized second law . . . applied to elastic materials

When a "system" . . .	. . . is acted upon by a "force". . .	. . . to change a "parameter" . . .	. . . the "system" opposes the change by a "resistance"	Equation
physical object	external push or pull (F)	acceleration (a)	object mass (m)	$F = ma$
elastic object	stress (σ)	strain (ε)	Young's modulus (E)	$\sigma = E\varepsilon$

From this discussion, you can see examples from many areas of engineering that are similar to Newton's second law. We often want to change something and find that it resists this change; this relationship is often linear. In all of these situations, we discover a coefficient that depends on the material encountered in the particular situation (mass, spring stiffness, circuit resistance, fluid viscosity, or Young's modulus).

Many other examples are not discussed here, such as Fourier's law of heat transfer, Fick's law of diffusion, and Darcy's law of permeability. As an engineering student, you can enhance your understanding of your coursework by attempting to generalize the knowledge presented in a single theory (such as Hooke's law) to other theories that may be presented in other courses. Many different disciplines of engineering are linked by common themes, and the more you can connect these theories across disciplines, the more meaningful your classes will become.

Combinations of Springs and Circuits

When connected, both springs and circuits form a resulting system that behaves like a single spring or single resistor. In a combination of springs, the stiffness of the spring configuration depends on the stiffness of each individual spring and on the configuration. This system stiffness is referred to as the effective spring constant (k_{eff}). In a network of circuits, the resistance of the network depends on the value of the individual resistors and on the configuration. This system resistance is referred to as the effective resistance (R_{eff}).

Springs in Parallel

When springs are attached in *parallel*, they must *displace the same distance* even though they may have different spring constants. The derivation below shows how this leads to an effective spring constant that is the sum of the individual spring constants in the system. Each spring is responsible for supporting a proportional amount of the force.

Writing Hooke's law for two springs each displacing the same distance (x):

$$F_1 = k_1 x \tag{a}$$

$$F_2 = k_2 x \tag{b}$$

Solve for F_1 in terms of F_2 since the displacement is the same:

$$F_1 = k_1 \frac{F_2}{k_2} = F_2 \frac{k_1}{k_2} \tag{c}$$

Writing Hooke's law as applied to the overall system:

$$F = k_{eff} x \tag{d}$$

Springs in parallel both displace the same distance.

The total force applied to the configuration (F) is the sum of the force supported by each spring:

$$F = F_1 + F_2 \tag{e}$$

Eliminating force (F) from Equation (e) with Equation (d):

$$k_{eff}\, x = F_1 + F_2 \tag{f}$$

Eliminating displacement (x) with Equation (b):

$$k_{eff}\frac{F_2}{k_2} = F_1 + F_2 \tag{g}$$

Substituting for F_1 with Equation (c):

$$k_{eff}\frac{F_2}{k_2} = F_2\frac{k_1}{k_2} + F_2 \tag{h}$$

Dividing Equation (h) by F_2:

$$\frac{k_{eff}}{k_2} = \frac{k_1}{k_2} + 1 \tag{i}$$

Multiplying Equation (i) by k_2 gives:

$$k_{eff} = k_1 + k_2 \tag{j}$$

A system of two springs in parallel will always be stiffer than either spring individually.

> A system of two springs in parallel will always be stiffer than either spring individually.

Springs in Series

When two springs are attached in *series*, the *force is the same for both springs*. The effective spring constant is derived below. The applied force affects each spring as though the other spring did not exist, and each spring can stretch a different amount.

Writing Hooke's law for two springs each under the same applied force (F):

$$F = k_1\, x_1 \tag{k}$$
$$F = k_2\, x_2 \tag{l}$$

Solve for x_1 in terms of x_2 since the force is the same:

$$x_1 = \frac{k_2}{k_1} x_2 \tag{m}$$

Writing Hooke's law as applied to the overall system:

$$F = k_{eff}\, x \tag{n}$$

> Springs in series are acted on by the same force for both springs.

The total distance stretched by the configuration (x) is the sum of the distance stretched by each spring:

$$x = x_1 + x_2 \tag{o}$$

Eliminating force (F) from Equation (n) with Equation (l):

$$k_2\, x_2 = k_{eff}\, x \tag{p}$$

Eliminating displacement (x) with Equation (o):

$$k_2 x_2 = k_{eff}\,(x_1 + x_2) \tag{q}$$

Substituting for x_1 with Equation (m):

$$k_2\, x_2 = k_{eff}\left(\frac{k_2}{k_1} x_2 + x_2\right) \tag{r}$$

Dividing Equation (r) by x_2

$$k_2 = k_{eff}\left(\frac{k_2}{k_1} + 1\right) \tag{s}$$

Dividing Equation (s) by k_2 gives

$$1 = k_{eff}\left(\frac{1}{k_1} + \frac{1}{k_2}\right) \tag{t}$$

Thus,

$$k_{eff} = \frac{1}{\left(\frac{1}{k_1} + \frac{1}{k_2}\right)} = \left(\frac{1}{k_1} + \frac{1}{k_2}\right)^{-1} \tag{u}$$

> A system of two springs in series will always be less stiff than either spring individually.

A system of two springs in series will always be less stiff than either spring individually.

These forms for springs in parallel and series generalize to any number of springs. For N springs in parallel, the effective spring constant is

$$k_{eff} = k_1 + k_2 + \cdots + k_{N-1} + k_N \tag{v}$$

For N springs in series, the effective spring constant is

$$k_{eff} = \left(\frac{1}{k_1} + \frac{1}{k_2} + \cdots + \frac{1}{k_{N-1}} + \frac{1}{k_N}\right)^{-1} \tag{w}$$

EXAMPLE 15-3

Find the displacement (x) in the spring combination shown, where Spring 1 (with a stiffness k_1) and Spring 2 (with a stiffness k_2) are connected in parallel, and the combination is then connected to Spring 3 (with a stiffness k_3) in series. Use the following values:

F = 0.2 N
k_1 = 10 N/m
k_2 = 5 N/m
k_3 = 8 N/m

First, we recognize that k_1 and k_2 are in parallel, so we can solve for an effective spring constant, using Equation (j).

$$k_{eff12} = k_1 + k_2 = 10\frac{N}{m} + 5\frac{N}{m} = 15\frac{N}{m}$$

The combination can then be redrawn to show k_{eff12} and k_3 in series.

Next, we solve for the effective spring constant using Equation (u).

$$k_{eff} = \left(\frac{1}{k_{eff12}} + \frac{1}{k_3}\right)^{-1} = \left(\frac{1}{15\frac{N}{m}} + \frac{1}{8\frac{N}{m}}\right)^{-1} = 5.2\frac{N}{m}$$

We can now solve for the displacement, using Hooke's law:

$$F = k_{eff}\, x$$

$$x = \frac{F}{k_{eff}} = \frac{0.2\,N}{5.2\frac{N}{m}} = 0.04\,m = 4\,cm$$

Comprehension Check 15-3

You have three springs, with stiffness 1 newton per meter [N/m], 2 newtons per meter [N/m], and 3 newtons per meter [N/m], respectively. How many unique combinations can be formed with these springs? Consider each spring alone, pairs of springs in both parallel and series, and all springs used at once.

Resistors in Series

When two resistors are connected in *series*, the *current through both of the resistors is the same*, even though the value of each resistor may be different. The derivation below shows the effective resistance of two resistors connected in series. Note that the voltage is applied to the entire system.

Writing Ohm's law for two resistors each with the same current:

$$V_1 = I\,R_1 \tag{A}$$
$$V_2 = I\,R_2 \tag{B}$$

Solving for V_1 in terms of V_2 since the current is the same:

$$V_1 = R_1\frac{V_2}{R_2} = V_2\frac{R_1}{R_2} \tag{C}$$

Writing Ohm's law as applied to the overall system:

$$V = I\,R_{eff} \tag{D}$$

The total voltage applied to the configuration (V) is the sum of the voltage applied to each resistor:

$$V = V_1 + V_2 \tag{E}$$

Eliminating voltage (V) from Equation (E) with Equation (D):

$$I\,R_{eff} = V_1 + V_2 \tag{F}$$

Eliminating current (I) from Equation (F) with Equation (B):

$$R_{eff}\frac{V_2}{R_2} = V_1 + V_2 \tag{G}$$

Substitution for V_1 with Equation (C):

$$R_{eff}\frac{V_2}{R_2} = V_2\frac{R_1}{R_2} + V_2 \tag{H}$$

Dividing Equation (H) by V_2:

$$R_{eff}\frac{1}{R_2} = \frac{R_1}{R_2} + 1 \tag{I}$$

> Resistors in series have the same current through both of the resistors.

A system of two resistors in series will always provide more resistance than either resistor individually.

Resistors in parallel have the same voltage through both of the resistors.

Multiplying Equation (I) by R_2:

$$R_{eff} = R_1 + R_2 \tag{J}$$

A system of two resistors in series will always provide more resistance than either resistor individually.

Resistors in Parallel

When two resistors are connected in *parallel*, the *voltage across both of the resistors is the same*. The current through each resistor may be different. The voltage is applied to the entire system.

Writing Ohm's law for two resistors each with the same voltage:

$$V = I_1 R_1 \tag{K}$$
$$V = I_2 R_2 \tag{L}$$

Solving for I_1 in terms of I_2 since the current is the same:

$$I_1 = \frac{R_2}{R_1} I_2 \tag{M}$$

Writing Ohm's law as applied to the overall system:

$$V = I\,R_{eff} \tag{N}$$

The total current (I) is the sum of the current flowing through each resistor:

$$I = I_1 + I_2 \tag{O}$$

Eliminating voltage (V) from Equation (N) using Equation (L):

$$I_2 R_2 = I\,R_{eff} \tag{P}$$

Eliminating current (I) using Equation (O):

$$I_2 R_2 = (I_1 + I_2)R_{eff} \tag{Q}$$

Substituting for I_1 using Equation (M):

$$I_2 R_2 = \left(\frac{R_2}{R_1} I_2 + I_2\right)R_{eff} \tag{R}$$

Dividing Equation (R) by I_2:

$$R_2 = R_{eff}\left(\frac{R_2}{R_1} + 1\right) \tag{S}$$

Dividing Equation (S) by R_2 gives:

$$1 = R_{eff}\left(\frac{1}{R_1} + \frac{1}{R_2}\right) \tag{T}$$

> A system of two resistors in parallel will always have less resistance than either resistor individually.

Thus,

$$R_{eff} = \frac{1}{\left(\frac{1}{R_1} + \frac{1}{R_2}\right)} = \left(\frac{1}{R_1} + \frac{1}{R_2}\right)^{-1} \tag{U}$$

A system of two resistors in parallel will always have less resistance than either resistor individually.

These forms for resistors in parallel and series generalize to any number of resistors. For N resistors in parallel, the effective resistance is

$$R_{eff} = \left(\frac{1}{R_1} + \frac{1}{R_2} + \cdots + \frac{1}{R_{N-1}} + \frac{1}{R_N}\right)^{-1} \tag{V}$$

This form, along with spring equation (w) shown earlier, is sometimes referred to as the "reciprocal of the sum of the reciprocals."

For N resistors in series, the effective resistance is

$$R_{eff} = R_1 + R_2 + \cdots + R_{N-1} + R_N \tag{W}$$

WARNING

Some of you may have seen a "simpler" form of the equation for two springs in series or two resistors in parallel.

For two springs in series: $k_{eff} = \dfrac{k_1 k_2}{k_1 + k_2}$

For two resistors in parallel: $R_{eff} = \dfrac{R_1 R_2}{R_1 + R_2}$

These forms are sometimes referred to as "the product over the sum."

These forms *do not generalize* to more than two elements.

If you have three or more elements, then you must use the "reciprocal of the sum of the reciprocals" form given earlier.

Human fibroblasts are connective tissue cells present in organs throughout the body. In this image, these cells can be seen spreading within a semi-interpenetrating network made of a polymer called polyethylene glycol diacrylate-hyaluronic acid (Pegda-HA). This material can be injected in a minimally invasive manner and cross-linked inside the body to form an insoluble gel with mechanical properties similar to many soft tissues in the human body. Such materials are being widely studied as "scaffolds" for cell transplantation in tissue engineering and regenerative medicine. The material degrades within 4–6 weeks, yielding physiological metabolites and water soluble polymers that are readily excreted through the kidneys.

Photo courtesy of K. Webb and J. Kutty

EXAMPLE 15-4

Find the current (I) in the circuit shown. Resistor 1 (with resistance R_1) and Resistor 2 (with resistance R_2) are connected in series, and the combination is then connected to Resistor 3 (with resistance R_3) in parallel. Use the following values:

$$V = 12 \text{ V}$$
$$R_1 = 7.5 \text{ k}\Omega$$
$$R_2 = 2.5 \text{ k}\Omega$$
$$R_3 = 40 \text{ k}\Omega$$

First, we recognize that R_1 and R_2 are in series, so we reduce R_1 and R_2 to a single effective resistor by using Equation (J).

$$R_{eff12} = R_1 + R_2 = 7.5 \text{ k}\Omega + 2.5 \text{ k}\Omega = 10 \text{ k}\Omega$$

Next, we can redraw the circuit so R_{eff12} and R_3 are in parallel. We solve for the effective resistance by using Equation (U).

$$R_{eff} = \left(\frac{1}{R_{eff12}} + \frac{1}{R_3} \right)^{-1} = \left(\frac{1}{10 \text{ k}\Omega} + \frac{1}{40 \text{ k}\Omega} \right)^{-1} = 8 \text{ k}\Omega$$

We can now solve the problem with Ohm's law:

$$V = I \, R_{eff}$$

$$I = \frac{V}{R_{eff}} = \frac{12 \text{ V}}{8 \text{ k}\Omega} = 0.0015 \text{ A} = 1.5 \text{ mA}$$

Comprehension Check 15-4

You have three resistors, with resistance 2 ohms [Ω], 2 ohms [Ω], and 3 ohms [Ω], respectively. How many unique combinations can be formed with these resistors? Consider each resistor alone, pairs of resistors both in parallel and in series, and all resistors used at once.

15.3   POWER FUNCTIONS

Trend	Equation	Data Form	Graphical Example
Power	$y = b\,x^m$	Positive m Value of zero at x = 0 Negative m Value of infinity at x = 0	Positive value of *m* — — Negative value of *m*

Power models take the form $y = b\,x^m$. Examples include

- Many geometric formulae involving areas, volumes, etc., such as the volume of a sphere (V) as a function of radius (r):

$$V = 4/3\,\pi\,r^3$$

- Distance (d) traveled by a body undergoing constant acceleration (a) over time (t), starting from rest:

$$d = a\,t^2$$

- Energy calculations in a variety of contexts, both mechanical and electrical, such as the kinetic energy (KE) of an object as a function of the object's velocity (v), where the constant (k) depends upon the object shape and type of motion:

$$KE = k\,m\,v^2$$

- Ideal gas law relationships, such as Boyle's law, relating volume (V) and pressure (P) of an ideal gas, holding temperature (T) and quantity of gas (n) constant:

$$V = (n\,R\,T)P^{-1}$$

General power model rules, given a power system of the form $y = b\,x^m + c$, assuming $x \geqslant 0$, are as follows:

- When $m = 1$, the model is a linear function.
- When $m = 0$, $y = b + c$, regardless of the value of x.
- When m is rational, the function will contain a rational exponent or may be described with a radical symbol ($\sqrt{}$). Certain rational exponents have special names (½ is "square root," ⅓ is "cube root").

- When m is an integer, the function will contain an integer exponent on the independent variable. Certain exponents have special names (2 is "squared," 3 is "cubed").
- When $0 < |m| < 1$ and $x < 0$, the function may contain complex values. In this chapter, we will only consider power law models where c is zero. In Chapter 16, we will discuss ways of dealing with data when the value of c is non-zero.

EXAMPLE 15-5

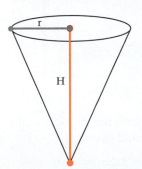

The volume (V) of a cone is calculated in terms of the radius (r) and height (H) of the cone. The relationship is described by the following equation:

$$V = \frac{\pi r^2 H}{3}$$

Given a height of 10 centimeters, calculate the volume of the cone when the radius is 3 centimeters.

$$V = \frac{\pi(3\text{ cm})^2(10\text{ cm})}{3} \approx 94.2\text{ cm}^3$$

> With a positive integer exponent, the volume of the cone increases as the independent variable (radius) increases. This observation is true with any power model with a positive integer exponent.

What is the volume of the cone when the radius is 8 centimeters?

$$V = \frac{\pi(8 \text{ cm})^2(10 \text{ cm})}{3} \approx 670 \text{ cm}^3$$

EXAMPLE 15-6

The resistance (R, $[g/cm^4 s]$) of blood flow in an artery or vein depends upon the radius (r, [cm]), as described by **Poiseuille's equation**:

$$R(r) = \frac{8\,\mu L}{\pi} r^{-4}$$

The viscosity of blood (μ, $[g/(cm\ s)]$) and length of the artery or vein (L, [cm]) are constants in the system. In studying the effects of a cholesterol-lowering drug, you mimic the constricting of an artery being clogged with cholesterol. You use the data you collect to create the following graph.

> With a negative integer exponent, the dependent variable (resistance) decreases as the independent variable (radius) increases. This trend is true for any power model with a negative integer exponent.

If the length of the artificial artery tested was 505 centimeters, what is the viscosity of the sample used to mimic blood, in units of grams per centimeter second $[g/(cm\ s)]$?

The constant "45" has physical meaning, found by comparison to the theoretical expression. In theory: $R(r) = \frac{8\mu L}{\pi} r^{-4}$ and from graph: $R = 45\ r^{-4}$

By comparison:

$$45\frac{g}{s} = \frac{8\mu L}{\pi} = \frac{8\,\mu(505\text{ cm})}{\pi}$$

$$\mu = 0.035\frac{g}{\text{cm s}}$$

Comprehension Check 15-5

The graph shows the ideal gas law relationship ($PV = nRT$) between pressure (P) and volume (V).

(a) What are the units of the constant (10)?

(b) If the tank is at a temperature of 300 kelvin and is filled with nitrogen (formula, N_2; molecular weight, 28 grams per mole), what is the amount of gas in the tank (n) in units of grams?

(c) If the tank is filled with 10 grams of oxygen (formula, O_2; molecular weight, 32 grams per mole), what is the temperature of the tank in units of degrees Celsius?

15.4 | EXPONENTIAL FUNCTIONS

Trend	Equation	Data Form	Graphical Example
Exponential	$y = b\,e^{mx}$	Defined value b (b ≠ 0) at x = 0 Positive m: asymptotic to 0 for large negative values of x Negative m: asymptotic to 0 at large positive values of x	Positive value of *m* — Negative value of *m*

Exponential models take the form $y = b\, e^{m\,x} + c$. Examples include

- The voltage (V) across a capacitor (C) as a function of time (t), with initial voltage (V_0) discharging its stored charge through resistance (R):

$$V = V_0\, e^{-t/(RC)}$$

- The number (N) of people infected with a virus such as smallpox or H1N1 flu as a function of time (t), given the following: an initial number of infected individuals (N_0), no artificial immunization available and dependance on contact conditions between species (C):

$$N = N_0\, e^{C\,t}$$

- The transmissivity (T) of light through a gas as a function of path length (L), given an absorption cross-section (s) and density of absorbers (N):

$$T = e^{-sNL}$$

- The growth of bacteria (C) as a function of time (t), given an initial concentration of bacteria (C_0) and depending on growth conditions (g):

$$C = C_0\, e^{gt}$$

Note that all exponents must be dimensionless, and thus unitless. For example, in the first equation, the quantity RC must have units of time.

Note that the intercept value (c) is zero in all of the above examples.

General Model Rules:

Given an exponential system of the form $y = b\, e^{m\,x} + c$

- When $m = 0$, $y = b + c$ regardless of the value of x.
- When $m > 0$, the model is a **growth function**. The minimum value of the growth model for $x \geqslant 0$ is $b + c$. As x approaches infinity, y approaches infinity.
- When $m < 0$, the model is a **decay function**. The value of the decay model approaches c as x approaches infinity. When $x = 0$, $y = b + c$.

What Is "e"?

An irrational number is a real number that cannot be expressed as the ratio of two integers. Pi (π) is an example.

The **exponential constant** "e" is an irrational number that can be truncated to 2.71828. It is defined as the base of the natural logarithm function. Sometimes, e is referred to as **Euler's number** or the **Napier constant.** The reference to Euler comes from the Swiss mathematician Leonhard Euler (pronounced "oiler," 1707–1783), who made vast contributions to calculus, including the notation and terminology used today. John Napier (1550–1617) was a Scottish mathematician credited with inventing logarithms and popularizing the use of the decimal point.

Growth Functions

Growth function

An exponential *growth function* is a type of function that increases without bound with respect to an independent variable. For a system to be considered an exponential growth function, the exponential growth model ($y = be^{mx} + c$) requires that m be greater than zero.

A more general exponential growth function can be formed by replacing the Napier constant with an arbitrary constant, or $y = b a^{mx} + c$. In the general growth function, a must be greater than 1 for the system to be a growth function. The value of a is referred to as the *base*, m is the *growth rate*, b is the *initial value*, and c is a *vertical shift*. Note that when a = 1 or m = 0, the system is reduced to $y = b + c$, which is a constant.

EXAMPLE 15-7

In 1965, Gordon E. Moore, co-founder of Intel Corporation, claimed in a paper that the number of transistors on an integrated circuit will double every two years. This idea by Moore was later referred to as **Moore's law**. The Intel 4004 CPU was released in 1971 as the first commercially available microprocessor. The Intel 4004 CPU contained 2,300 transistors. This system can be modeled with the following growth function.

$$T = T_0 2^{\frac{t}{2}}$$

In the equation, T_0 represents the initial number of transistors, and t is the number of years since T_0 transistors were observed on an integrated circuit. Predict the number of transistors on an integrated circuit in 1974 using the Intel 4004 CPU as the initial condition.

$$t = 1974 - 1971 = 3 \text{ years}$$

$$T = T_0 2^{\frac{t}{2}} = 2,300\left(2^{\frac{3}{2}}\right) = 2,300\left(2^{1.5}\right) \approx 6,505 \text{ transistors}$$

In 1974, the Intel 8080 processor came out with 4,500 transistors on the circuit.

Predict the number of transistors on integrated circuits in 1982 using the Intel 4004 CPU as the initial condition.

$$t = 1982 - 1971 = 11 \text{ years}$$

$$T = T_0 2^{\frac{t}{2}} = 2,300\left(2^{\frac{11}{2}}\right) = 2,300\left(2^{5.5}\right) \approx 104,087 \text{ transistors}$$

In 1982, the Intel 286 microprocessor came out with 134,000 transistors in the CPU.

Predict the number of transistors on integrated circuits in 2007 using the Intel 4004 CPU as the initial condition.

$$t = 2007 - 1971 = 36 \text{ years}$$

$$T = T_0 2^{\frac{t}{2}} = 2,300\left(2^{\frac{36}{2}}\right) = 2,300\left(2^{18}\right) \approx 603,000,00 \text{ transistors}$$

In 2007, the NVIDIA G80 came out with 681,000,000 transistors in the CPU.

No one really knows how long Moore's law will hold up. It is perhaps interesting to note that claims have consistently been made for the past 30 years that Moore's law will only hold up for another 10 years. Although many prognosticators are still saying this, some are not. There is, however, a limit to how small a transistor can be made. Any structure has to be at least one atom wide, for example, and as they become ever smaller, quantum effects will probably wreak havoc. Of course, chips can be made larger, multilayer structures can be built, new technologies may be developed (the first functional memristor was demonstrated in 2008), on so forth.

EXAMPLE 15-8

An environmental engineer has obtained a bacteria culture from a municipal water sample and allowed the bacteria to grow. After several hours of data collection, the following graph is created.

What was the initial concentration of bacteria?

In theory: $B = B_0 e^{gt}$ *and from graph:* $B = 10\, e^{0.2t}$

By comparison: $B_0 = 10$ *bacteria*

What was the growth constant (g) of this bacteria strain?

In theory: $B = B_{0,r} e^{gt}$ *and from graph:* $B = 10\, e^{0.2t}$

By comparison: g = 0.2 per hour. Recall that exponents must be unitless, so the quantity of (g t) must be a unitless group. To be unitless, g must have units of inverse time.

The engineer wants to know how long it will take for the bacteria culture population to grow to 30,000.

To calculate the amount of time, plug in 30,000 for B and solve for t:

$$30{,}000 = 10 e^{0.2t}$$

$$3000 = e^{0.2t}$$

$$\ln(3000) = \ln\left(e^{0.2t}\right) = 0.2\, t$$

$$t = \frac{\ln(3000)}{0.2\left[\frac{1}{h}\right]} = 40 \text{ hours}$$

Decay Functions

Decay function

A *decay function* is a type of function that decreases and asymptotically approaches a value. In the exponential decay model ($y = b\,e^{-m\,x} + c$), m is a positive value that represents the **decay rate**.

EXAMPLE 15-9

An electrical engineer wants to determine how long it will take for a particular capacitor in a circuit to discharge. The engineer wired a voltage source across a *capacitor* (C, farads) and a resistor (R, ohms) connected in series. After the capacitor is fully charged, the circuit is completed between the capacitor and resistor, and the voltage source is removed from the circuit. The product of R and C in a circuit like this is called the "time constant" and is usually denoted by the Greek letter tau ($\tau = RC$).

The time constant is the amount of time required for the circuit to discharge to approximately one-third of its initial value. Rather than 0.333, this value is actually $e^{-1} = 0.368$.

$$\tau = \frac{1}{R\,C}$$

The following equation can be used to calculate the voltage across a discharging capacitor at a particular time.

$$V = V_0 e^{-\tau} = V_0 e^{-\frac{t}{RC}}$$

Assuming a resistance of 100 kilo-ohms [kΩ], a capacitance of 100 microfarads [μF], and an initial voltage (V_0) of 20 volts [V], determine the voltage across the capacitor after 10 seconds.

$$V = 20\ [\text{V}]\ e^{-\frac{10\ \text{s}}{(100\ \text{k}\Omega)(100\ \mu\text{F})}} = 20[\text{V}]\ e^{-\frac{10\ \text{s}}{(100 \times 10^3\ \Omega)(100 \times 10^{-6}\ \text{F})}} \approx 7.36\ \text{V}$$

Assuming a resistance of 200 kilo-ohms [kΩ], a capacitance of 100 microfarads [μF], and an initial voltage (V_0) of 20 volts [V], determine the voltage across the capacitor after 20 seconds.

$$V = 20\ e^{-\frac{20\ \text{s}}{(200\ \text{k}\Omega)(100\ \mu\text{F})}} \approx 7.36\ \text{V}$$

Note that doubling the resistance in the circuit doubles the amount of time required to discharge the capacitor. In RC circuits, it is easy to increase the discharge time of a capacitor by increasing the resistance in the circuit.

Comprehension Check 15-6

The decay of a radioactive isotope was tracked over a number of hours, resulting in the following data. The decay of a radioactive element is modeled by the following equation, where C_0 is the initial amount of the element at time zero, and k is the decay constant of the isotope.

$$C = C_0 \, e^{-k\,t}$$

(a) What is the initial concentration of the isotope?
(b) Determine the decay constant of the isotope, including value and units.

$C = 35\,e^{-t}$

Concentration (C) [g]

Time (t) [h]

Picture of a single mortar shot. The creation of fireworks involves knowledge of chemistry (what materials to include to get the desired colors), physics and dynamics (what amounts of combustible charge should be included to launch the object properly), and artistry (what colors, shapes, patterns, and sounds the firework should emit such that it is enjoyable to watch). This picture is a close-up of the instant when a firework is detonating.

Photo courtesy of E. Fenimore

In-Class Activities

ICA 15-1

Mercury has a dynamic viscosity of 1.55 centipoises and a specific gravity of 13.6.

(a) What is the density of mercury in units of kilograms per cubic meter?
(b) What is the dynamic viscosity of mercury in units of pound-mass per foot second?
(c) What is the dynamic viscosity of mercury in units of pascal seconds?
(d) What is the kinematic viscosity of mercury in units of stokes?

ICA 15-2

SAE 10W30 motor oil has a dynamic viscosity of 0.17 kilograms per meter second and a specific gravity of 0.876.

(a) What is the density of the motor oil in units of kilograms per cubic meter?
(b) What is the dynamic viscosity of the motor oil in units of pound-mass per foot second?
(c) What is the dynamic viscosity of the motor oil in units of centipoise?
(d) What is the kinematic viscosity of the motor oil in units of stokes?

ICA 15-3

You have two springs each of stiffness 1 newton per meter [N/m] and one spring of stiffness 2 newtons per meter [N/m].

(a) List all possible stiffness combinations that can be formed with these springs. Your solution should include drawings and calculations of each unique configuration.
(b) How many unique combinations can be formed?
(c) What is the stiffest configuration?
(d) What is the least stiff configuration?

ICA 15-4

You have three resistors of resistance 30 ohm [Ω].

(a) List all possible resistance combinations that can be formed with these resistors. Your solution should include drawings and calculations of each unique configuration.
(b) How many unique combinations can be formed?
(c) What is the greatest resistance configuration?
(d) What is the least resistance configuration?

ICA 15-5

Four springs were tested, with the results shown graphically below. Use the graph to answer the following questions. Be sure to justify your answers as to *why* you are making the choice.

(a) Which spring is the stiffest?
(b) Which spring, if placed in parallel with Spring C, would yield the stiffest combination?
(c) Which spring, if placed in series with Spring C, would yield the stiffest combination?
(d) Rank the following combinations in order of stiffness:

Spring A and Spring D are hooked in parallel
Spring B and Spring C are hooked in series, then connected with Spring D in parallel
Spring A
Spring D

ICA 15-6

Four circuits were tested, with the results shown graphically below. Use the graph to answer the following questions. Be sure to justify your answers as to *why* you are making the choice.

(a) Which resistor gives the most resistance?

(b) What is the resistance of Resistor A?

(c) Which resistor, if placed in parallel with Resistor C, would yield the highest resistance?

(d) Which resistor, if placed in series with Resistor C, would yield the highest resistance?

(e) On the graph below, draw and label the resulting line for placing Resistor B and Resistor C in series. The location does not need to be exact but should approximate the resulting resistance.

(f) On the graph below, draw and label the resulting line for placing Resistor B and Resistor D in parallel. The location does not need to be exact but should approximate the resulting resistance.

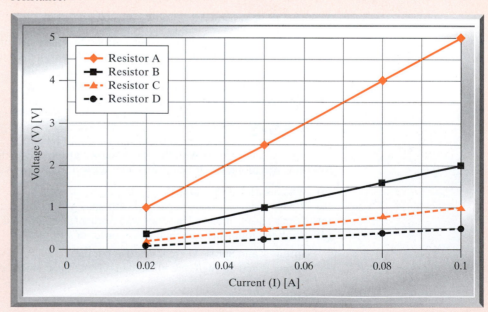

ICA 15-7

The graph shows the ideal gas law relationship ($P\,V = n\,R\,T$) between volume (V) and temperature (T).

(a) What are the units of the slope (0.0175)?

(b) If the tank has a pressure of 1.2 atmospheres and is filled with nitrogen (formula, N_2; molecular weight, 28 grams per mole), what is the amount of gas in the tank (n) in units of grams?

(c) If the tank is filled with 10 grams of oxygen (formula, O_2; molecular weight, 32 grams per mole), what is the pressure of the tank (P) in units of atmospheres?

ICA 15-8

An inductor is an electrical device that can store energy in the form of a magnetic field. In the simplest form, an inductor is a cylindrical coil of wire, and its inductance (L), measured in henrys [H], can be calculated by

$$L = \frac{\mu_0\, n^2 A}{\ell}$$

where

μ_0 = permeability of free space = $4\,\pi \times 10^{-7}$ [newtons per ampere squared, N/A^2]
n = number of turns of wire [dimensionless]
A = cross-sectional area of coil [square meters, m^2]
ℓ = length of coil [meters, m]
L = inductance [henrys, H] = [J/A^2] (One henry is one joule per ampere squared.)

Several inductors were fabricated with the same number of turns of wire (n) and the same length (ℓ), but with different diameters, thus different cross-sectional areas (A). The inductances were measured and plotted as a function of cross-sectional area, and a mathematical model was developed to describe the relationship, as shown on the graph below.

(a) What are the units of the slope (9.66)?
(b) For an inductor fabricated as described above, what is its diameter if its inductance is 0.2 henrys? Give your answer in centimeters.
(c) If the length of the coil (ℓ) equals 0.1 meter, how many turns of wire (n) are in the inductor?

ICA 15-9

The resistance of a wire (R, [ohm]) is a function of the wire dimensions (A = cross-sectional area, L = length) and material (ρ = resistivity) according to the relationship

$$R = \frac{\rho L}{A}$$

The resistance of three wires was tested. All wires had the same cross-sectional area.

Length (L) [m]	0.01	0.1	0.25	0.4	0.5	0.6
Wire #1	8.00E-05	8.00E-04	2.00E-03	3.50E-03	4.00E-03	4.75E-03
Wire #2	4.75E-05	4.80E-04	1.00E-03	2.00E-03	2.50E-03	3.00E-03
Wire #3	1.50E-04	1.70E-03	4.25E-03	7.00E-03	8.50E-03	1.00E-02

(a) Plot the data and fit a linear trendline model to each wire.
(b) From the following chart, match each wire (1, 2, and 3) with the correct material according to the results of the resistivity determined from the trendlines, assuming a 0.2 cm diameter wire was used.

Material	Resistivity (ρ) [Ω m] $\times 10^{-8}$
Aluminum	2.65
Copper	1.68
Iron	9.71
Silver	1.59
Tungsten	5.60

ICA 15-10

A piano, much like a guitar or a harp, is a stringed instrument. The primary difference is that a piano has a keyboard containing 88 keys that, when pressed, cause a small felt hammer to strike the strings (mostly 3) associated with that key. Each hammer can hit the strings harder or softer depending on the force applied to the key on the keyboard. In order to create the different pitches of sound, the strings for each key are different lengths as well as having different mass per unit length.

The equation for the fundamental frequency of a vibrating string is given by

$$f = \frac{\sqrt{T/\mu}}{2L}$$

where

f = frequency [Hz]
T = string tension [N]
μ = mass per unit length [kg/m]
L = string length [m]

In avant-garde music, the piano is sometimes played much like a harp, where the strings are directly manipulated (plucked, strummed, etc.) without the use of the keys on the keyboard. Assume we construct a special piano for use by an avant-garde musician in which each string has the same mass per unit length and the same length on the soundboard. The only difference is the string tension on each string on the soundboard. To test our design, we collect data using a single string on the sound board and create a graph of the observed frequency at different string tensions:

$f = 16.14\ T^{0.49}$

(a) What are the units of the coefficient (16.14)?
(b) If the observed frequency is 450 hertz, what is the string tension in newtons?
(c) If mass per unit length is 4.5 grams per meter, what is the length of the string in meters?
(d) If the length of the string is 0.7 meters, what is the mass per unit length in kilograms per meter?

ICA 15-11

To assist airplane manufacturers, NASA and other similar agencies around the world regularly release atmosphere models of Earth that are updated based on climate change around the world. Consider the graphs (Figures 1–3) of an atmosphere model of three different atmospheric layers (troposphere, lower stratosphere, upper stratosphere) and answer the following questions. These models are based on http://www.grc.nasa.gov/WWW/K-12/airplane/atmos.html.

(a) In Figure 1, what are the units of the value "2180.4" given in the trendline?
(b) In Figure 2, what are the units of "−5E05" given in the trendline?

(c) What is the pressure at an altitude of 0 feet? How does this compare to the definition of standard atmospheric pressure of 1 atmosphere?

(d) What is the altitude at an air pressure of 150 pounds-force per square foot?

(e) One aircraft manufacturer wants a single model for all 3 atmospheric levels given. What model, formed by combining the three models shown, would you recommend and why?

Figure 1 Earth Atmosphere Model: Troposphere.

Figure 2 Earth Atmosphere Model: Lower Stratosphere.

Figure 3 Earth Atmosphere Model: Upper Stratosphere.

CHAPTER 16

MATHEMATICAL MODELS

As we have already seen, a large number of phenomena in the physical world obey one of three basic mathematical models.

- Linear: $y = mx + b$
- Power: $y = bx^m + c$
- Exponential: $y = be^{mx} + c$

As we have mentioned previously, Excel can determine a mathematical model (trendline equation) for data conforming to all three of these model types, with the restriction that the constant c in the power and exponential forms must be 0.

Here we consider how to determine the best model type for a specific data set, as well as learning methods of dealing with data that fit a power or exponential model best but have a non-zero value of c.

Linear

Power

Exponential

16.1   SELECTING A TRENDLINE TYPE

When you determine a trendline to fit a set of data, in general you want the line, which may be straight or curved, to be as close as is reasonable to most of the data points. *The objective is **not** to ensure that the curve passes through every point.*

To determine an appropriate model for a given situation, we use five guidelines, presented in general order of importance:

1. Do we already know the model type that the data will fit?
2. What do we know about the behavior of the process under consideration, including initial and final conditions?

3. What do the data look like when plotted on graphs with logarithmic scales?
4. How well does the model fit the data?
5. Can we consider other model types?

GUIDELINE 1: Determine if the Model Type is Known

If you are investigating a phenomenon that has already been studied by others, you may already know which model is correct or perhaps you can learn how the system behaves by looking in appropriate technical literature. In this case, all you need are the specific values for the model parameters since you already know the form of the equation. As we have seen, Excel is quite adept at churning out the numerical values for trendline equations.

If you are certain you know the proper model type, you can probably skip guidelines 2 and 3, although it might be a good idea to quantify how well the model fits the data as discussed in guideline 4. For example, at this point you should know that the extension of simple springs has a linear relationship to the force applied. As another example, from your study of the ideal gas law, you should know that pressure is related to volume by a power law model (exponent $= -1$).

At other times, you may be investigating situations for which the correct model type is unknown. If you cannot determine the model type from experience or references, continue to guideline 2.

GUIDELINE 2: Evaluate What is Known About the System Behavior

The most important thing to consider when selecting a model type is whether the model makes sense in light of your understanding of the physical system being investigated. Since there may still be innumerable things with which you are unfamiliar, this may seem like an unreasonable expectation. However, by applying what you *do* know to the problem at hand, you can often make an appropriate choice without difficulty.

When investigating an unknown phenomenon, we typically know the answer to at least one of three questions:

1. How does the process behave in the initial state?
2. How does the process behave in the final state?
3. What happens to the process between the initial and final states—if we sketch the process, what does it look like? Does the parameter of interest increase or decrease? Is the parameter asymptotic to some value horizontally or vertically?

EXAMPLE 16-1

Suppose we don't know Hooke's law and would like to study the behavior of a spring. We hang the spring from a hook, pull downward on the bottom of the spring with varying forces, and observe its behavior. We know initially the spring will stretch a little under its own weight even before we start pulling on it, although in most cases this is small or negligible. As an extreme case, however, consider what would happen if you hang one end of a Slinky® from the ceiling, letting the other end fall as it will. As we pull on the spring, we realize the harder we pull, the more the spring stretches. In fact, we might assume that in a simple world, if we pull twice as hard, the spring will stretch twice as far, although that might not be as obvious. In words we might say,

The distance the spring stretches (x) is directly proportional to the pulling force (F), or we might express the behavior as an equation:

$$x = kF + b$$

where *b* is the amount of stretch when the spring is hanging under its own weight. This is what we mean by using an "expected" form. Always remember, however, that what you "expect" to happen may be in error.

In addition, suppose we had tested this spring by hanging five different weights on it and measuring the stretch each time. After plotting the data, we realize there is a general trend that as the weight (force) increases, the stretch increases, but the data points do not lie exactly on a straight line. We have two options:

- If we think our assumption of linear behavior may be in error, we can try nonlinear models.
- Or we can use a linear model, although the fit may not be as good as one or more of the nonlinear models.

To bring order to these questions, we should ask the following sequence of questions:

- - Dependent variable constant
- - - Independent variable constant

Is the system linear?

Linear systems have the following characteristics. If any of these is not true, then the system is not linear.

1. As the independent variable gets larger, the dependent variable continues to increase (positive slope) or decrease (negative slope) without limit. (See item 4 below.)
2. If the independent variable can become negative, as it continues negative, the dependent variable continues to decrease (positive slope) or increase (negative slope) without limit unless one of the variables is constant. (See item 4 below.)
3. The rate of increase or decrease is constant; in other words, it will not curve upward or downward but is a straight line.
4. There are no horizontal or vertical asymptotes unless the dependent variable is defined for only *one* value of the independent variable *or* if the dependent variable is the same value for *all* values of the independent variable.

Examples illustrating if a system is linear:

- You are driving your car at a constant speed of 45 miles per hour. The longer you drive, the farther you go, without limit. In addition, your distance increases by the same amount each hour, regardless of total time elapsed. This is a linear system.
- You observe the temperature of the brake disks on your car to be slowly decreasing. If it continued to decrease without limit, the temperature would eventually be less than absolute zero (an impossibility); thus, it is not linear. Also, it seems reasonable that the temperature will eventually approach the surrounding air temperature; thus, there is a horizontal asymptote.

If the system is not linear, is there a vertical asymptote?

If there *is* a vertical asymptote, it will also have a horizontal asymptote. This is a power law model with a negative exponent.

If there is not a vertical asymptote, is there a horizontal asymptote?

If there is a horizontal asymptote (but not a vertical one), then the model is exponential. If the horizontal asymptote occurs for positive values of the independent variable, then the exponent is negative. If the horizontal asymptote occurs for negative values of the independent variable, then the exponent is positive.

What if there is not a horizontal asymptote or a vertical asymptote?

It is a power law model with a positive exponent. Such models can have a variety of shapes. This sequence of questions can be represented pictorially as shown below.

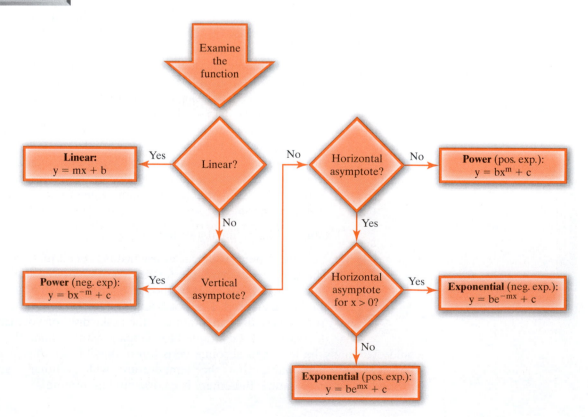

GUIDELINE 3: Convert Axes to a Logarithmic Scale

If the logarithm of the dependent or independent variable is plotted instead of the variable itself, do the modified data points appear to lie on a straight line?

To see how logarithmic axes are constructed, let's consider a simple case. Plotting the data points below gives the graph shown below.

x	y
2	0.1
4	1
6	10
8	100

Now let's take the logarithm (base 10) of the independent (vertical) variable and plot.

x	y	Log y
2	0.1	−1
4	1	0
6	10	1
8	100	2

Finally, we simply change the labels on the vertical axis of the second chart back to the original values of y from which each logarithmic value was obtained.

Note that the vertical positions of the data points are determined by the logarithms of the values, but the numeric scale uses the actual data values.

A couple of points to note about the use of logarithmic scales:

- The original data would fit an exponential model ($y = 0.01\,e^{1.15x}$), and when plotted on a logarithmic vertical axis, the data points appear in a straight line.

- The logarithmic axis allows us to more easily distinguish between the values of the two lowest data points, even though the data range over three orders of magnitude. (On the original graph, 0.1 and 1 were almost in the same vertical position.)
- Note that you *do not* have to calculate the logarithms of the data points. You simply plot the actual values on a logarithmic scale.

Logarithm graphs are discussed in more detail in Section 16.2. How does this help us determine an appropriate model type?

- Plot the data using normal (linear) scales for both axes. If the data appear to lie more or less in a straight line, a linear model is likely to be a good choice.
- Plot the data on a logarithmic vertical scale and a normal (linear) horizontal scale. If the data then appear to lie more or less in a straight line, an exponential model is likely to be a good choice.
- Plot the data with logarithmic scales for *both* axes. If the data then appear to lie more or less in a straight line, a power law model is likely to be a good choice.
- Although not covered in this course, you could plot the data on a logarithmic horizontal scale and a normal (linear) vertical scale. If the data then appear to lie more or less in a straight line, a logarithmic model is likely to be a good choice.

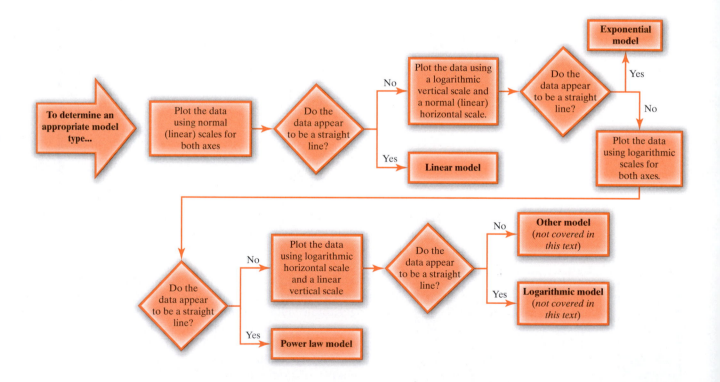

GUIDELINE 4: Consider the R² Value

When a trendline is generated in Excel, the program can automatically calculate an **R² value**, sometimes called the **coefficient of determination**. The R^2 value is an indication of the variation of the actual data from the equation generated—in other words, it is a measure of how well the trendline fits the data. The value of R^2 varies between 0 and 1. If the value of R^2 is exactly equal to 1, a perfect correlation exists between the data and the trendline, meaning that the curve passes exactly through all data points. The farther R^2 is from 1, the less confidence we have in the accuracy of the model generated. When fitting a trendline to a data set, we always report the R^2 value to indicate how well the fit correlates with the data.

In reality, a fit of $R^2 = 1$ is rare, since experimental data are imprecise in nature. Human error, imprecision in instrumentation, fluctuations in testing conditions, and natural specimen variation are among the factors that contribute to a less-than-perfect fit. *The best R^2 value is not necessarily associated with the best model and should be used as a guide only*. Once again, making such decisions becomes easier with experience.

When displaying the equation corresponding to a trendline, you may have already noticed how to display the R^2 value.

✎ **To display an R^2 value:**

- Right-click the trendline or choose the trendline, then choose **Format > Format Selection**.
- In the **Format Trendline** window that opens, from the **Trendline** Options tab, check the box for **Display R-squared value on chart**. Click **Close**.

⌘ **Mac OS:** To show the R^2 value on a Mac, double-click the trendline. In the window that opens, click **Options** and select **Display R-squared value**. Click **OK**.

Try different models and compare the R^2 values. This is sort of a "trial and error" approach, and to rely on it exclusively can be risky. If you have no idea what type of model to use, this may be a reasonable step, however.

- If one of the R^2 values is considerably smaller than the others, say, more than 0.2 less, then that model very likely can be eliminated.
- If one of the R^2 values is considerably larger than the others, say, more than 0.2 greater, then that model very likely is the correct one.

In any case, you should always consider Guidelines 1 through 3 above to minimize the likelihood of error.

WARNING!

While practicing with trendlines in the preceding chapters, you may have noticed a choice for polynomial models. Only rarely would this be the proper choice, but we mention it here for one specific reason—a polynomial model can always be found that will perfectly fit any data set. In general, if there are N data points, a polynomial of order N − 1 can be found that goes exactly through all N points. (Excel can calculate polynomials up to 6th order.) For example, a data set with 5 data points is plotted below. A 4th order polynomial can be found that perfectly fits the data. Let's consider a simple spring stretching example to illustrate why a perfect fit to the data is not necessarily the correct model.

The graph shows the five data points for spring displacement as a function of force. Note that as force increases, displacement increases, but the points are certainly not in a straight line. Also shown is a 4th order polynomial model that goes through every point—a perfect fit. This, however, is a terrible model.

$$x = 0.0007\,F^4 - 0.019\,F^3 + 0.162\,F^2 - 0.47\,F + 0.51$$

Presumably you agree that as force increases, displacement *must* increase as well. The polynomial trendline, however, suggests that as force increases from about 7 to 10 newtons, the displacement *decreases*.

Always ask yourself if the model you have chosen is obviously incorrect, as in this case. We do not use polynomial models in this book, and so discuss them no further.

THE THEORY OF OCCAM'S RAZOR

It is vain to do with more what can be done with less.

or

Entities are not to be multiplied beyond necessity.

—*William of Occam*

It is probably appropriate to mention Occam's Razor at this point. Those who choose to pursue scientific and technical disciplines should keep the concept of Occam's Razor firmly in mind. **Occam's Razor refers to the concept that the simplest explanation or model to describe a given situation is usually the correct one.** It is named for William of Occam, who lived in the first half of the 14th century and was a theologian and philosopher.

EXAMPLE 16-2

The velocity of a ball was recorded as it rolled across a floor after being released from a ramp at various heights. The velocities were then plotted versus the release heights. We want to fit a trendline to the data. We start with the simplest form, a linear fit, shown on the left. We know that if the ramp is at a height of zero, the ball will not roll down the ramp without any external forces. The linear fit yields an intercept value of 0.6, indicating that the ball will have an initial velocity of 0.6 m/s when the ramp is horizontal, which we know to be untrue. It seems unlikely experimental variation alone would generate an error this large, so we try another model.

We choose a power fit, shown in the center. With an R^2 value of 0.86, the equation fits the data selection well, but is there a better fit? Using the same data, we try a third-order polynomial to describe the data. The polynomial model, which gives a perfect fit, is shown on the bottom with an R^2 value of 1. While the polynomial trendline gave the best fit, is this really the correct way to describe the data? Recall that in theory the potential energy of the ball is transformed into kinetic energy according to the conservation of energy law, written in general terms as

$$PE_{initial} = KE_{final} \quad or \quad m\,g\,H = \frac{1}{2}\,m\,v^2$$

Therefore, the relationship between velocity and height is a relationship of the form

$$v = (2\,g\,H)^{\frac{1}{2}} = (2\,g)^{\frac{1}{2}} H^{\frac{1}{2}}$$

The relation between velocity and height is a power relationship; velocity varies as the square root of the height. The experimental error is responsible for the inaccurate trendline fit. In most instances, the polynomial trendline will give a precise fit but an inaccurate description of the phenomenon. **It is better to have an accurate interpretation of the experimental behavior than a perfect trendline fit!**

GUIDELINE 5: Should We Consider Model Types Not Covered Here?

Many phenomena may be accurately characterized by a linear model, power law model, or exponential model. However, there are innumerable systems for which a different model type must be chosen. Many of these are relatively simple, but some are mind-bogglingly complicated. For example, modeling electromagnetic waves (used for television, cell phones, etc.) or a mass oscillating up and down while hanging from a spring requires the use of trigonometric functions.

You should always keep in mind that the system or phenomenon you are studying may not fit the three common models we have covered in this book.

NOTE ON ADVANCED MATH

Actually, sinusoids (sine or cosine) can be represented by exponential models through a mathematical trick first concocted by Leonhard Euler, so we now refer to it as Euler's identity. The problem is that the exponents are imaginary (some number times the square root of −1).

Euler's identity comes up in the study of calculus, and frequently in the study of electrical or computer engineering, and early in the study of electric circuits.

Euler's identity can be expressed in several different forms. The basic identity can be stated as

$$e^{i\pi} = -1$$

where i is the square root of −1.

Another form often used in electrical engineering is

$$\cos\theta = 0.5\,(e^{i\theta} + e^{-i\theta})$$

16.2 INTERPRETING LOGARITHMIC GRAPHS

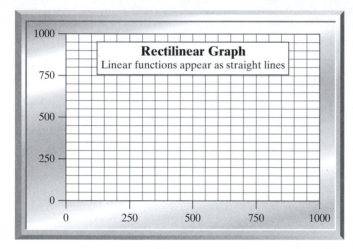

A "regular" plot, shown on a graph with both axis at constant-spaced intervals, is called **rectilinear**. When a linear function is graphed on rectilinear axis, it will appear as a straight line. Often, it is convenient to use a scale on one or both axes that is not linear, where values are not equally spaced but instead "logarithmic," meaning that powers of 10 are equally spaced. Each set of 10 is called a **decade** or **cycle**. A logarithmic scale that ranges either from 10 to 1000 would be 2 cycles, 10 to 100 and 100 to 1000. Excel allows you to select a logarithmic scale for the abscissa, the ordinate, or both.

If one scale is logarithmic and the other linear, the plot is called **semilogarithmic**, or **semilog**. Note in the figure that the abscissa has its values equally spaced and so is a linear scale. However, the ordinate has powers of 10 equally spaced and thus is a logarithmic scale.

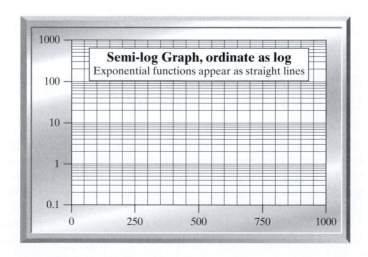

If both scales are logarithmic, the plot is called **full logarithmic** or **log-log**. Note in the figure that both axes have powers of 10 equally spaced.

There are four different combinations of linear and logarithmic axes, each corresponding to one of four specific trendline types that will appear linear on that particular graph type. If the plotted data points are more or less in a straight line when plotted with a specific axis type, the corresponding trendline type is a likely candidate, as discussed earlier.

Once the data are plotted as logarithmic, how do you read data from this graph? This is perhaps best shown through examples.

DEVILISH DERIVATION

Consider a power law model: y = b x^m

Now take the logarithm of both sides of the equation.

$$\log y = \log (b\, x^m) = \log b + \log x^m = \log b + m \log x$$

Using the commutative property of addition, you can write

$$\log y = m \log x + \log b$$

Since b is a constant, log b is also a constant. Rename log b and call it b′.

Since x and y are both variables, log x and log y are also variables. Call them x′ and y′, respectively.

Using the new names for the transformed variables and the constant b:

$$y' = m\, x' + b'$$

This is a linear model! Thus, if the data set can be described by a power law model and you plot the logarithms of both variables (instead of the variables themselves), the transformed data points will lie on a straight line. The slope of this line is m, although "slope" has a somewhat different meaning than in a linear model.

EXAMPLE 16-3

An unknown amount of oxygen, kept in a piston type container at a constant temperature, was subjected to increasing pressure (P), in units of atmospheres; as the pressure (P) was increased, the resulting volume (V) was recorded in units of liters. We have found that a log-log plot aligns the data in a straight line. Using the figure, determine the mathematical equation for volume (V) in units of liters, and of a piston filled with an ideal gas subjected to increasing pressure (P) in units of atmospheres.

Since the graph appears linear on log-log paper, we can assume a power law relationship exists of the form:

$$V = b\,P^m$$

For illustration, a line has been sketched between the points for further clarification of function values.

To determine the power of the function (m), we can estimate this by the number of decades of "rise" (shown as a vertical arrow) divided by the decades of "run" (shown as a horizontal arrow):

$$\text{slope} = \frac{\text{Change in decades of Volume}}{\text{Change in decades of Pressure}}$$

$$= \frac{-1 \text{ decade}}{1 \text{ decade}} = -1$$

To establish the constant value (b), we estimate it as the ordinate value when the abscissa value is 1, shown in the shaded circle. When the pressure is 1 atmosphere, the volume is 50 liters.

The resulting function:

$$V = 50\ P^{-1}$$

EXAMPLE 16-4

When a body falls, it undergoes a constant acceleration. Using the figure, determine the mathematical equation for distance (d), in units of meters, of a falling object as a function of time (t), in units of seconds.

Since the graph appears linear on log-log paper, we can assume a power law relationship exists of the form:

$$d = bt^{m}$$

For illustration, a line has been sketched between the points for further clarification of function values.

To establish the power of the function (m), we estimate it as the number of decades of "rise" (shown as vertical arrows) divided by the decades of "run" (shown as a horizontal arrow):

$$\text{slope} = \frac{\text{Change in decades of Distance}}{\text{Change in decades of Time}} = \frac{2 \text{ decade}}{1 \text{ decade}} = 2$$

To establish the constant value (b), we estimate it as the ordinate value when the abscissa value is 1, shown in the shaded circle. When the time is 1 second, the distance is 5 meters.

The resulting function: $d = 5t^2$

   This matches well with the established theory, which states

$$d = \frac{1}{2}gt^2$$

The value of ½ g is approximately 5 m/s².

DEVILISH DERIVATION

Consider an exponential model: $y = b\,e^{mx}$

Now take the logarithm of both sides of the equation.

$$\log y = \log(b\,e^{mx}) = \log b + \log e^{mx} = \log b + (mx)\log e$$

Using the commutative property of addition, you can write:

$$\log y = m(\log e)x + \log b$$

Since b is a constant, log b is also a constant. Rename log b and call it b′. Since y is a variable, log y is also a variable; call it y′.

Using the new names for the transformed variable y and the constant b:

$$y' = m(\log e)x + b'$$

This is a linear model! Thus, if the data set can be described by an exponential law model, and you plot the logarithm of y (instead of y itself) versus x, the transformed data points will lie on a straight line. The slope of this line is m(log e), but again, "slope" has a somewhat different interpretation. The term (log e) is a number, approximately equal to 0.4343; the slope = 0.4343 m.

EXAMPLE 16-5

A chemical reaction is being carried out in a reactor; the results are shown graphically in the figure. Determine the mathematical equation that describes the reactor concentration (C), in units of moles per liter, as a function of time spent in the reactor (t), in units of seconds.

Since the graph appears linear on semi-log paper where the ordinate is logarithmic, we can assume an exponential law relationship exists of the form:

$$C = be^{mt}$$

For illustration, a line has been sketched between the points for further clarification of function values.

Since this is an exponential function, to determine the value of m, we must first determine the slope:

$$\text{slope} = \frac{\text{Change in decades of Concentration}}{\text{Change in Time}}$$

$$= \frac{-1 \text{ decade}}{21.5 \text{ s} - 10 \text{ s}} = -0.087 \text{ s}^{-1}$$

The value of m is then found from the relationship: slope = m (log e).

$$m = \frac{\text{slope}}{\log(e)} = \frac{-0.087 \text{ s}^{-1}}{0.4343} = -0.2 \text{ s}^{-1}$$

When time = 0 seconds, the constant (b) can be read directly and has a value of 6 [mol/L]. The resulting function:

$$C = 6e^{-0.2t}$$

EXAMPLE 16-6

The data shown graphically in the figure describe the discharge of a capacitor through a resistor. Determine the mathematical equation that describes the voltage (V), in units of volts, as a function of time (t), in units of seconds.

Since the graph appears linear on semi-log paper where the ordinate is logarithmic, we can assume an exponential law relationship exists of the form:

$$V = be^{mt}$$

For illustration, a line has been sketched between the points for further clarification of function values.
Since this is an exponential function, to determine the value of m, we must first determine the slope:

$$\text{slope} = \frac{\text{Change in decades of Voltage}}{\text{Change in Time}}$$

$$= \frac{-1 \text{ decade}}{36 \text{ s} - 10 \text{ s}} = -0.038 \text{ s}^{-1}$$

The value of m is then found from the relationship: slope = m (log e).

$$m = \frac{\text{slope}}{\log (e)} = \frac{-0.038 \text{ s}^{-1}}{0.4343} = -0.89 \text{ s}^{-1}$$

When time = 0 seconds, the constant (b) can be read directly and has a value of 30 volts. The resulting function:

$$V = 30 \, e^{-0.89 \, t}$$

16.3 CONVERTING SCALES TO LOG IN EXCEL

✎ **To convert axis to logarithmic:**

- Right-click the axis to convert to logarithmic. The Format Axis window will appear.
- Click **Axis Options**, then check the box for **Logarithmic scale**.

Alternatively:

- Click the chart. In the tool bar, select **Layout > Axes** and choose the axis to convert.
- In the corresponding menu, select **Show Axis with Log Scale**.

⌘ **Mac OS:** Double-click on the axis you want to convert to logarithmic. The Format Axis window will appear.

Click **Scale** in the list on the left side of the window, and then click the checkbox near the bottom that says "Logarithmic scale."

16.4 DEALING WITH LIMITATIONS OF EXCEL

As we have mentioned earlier, Excel will not correctly calculate a trendline for a power or exponential model containing a vertical offset. In other words, it can calculate appropriate values for b and m in the forms

$$y = bx^m + c \qquad \text{or} \qquad y = be^{mx} + c$$

only if c = 0. Note that if the data inherently has a vertical offset, Excel may actually calculate a trendline equation, but the values of b and m will not be accurate.

In addition, if any data value, dependent or independent, is less than or equal to zero, Excel cannot calculate a power law model. If one or more dependent variable data points is less than or equal to zero, exponential models are unavailable.

In the real world, there are numerous systems best modeled by either a power or exponential model with a non-zero value of c or with negative values, so we need a method for handling such situations.

Case 1: Vertical Asymptote

Since a vertical asymptote implies a power model with a negative exponent, there will be a horizontal asymptote as well. If the horizontal asymptote is not the horizontal axis (implying a vertical offset), Excel will calculate the model incorrectly or not at all. The object here is to artificially move the asymptote to the horizontal axis by subtracting the offset value from every data point. If you have a sufficient range of data, you may be able to extract the offset from the data.

For example, if the three data points with the largest values of x (or smallest if the asymptote goes to the left) have corresponding y values of 5.1, 5.03, and 5.01, the offset is likely to be about 5. (This assumes there are other values in the data set with considerably different y values.) You can also try to determine from the physical situation being modeled at what non-zero value the asymptote occurs. In either case, simply subtract the offset value from the vertical component of *every* data point, plot this modified data, and determine a power trendline. Once the trendline equation is displayed, edit it by adding the offset to the power term. (Note that you subtract the value from the data points but add it to the final equation.)

Case 2: No Horizontal Asymptote

Assuming it has been established that the model should not be linear, Case 2 implies a power model with a positive exponent. If you have a data point for x = 0, the corresponding y value should be very close to the vertical offset. Also, you may be able to determine the offset value by considering the physical situation. In either case, proceed as in case 1, subtracting the offset from every data point, etc.

Case 3: Horizontal Asymptote, No Vertical Asymptote

Case 3 implies an exponential model. The object, as in Case 1, is to artificially move the asymptote to the horizontal axis. Also as in Case 1, you may be able to determine the offset by considering the physical system or by looking at the data points with the largest or smallest values. Again, subtract the determined value from every data point, etc.

NOTE

Sometimes, due to inaccuracy of measurement, one or more of the data points near the asymptote may be negative after the offset value is subtracted, and Excel will be unable to process the data. You can circumvent this either by deleting such data points or by making the vertical component a very small positive value.

EXAMPLE 16-7

The following data were collected in an experiment. We wish to determine an appropriate model for the data.

As the independent variable gets larger, the dependent variable appears to be approaching 10. This is even more apparent when were graphed. Subtracting this assumed offset from every data point gives a new column of modified dependent data.

Independent	Dependent	Modified Dependent
0	25.0	15.0
12	18.0	8.0
24	14.0	4.0
36	12.5	2.5
48	11.0	1.0
60	10.5	0.5
72	10.4	0.4
84	10.2	0.2
96	10.1	0.1

In chart: $y = 14.3\ e^{-0.05x}$

Since we subtracted 10 from every data point, we need to correct the equation by adding 10, giving $y = 14.3\ e^{-0.5x} + 10$.

Case 4: A Few Values with Small Negative Value, Most Positive

In Case 4, the negative values may be a result of measurement inaccuracy. Either delete these points from the data set, or change the negative values to a very small positive value.

Case 5: Many or All Data Points Negative

If the independent values are negative, try multiplying every independent value by −1. If this works, then make the calculated value of b negative after the trendline equation is calculated. You may have to apply some of the procedures in the previous cases after negating each data value.

Negative dependent values may simply be a negative offset to the data. If you can determine the asymptote value, ask if essentially all values are greater than the asymptote value. If so, it is probably just an offset. If not, then multiply every dependent data value by −1, and proceed in a manner similar to that described in the preceding paragraph.

EXAMPLE 16-8

The following data were collected during an experiment. We wish to determine an appropriate model for the data.

You notice that all of the dependent values are negative. Since the value when the independent variable is zero is non-zero, this is likely to be an exponential model, although it could possibly be a power law model with an offset.

Independent	Dependent	Modified Dependent
0	−25	25
5	−37	37
10	−45	45
15	−58	58
20	−85	85
25	−106	106
30	−154	154

Following the first assumption (that the model is exponential), multiply every data point by −1, plot the inverted data, and fit an exponential trendline. This seems to fit the data quite well. Finally, negate the trendline equation to obtain the true model, giving y = −25.6 e$^{0.06x}$.

If All Else Fails

If you are convinced that the model is exponential or power with an offset but you cannot determine its value, consider making further measurements for larger or smaller values of the independent variable. Particularly in data that has a horizontal asymptote, further measurements may make the value of the asymptote more obvious. In the first chart below, the value of the asymptote is not clear. By extending the measurements in the direction of the asymptote (positive in this case), it is clear that the asymptote has a value of 2. Note also that it becomes much clearer that the data are not linear.

EXAMPLE 16-9

The data shown describe the discharge of a capacitor through a resistor. Before the advent of microprocessors, intermittent windshield wipers in automobiles often used such circuits to create the desired time delay. We wish to determine an appropriate model for the data.

Time (t) [s]	2	10	22	39
Voltage (V) [V]	24	13	5	1

$V = 30\,e^{-0.086\,t}$
$R^2 = 0.997$

$V = -0.6\,t + 21$
$R^2 = 0.87$

$V = 67\,t^{-1}$
$R^2 = 0.815$

- *Select the data series and create a linear trendline, being sure to display the equation and the R^2 value.*
- *Without deleting the first trendline, click one of the data points again— be sure you select the points and not the trendline—and again add a trendline, but this time choose a power trend.*
- *Repeat this process for an exponential trend.*

You should now have a chart with three trendlines. Things to note:

- *Neither the linear nor power trendlines are very good compared to the exponential line, and both have an R^2 value less than 0.9. These are probably not the best choice.*

- *The exponential model fits the data very closely and has an R^2 value greater than 0.95; thus, it is probably the best choice.*

As a model check, compare the graph by using logarithmic scales. If the model is exponential, the data should appear linear on a semilogarithmic plot with the ordinate shown as logarithmic.

Based on this analysis, you would choose the exponential model. As it turns out, the exponential model is indeed the correct one, being the solution to a differential equation describing the capacitor's behavior. Most students learn about this in second semester physics, and some study it in much more depth in electrical and computer engineering courses.

EXAMPLE 16-10

These data describe the temperature of antifreeze (ethylene glycol) in the radiator of a parked car. The temperature of the surrounding environment is $-20\,°F$. The initial temperature (at t = 0) is unknown.

Time (t) [min]	10	18	25	33	41
Temperature (T) [°F]	4.5	1.0	−2.1	−4.6	−6.4

- Determine an appropriate model type for these data.
- Determine the vertical offset of the data.
- Plot the modified data and generate the correct trendline equation to describe this data.

It seems reasonable that the temperature will be asymptotic to the surrounding temperature (−20 °F) as time goes on. Also, there is no known mechanism whereby the temperature could possibly go to infinity for any finite value of time, so there is no vertical asymptote. This indicates an exponential model with a negative exponent. Since the asymptote is at −20 °F, subtract −20 (i.e., add 20) from every data point before plotting.

Time (t) [min]	10	18	25	33	41
Temperature (T) [°F]	4.5	1.0	−2.1	−4.6	−6.4
Offset Temperature (T_o) [°F]	24.5	21.0	17.9	15.4	13.6

Since you subtracted −20 from every data point, you should add −20 to the trendline equation, giving

$$T = 29.5\,e^{-0.019\,t} - 20$$

EXAMPLE 16-11

Assume the car in Example 16-10 is cranked up and driven 50 feet into a garage. The temperature inside the garage is 5°F. These data describe the temperature of antifreeze in the radiator after it is driven into the garage and the motor turned off.

Time (t) [min]	5	13	25	34	51
Temperature (T) [°F]	−13.0	−10.0	−6.8	−4.5	−1.5

The antifreeze temperature will increase asymptotically to 5 °F. Since all of the data points are less than this, multiply through by −1 before applying the offset of 5 °F. Also, be rather careful here. Since you are dealing with the negatives of the temperatures, you must deal with the negative of the offset as well. Thus, multiply each temperature by −1, then subtract the negative of the asymptote value of 5 °F (add 5 to each value). This approach gives the following.

Time (t) [min]	5	13	25	34	51
Temperature (T) [°F]	−13.0	−10.0	−6.8	−4.5	−1.5
Adjusted Temperature (T_A) [°F]	18.0	15.0	11.8	9.5	6.5

$$T_A = 20\,e^{-0.022\,t}$$
$$R^2 = 0.99$$

Since you are working with the negatives of the values, put a minus by the exponential term. Since you subtracted −5 from the data, you might initially think to add −5 to the equation. However, since you used the negatives of the values, add 5 as an offset in the equation since the two negatives cancel. This gives the final equation

$$T = 5 - 20\,e^{-0.022\,t}$$

In-Class Activities

ICA 16-1

Capillary action draws liquid up a narrow tube against the force of gravity as a result of surface tension. The height the liquid will move up the tube depends on the radius of the tube. The following data were collected for water in a glass tube in air at sea level. Show the resulting data and trendline, with equation and R^2 value, on the appropriate graph type (rectilinear, semilog, or log-log) to make the data appear linear.

Radius (r) [cm]	0.01	0.05	0.10	0.20	0.40	0.50
Height (H) [cm]	14.0	3.0	1.5	0.8	0.4	0.2

ICA 16-2

Several reactions are carried out in a closed vessel. The following data are taken for the concentration (C) of compounds A, B, and C [grams per liter, g/L] as a function of time (t) [minutes, min], from the start of the reaction. Show the resulting data and trendlines, with equation and R^2 value, on the appropriate graph type (rectilinear, semilog, or log-log) to make the data appear linear.

Time (t) [min]	2	5	8	15	20
Concentration of A (C_A) [g/L]	1.1	1.3	1.4	1.6	1.7
Concentration of B (C_B) [g/L]	1.3	1.5	1.7	2.0	2.3
Concentration of C (C_C) [g/L]	1.6	3.0	4.5	6.5	8.0

ICA 16-3

An environmental engineer has obtained a bacteria culture from a municipal water sample and allowed the bacteria to grow. The data are shown below. Show the resulting data and trendline, with equation and R^2 value, on the appropriate graph type (rectilinear, semilog, or log-log) to make the data appear linear.

Time (t) [h]	0	2	3	5	6	7	9	10
Concentration (C) [ppm]	0	21	44	111	153	203	318	385

ICA 16-4

In a turbine, a device used for mixing, the power required by the turbine depends on the size and shape of impeller. In the lab, you have collected the following data. Show the resulting data and trendline, with equation and R^2 value, on the appropriate graph type (xy scatter, semilog, or log-log) to make the data appear linear.

Diameter (D) [ft]	0.5	0.75	1	1.5	2	2.25	2.5	2.75
Power (P) [hp]	0.004	0.04	0.13	0.65	3	8	18	22

ICA 16-5

Being quite interested in obsolete electronics, Angus has purchased several electronic music synthesis modules dating from the early 1970s and is testing them to find out how they work. One module is a voltage-controlled amplifier (VCA) that changes the amplitude (loudness) of an audio signal by changing a control voltage into the VCA. All Angus knows is that the magnitude of the control voltage should be less than 5 volts. He sets the audio input signal to an amplitude of 1 volt, then measures the audio output amplitude for different control voltage values. The table below shows his data. Show the resulting data and trendline, with equation and R^2 value, on the appropriate graph type (xy scatter, semilog, or log-log) to make the data appear linear.

Control Voltage (V) [V]	−4.0	−2.5	−1.0	0.0	1.0	2.5	4.0
Output Amplitude (A) [V]	0.116	0.324	0.567	0.962	1.690	3.320	7.270

ICA 16-6

Referring to the previous ICA, Angus is also testing a voltage-controlled oscillator. In this case, a control voltage (also between −5 V and +5 volts) changes the frequency of oscillation in order to generate different notes. The table below shows his measurements. Show the resulting data and trendline, with equation and R^2 value, on the appropriate graph type (xy scatter, semilog, or log-log) to make the data appear linear.

Control Voltage (V) [V]	−4.0	−2.5	−1.0	0.0	1.0	2.5	4.0
Output Frequency (F) [Hz]	28	99	227	539	989	3110	8130

ICA 16-7

A growing field of inquiry that poses both great promise and great risk for humans is nanotechnology, the construction of extremely small machines. Over the past couple of decades, the size that a working gear can be made has consistently gotten smaller. The table shows milestones along this path.

Years from 1967	0	5	7	16	25	31	37
Minimum Gear Size [mm]	0.8	0.4	0.2	0.09	0.007	2E-04	8E-06

(a) Show the resulting data and trendline, with equation and R^2 value, on the appropriate graph type (xy scatter, semilog, or log-log) to make the data appear linear. What do you think is the limiting factor on the smallest size that gears can eventually be manufactured?

(b) According to this model, how many years does it take (from any point in time) for the minimum size to be cut in half?

(c) According to the model, during what year will the smallest gear be one-tenth the size of the smallest gear in 2009?

ICA 16-8

If an object is heated, the temperature of the object will increase. The thermal energy (Q) associated with a change in temperature (ΔT) is a function of the mass of the object (m) and the specific heat (C_p), which is the rate at which the object will lose or gain energy. Specific heat is a material property, and values are available in literature. In an experiment, heat is applied to the end of an object, and the temperature change at the other end of the object is recorded. An unknown material is tested in the lab, yielding the following results.

Heat Applied (Q) [J]	2	8	10	13	18	27
Temp Change (ΔT) [K]	1.5	6.0	7.0	9.0	14.0	22.0

(a) Show the resulting data and trendline, with equation and R^2 value, on the appropriate graph type (xy scatter, semilog, or log-log) to make the data appear linear.

(b) If the material was titanium, what mass of sample was tested?

(c) If a 4-gram sample was used, which of the following materials was tested?

Material	Specific Heat Capacity (C_p) [J/(g K)]
Aluminum	0.91
Copper	0.39
Iron	0.44
Lead	0.13
Molybdenum	0.30
Titanium	0.54

ICA 16-9

Biosystems engineers often need to understand how plant diseases spread in order to formulate effective control strategies. The rate of spread of some diseases is more or less linear, some increase exponentially, and some don't really fit any standard mathematical model.

Grey leaf spot of corn is a disease (caused by a fungus with the rather imposing name of *Cercospora zeae-maydis*) that causes chlorotic (lacking chlorophyll) lesions and eventually necrotic (dead) lesions on corn leaves, thus reducing total photosynthesis and reducing yield. In extremely severe cases, loss of the entire crop can result.

During a study of this disease, the number of lesions per corn leaf was counted every 10 days following the initial observation of the disease, which we call day 0. At this time, there was an average of one lesion on every twenty leaves, or 0.05 lesions per leaf. The data collected during the growing season are tabulated.

Day	Lesions per Leaf
0	0.05
20	0.10
30	0.20
40	0.26
60	0.60
80	1.30
90	2
110	4
120	6
140	17
150	20
170	40
190	112
200	151

(a) Show the resulting data and trendline, with equation and R^2 value, on the appropriate graph type (xy scatter, semilog, or log-log) to make the data appear linear.
(b) According to the model, how many lesions were there per leaf at the start of the survey?
(c) How many lesions are there per leaf after 97 days?
(d) If the model continued to be accurate, how many days would be required to reach 250 lesions per leaf?

ICA 16-10

A **pitot tube** is a device used to measure the velocity of a fluid, typically, the airspeed of an aircraft. The failure of a pitot tube is credited as the cause of Austral Líneas Aéreas flight 2553 crash

in October 1997. The pitot tube had frozen, causing the instrument to give a false reading of slowing speed. As a result, the pilots thought the plane was slowing down, so they increased the speed and attempted to maintain their altitude by lowering the wing slats. Actually, they were flying at such a high speed that one of the slats ripped off, causing the plane to nosedive; the plane crashed at a speed of 745 miles per hour [mph].
In the pitot tube, as the fluid moves, the velocity creates a pressure difference between the ends of a small tube. The tubes are calibrated to relate the pressure measured to a specific velocity. This velocity is a function of the pressure difference (ΔP [=] Pa) and the density of the fluid (ρ [=] kg/m^3).

$$ v = \left(\frac{2}{\rho}\right)^{0.5} p^m $$

Pressure (P) [Pa]	50,000	101,325	202,650	250,000	304,000	350,000	405,000	505,000
Velocity Fluid A (v_A) [m/s]	11.25	16.00	23.00	25.00	28.00	30.00	32.00	35.75
Velocity Fluid B (v_B) [m/s]	9.00	12.50	18.00	20.00	22.00	24.00	25.00	28.00
Velocity Fluid C (v_C) [m/s]	7.50	11.00	15.50	17.00	19.00	20.00	22.00	24.50

(a) Show the resulting data and trendline, with equation and R^2 value, on the appropriate graph type (xy scatter, semilog, or log-log) to make the data appear linear.
(b) First determine what the units of the constant in front of the pressure variable must be to yield velocity when multiplied, and then determine how to obtain those units from density. Determine the value and units of the density for each fluid.
(c) From the following chart, match each fluid (A, B, and C) with the correct name according to the results of the density determined from the trendlines.

Fluid	Specific Gravity
Acetone	0.79
Citric Acid	1.67
Glycerin	1.26

ICA 16-11

For the graph shown, determine the equation of the trendline. The trendline will fit one of the following forms. In each case, the constants "m" and "b" may be positive or negative. You must choose the correct trendline equation from choices (A)–(C) below, replacing the variables "y" and "x" with those used in the graph, and determine the value and units of the constants (m and b).

(A) $y = m\,x + b$
(B) $y = b\,x^m$
(C) $y = b\,e^{mx}$

ICA 16-12

For the graph shown, determine the equation of the trendline. The trendline will fit one of the following forms. In each case, the constants "m" and "b" may be positive or negative. You must choose the correct trendline equation from choices (A)–(C) below, replacing the variables "y" and "x" with those used in the graph, and determine the value and units of the constants (m and b).

(A) $y = m\,x + b$
(B) $y = b\,x^m$
(C) $y = b\,e^{mx}$

As a reminder, Reynolds Number is discussed in Chapter 12.

ICA 16-13

When discussing the flow of a fluid through a piping system, we say that friction occurs between the fluid and the pipe wall due to viscous drag. The loss of energy due to the friction of fluid against the pipe wall is described by the friction factor. The *Darcy friction factor* (f) was developed by Henry Darcy (1803–1858), a French scientist who made several important contributions to the field of hydraulics. The friction factor depends on several other factors, including flow regime, Reynolds' number, and pipe roughness. The friction factor can be determined in several ways, including from the Moody diagram (shown below).

Olive oil having a specific gravity = 0.914 and a viscosity = 100.8 centipoise is draining by gravity from the bottom of a tank. The drain line from the tank is a 4-inch diameter pipe made of commercial steel (pipe roughness, ε = 0.045 millimeters). The velocity is 11 meters per second. Determine the friction factor for this system, using the following process:

Step 1: Determine the Reynolds number: $\mathrm{Re} = \dfrac{\rho \mathrm{VD}}{\mu}$

Step 2: Determine flow regime.

- If the flow is laminar (Re ≤ 2000), proceed to Step 4.
- If the flow is turbulent or transitional (Re > 2,000), continue with Step 3.

Step 3: Determine the relative roughness ratio: ε/D

Step 4: Determine the Darcy friction factor (f) from the diagram.

ICA 16-14

Repeat ICA 16-13 with the following conditions:

Olive oil having a specific gravity = 0.914 and a viscosity = 100.8 centipoise is draining by gravity from the bottom of a tank. The drain line from the tank is a 4-inch diameter pipe made of commercial steel (pipe roughness, ε = 0.045 millimeters). The velocity is 1.5 meters per second. Determine the friction factor for this system.

ICA 16-15

Repeat ICA 16-13 with the following conditions:

Lactic acid, with a specific gravity = 1.249; dynamic viscosity = 40.33 centipoise, is flowing in a 1½-inch diameter galvanized iron pipe at a velocity of 17 meters per second. Assume the pipe roughness (ε) of galvanized iron is 0.006 inches [in].

ICA 16-16

Repeat ICA 16-13 with the following conditions:

Lactic acid, with a specific gravity = 1.249; dynamic viscosity = 40.33 centipoise, is flowing in a 1½-inch diameter galvanized iron pipe at a velocity of 1.5 meters per second. Assume the pipe roughness (ε) of galvanized iron is 0.006 inches [in].

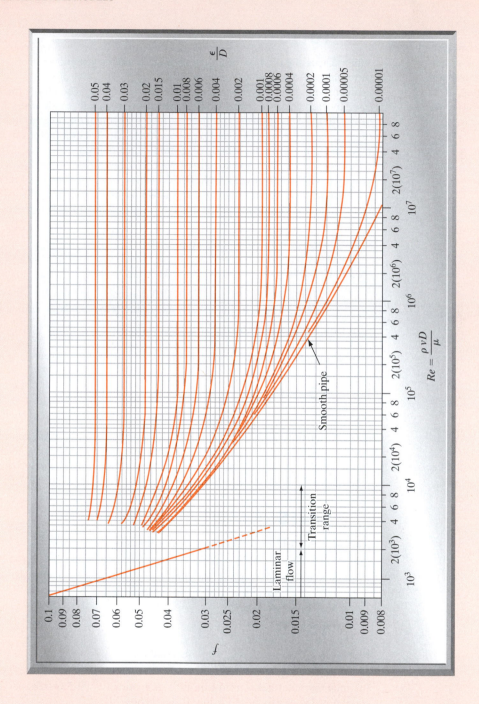

1. The following functions could describe the oscillations in electrical circuits or the vibrations of machines or structures. Plot these two functions on the same graph for time (t) from 0 to 1 second using 0.02 second intervals. Your worksheet should be constructed using absolute references for a, b, c, and d.

$$x = 10\,e^{-at}\sin(bt + 2) \qquad\qquad y = 7\,e^{-ct}\cos(dt - 3)$$

   (a) Suppose $a = 0.5$/s; $b = 3$ rad/s; $c = 0.4$/s; and $d = 5$ rad/s.
   (b) Suppose $a = 0.2$/s; $b = 2$ rad/s; $c = 0.5$/s; and $d = 2$ rad/s.
   (c) Suppose $a = 1.0$/s; $b = 6$ rad/s; $c = 1.0$/s; and $d = 6$ rad/s.
   (d) Suppose $a = 0.8$/s; $b = 6$ rad/s; $c = 0.4$/s; and $d = 2$ rad/s.

2. The height and speed of a projectile (such as a thrown ball) launched with an initial speed v_0 at an angle θ to the horizontal are given by the following functions. Plot these two functions on the same graph for time from 0 to 5 seconds by 0.1 second intervals. Your worksheet should be constructed using absolute references for the initial velocity and angle.

$$H = v_0\,t\sin\theta - 0.5\,gt^2 \qquad\qquad v = \sqrt{v_0^2 - 2\,v_0\,gt\sin\theta + (gt)^2}$$

   (a) Suppose that $\theta = 40°$ and $v_0 = 20$ meters per second.
   (b) Suppose that $\theta = 60°$ and $v_0 = 30$ meters per second.
   (c) Suppose that $\theta = 70°$ and $v_0 = 20$ meters per second.
   (d) Suppose that $\theta = 80°$ and $v_0 = 15$ meters per second.

3. In the design of a hard drive, the faster the disk spins, the faster information can be read from and written to the disk. In general, the more information stored on the disk, the larger the diameter of the disk must be. Unfortunately, the larger the disk, the lower the maximum rotational speed must be to avoid stress-related failures. Assume that the minimum allowable rotational speed (S) of the new hard drive is 6000 revolutions per minute [rpm], and assume that the rotational speed must meet the criterion $S < 12000 - 150\,D^2$, where D is the diameter of the disk in inches.

   (a) Create a graph of these two inequalities with rotational speed on the ordinate and diameter on the abscissa. The values on the vertical axis should range from 0 to 12000 revolutions per minute [rpm], and the values on the horizontal axis should range from 0 to 7 inches [in].
   (b) Graphically determine the range of allowable rotational speeds for a 4-inch-diameter disk.
   (c) Graphically determine the largest diameter disk that meets the design criteria.

4. Suppose you are directed to design a pumping system to safely discharge a toxic industrial waste into a municipal reservoir. The concentrated wastewater from the plant will be mixed with fresh water from the lake, and this mixture is to be pumped into the center of the lake. You realize that the more water you mix with the waste, the more dilute it will be and thus

will have a smaller impact on the fish in the lake. On the other hand, the more water you use, the more it costs in electricity for pumping. Your objective is to determine the optimum amount of water to pump so the overall cost is at a minimum.

- Assume that the cost of pumping is given by the expression $C_{pump} = 10 * Q^2$. The cost C_{pump} [$/day] depends on the pumping rate Q [gallons per minute or gpm] of the water used to dilute the industrial waste.
- Now suppose that a study has found that as more and more water is used to dilute the waste, the fish loss C_{fish} [$/day] can be expressed as: $C_{fish} = 2{,}250 - 150 * Q$.

With this information, construct a graph with pumping rate on the abscissa and cost on the ordinate, showing the pumping cost, the fish loss cost, and total cost. For the scale, plot 0 to 15 gallons per minute [gpm] for flow rate and $0 to $2250 for cost. Determine both the minimum total cost and the corresponding flow rate.

5. In a simple electric circuit, the current (I) must remain below 40 milliamperes (I < 40 mA), and must also satisfy the function $I > 10^{-6} e^{25\,V}$, where V is the voltage across a device called a diode.

   (a) Create a graph of these two inequalities with current on the ordinate and voltage on the abscissa. The values on the vertical axis should range from 0 to 50 milliamperes, and the values on the horizontal axis should range from 0 to 0.8 volts.
   (b) Graphically determine the maximum allowable voltage across the diode.

6. A company that fabricates small, custom machines has been asked to generate a machine that throws darts at a dart board as precisely and accurately as possible. To assess the precision and accuracy of each proposed design, the engineers build a model and record the distance from the bull's-eye of the dart board to the location of each dart thrown—both the straight line distance (A), as well as the horizontal (B) and vertical (C) distances are recorded separately with regard to the bull's-eye, as demonstrated in the figure. The engineers throw 15 darts with their prototype machine and record the 3 data points for each dart.

   Using the data collected for a design, create a histogram and a CDF for the straight-line distance (A), as well as the horizontal (B) and vertical (C) distances and determine which graph or graphs are better for assessing the performance of the design. Justify your answer with a few sentences about why you selected the graph or graphs.

Dart #	Straight Distance (A) [in]	Horizontal Distance (B) [in]	Vertical Distance (C) [in]
1	0.703	0.400	−0.578
2	2.740	1.976	1.898
3	1.555	−1.484	0.466
4	0.387	−0.387	0.010
5	1.180	−1.147	−0.278
6	1.424	−1.418	0.132
7	0.899	0.835	0.335
8	2.069	0.330	2.043
9	0.547	0.514	0.187
10	3.451	1.566	3.076
11	2.532	2.361	−0.912
12	3.070	1.602	2.619
13	1.386	0.176	1.375
14	1.637	1.458	−0.744
15	0.888	−0.822	0.337

7. Essentially all manufactured items are made to some "tolerance," or how close the actual product is to the nominal specifications. For example, if a company manufactures hammers, one customer might specify that the hammers should weigh 16 ounces. With rounding, this means that the actual weight of each hammer meets the specification if it weighs between 15.5 and 16.5 ounces. Such a hammer might cost 10 dollars. However, if the U.S. military specifies that an essentially identical hammer should have a weight of 16.000 ounces, then in order to meet specifications, the hammer must weigh between 15.9995 ounces and 16.0005 ounces. In other words, the weight must fall within a range of one-thousandth of an ounce. Such a hammer might cost $1000.

   With that preamble, let's move on to the problem.

   You have purchased a "grab bag" of 100 supposedly identical capacitors. You got a really good price, but there are no markings on the capacitors. All you know is that they are all the same nominal value. You want to discover not only the nominal value but also the tolerance: are they within 5% of the nominal value, or within 20% ... ? You set up a simple circuit with a known resistor and each of the unknown capacitors. You charge each capacitor to 10 volts, then use an oscilloscope to time how long it takes for each capacitor to discharge to 2 volts. In a simple RC (resistor-capacitor) circuit, the voltage (V_C) across a capacitor (C) discharging through a resistor (R) is given by

   $$V_c = V_o e^{-t/(RC)}$$

   where t is time in seconds and V_o is the initial voltage across the capacitor.

   After measuring the time for each capacitor to discharge from 10 V to 2 V, you scan the list of times to find the fastest and slowest. Since the resistor is the same in all cases, the fastest time corresponds to the smallest capacitor in the lot, and the slowest time to the largest. The fastest time was 3.3 microseconds and the slowest was 3.7 microseconds. For the two capacitors, you have the two pairs of data points.

   Plot these points in Excel on the same graph, using time as the independent variable. Fit an exponential trendline to the data.

Time for C_1 (sec)	Voltage of C_1
0	10
3.3×10^{-6}	2

Time for C_2 (sec)	Voltage of C_2
0	10
3.7×10^{-6}	2

   (a) Referring to the known form of the response, explain why the faster time corresponded to the smaller capacitor.

   (b) The trendline for each capacitor had an R^2 value of 1, a perfect fit of trendline to data. Explain why this happened.

   (c) Assuming that for these measurements you chose a precision resistor with a value of R = 1000.0 ohms, determine the capacitance of the largest and smallest capacitors.

   (d) You selected the fastest and slowest discharge times from a set of 100 samples. Since you had a fairly large sample set, it is not a bad assumption, according to the laws of large numbers, that these two selected data sets represent capacitors near the lower and higher end of the range of values within the tolerance of the devices. Assuming that the nominal value is the average of the minimum and maximum allowable values, what is the nominal value of the set of capacitors?

   (e) What is the tolerance, in percent, of these devices? As an example, if a nominal 1 μF (microfarad) capacitor had an allowable range of 0.95 μF < C < 1.05 μF, the tolerance would be 5%.

   (f) If standard tolerances of capacitors are 5%, 10%, and 20%, to which of the standard tolerances do you think these capacitors were manufactured? If you pick a smaller tolerance than you calculated, justify your selection. If you picked a higher tolerance, explain why the tolerance is so much larger than the measured value.

8. You will be given an Excel spreadsheet, titled "Baseball Data," that contains salary data from major league baseball in 2005. Use the data provided to determine

   (a) Salary mean, median, and standard deviation for all players.
   (b) Salary mean, median, and standard deviation for the Arizona Diamondbacks.
   (c) Salary mean, median, and standard deviation for the Atlanta Braves.
   (d) Salary mean, median, standard deviation for all pitchers. (HINT: sort first!)
   (e) Salary mean, median, standard deviation for all outfielders. (HINT: sort first!)
   (f) Which team had the highest average salary?
   (g) Which team had the lowest average salary?
   (h) Which position had the highest average salary? (HINT: sort first!)
   (i) Which position had the lowest average salary? (HINT: sort first!)
   (j) Draw a histogram and CDF in Excel for all players.
   (k) What percentage of players earned more than $1 million?
   (l) What percentage of players earned more than $5 million?

9. The information for Problem 9 was taken from the report of the EPA on the U.S. Greenhouse Gas Inventory (http://yosemite.epa.gov/oar/globalwarming.nsf).

   Greenhouse gas emission inventories are developed for a variety of reasons. Scientists use inventories of natural and anthropogenic emissions as tools when developing atmospheric models. Policy makers use inventories to develop strategies and policies for emissions reductions and to track the progress of those policies. Regulatory agencies and corporations rely on inventories to establish compliance records with allowable emission rates. Businesses, the public, and other interest groups use inventories to better understand the sources and trends in emissions.

   The Inventory of U.S. Greenhouse Gas Emissions and Sinks supplies important information about greenhouse gases, quantifies how much of each gas was emitted into the atmosphere, and describes some of the effects of these emissions on the environment.

   In nature, carbon is cycled between various atmospheric, oceanic, biotic, and mineral reservoirs. In the atmosphere, carbon mainly exists in its oxidized form as carbon dioxide. Carbon dioxide (CO_2) is released into the atmosphere primarily as a result of the burning of fossil fuels (oil, natural gas, and coal) for power generation and in transportation. It is also emitted through various industrial processes, forest clearing, natural gas flaring, and biomass burning.

   The EPA website provides data on emissions. The data found in the file "Carbon Dioxide Emissions Data" were taken from this website for the year 2001 for all 50 states and the District of Columbia.

   (a) Use the data on all 50 states (+DC) provided in "Carbon Dioxide Emissions Data" to create a histogram with an appropriate bin size; use Excel.
   (b) Determine the mean and median of the data. Indicate each of these values on your graph for part (a).
   (c) Which value more accurately describes the data? Indicate your choice (mean or median) and the value of your choice. Justify your answer in words.

10. The following set of data was collected by Ed Fuller of the NIST Ceramics Division in December 1993. The data represent the polished window strength, measured in units of ksi [kilopounds per square inch], and were used to predict the lifetime and confidence of airplane window design. Use the data set to generate a histogram and CDF; use Excel (http://www.itl.nist.gov/div898/handbook/eda/section4/eda4291.htm).

11. Choose *one* of the following options and collect the data required. For the data source you select, do the following:

   ■ Construct a histogram, including justification of bin size.
   ■ Determine the mean, median, variance, and standard deviation values.
   ■ Construct a cumulative distribution function.

   (a) On a campus sidewalk, mark two locations 50 feet apart. As people walk along, count how many steps they take to go the 50 feet. Do this for 125 individuals.

(b) Select 250 words at random from a book (fiction). Record the number of letters in each word. Alternatively, you can count and record the words in 250 sentences.

(c) Go to one section of the library, and there record the number of pages in 125 books in that same section.

(d) Interview 125 people to determine how far their home is, in miles, from the university.

12. During the month of November, the HVAC system in your apartment appeared to be broken. To prove this, you record the following daily high temperatures in degrees Fahrenheit [°F], taken every other day:

58	54	60	58	55	55	60	60
86	80	85	85	82	85	91	93

(a) Calculate the mean and median of the data.

(b) Draw the associated histogram and CDF for the data; plot temperature on the abscissa and days on the ordinate.

(c) When you take your complaint to the apartment manager, he fails to see the problem; according to his heating bill, your apartment had an average temperature of 72°F. Show both the mean and median on the graph drawn for part (b). Which is a better presentation of the data, part (a) or part (b)? Justify.

13. The drag force on a sphere moving through a fluid is given by

$$F_D = \frac{1}{2}\rho\, v^2\, A_p\, C_d$$

The variables used are fluid density, ρ; coefficient of drag, C_d (dimensionless); velocity, v; and the drag force, F_D. The area of the object the force acts upon is A_p, and so for spheres is given by the area of a circle.

The Reynolds number in this situation is written as

$$Re = \frac{D_p\rho v}{\mu}$$

where D_p is the diameter of the object the force acts upon and μ is the viscosity of the fluid.

At small Reynolds numbers (Re < 10), the drag coefficient of a sphere can be determined by a dimensionless relationship known as Stoke's law.

(a) Analyze the data shown to determine two dimensionless parameters: Reynolds number (Re) and coefficient of drag (C_d).

(b) Be sure to *watch the units*! Create a graph of the resulting coefficient of drag (C_d, on the ordinate) versus the Reynolds number (Re, on the abscissa).

(c) Fit a trendine to the data. Linearize the data by creating a rectilinear, semilog, or log-log chart. What is the relationship between C_d and Re?

14. An environmental engineer has obtained a bacteria culture from a municipal water sample and has allowed the bacteria to grow.

Time (t) [min]	1	2	4	6	7	9	10
Concentration (C) [ppm]	9	15	32	63	102	220	328

(a) Plot the data and fit an appropriate trendline model, using the possible choices of linear, power, or exponential.

(b) Suppose the growth of bacteria is governed by a process described by the following equations: Type A bacteria, $C_A = a + bt$; Type B bacteria, $C_B = ae^{bt}$; and Type C bacteria, $C_C = at^b$. What type of bacteria is the engineer dealing with? Justify your answer using logarithmic plots, and determine the values of a and b for the model chosen.

15. An environmental engineer has obtained a bacteria culture from a municipal water sample and allowed the bacteria to grow.

Time (t) [min]	1	2	4	6	7	9	10
Concentration (C) [ppm]	12	17	27	37	42	52	57

(a) Plot the data and fit an appropriate trendline model, using the possible choices of linear, power, or exponential.

(b) Suppose the growth of bacteria is governed by a process described by the following equations: Type A bacteria, $C_A = a + bt$; Type B bacteria, $C_B = a e^{bt}$; and Type C bacteria, $C_C = a t^b$. What type of bacteria is the engineer dealing with? Justify your answer using logarithmic plots, and determine the values of a and b for the model chosen.

16. An environmental engineer has obtained a bacteria culture from a municipal water sample and allowed the bacteria to grow.

Time (t) [min]	1	2	4	6	7	9	10
Concentration (C) [ppm]	0.5	1.5	4.0	7.3	9.0	13.5	16.0

(a) Plot the data and fit an appropriate trendline model, using the possible choices of linear, power, or exponential.

(b) Suppose the growth of bacteria is governed by a process described by the following equations: Type A bacteria, $C_A = a + bt$; Type B bacteria, $C_B = a e^{bt}$; and Type C bacteria, $C_C = a t^b$. What type of bacteria is the engineer dealing with? Justify your answer using logarithmic plots, and determine the values of a and b for the model chosen.

17. A capacitor is an electrical device that can store energy in the form of an electric field. A capacitor is simply two conducting plates (usually metal) separated by an insulator, with a wire connected to each of the two plates. For a simple capacitor with two flat plates, the capacitance (C) [F, farads] can be calculated by

$$C = \varepsilon_r \varepsilon_0 A/d$$

where

$\varepsilon_0 = 8.854 \times 10^{-12}$ F/m (the permittivity of free space in farads per meter)
ε_r = relative static permittivity, a property of the insulator [dimensionless]
A = area of overlap of the plates [meters squared, m^2]
d = distance between the plates [meters, m].

Capacitance (C) [nF] vs Plate Area (A) [m^2]; C = 19.15 A

Several experimental capacitors were fabricated with different plate areas (A), but with the same inter-plate distance (d = 1.2 mm) and the same insulating material, and thus the same relative static permittivity (ε_r). The capacitance of each device was measured and plotted versus the plate area. A mathematical model to describe the data was then obtained. The graph and equation are shown below. The numeric scales were deliberately omitted.

(a) What are the units of the slope (19.15)?

(b) If the capacitance is 2 nanofarads [nF], what is the area (A) of the plates?

(c) What is the relative static permittivity of the insulating layer?

(d) If the distance between the plates were doubled, how would the capacitance be affected?

18. The viscosity of fluids depends on temperature. For many fluids, the higher the temperature, the lower the viscosity or the easier it becomes to move the fluid. Use the two graphs below to answer the following questions.

(a) What is the dynamic viscosity and kinematic viscosity of kerosene at 50°F? What is the specific gravity of kerosene?

(b) What is the dynamic viscosity and kinematic viscosity of glycerin at 100°F? What is the specific gravity of glycerin?

(c) What is the dynamic viscosity and kinematic viscosity of gasoline at 50°F? What is the specific gravity of gasoline?

(d) What is the dynamic viscosity and kinematic viscosity of crude oil at 150°F? What is the specific gravity of crude oil?

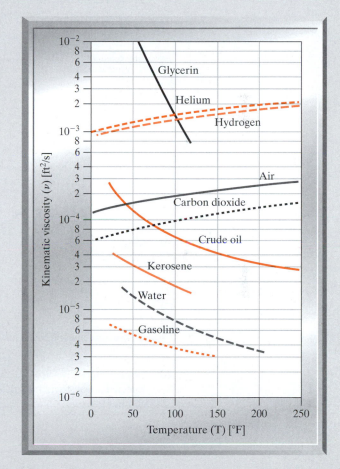

19. The relationship of the power required by a propeller (shown as the power number, on the ordinate) and the Reynolds number (abscissa) is shown in the graph below. For a propeller, the Reynolds number (Re) is written slightly differently, as

$$Re = \frac{D^2 n\rho}{\mu}$$

where D is the blade diameter [meters] and n is the shaft speed [hertz]. The power number (N_p) is given by the following, where P is the power required [watts].

$$N_p = \frac{P}{\rho n^2 D^5}$$

Use the chart below to answer the following questions:

(a) If the Reynolds number is 120, what is the power number for a system described by Curve A?

(b) If the power number (N_p) is 30, what is the Reynolds number for a system described by Curve C?

(c) If the Reynolds number is 4,000, what is the power (P) required in units of watts at a shaft speed (n) of 0.03 hertz? Assume the system contains acetone, with a kinematic viscosity of 0.419 stokes. (Hint: use the Reynolds number of the system to first calculate the diameter, then find the power number, then calculate the power.)

(d) If the power number (N_p) is 5, what is the diameter (D) of the blade in units of centimeters at a shaft speed (n) is 0.02 hertz? Assume the system contains brine, with a kinematic viscosity of 0.0102 stokes. (Hint: find the Reynolds number of the system first, then calculate the diameter.)

20. When a fluid flows around an object, it creates a force, called the drag force, that pulls on the object. The coefficient of drag (C_d) is a dimensionless number that describes the relationship between the force created and the fluid and object properties, given as

$$C_d = \frac{F_D}{\frac{1}{2}\rho v^2 A_p}$$

where F_D is the drag force, ρ is the fluid density, and v is the velocity of the object relative to the fluid. The area of the object the force acts upon is A_p, and for spheres is given by the area of a circle. The Reynolds number in this situation is written as

$$Re = \frac{D_p \rho v}{\mu}$$

where D_p is the diameter of the object the force acts upon.

(a) If the Reynolds number is 500, what is the coefficient of drag?

(b) If the coefficient of drag is 2, what is the Reynolds number?

Ethylene glycol has a dynamic viscosity of 9.13 centipoise and a specific gravity of 1.109.

(c) If the fluid flows around a sphere of diameter 1 centimeter travelling at a velocity of 2.45 centimeters per second, determine the drag force on the particle in units of newtons. (Hint: determine the Reynolds number of the system, then find the coefficient of drag, then calculate the force.)

(d) If a coefficient of drag of 100 is produced, what is the diameter of the particle? Assume the fluid is moving at 1 centimeter per second. (Hint: determine the Reynolds number of the system, then calculate the diameter.)

21. When a fluid flows around an object, it creates a force, called the drag force, that pulls on the object. The coefficient of drag (C_d) is a dimensionless number that describes the relationship between the force created and the fluid and object properties, given as

$$C_d = \frac{F_D}{\frac{1}{2}\rho v^2 A_p}$$

where F_D is the drag force, ρ is the fluid density, and v is the velocity of the object relative to the fluid. The area of the object the force acts upon is A_p, and for spheres is given by the area of a circle. The Reynolds number in this situation is written as

$$Re = \frac{D_p \rho v}{\mu}$$

where D_p is the diameter of the object the force acts upon.

(a) If the Reynolds number is 3,000, what is the coefficient of drag?

(b) If the coefficient of drag is 5, what is the Reynolds number?

Methyl ethyl ketone (MEK) has a specific gravity of 0.805 and a dynamic viscosity of 0.0043 grams centimeter second.

(c) If MEK flows around a sphere of diameter 0.5 centimeter travelling at a velocity of 1.5 centimeters per second, determine the drag force on the particle in units of newtons. (Hint: determine the Reynolds number of the system, then find the coefficient of drag, then calculate the force.)

(d) If a coefficient of drag of 20 is produced, what is the diameter of the particle? Assume the fluid is moving at 15 centimeters per second. (Hint: determine the Reynolds number of the system, then calculate the diameter.)

22. A phase diagram for carbon and platinum is shown. Assuming the lines shown are linear, we can say the mixture has the following characteristics:

- Below 1700 °C, it is a mixture of solid platinum and graphite.
- Above 1700 °C, there are two possible phases: a liquid (L) phase and a liquid (L)+ graphite phase. The endpoints of the division line between these two phases are labeled on the diagram. Develop a worksheet to determine the phase of a mixture, given the temperature and carbon content. A sample worksheet is shown below.

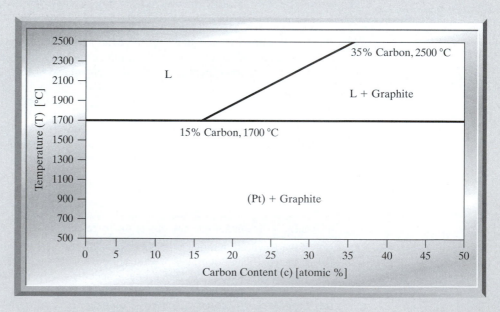

(a) Write the equation to describe the temperature of the dividing line between the liquid (L) region and the liquid (L) + graphite region in Column C. Reference the carbon content found in Column B as needed. Add any absolute reference cells you feel are needed to complete this calculation.

(b) Write the conditional statement to determine the phase in Column D. For simplicity, call the phases Pt + G, L, and L + G.

	A	B	C	D	E
1					
2					
3	Maximum Temperature for Pt + G			1700	[°C]
4					
5					
6					
7	Temperature	Carbon Content	Temp between	Phase	
8	(T) [°C]	(c) [%]	L & L+G		
9	1220	30			
10	1300	34			
11	1150	45			
12	1310	56			

23. A simplified phase diagram for cobalt and nickel is shown. Assuming the lines shown are linear, we can say the mixture has the following characteristics:

- Above 1500°C, it is a liquid.
- Below 1500°C, there are two possible phases: face-centered cubic (FCC) phase and hexagonal close-packed (HCP) phase. Develop a worksheet to determine the phase of a mixture, given the temperature and nickel content. A sample worksheet is shown below.

(a) Write the mathematical equation to describe the dividing line between the HCP region and the FCC region in Column C. Reference the nickel content found in Column B as needed. Add any absolute reference cells you feel are needed to complete this calculation.

(b) Write the conditional statement to determine the phase in Column D. For simplicity, call the phases HCP, FCC, and L.

	A	B	C	D	E	F
1						
2						
3		Minimum Temperature for Liquid			1500	[°C]
4						
5	Temperature	Nickel Content	Temp Between	Phase		
6	(T) [°C]	(w) [%]	HCP & FCC			
7	250	12				
8	1015	43				
9	125	49				
10	360	27				
11	108	68				
12	200	40				

24. The phase diagram below for the processing of a polymer relates the applied pressure to the raw material porosity.

- Region A or B = porosity is too high or too low for the material to be usable.
- Region C = combinations in this region yield material with defects, such as cracking or flaking.
- Region D = below a pressure of 15 psi the polymer cannot be processed.
- Region E = optimum region to operate.

There are often multiple ways to solve the same problem; here we look a few alternative ways to determine the phase of the material and the processibility of the material.

(a) Method One: Conditional Statements

Begin on the worksheet titled Conditionals.

In Column C, develop the equation for the line dividing the phases of Region E and Region C. Assume it was written in cell C9 and copied down Column C.

In Column D, write an expression to determine the phase of the material (Phase A–Phase E).

In Column E, write an expression to determine if the material is processible. When the conditions of Phase E are met, the cell should be highlighted by conditional formatting.

	A	B	C	D	E
1					
2					
3	Porosity Upper Limit [%]			85	
4	Porosity Lower Limit [%]			15	
5	Pressure Limit [psi]			15	
6					
7	Pressure	Porosity	Porosity between	Phase	Is Material Able
8	(P) [psi]	(ε) [%]	C and E		to be Processed?
9	13	68			
10	26	39			
11	10	43			
12	15	82			
13	22	47			
14	16	84			
15	7	73			
16	25	45			

(b) Method Two: Data Validation

Begin on the worksheet titled Validation.

In Column A and Column B, use data validation to restrict the user from entering values outside the valid parameter ranges: Pressure: 0–35 psi and Porosity: 0–100%.

In Column C, develop the equation for the line dividing the phases of Region E and Region C.

In Column D, write an expression to determine the phase of the material (Phase A–Phase E).

In Column E, write an expression to determine if the material is processible. When the conditions of Phase E are met, the cell should be highlighted by conditional formatting.

	A	B	C	D	E
1					
2					
3	Porosity Upper Limit [%]			85	
4	Porosity Lower Limit [%]			15	
5	Pressure Limit [psi]			15	
6					
7	Pressure	Porosity	Porosity between	Phase	Porosity
8	(P) [psi]	(ε) [%]	C and E		Check
9	50	68			
10	Processig Pressure / Value must fall between 0 - 35 psi	Processing Pressure			
11		Value must lie between 0 - 35 psi. Please reenter a valid pressure reading.			
12					
13		Retry Cancel Help			
14					

(c) Method Three: Slider Bars

Begin on the worksheet titled Slider Bars.

In Column A and Column B, use slider bars to control the entry values for the parameters: Pressure: 0–35 psi and Porosity 0–100%.

In Column C, develop the equation for the line dividing the phases of Region E and Region C.

In Column D, write an expression to determine the phase of the material (Phase A–Phase E).

	A	B	C	D	E
1					
2					
3	Porosity Upper Limit [%]			85	
4	Porosity Lower Limit [%]			15	
5	Pressure Limit [psi]			15	
6					
7	Pressure	Porosity	Porosity between	Phase	Porosity
8	(P) [psi]	(ε) [%]	C and E		Check
9	14	36			
10					
11					
12					
13					
14					

In Column E, write an expression to determine if the material is processible. When the conditions of Phase E are met, the cell should be highlighted by conditional formatting.

(d) Can you write an expression in Column F to tell the user why the material was rejected? For example, under the conditions of Pressure = 25 psi and Porosity = 40%, the statement might say "Porosity too low."

(e) Can you write an expression in Column F to tell the user why the material was rejected and how to fix the problem to make the product useful? For example, under the conditions of Pressure = 25 psi and Porosity = 40%, the statement might say "Porosity too low, increase to at least a value of 55%."

25. The following phase diagram is for salt water. There are four possible phases, which depend on the temperature and the sodium chloride content (NaCl).

- Ice and SC = Mixed ice and salt crystals
- Ice and SW = Ice and saltwater
- SW = Saltwater
- SW and SC = Saltwater and salt crystals.

There are often multiple ways to solve the same problem; here we look a few alternative ways to determine the phase of the mixture.

(a) Method One: Conditional Statements

Begin on the worksheet titled Conditionals.

In Column C, develop the equation for the line dividing the phases of the ice–saltwater mix and the saltwater. Assume it was written in cell C11 and copied down.

In Column D, develop the equation for the line dividing the phases of the saltwater and the saltwater–salt crystals mix. Assume it was written in cell D11 and copied down.

In Column E, write an expression to determine the phase of the mixture. Use conditional formatting to highlight the various phases. Provide a color key.

	A	B	C	D	E
1					
2					
3					
4					
5					
6					
7	Upper Limit of Mixed Ice and Salt Crystals			-21	[°C]
8					
9			Dividing Temp [°C]		
10	NaCl [%]	Temp [°C]	Ice and SW to SW	SW to SW and SC	Phase
11	83	-25			
12	73	2			
13	30	12			
14	56	-12			
15	90	31			
16	39	23			
17	33	-16			
18	35	5			
19	16	-10			
20	87	-25			
21	28	30			
22	36	-5			
23	10	22			
24	17	-5			
25	22	-23			
26	60	2			
27					

(b) **Method Two: Data Validation**

Begin on the worksheet titled Validation.

In Column A and Column B, use data validation to restrict the user from entering values outside the valid parameter ranges: NaCl (%): 0–100%; Temp [°C]: −35°C to 35°C.

In Column C, develop the equation for the line dividing the phases of the ice–saltwater mix and the saltwater.

In Column D, develop the equation for the line dividing the phases of the saltwater and the saltwater–salt crystals mix.

In Column E, write an expression to determine the phase of the mixture. Use conditional formatting to highlight the various phases. Provide a color key.

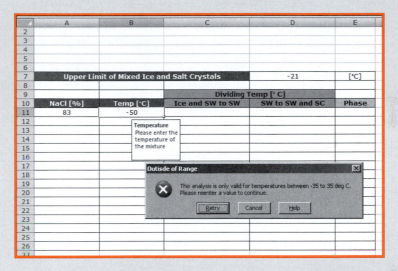

(c) **Method Three: Slider Bars**

Begin on the worksheet titled Slider Bars.

In Column A and Column B, use slider bars to control the entry values for the parameters: NaCl (%): 0–100%; Temp: [°C] −35°C to 35°C. Since slider bars will only operate with positive numbers, the temperature calculation will require an extra step.

In Column C, develop the equation for the line dividing the phases of the ice–saltwater mix and the saltwater.

In Column D, develop the equation for the line dividing the phases of the saltwater and the saltwater–salt crystals mix.

In Column E, write an expression to determine the phase of the mixture. Use conditional formatting to highlight the various phases. Provide a color key.

> This problem requires iteration. For more information, refer to Chapter 14.3 on Automated Calculations.

26. In this problem, we model the trajectory of a skydiver's jump from an airplane. The skydiver will free-fall until reaching a specified height above the ground. At that point the skydiver will pull the rip cord to release the parachute slowing the skydiver's descent. Create a worksheet as follows.

- Cells B5–B10 and B16–B18 contain the numbers shown in the figure below.
- Cell B12 contains the time elapsed since the skydiver jumped from the airplane. To increment the value in Cell B12, we hold down the F9 button. That repeatedly adds the time increment value in B10 to the previous value in B12. We can reset the value in B12 by selecting B12, pressing the F2 key and then Enter. This resets the "clock."

⌘ **Mac OS:** Excel 2008 does not include the iteration function. Either run Excel 2004 for Mac or run Windows and use Excel 2007.

- Cell B20 contains the skydiver's comments. Upon reaching the ground, the skydiver's comment should change from "Whee!" to "Down Safely."
- Cell F5 contains an equation to calculate the time for the parachute to open (t_{chute}), which is function of the difference between the jump height (H_{jump}) and the height at which the parachute opens (H_{chute}) as well as of the acceleration due to gravity.

$$t_{chute} = \sqrt{\frac{2(H_{jump} - H_{chute})}{g}}$$

- Cell F8 contains an equation to calculate the airplane's horizontal position. This can be calculated very simply from the airplane's speed and the elapsed time.
- Cell F9 contains the airplane's elevation, which remains the same as the skydiver's jump height, that is, the airplane continues in level flight after the skydiver jumps.
- Cell F12 contains an equation to calculate the skydiver's horizontal position. Since we are neglecting the effects of wind drag on the skydiver's horizontal travel, this can be calculated very simply from the skydiver's horizontal speed, which is the same as the airplane's speed, and the time elapsed since leaving the airplane.

- Cell F13 contains an equation to calculate the skydiver's elevation above the ground assuming that the parachute has not opened. This is calculated with the following free-fall trajectory formula

$$H_{nochute} = H_{jump} - \frac{g\,t^2}{2}$$

- Cell F14 contains an equation to calculate the skydiver's elevation above the ground. This will equal the height calculated in F13 whenever that height is greater than or equal to the height to open the parachute, specified in B7. After the height in F13 falls below the height specified in B7, the height to plot is calculated with the following formula

$$Y_{diver} = H_{chute} - v_{chute}(t - t_{chute})$$

- Cell F17 contains an equation to calculate the horizontal position of the parachute. This will equal the horizontal position of the skydiver if the diver's height is less than or equal to the height to open the parachute. Before the diver's height reaches the height to open the chute, the horizontal position of the parachute is set to its initial value typed into B16 (which is off the chart, set at 1500). This requires that the axes scales be set to a maximum of 1000 feet.
- Cell F18 contains an equation to calculate the elevation of the parachute. This will equal the elevation of the skydiver plus the height of the parachute above the skydiver after the parachute has been opened. Until the elevation to plot falls below the height at which the chute opens, the elevation of the parachute is set to its initial value typed into B16 (which is off the chart).
- The proper plot chart contains a legend and three data series. The legend is used as a text box. The three series are:

  - The skydiver position with x = F12 and y = F14. The marker option is set as a solid, grey triangle (size 12). The series name is set to cell B20. This will display in the Legend whatever text is in cell B20, which will be "Whee!" or "Down Safely" depending on whether the skydiver has reached the ground.
  - The parachute position with x = F17 and y = F18. The marker option is set to a hollow, black circle (size 20). The default series name, Series 2, should be deleted from the legend.
  - The airplane position with x = F8 and y = F9. The marker option is set to a large (size 40) grey horizontal dash. The default series name, Series 3, should be deleted from the legend.

- Set your worksheet to calculate iteratively so that holding down the F9 button will cause the elapsed time to advance, causing the skydiver and airplane (and later the parachute) to move across the graph.

27. Rework Question 26, using slider bars instead of iteration.
28. We throw an egg horizontally from the top of a cliff. We want to watch as the egg drops and finally goes splat. The equations of motion for the egg (neglecting aerodynamic drag) are

$$x = v\,t \qquad \text{and} \qquad y = y_0 - \frac{1}{2}\,g\,t^2$$

Here, x is the horizontal distance, y_0 the initial height (height of the cliff), g the acceleration of gravity, t the elapsed time, and v the initial velocity (horizontal).

   If we iterate on time, we can calculate the position of the egg at any time (x and y position or coordinates). Create a worksheet like the one shown below. Fill in all cells except B10 and B16, which require conditional (IF) statements. Use data validation to limit the initial height to 100 feet.

⌘ **Mac OS:** Excel 2008 does not include the iteration function. Either run Excel 2004 for Mac or run Windows and use Excel 2007.

The program will start with just a single data point on the plot (maybe a circle point to represent the egg). Once the egg hits the ground, we want a large diamond for the yolk to show up. This will be a second point.

In cell B10, write an IF statement to show the sum of B9 + B10 until the egg hits the ground (i.e., the y value goes negative), and then it will use B10. This means that after the egg has reached the ground, the actual value of time will cease to change.

Construct a graph to plot the single x and y values for the egg. Fix both axis scales as 0 to 100 feet with a major unit of 20. This will keep the scale the same as the egg falls.

Now add one more data series to the graph where the egg hits the ground. It will have a y coordinate of zero and an x coordinate of the horizontal position when the egg hits (and time ceases to change). Write an IF statement in cell B16 to check whether the egg has hit. If it has not hit the ground yet, specify an x coordinate of 1000, so that the point does not appear on the graph. If the egg has hit, specify the value in cell B13. Select a large diamond marker shape.

Now you should be able to run the program and watch the egg fall and go splat. Try different speeds and different cliff heights (initial height).

29. Rework Question 28, using slider bars instead of iteration.

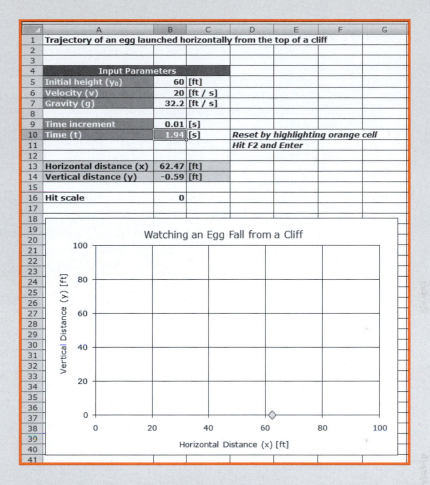

	A	B	C	D	E	F	G
1	Trajectory of an egg launched horizontally from the top of a cliff						
2							
3							
4	Input Parameters						
5	Initial height (y_0)	60	[ft]				
6	Velocity (v)	20	[ft / s]				
7	Gravity (g)	32.2	[ft / s]				
8							
9	Time increment	0.01	[s]				
10	Time (t)	1.94	[s]	Reset by highlighting orange cell			
11				Hit F2 and Enter			
12							
13	Horizontal distance (x)	62.47	[ft]				
14	Vertical distance (y)	-0.59	[ft]				
15							
16	Hit scale	0					
17							

30. When discussing the flow of a fluid through a piping system, we say that friction occurs between the fluid and the pipe wall due to viscous drag. The loss of energy due to the friction of fluid against the pipe wall is described by the friction factor. The Darcy friction factor (f) was developed by Henry Darcy (1803–1858), a French scientist who made several important contributions to the field of hydraulics. The friction factor depends upon several other factors, including flow regime, Reynolds number, and pipe roughness. The friction factor can be determined in several ways, including with the Moody diagram (shown below, discussed in ICA 16-13) and several mathematical approximations presented here.

In the laminar flow range, the *Darcy friction factor* can be determined by the following formula, shown as the linear line on the Moody diagram.

$$f = \frac{64}{Re}$$

In the turbulent range, the friction factor is a function of the Reynolds number and the roughness of the pipe (ε). For turbulent flow smooth pipes (where the relative roughness ratio (ε/D) is very small), the *Blasius formula* can be used to calculate an approximate value for the Darcy friction factor.

$$f = 0.316\,(Re)^{-1/4}$$

This simple formula was developed by Paul Richard Heinrich Blasius (1883–1970), a German fluid dynamics engineer. Later, a more accurate but more complex formula was developed in 1939 by C. F. Colebrook. Unlike the Blasius formula, the Colebrook formula directly takes into account the pipe roughness.

The *Colebrook formula* is shown below. Notice that both sides of the equation contain the friction factor, requiring an iterative solution.

$$\frac{1}{\sqrt{f}} = -2\log\left(\frac{\varepsilon/D}{3.7} + \frac{2.51}{Re\sqrt{f}}\right)$$

To begin the iteration, the Colebrook calculation must have an initial value. Use the Blasius approximation as the initial guess to begin the iterative calculation utilizing a conditional statement.

Prepare an Excel worksheet to compute the friction factor from the input parameters shown below:

Input Parameters:

- Fluid: should be chosen from a drop-down list using the material properties listed below. Used with the LOOKUP function to determine:

  - Density (ρ) [grams per cubic centimeter]
  - Viscosity (μ) [centipoises]

- Volumetric flow rate (Q) [gallons per minute]
- Diameter (D) [inches]
- Type of pipe: should be chosen from a drop-down list using the properties listed below. Used with the LOOKUP function to determine:
  - Pipe roughness (ε) [millimeters]

	A	B	C	D	E	F	G
1							
2	*Enter input values in orange cells.*						
3	*Press F9 button to calculate. If in turbulent conditions, press F9 until the Colebrook value does not change.*						
4							
5							
6	Fluid	Water		Pipe Type	Commercial Steel		
7	Density	Viscosity	Volumetric Flowrate	Diameter	Roughness	Reynolds Number	Flow Regime
8	(ρ) [g/cm3]	(μ) [cP]	(Q) [gpm]	(D) [in]	(ε) [mm]	(Re) [-]	
9	1	1.002	500	6	0.045	263,020	Turbulent
10							
11	Darcy Friction Factor (f) [-]						
12	Laminar	Colebrook					
13		0.0172					
14							

Pipe	Equivalent Roughness (ε) [mm]
Cast Iron	0.26
Commercial Steel	0.045
Concrete	1.5
Drawn Tubing	0.0015
Galvanized Iron	0.15
Riveted Steel	4.5
Smooth Pipe	0
Wrought Iron	0.045

Output Parameters:

- Be sure to include the appropriate unit conversions
- Reynolds number
- Flow regime (laminar, transitional, or turbulent)
- Only the correct Darcy friction factor (one of these two values) should be displayed.

  - For laminar flow, use the equation: $f = 64/Re$
  - For turbulent flow determined with the Colebrook formula.

Compound	Specific Gravity	Viscosity [cP]
Acetone	0.791	0.3311
Acetic Acid	1.049	1.2220
Benzene	0.879	0.6470
Iso-Butyl Alcohol	0.802	3.9068
Castor Oil	0.960	1027
Ethyl Alcohol	0.789	1.1943
Glycerol	1.261	4220
Lactic Acid	1.249	40.3300
Methyl Ethyl Ketone	0.805	0.4284
Olive Oil	0.914	101
Pentane	0.626	0.2395
Phenol	1.071	12.7440
Tetrachlorothylene	1.466	0.8759
Turpentine	0.860	1.4870
Water	1.000	1.0020
m-Xylene	0.864	0.6475

31. Most resistors are so small that the actual value would be difficult to read if printed on the resistor. Instead, colored bands denote the value of resistance in ohms. Anyone involved in constructing electronic circuits must become familiar with the color code, and with practice, one can tell at a glance what value a specific set of colors means. For the novice, however, trying to read color codes can be a bit challenging.

You are to design a worksheet similar to the one shown, allowing the user to enter a resistance value and automatically show the color code for that resistance. Note the cells below "First Digit," "Second Digit," and "Number of Zeros" should actually take on the appropriate colors in addition to showing the numerical value. The cells below should use the LOOKUP function to show the color names. The colors assigned to each digit are on the right of the worksheet for convenience.

To enter the resistance, use data validation along with the table on the left. The resistance will be entered in two parts, the first two digits, and a power of 10 by which those digits will be multiplied. Include an appropriate input message when the cell is selected and an appropriate warning if the user enters an invalid number. The user should be able select from a drop-down menu. Note that you must use the two input values to calculate the total resistance, the first digit, the second digit, and the number of zeros.

As examples, a resistance of 4700 ohms (Ω) has first digit 4 (yellow), second digit 7 (violet) and 2 zeros following (red). A resistance of 56 Ω would be 5 (green), 6 (blue), and 0 zeros (black); 1,000,000 Ω is 1 (brown), 0 (black), and 5 zeros (green). Particularly note that if the second digit is zero, it does not count in the multiplier value. There are numerous explanations of the color code on the web if you need further information or examples.

The worksheet shown is available online except for the nine cells in the center containing the desired resistance values and the color code for that resistance. (Unfortunately, this text is not printed in full color, but the online version will show the colors in all their glory!)

HINT

Without giving too much away, try typing "Round" into Excel help, click on the link to the ROUND function, then scroll down to the list of related functions. One or more of those might be useful.

Resistor Color Code Calculator

| Resistor Values | | | Enter Values Below | | | | | | | Color Code | |
Standard Values	Multipliers		First Two Digits	Multiplier	Resistance	First Digit	Second Digit	Number of zeros		Value	Color
10	1		68	1,000	68000	6	8	3		0	Black
12	10					Blue	Grey	Orange		1	Brown
15	100		Please enter first two digits of resistance value or select from drop-down menu.							2	Red
18	1,000									3	Orange
22	10,000									4	Yellow
27	100,000									5	Green
33	1,000,000									6	Blue
39										7	Violet
47										8	Grey
56										9	White
68											
82											

32. A piano, much like a guitar or a harp, is a stringed instrument. The primary difference is that a piano has a keyboard containing 88 keys that, when pressed, cause a small felt hammer to strike the strings (mostly 3) associated with that key. Each hammer can hit the strings harder or softer depending on the force applied to the key on the keyboard. In order to create the different pitches of sound, the strings for each key are different lengths as well as having different mass per unit length.

In avant-garde music, the piano is sometimes played much like a harp, where the strings are directly manipulated (plucked, strummed, etc.) without the use of the keys on the keyboard. Assume we construct a special piano for use by an avant-garde musician in which each string has the same mass per unit length and the same tension on the soundboard. The only difference is the length of each string on the soundboard. To test our design, we collect the following data on the observed frequency at different string lengths:

Length (L) [m]	0.41	0.48	0.61	0.69	0.86	0.97	1.1	1.3	1.5
Frequency (f) [Hz]	1500	1300	1000	880	740	650	600	490	400

The equation for the fundamental frequency of a vibrating string is given by

$$ f = \frac{\sqrt{T/\mu}}{2L} $$

where   f = frequency [Hz]
T = string tension [N]
μ = mass per unit length [kg/m]
L = string length [m]

(a) Using the equation above, express the unit hertz [Hz] in terms of base SI units. You must prove this relationship using the equation given.
(b) Is the relationship between frequency and length linear, power, or exponential?
(c) Create a graph of the observed frequency data, including the trendline and equation generated by Excel.
(d) If the tension was reduced to half of its original value, would the frequency increase or decrease and by how much?
(e) If the tension on the strings is 200 newtons, what is the mass per unit length in grams per meter?
(f) If the mass per length of the string is 0.5 grams per meter, what is the tension in newtons?

33. Guitars are constructed in a manner to allow musicians to place their fingers on different arrangements of strings to create chords. Chords are arrangements of 3 or more notes (or pitches) played at the same time. (There is some disagreement among musicologists concerning whether 2 notes played simultaneously should be classified as a chord.) On a standard electric guitar, there are typically 6 strings strung from the nut down the neck of the guitar with 21 frets set in the neck terminating at the bridge on the guitar. On a 6-string, 21-fret guitar, there are 126 different locations that can generate a pitch plus the 6 pitches generated by playing a string without pressing a fret. However, each fretted string on a guitar can take on one of 12 different semitones. As a guitarist moves their finger from lower to higher frets on a string on a guitar, each note changes by increasing 1 semitone.

The 12-tone semitone transitions are shown in the chart below:

From:	To:
A	A#/Bb
A#/Bb	B
B	C
C	C#/Db
C#/Db	D
D	D#/Eb
D#/Eb	E
E	F
F	F#/Gb
F#/Gb	G
G	G#/Ab
G#/Ab	A

In other words, if we are on the 5th string playing a B on the 10th fret, moving to the 11th fret would play a C.

Fill in the provided Excel worksheet using LOOKUP statements to the semitone transition table for a standard tuning guitar (E-A-D-G-B-E) to show where each note will be located on the guitar fretboard. The worksheet should be set up so that it can be easily modified for different guitar tunings:

(a) What will the resulting notes on the 5th fret be if an "open tuning," typical of blues and slide guitar music, if the open G tuning were used (G-B-D-G-B-D)?

(b) What will the resulting notes on the 8th fret be if the D-A-D-G-A-D tuning were used (e.g., Led Zeppelin's "Kashmir")?

(c) Use conditional formatting to highlight the sharps and flats (notes with # or b) in light yellow.

(d) Use conditional formatting to highlight all entries with the same note name as one of the base tuning notes within the column by making the text white on a black background. For standard tuning, this would highlight all occurrences of E in the first column, A in the second column, etc.

(e) Use data validation to allow the user to choose the notes from a drop-down list.

34. As part of an electronic music synthesizer, you need to build a gizmo to convert a linear voltage to an exponentially related current. Specifically, you want the current to double every time you increase the voltage by one volt. You decide to use the well-known exponential relationship between current and voltage in a bipolar junction transistor as the basis for this converter. You build three prototype circuits and make several measurements of voltage and current in each. The collected data is given in the table below.

Circuit A		Circuit B		Circuit C	
Voltage (V_A) [V]	Current (I_A) [mA]	Voltage (V_B)[V]	Current (I_B) [mA]	Voltage (V_C) [V]	Current (I_C) [mA]
−2.7	0.28	−2.7	0.11	0	0.79
−0.4	1.05	−1.5	0.36	0.5	1.59
0	1.74	0	1.34	1.4	5.41
1.2	3.17	0.8	2.37	2.3	20.28
2.9	7.74	2.6	14.53	2.9	41.44

(a) Plot this data, and determine an appropriate mathematical model for each circuit. If the model chosen is not linear, adjust the axis to the appropriate log scales to make the data appear linear. Justify your choice of model.

(b) Which circuit is closest to achieving the goal of doubling the current for each one volt increase?

(c) Estimate from these data sets the value that should appear in the exponent if the current is to double for each increase of one volt.

(d) Calculate the value that should appear in the exponent if the current is to double for each increase of one volt. Note that you do NOT need the data above nor the chart you created to do this calculation.

PART 5

PUNCTILIOUS PROGRAMMING

The first 90% of the code accounts for 10% of the development time. The remaining 10% of the code accounts for the other 90% of development time.

TOM CARGILL

Computers are controlled by software that can be designed in a variety of programming languages. Computer programs are a translation of what you want to accomplish into something the computer can understand, so the term "programming language" is particularly appropriate. Some computer programs are installed permanently or temporarily on computer chips, and others are installed on a variety of other media, such as hard drives or removable media like CD-ROMs.

Computers relentlessly produce a particular result given a particular set of input conditions. This can be frustrating when you make a simple mistake in a computer program—the computer will do exactly what you tell it to do, even if your mistake would be obvious to a person.

> **If a computer always produces the same result every time given the same input conditions, then why does my computer crash sometimes when I'm doing something that should work?**
>
> **HERE'S WHY**
>
> *The computers you use are simultaneously running a large number of complicated computer programs, including the operating system, background programs, and whatever programs you have started intentionally. Sometimes these programs compete for resources, causing a conflict. Other times, programs are complicated enough that the "input conditions," including the configuration of data in memory and on the hard disk, the time on the system clock, and other factors that change all the time while the computer is running create a combination of circumstances that the programmers never anticipated and so did not include programming code to handle, and the system crashes.*

The biggest difference between a computer and a person is that you can ask a person open-ended questions—questions like design questions that can have many answers. Computers can only process questions that have a single answer.

This makes the process of programming a computer a bit like trying to ask another person to solve a problem when they are on the other side of a wall and you can only communicate by passing them slips of paper asking questions that can have only one answer and waiting for the person to pass back a slip of paper with the answer on it.

WISE WORDS: WHAT IS YOUR FAVORITE PART OF YOUR CURRENT JOB?

The sheer hour-to-hour diversity of projects and work that I get to see. I learn every day, and that's a blast.

E. Basta, CME

I work with many different companies and get to continually learn new industries, new technologies, and ways companies tackle similar challenges differently. Each engagement typically lasts three months.

M. Ciuca, ME

Finding ways to improve our parts so that our customers can run their hard drives at higher speeds and with greater densities.

E. D'Avignon, CpE

Every day, I get to see a 2-dimensional plan be transformed into something 3D and real. It is exciting to see the progress of a schedule on a daily basis, and exciting to deal with the problems that come about when plans don't exactly work. Coming up with solutions to problems and helping the progress of a building or structure is what keeps me on my toes.

L. Edwards, CE

I love the people I work with. They make work fun. I also love the feeling of seeing a product on the shelf in a store and knowing I was a part of getting it there. I also love the fact that I am never really doing the same thing twice. There is always a new and different challenge, so the job is never boring.

S. Forkner, ChE

Mentoring my engineering staff. Seeing each of them grow and develop professionally is an indescribable feeling.

J. Huggins, ME

I am always using what I learned in engineering school and continually learning new technology. I love that I am consistently stimulated by new challenges, whether technical or legal.

M. Lauer, Enviro E

I love working on the next rocket to the Moon; I've enjoyed learning a new discipline and meeting the people involved.

E. Styles, CpE

The flexibility and responsibility. As an engineer, you are already assumed to be "a cut above" everyone else. Management typically gives an engineer a project goal, and the engineer develops it. There is very little hand-holding.

A. Thompson, EE

SOME ADVANTAGES OF COMPUTERS

Given our description of how computers work, it may sound to some as if computers are too simple to be useful. The value of programming is linked to a few important characteristics of computers.

Calculation speed: Although computers can only answer analysis questions, they can answer such questions very quickly—in small fractions of a second. Computer programs can therefore ask the computer a lot of questions in a short time, and thus find the answer to more complicated problems by breaking down the complicated question into a series of simple questions.

Information storage: In "Memory: Science Achieves Important New Insights into the Mother of the Muses" (*Newsweek*, September 29, 1986), Sharon Begley estimates that the mind can store an estimated 100 trillion bits of information. The typical computer has a small amount of storage compared to that, but computers are gaining. Where computers have a bigger advantage is that new information can be incorporated in a fraction of the time it takes a human to learn it.

Information recall: Computers have nearly 100% recall of information, limited only by media failures. The human brain can be challenged to recall information in exactly the same form as it was stored.

HOW I LEARNED TO PROGRAM . . . AN INTJ APPROACH

By: David Bowman

Learning how to program a computer is different from learning a new concept or theory. Especially for someone not typically interested in going to school to graduate as a programmer, it may seem like learning to program a computer might be a waste of time. However, computers are so integrated with every aspect of engineering, business, and research, it is critical for any engineer to have some degree of experience with how a computer can be beneficial for solving problems.

As a computer engineer, I realized the value of programming but was more interested in the hardware side of computer engineering. Regardless, I had to learn how to program a computer just like every other student entering the discipline. It is important to recognize that learning to program has very little in common with learning something like calculus. Programming is a skill that's part of the bigger picture of engineering problems, like soldering is an important skill for being able to wire a circuit. Here are a few tips to guide you while you are learning this new skill.

Learning to Write Computer Programs

- **Spend more time planning than typing.** Regardless of what type of program or function you're writing, before you even begin to type any code, you should have a plan. That plan should include a scope definition, written algorithm, flowchart, and a tentative list of any variables, functions, or other built-in controls you might need. When I write a new program from scratch, I like to draw a few pictures on a whiteboard to get an idea of how my program and functions will interact.
- **Don't study. Practice!** Studying programming doesn't really make sense. Instead, you should practice writing code.
- **Don't get overwhelmed.** Pace yourself! There is almost no chance that you will sit down and write a program from start to finish and have that program work 100% without the need to fix problems in your code. Sometimes the problems are just typos or syntax errors, but sometimes the problems represent a higher misconception

about the problem. It's always OK to step back, reevaluate the algorithm, and start again from scratch.

- **Recycling isn't just for cans and newspapers.** If you wrote a clever piece of code, remember it—if you ever need to solve a problem that involves all or part of a program you've written in the past, there's no need to reinvent the wheel. If you write a piece of code that you think will be helpful with most programs, turn it into a function and use it when you need it.

- **Documentation is almost more important than the code itself.** As you're writing your code, pay attention to including comments throughout the code. Those comments will prove to be invaluable later as you, your boss, your coworker, or anyone else who's reading your code needs to change the functionality of the program. Your comments should be clear enough so that someone who doesn't understand programming can pick up your code and comprehend what's going on.

- **Keep it simple, keep it clean.** Only create variables when you absolutely need to in the code. Don't recalculate a value that you stored in a variable earlier in the code. Keep everything properly indented so that you can follow the structure of your program. If your own code is difficult for you to understand, it's going to be difficult for someone else to understand.

Developing a Program from Scratch

Step 1: Clearly define what you want your program to do. Come up with a problem statement that says exactly what your code is supposed to accomplish. In addition, define the scope of the problem.

Step 2: Create an algorithm. Create a written algorithm or flowchart that defines each step in the sequence for solving the problem.

Step 3: Create a blank file with a header and comments. Write a well documented header for the program; include the problem statement, documentation on all variables needed in the code, a creation and "last modified" date, and the names of each contributor to the program. Populate the file with comments. As a rule of thumb, each step in a written algorithm should be a comment in the code.

Step 4: Write the program. Write the code necessary to solve the problem. If necessary, develop any extra functions needed to be called from the main program.

Step 5: Test the program. Test the program to see if all bugs have been removed. Bugs might include syntax errors, errors in formula calculations, misuse of variables, or other less obvious problems. If the problems are not immediately identified, trace through the code by hand to identify where the problem is occurring.

HOW I LEARNED TO PROGRAM . . . AN ENFP APPROACH

by Matthew Ohland

I was a bit nervous about programming at the beginning—my sister attended engineering school six years ahead of me, and I had seen one of her computer programs—a row of punch cards about two feet long that tracked the population of fish in a pond. My classmates and I were among the first to have the chance to experience personal computers in junior high school—the TRS-80 was the popular model at the time. I got a little programming experience there, but I never had the time (or took the time) to see them as powerful tools—they were more a curiosity. Sensing the onset of the revolution in computing, I took a summer class in computing in the summer between high school graduation and starting college. I believe the three main things I learned were:

1. **Structured programming is better**—it's easier to read programs written without "goto" and "break" statements allowing the flow to jump around like BASIC did.
2. Procedural programming is like other designed systems I was familiar with (like a car), which have well-defined subsystems. This made sense to me, and made it clearer that **computer programs were something that people *designed*.**
3. **People should *never* use their spouse's name for a password**, especially not if they are teaching a computer programming class.

While I had learned to break computer programs down into procedures, my programming was pretty linear in format, and I didn't really understand the power of computing. In college, I learned FORTRAN in my second semester as part of a statics class. Even then, I resisted using iteration to generate an approximate solution. We had to write a program to determine the maximum stress in a system, which required computing the stress in different orientations iteratively. Since the problem was based on trigonometric principles, it was possible to solve it without iteration, so that's what I did. On the one hand, my program was much shorter and more efficient than anyone else's. On the other hand, I completely missed the point of the assignment, which was to learn how to have a computer guess at a lot of configurations quickly and then establish criteria to describe an optimal solution. Sometimes in the early stages of learning to program, an assignment may seem trivial or contrived, but there is often an underlying point and broader learning outcome to the assignment.

When I program, I need to visualize the whole program. As an experience programmer, if the program is simple enough, I can see it all in my head, and I just start typing. If the program is longer, or if I am collaborating with someone else, I draw a graphical algorithm showing what information the program needs, what outputs the program is generating, and finally what has to happen in between to connect the two.

I have learned to program in a few languages—Pascal, FORTRAN, MATLAB, and Maple. What's really interesting is that I have found that I can read code written in programming languages that I've never learned. Learning programming has helped me develop logical thinking patterns that seem have universal applicability—not just in programming, but also more generally in any design process.

LEARNING OBJECTIVES

The overall learning objectives for this part include:

- Defining the scope of a problem and creating a written or graphical algorithm to solve the problem;
- Performing basic matrix operations;
- Reading and interpreting MATLAB programs written by others;
- Writing a program or function using MATLAB and including I/O, plots, conditionals, and loops;
- Design formatted output in a MATLAB program;
- Debugging a program to identify different types of errors.

WISE WORDS: WOULD YOU CONSIDER YOUR CURRENT POSITION TO BE "PURE ENGINEERING," A "BLEND OF ENGINEERING AND ANOTHER FIELD," OR "ANOTHER FIELD"?

I always feel my work is not "pure engineering", but rather often a blend of engineering, sales, accounting, research, inspection, and maintenance.

E. Basta, CME

I would consider my career in another field from engineering, however, highly reliant on my engineering background. As a management consultant, I have to break down complex problems, develop hypotheses, collect data I believe will prove or disprove the hypotheses, and perform the analysis. My focus area is companies who develop highly engineered products.

M. Ciuca, ME

My position is mostly pure engineering.

E. D'Avignon

I work in a blend of engineering and business. I spend most of my time working on business-related activities—forecasting, variance reporting, and timing/work decisions—but I also have to work closely with our field engineers and understand our project scopes. I use both my business and engineering knowledge on a daily basis—without each, I would not be able to succeed at my job.

R. Holcomb, IE

It is definitely a blend of engineering and law with a heavy dose of technical writing. It takes the thinking of an engineer or scientist to truly comprehend the inventions and the skill of a writer to convey the inventor's ideas in written and image terms that others will understand (including juries of lay people). It takes the thinking of a lawyer to come up with creative strategies and solutions when faced with a certain set of facts.

M. Lauer, Enviro E

My current position is definitely a blend of engineering and at least one other field, but more like five other fields. I definitely use my engineering background in the way I think, the way I analyze data, how I approach problems, and how I integrate seemingly unrelated information together. The project management skills that I learned in engineering are helpful, too.

B. Holloway, ME

Even though my boss calls Hydrology "Voodoo Engineering," it is pure engineering.

J. Meena, CE

A blend of mechanical/aerospace engineering and human factors engineering—and management.

R. Werneth, ME

CHAPTER 17
ALGORITHMS, PROGRAMS, AND FUNCTIONS

Learning to create effective algorithms is a crucial skill for any aspiring engineer. In general, an **algorithm** is a well-defined sequence of instructions that describe a process. Algorithms can be observed in everyday life through oral directions ("Simon Says: raise your right hand"), written recipes ("Bake for 15 minutes at 350 degrees Fahrenheit"), graphical assembly instructions, or other graphical cues. As an engineer, writing any algorithm requires a complete understanding of all the necessary actions and decisions that must occur to complete a task.

Before writing an algorithm, you must answer a few questions before attempting to design the process. To even begin thinking of a strategy to describe a process, you must have carefully defined the scope of the problem. The **scope** of an algorithm is the overall perspective and result that the algorithm must include in its design.

For example, if we are required to "sum all numbers between 1 and 5," before thinking about an approach to solve the problem, we must first determine if the scope is properly defined. Does the word "between" imply that 1 and 5 are included in the sum? Do "numbers" include only the integer values? What about the irrational numerical values? Clearly, we observe that we cannot properly define the scope of the charge to add all numbers between 1 and 5.

Likewise, imagine you were charged to design a device that transports people from Atlanta, Georgia, to Los Angeles, California. How many people must the device transport? Does the device need to travel on land? Should it travel by air? Should it travel by water? Does the device require any human interaction?

This section covers two methods of defining a process: with **written algorithms** and with **graphical algorithms**. Both methods require your properly identifying the scope of the problem and identifying all of the necessary **input** and **output** of the process.

17.1   SCOPE

One of the most difficult steps in designing an algorithm is properly identifying the entire scope of the solution. Like solving a problem on paper involving unit conversions and equations, it's often necessary to state all of the known and unknown variables in order to determine a smart solution to the problem. If information is left out of the problem, it might be necessary to state an assumption in order to proceed with a solution. After all variables and assumptions about the problem have been identified, it is then possible to create a sequence of actions and decisions to solve the problem.

To clearly understand the scope of the problem, we often find it helpful to formally write out the known and unknown information, as well as state any assumptions necessary to solve the problem. In the following examples, notice that as the problem statements become more and more refined, the number of necessary assumptions decreases and eventually disappears.

EXAMPLE 17-1

Problem: Sum all numbers between 1 and 10.

Known:

- *The minimum value in the sum will be 1.*
- *The maximum value in the sum will be 10.*

Unknown:

- *The sum of the sequence of numbers.*

Assumptions:

- *We will only include the whole number values (e.g., 1, 2, 3, …) in the sum.*
- *The sum will include the starting value of 1 and the ending value of 10.*

EXAMPLE 17-2

Problem: Sum all numbers between (and including) 1 and 10.

Known:

- *The minimum value in the sum will be 1.*
- *The maximum value in the sum will be 10.*

Unknown:

- *The sum of the sequence of numbers.*

Assumptions:

- *We will only include the whole number values (e.g., 1, 2, 3, …) in the sum.*

EXAMPLE 17-3

Problem: Sum all whole numbers between (and including) 1 and 10.

Known:

- *The minimum value in the sum will be 1.*
- *The maximum value in the sum will be 10.*

Unknown:

- *The sum of the sequence of numbers.*

Assumptions:

- *[None]*

Comprehension Check 17-1

For the problem statement, list all knowns, unknowns, and assumptions.
Problem: Sum all even numbers between (and including) 2 and 20.

Comprehension Check 17-2

For the problem statement, list all knowns, unknowns, and assumptions.
Problem: Multiply all powers of 5 between (and including) 5 and 50.

17.2   WRITTEN ALGORITHMS

A written algorithm is a narrative set of instructions required to solve a problem. In everyday life, we encounter written algorithms in the form of oral instructions or written recipes. However, it is extremely common for humans to "fill in the blanks" on a poorly written algorithm. Imagine you are handed a strongly guarded family recipe for tacos. One of the steps in the archaic recipe is to "cook beef on low heat until done." To the veteran cook, it's apparent that this step requires cooking the prepared ground beef on a stove-top in a sauce pan for approximately 10 minutes on a burner setting of 2–3. To a first-time cook, the step is poorly defined and could result in potentially inedible taco meat.

Engineers and Written Algorithms

As an engineer, to write effective algorithms you must ensure that every step you include in a written algorithm must not be subject to misinterpretation. It is helpful to write an algorithm as if it were to be read by someone completely unfamiliar with the topic. Each step in the written algorithm should be written such that the stepwise scope is properly defined. The **stepwise scope** is all of the known and unknown information at that point in the procedure. If a step in an algorithm contains an assumption, you must formally declare it before proceeding with the next step. By ensuring that the stepwise scope is well defined, you ensure that your algorithm will not be subject to misinterpretation.

All written algorithms should be expressed sequentially. The most effective algorithms are written with many ordered steps, wherein each step contains one piece of information or procedure. While the author of an algorithm may consider each step in an algorithm to be "simple," it might not be trivial to an external interpreter. When writing an algorithm, it's helpful to assume that the reader of your algorithm can only perform small, simple tasks. Assume that your algorithm can be interpreted by a computer. A computer can execute small tasks efficiently and quickly, but unlike a human, a computer cannot fill in the blanks with information you intended the reader to assume.

Decision-making can be expressed in a written algorithm. Assume you are designing a process to determine if the value read from a temperature sensor in a vehicle indicates it is unsafe for operation. To express the decision in a written algorithm, phrase your decisions in questions that have a "yes" or "no" response.

> **NOTE**
>
> All decisions in a written algorithm must be YES or NO questions.

Format of Written Algorithms

The first step in writing any algorithm is defining the scope of the problem. After you define the scope of the problem, create an *ordered* or *bulleted list* of actions and decisions. Imagine taking an English class and writing a research report on the influence of

19th-century writers on modern-day fiction authors. Before writing the paper, you would create an outline to ensure that your topics have connectivity and flow. Just like the outline of an English paper, an algorithm is best expressed as a sequential list rather than as complete paragraphs of information.

If a decision is required in the algorithm, indent the actions to indicate the action is only associated with the particular condition.

EXAMPLE 17-4

Problem: Create an algorithm to convert a temperature from relative units [°F or °C] to the corresponding absolute units [K or °R].

Known:

- *Temperature in relative units (degrees Celsius or degrees Fahrenheit).*

Unknown:

- *Temperature in absolute units (kelvin or degrees Rankine).*

Assumptions:

- *Since the problem does not explicitly state the temperature to be converted, assume that the interpreter of the algorithm will input the temperature and units.*

Algorithm:

1. *Input the numeric value of the temperature.*
2. *Input the units of the numeric value of the temperature.*
3. *Ask if the input unit is degrees Fahrenheit.*

   a. *If yes, convert the value to degrees Rankine.*
   b. *If no, convert the value to kelvin.*

4. *Display the converted value and absolute unit.*
5. *End the process.*

EXAMPLE 17-5

Problem: Create an algorithm to calculate the sum of a sequence of whole numbers, given the upper and lower bounds of the sequence.

Known:

- *Upper bound of whole number sequence.*
- *Lower bound of whole number sequence.*

Unknown:

- *Sum of all whole numbers between the upper and lower bound.*

Assumptions:

- *Since the problem does not explicitly state the upper and lower bounds, assume the interpreter will ask for the values.*
- *Include the boundary values in the summation.*

Algorithm:

1. *Input the lower bound of the sequence.*
2. *Input the upper bound of the sequence.*
3. *If the lower bound is larger than the upper bound,*
   - a. *Warn the user that the input is invalid.*
   - b. *End the process.*
4. *If the upper bound is larger than the lower bound,*
   - a. *Create a variable to keep track of the sum (S).*
   - b. *Create a variable to keep track of the location in the sequence (I).*
   - c. *Set the initial value of S to be zero.*
   - d. *Set the initial value of I to be the lower bound.*
   - e. *If the value of I is less than or equal to the upper bound,*
      - i. *Add I to the current value of S.*
      - ii. *Add one to the current value of I.*
      - iii. *Return to step 4.e. and ask the question again.*
   - f. *If the value of I is greater than the upper bound,*
      - i. *Display the sum of the sequence (S).*
5. *End the process.*

In step 4.e.iii, we required that the interpreter return to an earlier step in the algorithm after changing the values of our variables. This allows us to create a **feedback loop** necessary to calculate the sequence of values. A feedback loop is a return to an earlier location in an algorithm with updated values of variables. It's important to note that if we failed to update the variables, the feedback loop will never terminate. A nonterminating feedback loop is also known as an **infinite feedback loop**.

Comprehension Check 17-3

Create a written algorithm to multiply all integer powers of 5, 5^x, for x between (and including) 5 and 50.

17.3 GRAPHICAL ALGORITHMS

To visualize a process, a graphical representations of algorithms is used instead of a written algorithm. A **flowchart** is a graphical representation of a written algorithm that describes the sequence of actions, decisions, and path of a process. Designing a flowchart forces the author of the algorithm to create small steps that can be quickly evaluated by the interpreter and enforces a sequence of all actions and decisions. Flowcharts are used by many different disciplines of engineering to describe different types of processes, so learning to create and interpret flowcharts is a critical skill for a young engineer. In fact, in the United States, any engineers that discover a new innovative algorithm can submit their concept to a patent office by representing the process in terms of a flowchart.

Three different shapes are used in the creation of flowcharts in this book; a number of other widely used operators are encountered across the world. In this book, we describe all actions with rectangles, all decisions with diamonds, and all connections between shapes with directional arrows.

Rules for Creating a Proper Flowchart

■ The flowchart must contain a START rectangle to designate the beginning of a process.
■ All actions must be contained within rectangles.
■ All decisions must be contained within diamonds.
■ All shapes must be connected by a one-way directional arrow.
■ The flowchart must contain an END rectangle to designate the end of a process.

Actions

Actions are any executable steps in an algorithm that do not require a decision to be made. Based on this definition, any defined variables, calculations, and input or output commands would all be contained within action rectangles.

All simple actions are contained within a single rectangle on the flowchart. For each rectangle, two arrows are always associated with the shape, with two exceptions. The inward arrow to the rectangle represents the input to the action. It is assumed that any variables defined in the stepwise scope of an action rectangle are accessible and can be used in the action. The outward arrow from the rectangle represents the output of the action. If any new variables or calculations are performed within the rectangle, those values are passed along to the next shape's stepwise scope.

■ Exception One: The START rectangle represents the beginning of the flowchart and does not contain an inward arrow. An oval shape is also commonly used to represent the start of an algorithm.
■ Exception Two: The END rectangle represents the end of the flowchart and does not contain an outward output arrow. An oval shape is also commonly used to represent the end of an algorithm.

Decisions

Decisions are any executable steps in an algorithm that require the answer to a question with "yes" or "no." All decisions in a flowchart must be represented within a diamond shape.

For each diamond on a flowchart, three arrows are always associated with the shape. The inward arrow to the diamond represents the input to the decision. It is assumed that any variables defined in the stepwise scope of a decision diamond are accessible and can be used in the decision. The two outward arrows that exit decision diamonds represent the conditional branch based on the outcome of the question asked within the diamond. If the outcome of the decision is true, the flow of the process will follow the "Yes" branch; otherwise, it will follow the "No" branch. Since no new variables are created in a decision diamond, the stepwise scope that enters the decision diamond is passed on to the next shape of each conditional branch.

EXAMPLE 17-6

Problem: Create an algorithm to convert a temperature from relative units [°F or °C] to the corresponding absolute units [K or °R].

Known:

- *Temperature in relative units (degrees Celsius or degrees Fahrenheit).*

Unknown:

- *Temperature in absolute units (kelvin or degrees Rankine).*

Assumptions:

- *Since the problem does not explicitly state the temperature to be converted, the interpreter of the algorithm will ask for the temperature and units.*

Flowchart:

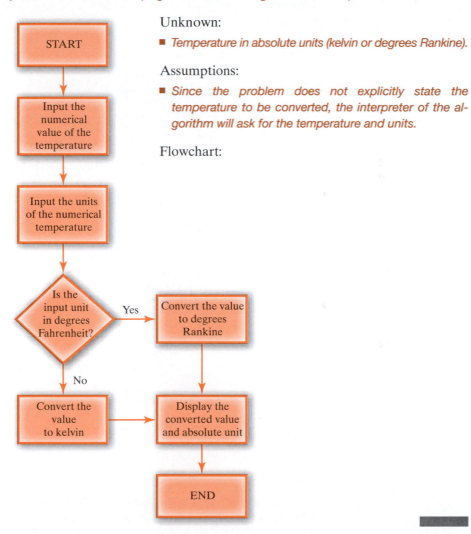

EXAMPLE 17-7

Problem: Create an algorithm to calculate the sum of a sequence of whole numbers, given the upper and lower bounds of the sequence.

Known:

- *Upper bound of whole number sequence.*
- *Lower bound of whole number sequence.*

Unknown:

- *Sum of all whole numbers between the upper and lower bound.*

Assumptions:

- *Since the problem does not explicitly state the upper and lower bounds, the interpreter will ask for the values. Include the boundary values in the summation.*

Flowchart:

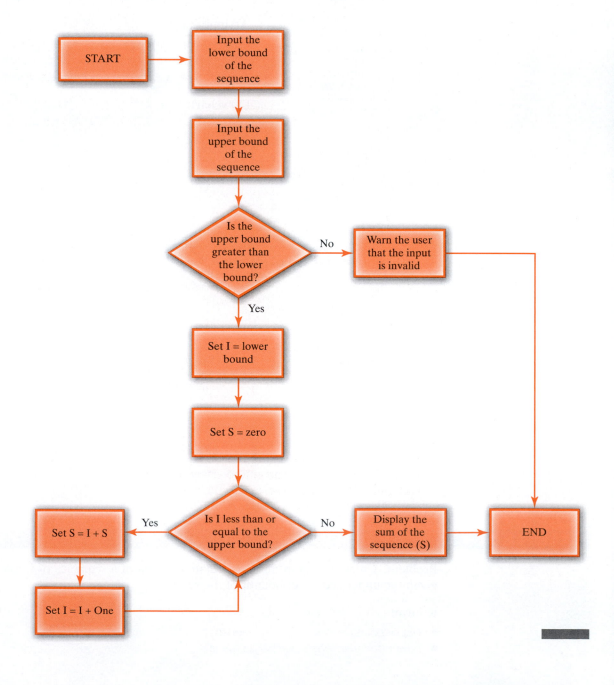

Comprehension Check 17-4

Create a graphical algorithm to multiply all integer powers of 5, 5x, for x between (and including) 5 and 50.

Flowchart Creation in Microsoft Word

Open a new Microsoft Word document. Click the **Insert** ribbon at the top of the Microsoft Word editor window. Click the **Shapes** drop-down menu in the Illustrations box. You will need to use the rectangle and diamond shapes under the Flowchart section as well as the directional arrows under the Lines section.

⌘ **Mac OS:** Access the flowchart shapes by clicking the Object Palette button near the top of the Formatting Palette, then clicking the Shapes button.

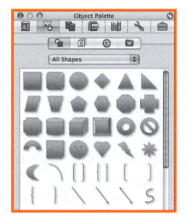

⬎ **To add an action:** Click the Rectangle tool from the Shapes menu and click-and-drag in the body of the document to draw a rectangle. Right-click the rectangle and click Add Text to add text on top of the rectangle.

⌘ **Mac OS:** Control-click or two-finger tap to access the **Add Text** menu item.

♻ **To add a decision:** Click the Diamond tool from the Shapes menu and click-and-drag in the body of the document to draw a diamond. Right-click the diamond and click Add Text to add text on top of the rectangle.

♻ **To insert the YES and NO labels,** use the Text Box option under the Basic Shapes menu. To remove the border, right-click the text box and click Format Text Box. On the Color and Lines tab, click the Color drop-down menu under the Lines section and select No Color. Click OK.

⌘ **Mac OS:** In the main menu, click **Insert > Text Box**. The default is probably "no border," but the border can be modified in the formatting palette.

♻ **To add an arrow:** Click the single direction arrow from the Shapes menu and starting from the source, click-and-drag to the destination.

17.4 PROGRAMS IN MATLAB

Programming is the process of expressing an algorithm in a language that a computer can interpret. To correctly automate a process on a computer, a programmer must be able to correctly speak the language both grammatically and semantically. In this book, we communicate with a computer by using the MATLAB programming language.

Programs are a set of instructions given to a computer to perform a specific task. In this chapter, we create MATLAB program files, or **M-Files**, to automate our processes.

Notes About Programming in General

Certain principles apply to programming, regardless of what language is being used. Entire textbooks have been written on this subject, but a few important concepts are addressed here.

> For each rectangle and diamond in a flowchart, there should be a few comments in the source code where that action occurs.

■ **Programming style:** Just like technical reports should follow a particular format to ensure they can be understood by someone reading them, all programs should have a few common elements of programming style, particularly the inclusion of comments that help identify the source of the program and what it does. For any program, a proper header should be written describing the scope of the problem, including a problem statement and definition of all input and output variables used in the program. Properly commenting the source code is also critical for ensuring that someone else can follow your work on the program. *As a rule of thumb, for each rectangle and diamond in a flowchart, there should be a few comments in the source code where that action occurs.*

■ **Program testing:** When you write a program, testing it is a critical step in making sure that it does what you expect. Lots of things can go wrong.

• *Syntax errors*: These violate the spelling and grammar rules of the programming language. Compilers (programs that interpret your computer program) generally identify their location and nature.

- *Runtime errors*: These occur when an inappropriate expression is evaluated during program execution. Such errors may occur only under certain circumstances. Examples of runtime errors are division by zero, the logarithm of zero, or the wrong kind input, such as a user's entering letters when numbers are expected. Computer programs should anticipate runtime errors and alert users to the error (rather than having the program unexpectedly terminate).

- *Formula coding errors*: It is not uncommon to enter a formula incorrectly in a form that does not create either a syntax or runtime error. Evaluating even a few simple calculations by hand and comparing them to the values calculated by the program will usually reveal this kind of error.

- *Formula derivation errors*: This difficult error to discover in programming occurs when the computer is programmed correctly but the programming is based on a faulty conceptualization. This kind of error is difficult to figure out because programmers generally assume that a program that executes without generating any errors is correctly written. This situation underscores the importance of checking a solution strategy before starting to write a computer program.

■ **Keeping track of units:** Some computer programs can be designed to have the user keep track of units as long as they use consistent units. Other computer programs require input data to be in a particular set of units. Regardless of which approach you use, the way in which units are managed must be clear to the user of the program and in the program's comments.

Why are units important? Read the LESSONS OF THE MARS CLIMATE ORBITER from page 246.

■ **Precision:** All computers have finite precision, the maximum number of significant digits they can manage. Some computer programs further restrict the precision. The issue of finite precision affects the possibility of two numbers being exactly equal. As a result, it is always necessary to consider how close two numbers must be in order to be considered equal rather than test for equality.

■ **Numerical vs. symbolic solutions:** In algebra, we learn that the equal sign means that whatever is on the left side of the equation identically equals whatever is on the right. This is known as symbolic manipulation. Some computer programs (e.g., Maple and MathCAD) specialize in manipulating symbolic expressions. Although MATLAB can manipulate symbolic expressions, it is primarily used for its numerical capabilities. Numerical computer packages and computer programming languages in general use the equal sign to assign the right-hand value to the variable on the left-hand side. The expression on the right of the "=" is evaluated, and then placed in the variable on the left.

■ **Documentation:** In any programming language, there is usually a special character or pair of characters that tell the program interpreter to ignore everything to the right of the special character to allow the programmer to write human readable notes within the code of the program. These notes are commonly referred to as "comments" within the code. A great deal of controversy surrounds how much documentation is necessary in any program, but this text proposes and follows the following guideline to properly document a program:

For every numbered item in a written algorithm, you should include that item as a comment in your code.

For every numbered item in a written algorithm, you should include that item as a comment in your code.

In addition, at the top of each program, it is extremely helpful to at least write a problem statement and document any variables used within the program. In a function, it is helpful to cluster your description of variables into three groups: input variables, output variables, and function variables.

Notes Particular to MATLAB

Before proceeding with the material on MATLAB, it would be wise to review the material on matrices in Appendix A. Since variables in MATLAB are inherently matrices, it is critical that you understand not only some basic matrix concepts but also the specific notation used in this text (as well as MATLAB) when referring to matrices.

Certain details specific to MATLAB are important to note early on.

■ **The MATLAB interface:** The MATLAB program window has a number of subwindows. We are not concerned with the Launch Pad or Command History windows, but two windows are of interest.

clc = clear the
Command Window

• The *Command Window* can be used as a powerful but awkward calculator. The button at the top-right side of the Command Window menu bar will "undock" the Command Window from the MATLAB program window. The ✕ button will close the Command Window. To restore the Command Window, select **Desktop > Command Window**. To clear any text from the Command Window, enter the special reserved MATLAB expression `clc` in the Command Window.

Word files = .docx
Excel files = .xlsx
Powerpoint
  files = .pptx
Matlab files = .m

• Although you can enter MATLAB statements in the Command Window, the *Editor/Debugger window* is the interface of choice for developing MATLAB programs. The editor stores your commands in M-files. This permanent record of your commands allows you to make small changes to either the program or inputs and to re-execute a complicated set of commands. The editor also assists in the formatting of MATLAB programs, using both indentation and color to distinguish program elements.

■ **Naming programs, functions, and variables in MATLAB:** In MATLAB, program names, function names, and variable names must all follow certain rules to prevent unintended side effects:

• All names must consist only of letters, numbers, and the underscore character (displayed by holding "Shift" and typing a hyphen).
• Names can *only* begin with alphabetic letters (no numbers or special characters).
• Names cannot be longer than 63 characters.
• Names in MATLAB are case-sensitive: `Frog` and `frog` are different.
• Names cannot have the same name as any other identifier (program name, function name, built-in function, built-in constant, reserved word, etc.), because MATLAB's order of execution will prevent the program from running in the intended order.

Comprehension Check 17-5

Which of the following are valid MATLAB variable names?

(a) `my name`   **(d)** `m`   **(g)** `my_var`   **(j)** `for`
(b) `length`   **(e)** `m6`   **(h)** `@clemson`   **(k)** `HELLO[]`
(c) `MyLength`   **(f)** `4m`   **(i)** `my.variable`

Order of Execution:
- Variable
- Built-In Function
- Program or Function in Current Directory
- Program or Function in Current Path

- **Order of execution:** Order of execution is the sequence MATLAB goes through when you type a name in either the Command Window or in a program or function. If you type `Frog` in the Command Window,

  1. MATLAB checks if `Frog` is a variable, and displays its value if it is.
  2. MATLAB checks if `Frog` is one of MATLAB's built-in functions, and executes that function if it is.
  3. MATLAB checks if `Frog` is a program or function in its current directory, and executes the program or function if it is.
  4. MATLAB checks if `Frog` is a program anywhere in its path, and executes the program or function if it is. The location where you store your MATLAB programs or functions must be included in MATLAB's path. You can check where MATLAB is looking for programs and functions by typing `path` in the Command Window. Any M-files in the directories listed will be visible to MATLAB. If you try to execute a M-file outside any of the directories listed, MATLAB will ask if you want to switch current directories or add that directory to the list of paths MATLAB can see.

- **`help` and `lookfor`:** In addition to a standard Windows help utility, the first line of MATLAB's help files can be searched for a keyword with the `lookfor` command. To display the entire contents of the help documentation available for a function, type `help` followed by a space and the function name. It is recommended that both `lookfor` and `help` be used *only* in the Command Window rather than those calls being incorporated in an M-file.

clear = clear workspace variables

- **The Workspace:** MATLAB's Workspace is the place where all of the variables are visible to the user or where programs are located. In the main interface of MATLAB, all variables created by the user directly in the Command Window or created as a result of executing a program are stored and displayed in the Workspace tab. To clear all the variables from the workspace, type the command `clear` into the Command Window. To remove a specific variable from the workspace, type the command `clear` followed by a space and the name of the variable.

Local variables are only visible to the program or function using them.

Global variables are visible within any program or function.

- One important fact to note about variables is that the variables in the workspace are only available within MATLAB's Command Window and to any running program. Functions, explained later, create a private workspace and cannot access any of the variables available in MATLAB's main workspace. The only way to access any variables in MATLAB's main workspace is to "pass" the variable to the program as function input. The phenomenon of having "main workspace" variables inaccessible by functions is because functions exist within a different scope from the programs. The variables created within the "private workspace" of a function are commonly referred to as **local variables**, since they are destroyed after the function executes and do not appear in the "main workspace" unless they're passed back as function output. Special constants like "pi" exist within the scope of programs. Variables that are visible within any scope are commonly referred to as **global variables**.

  In the diagram below, consider the two inner boxes to be programs (Prgm1) or functions (Func1) and the outer box represents all of the other variables in MATLAB's Workspace. Since we do not include any commands to clear the variables from the workspace at the top of the program Prgm1, the program is able to "see" and use all of the variables within MATLAB's workspace. Additionally, it will create a new variable in MATLAB's workspace (`c`) after we execute program Prgm1.

  In contrast, the function can't "see" or use any of the variables in MATLAB's Workspace and creates its own separate workspace to execute the code inside the function. In order to run the function, the programmer must provide the values of variables e and g by "passing" the variables into the function, the code calculates the value d, and returns d as function output. Assume that we store the output of the function in the variable h when the function returns to the MATLAB workspace. After the function has executed, the separate workspace is destroyed, along with variables d and k, which were never stored in MATLAB's main workspace.

what = display a list of programs in current directory.

who = display a list of variables in current workspace.

- **what** and who: To display a list of all MATLAB programs in the current directory, type what into the Command window. To display all variables currently in MATLAB's workspace, type who in the Command Window. Remember that all variables automatically clear after you exit MATLAB.
- **Special constants:** In addition to the names of special functions, MATLAB reserves certain names as special constants, such as pi. To search for a special constant, use the lookfor command.

% = comment

- **Comments:** For documenting or commenting any code in MATLAB, a special character is used to "comment" out text: the percent symbol (%). The comment character can appear anywhere in a MATLAB program or function, but preferably either at the beginning of a line or middle of a line of MATLAB code. Everything after the percent sign to the end of the line is ignored by MATLAB, so you as a programmer can "comment out" any broken lines of code during code testing in addition to providing comments. Program or function file headers will usually have 6–10 lines of comments at the top of every program and more throughout the rest of the program or function.

EXAMPLE 17-8

We want to create a MATLAB program to convert Cartesian coordinates (x, y) to polar coordinates (r, θ). Before we can even attempt to write the program, we need to recall some information about the coordinate systems.

We know the following:

- In the Cartesian coordinate system, the coordinates (x, y) represent how "far away" a particular data point is from the origin (0, 0) by discussing the horizontal and vertical distance with respect to the origin. The x-value represents the horizontal distance and the y-value represents the vertical distance.
- In the polar coordinate system, the coordinates (r, θ) represent the exact distance from the point to the origin (r) as well as the angle of elevation (θ) with respect to the origin (0, 0) in units of radians.

- The conversion from polar coordinates to Cartesian coordinates is

$$x = r \cos(\theta)$$
$$y = r \sin(\theta)$$

where θ the angle of elevation in radians.

- The conversion from Cartesian coordinates to polar coordinates is

$$r = \sqrt{x^2 + y^2} \qquad \theta = \tan^{-1}\left(\frac{y}{x}\right)$$

We have the equations necessary to write a program to convert Cartesian coordinates to polar coordinates, and vice versa. Now, we need to ask MATLAB for more information on doing things like calculating a square root and calculating an inverse tangent of a value.

In MATLAB's Command Window, we search for the built-in functions to calculate a square root and an inverse tangent. After we type `lookfor square root` into the Command Window, we notice that MATLAB gives us an error message. With the `lookfor` command, MATLAB is expecting us to provide only one keyword, so we decide try our search again, this time including only the word "square."

Because MATLAB is searching through every built-in function, we need a way to tell MATLAB to stop, because when the function `sqrt` appears, it seems to be the exact function we need. To terminate a MATLAB command early, we press CTRL + C (that is, the CTRL key and the "C" key at the same time) on our keyboard, and the prompt will return to the Command Window. Control + C can also be used to stop a program caught in an infinite loop or that otherwise wandered away and got lost.

To ask for more information on the function `sqrt`, we type `help sqrt` to double-check that the function is going to perform our intended calculation.

> CTRL + C = stops MATLAB from executing the current task.

Repeating this process to find the inverse tangent function, we determine that `atan` is the correct function necessary to calculate the inverse tangent. We note, from the help documentation, that the `atan` function will return the inverse tangent in radians.

To create a program, we click the New M-File button on MATLAB's main screen.

After creating a new blank M-file, we need to write out our problem statement and start keeping track of any variables we create within the program.

Next, we start writing our code. In a MATLAB program, it is good practice to start the program by clearing all of the variables from the workspace (`clear`) and clearing the Command Window (`clc`). That way, we guarantee that if we run our program on a different computer or run it after restarting MATLAB, our program will create all the variables we need to solve the problem and not rely on the workspace having variables predefined before code execution.

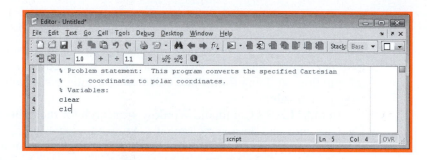

Now we're ready to start our program. Let's look at the step-by-step written algorithm necessary to solve this problem:

1. Create a variable to contain the x coordinate as a variable named "x".
2. Create a variable to contain the y coordinate as a variable named "y".
3. Calculate the radius by taking the square root of the sum of the squares of x and y and store the result in a new variable called "r".
4. Calculate the angle by taking the inverse tangent of the result of dividing y by x and store the result in a new variable called "Theta".

In the written algorithm, the first thing we have to do is to manually assign the values of x and y. For this example, we assign x to contain the value of 5 and y to contain the value of 10, but in later chapters, we show how the user of a program can enter values.

Note that MATLAB highlights the = symbol in the editor. When we move the cursor over the assignment operator (=), a pop-up message tells us that we can suppress the output of the assignment expression by using a semicolon. For now, we omit the semicolon and observe MATLAB's behavior when we leave the code unsuppressed.

Next, we type the equations to calculate the radius and the angle, using the built-in MATLAB functions we discovered with `lookup` and `help`.

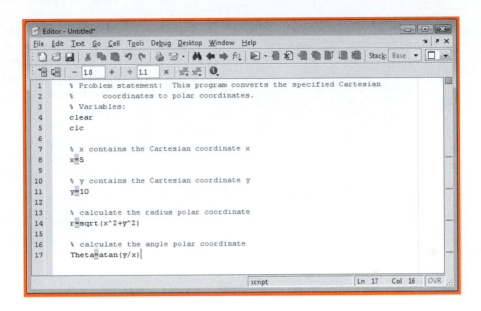

Next, we save the program in MATLAB's path with a file name that follows the program-naming convention. We choose the name "CartesianToPolar" since that name nicely describes the purpose of the program.

To save a program, we click the Save icon in the Editor window.

Alternatively, we can go to **File > Save** in the Editor menu. When the **Save File As** window appears, we type the name of the program or function and click **Save**.

⌘ **Mac OS:** MATLAB is one of the few programs that does *not* modify the main menu at the top of the screen. Instead, its main menu is at the top of the MATLAB window, just as in the Windows implementation. This is because MATLAB is actually running under UNIX®.

CAUTION

Remember, MATLAB programs and functions can *not* have any spaces in their file name, so be sure that a file is named without spaces, for example, CartesianToPolar.m— *not* "Cartesian To Polar.m".

Running the Program

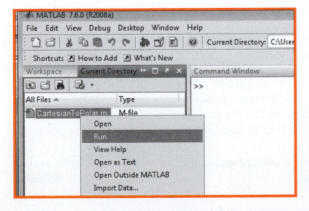

There are a number of different ways to run a MATLAB program located within MATLAB's path.

Method 1: In the main MATLAB window (not the Editor,) type the name of the program in the Command Window and press **Enter**.

Method 2: In the main MATLAB window (not the Editor,) right-click the program you want to run while in the Current Directory box and select Run.

⌘ **Mac OS:** Control-click or two-finger tap.

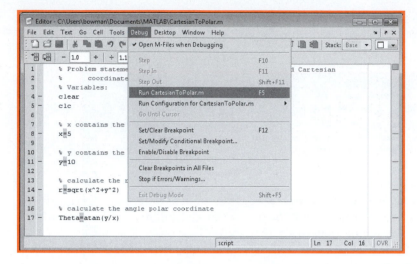

Method 3: With the program loaded in the Editor window, go to the Debug menu and click the option Run *file name*.

Method 4: With the program loaded in the Editor window, click the Save and Run button in the toolbar.

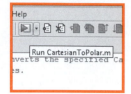

Method 5: With the program loaded in the Editor window, click the F5 key on the keyboard to automatically save any changes to the file and run the program.

⌘ **Mac OS:** This option is not available because the function keys have different uses.

Regardless of which method you choose to run the program, the output from the program will appear in the Command Window.

> To supress output, end the code line with a semicolon (;).

We see that MATLAB spits out the results of every assignment and calculation performed in the program. Let's assume we want to show only the output from the calculations of r and Theta, and not the output of the assignment of x and y. To do this, we add a semicolon (;) at the end of each line of code we do not want showing up in the output of the program.

Now when we save and run our modified program, we will see only the output from the calculations of r and Theta.

Programming Features in Our Coverage of MATLAB

MATLAB has many advanced programming features that make sophisticated numerical analysis, including simulation, possible. In this book, you will learn basic programming structures to solve engineering analysis problems.

- **Arrays.** These offer the opportunity to store data in their natural structure: one-dimensional arrays for vectors, two-dimensional arrays for matrices and paired data, and so on. Arrays are also more efficient for data storage.
- **Input and output.** By gathering input, the program can solve a different problem each time the program is executed. Students can also produce formatted output to communicate with the user.
- **Conditional statements.** These allow a program to take different paths for various user input, intermediate calculations, or final results. Conditional statements allow the programmer to ask questions in order to make decisions.
- **Looping.** Repetition is critical in breaking down complicated problems into many simple calculations.
- **Plotting.** Graphs communicate data relationships in a variety of forms.

Data and Program Structures

MATLAB actually views all data structures as arrays—we discuss the importance and the implications of this later. For now, consider the following data and program structures and how they are used.

Numbers

Numbers are also referred to as scalars, as opposed to vectors. There are many number types—integers and floating-point numbers of different levels of precision. Double-precision floating-point numbers (16 digits stored, four of which are displayed) are the default. The display format can be manipulated independently of what kind of number is stored—for example, integers can be displayed to a specified number of decimal, or floating-point numbers can be displayed in scientific notation.

Comprehension Check 17-6

Type the following expressions into the MATLAB Command Window and observe how MATLAB stores the variables in the workspace:

```
A=1          B=1.2        C=1e-10      D=7i+2        E=2;
```

What effect does the semicolon have on the expression?

Text

Strings of text, or character strings (including letters, numbers, and other special characters such as a question mark) can also be stored. Text strings must be placed in quotes to avoid being mistaken for variable or function names (or causing errors where the strings have no definition as a variable or function). For example, `Name = 'Jonas Doe'` will store the text in quotes in the variable `Name`. Note that single quotes are used.

Comprehension Check 17-7

Type the following expressions into the MATLAB Command Window and observe how MATLAB stores the variables in the workspace:

```
A='Hello'       B='Hello World!'          C='It''s 3:00 PM'
```

What effect did two single quotes in the expression have on the output?

Arrays

For more information on matrices, refer to Appendix A.

Also referred to as matrices, arrays are used to store blocks of numbers or text. When a row is entered, spaces or commas separate columns. For example, the row vector `[1 2 3]` can be entered as it appears or as `[1, 2, 3]`. Similarly, use a semicolon or the Enter key to separate columns. For example, you can enter the column vector

$$\begin{bmatrix} 1 \\ 2 \\ 3 \end{bmatrix}$$ as `[1; 2; 3]` or [1<ENTER>2<ENTER>3].

Comprehension Check 17-8

Type the following expressions into MATLAB and observe how MATLAB stores the variables in the workspace:

```
A = [1 2 3]         B=[1,2,3]        C=[2;4;6]       D=[2:6]

E = [1 2 3; 4 5 6]  F=[1,2,3;4,5,6]  G=[1:3;4,5,6]   H=[1:3;4:6]
```

What does the colon operator do if it is used inside square brackets?
What does the comma do if it is used inside square brackets?
What is the difference between a blank space and the comma inside square brackets?

Programs

A program is a series of executable MATLAB statements in a special text file with the suffix ".m" instead of ".txt" or ".docx" or ".xlsx", and so forth. Files with the ".m" suffix are commonly referred to in the text as "M-files." The statements in an M-file are executed as if each line were typed into the MATLAB Command Window.

Variables

Special placeholders, called variables, inside MATLAB contain information like text and numbers. MATLAB has built-in variables such as `pi`, so take care to properly name any variable you use in MATLAB. Also note that you should *not* place a value into such variables (e.g., do not write `pi=3.14`).

Built-in Functions

MATLAB has many built-in functions, some of which you will recognize immediately: `sqrt`, `abs`, `log`, `sin`, `cos`, `tan`, etc. These common functions each have a single argument and a single returned value and are called in the traditional manner: `sqrt(d)`, `abs(t)`, `sin(x)`. Some built-in functions act on multiple arguments, such as `plot(x,y)`—discussed in detail later—and some return multiple output variables.

Matrix Operations in MATLAB

Review the sections on matrices in Appendix A, particularly if some of the following reads like an extraterrestrial language to you.

One of the most powerful features of MATLAB is its ability to directly manipulate matrices. Many things that MATLAB can do with a single statement would require numerous lines of code, nested loops, etc., in most other computer languages. A few of these matrix operations are shown below.

- *Assigning* part of an array to another array
  - `a=b(2,:)` puts the second row of `b` into `a`.
  - `c=d(:,3:5)` puts the third through fifth columns of `d` into `c`.
- *Transposing* matrices—converting rows to columns, and vice versa (the "glitch" (') or single quote operator means transpose)
  - $[1; 2; 3]' = [1\ 2\ 3]$
  - $[0\ 8\ 6; 3\ 5\ 7]' = [0\ 3; 8\ 5; 6\ 7]$
- *Incrementing* a vector or an array (increasing all terms by a scalar)
  - If `TempC` is an array of Celsius temperatures, `TempK = TempC + 273.15` converts all array elements to units of kelvin
- *Scaling* a vector or an array (multiplying all terms by a scalar)
  - If `LengthInches` is an array of lengths in inches, `LengthCentimeters = LengthInches * 2.54`.
- *Adding* and *subtracting* arrays
  - $[1\ 2\ 3; 4\ 5\ 6] - [8\ 7\ 6; 4\ 2\ 0] = [-7\ -5\ -3; 0\ 3\ 6]$
  - If `A` is a vector distance traveled on day 1 and `B` is a vector distance traveled on day 2, then the vector addition `A + B` is the total vector distance traveled.
- *Repeating* a basic operation on every element of the array
  - `sqrt(A)` takes the square root of all elements of array `A`.
- *Operating term-by-term*, as when array operators (preceded by a period, as in `.*`) cause elements of one array to operate on the corresponding elements in another array
  - $[1\ 3; 5\ 8] .* [0\ 1; 4\ 2] = [0\ 3; 20\ 16]$
  - If `distance` is a vector of distances traveled by a bicyclist in each leg of a cycling event and `speed` is a vector of the speed the cyclist travels for each leg, then `time = distance ./ speed` computes a vector that contains the time it takes the cyclist to complete each leg.
  - If `accounts` is a vector of the current balance of a number of accounts and `rates` is a vector of the interest rates paid to each of those accounts, then `interest = accounts .* rates` calculates the interest accrued to each account.
  - If `radius` is a vector of circle radii, then `area = pi * radius .^ 2` is a vector of the circle areas.

> **NOTE**
>
> Some MATLAB operations do *not* perform element-by-element operations if the dot is omitted, but instead perform matrix operations. For example, `A * B` multiplies array `A` by array `B` using matrix multiplication. This is very different from the element-by-element multiply of `A .* B`. See Appendix A for more details.

> **ARGUMENTS**
>
> In computer programming, and often in math as well, the word "argument" does *not* refer to an altercation or vehement disagreement. In this context, it means the information that is given to a function as input to be processed. If you use the function `sqrt` to find the square root of 49, you would write `sqrt(49)`: 49 is the argument of sqrt—the value upon which the `sqrt` function performs its calculations. Functions may have zero arguments (e.g., `rand` will calculate a single random number between 0 and 1—no input is required), one argument (e.g., `sqrt(4)`), or two or more arguments (e.g., `power(a,b)` raises `a` to the power `b`).

17.5 | DEBUGGING MATLAB CODE

If a program has syntax errors, MATLAB (and other compilers) will give you useful feedback regarding the type and location of the error to help you fix, or **debug**, your program. If this information is not sufficient for you to understand the problem, check with others who might have seen this error previously—some errors are particularly common. If you still cannot identify the error, you might need to use a more formal process to study the error. For other kinds of errors, this formal process is essential for diagnosing the problems. When you first write a program, you must test the output of the program with a set of inputs for which you know result. Such test cases would include:

- Simple cases for which you can quickly compute the expected output;
- Cases provided by an instructor or textbook with a published solution;
- Cases for which results have already been produced by a previously tested program that does the same thing;
- Test cases that are customarily used to test programs.

When the output of a program is different from what you expect, the debugging process begins. It is common to debug shorter programs simply by reading them and writing a few notes. Longer programs may require the use of the MATLAB Debugger, which is available with the MATLAB Editor. Whether you are using the MATLAB Debugger or are debugging "by hand" or in the Command Window, the same techniques apply.

Why do we say that computers (and programs) have bugs? Grace Murray Hopper is quoted in the April 16, 1984, issue of Time Magazine as saying, "From then on, when anything went wrong with a computer, we said it had bugs in it"—referring to when a 2-inch-long moth was removed from an experimental computer at Harvard in 1947.

Preparing for Debugging

When using the Editor/Debugger, open the program for debugging. If it is open, make sure changes are saved—MATLAB will run the saved version without including recent changes. If you are debugging from a printed program and output, make sure that the printed output came from the version you are reading. In preparing to debug a program, it is critical to be able to reproduce the conditions that caused the bug in the first place. If the bug occurs in processing data from a large data set, you may have to split the data set to find the specific data that trigger the bug.

Setting Breakpoints

A breakpoint stops the program at the specified location to allow you to examine the content before the program continues.

When using the Editor/Debugger, establish **breakpoints** that allow you to check your agreement with the program at various stages. A breakpoint stops the program at the specified location to allow you to examine the contents of variables, etc., before the program continues. This will help you find the specific location when the program does something you weren't expecting. When debugging shorter programs, write down everything the program does or make all program results display in the Command Window by removing semicolons that were used to suppress output to the Command Window.

MATLAB has three types of breakpoints—standard (set by location in the program), conditional (triggered at a specified location by specified conditions), and error (triggered by a particular error). We focus on standard breakpoints.

Breakpoints are normally shown in red. If they are shown in grey, either the file has not been saved since changes were made to it or there is a syntax error in the file. All breakpoints remain in a file until you clear (remove) them or until they are automatically cleared.

Clear a breakpoint after determining that a line of code is not causing a problem. You can conduct limited testing without setting a breakpoint by using the Go Until Cursor menu selection. When `clear name` or `clear all` appears as a statement in an M-file that you are debugging, it clears the breakpoints.

Stepping Through a Program

When using the Editor/Debugger, you must start by running the program. You can choose to step through the program one line at a time or have it continue until a breakpoint is encountered. A green arrow will point to the program line where execution will begin if you elect to continue the program. When stepping through a program, you can choose to step through functions line by line or to simply return their results immediately.

Examining Values

When the program is paused at a breakpoint or when you are stepping through the program a line at a time, you can view the value of any variable to see whether a line of code has produced the expected result. If the result is as expected, continue running or step to the next line. If the result is not as expected, then that line, or a previous line, contains an error. You can examine the value of a variable in the Workspace window in the main MATLAB program window or by holding the mouse over the variable in the Editor/Debugger. It is also possible (again, when the program is paused) to select either a variable or an expression and select "Evaluate Selection" from the context menu.

Correcting Problems and Ending Debugging

If a problem is difficult to diagnose, one method to help define or correct problems is to change the value of a variable in a paused program to see if continuing with the new value produces expected results. A new value can be assigned through the Command Window, Workspace browser, or Array Editor. Do not make changes to an M-file while MATLAB is in debug mode—it is best to exit debug mode before editing an M-file. If you edit an M-file while in debug mode, you can get unexpected results when you run the file.

17.6 FUNCTIONS IN MATLAB

Functional programming allows the designer of an algorithm to recycle a segment of code to reduce the amount of redundant code. Decreasing code redundancy allows for cleaner source code that easier to read. **Functions** are a special class of programs that require the user to pass in input variables and capture output variables.

User-defined functions are similar to programs in that they contain a series of executable MATLAB statements in M-files that exist independently of any program we create in MATLAB. To allow any MATLAB program to call a user-defined function, we must follow a few guidelines to ensure that our function will run properly.

Function Creation Guidelines

■ The first line in a function must be a function definition line:

```
function [output_variables] = function_name(input_variables)
```
■ `output_variables`: A comma-separated list of output variables in square brackets
■ `input_variables`: A comma-separated list of input variables in parentheses
■ `function_name`: Function names cannot be the same as a built-in function, user-created variables, or any reserved word in MATLAB. Avoid creating variables with the same name as a user-defined function because you will not be able to call your function.

■ **The name of the function must be the same as the file name**. For example, if the function name is `sphereVolume`, the function must be saved in an M-file named `sphereVolume.m`, where `.m` is just the extension of the M-file. In addition to the aforementioned function-naming rules, the name of the function must also follow all program-naming rules, including omitting spaces in the file name.
■ **CAUTION**: To run a function, you must call it from MATLAB's Command Window by passing in input variables or properly calling it from another program or function. Running the function using the "Run" commands in the Editor/Debugger window will report errors to the Command Window.
■ Do *not* run the `clear` command after the function header because it will clear the input variables. The `clear` and `clc` commands should only appear at the top of programs, not functions!

Function Structure and Use

The following examples demonstrate the structure of different functions and how MATLAB programs can use the functions to reduce code redundancy.

EXAMPLE 17-9

Assume we are required to create a function, `areaCircle`, to calculate the area of a circle given the radius. The function should accept one input (radius) and return one output (area). Furthermore, we are told to suppress all calculations within the function.

Source Code

```
% Problem Statement: This function calculates the area of a circle
% Input:
%   r - Radius [any length unit]
% Output:
%   A - Area [units of input^2]
function [A] = areaCircle(r)
% Calculation of the area of a circle
A = pi*r^2;
```

Typical Usage (can be typed in Command Window or can be a separate program):

You can call the function by typing the name of the function and passing in a value for the radius. Note that the result is stored in a variable called "`ans`" when you don't specify a variable to capture the output.

```
>> areaCircle(3)
ans =
   28.2743
```

You can call the function by typing the name of the function and passing in a variable.

```
>> Rad=4
Rad =
      4
>> CircleA=areaCircle(Rad)
CircleA =
   50.2655
```

Comprehension Check 17-9

What is the output when you "pass in" the value 10 to the function in Exercise 17-9 above?

Assume you want to store the result of the previous function call in a variable called `Dogs`. What is the command you would type in MATLAB?

EXAMPLE 17-10

Assume we are required to create a function, `volumeCylinder`, to calculate the volume of a cylinder given the radius. The function should accept two inputs (radius, height) and return one output (volume). For this example, you may assume that the user will input the radius and height in the same units. Any calculations in the functions must be suppressed.

Source Code

```
% Problem Statement: This function calculates the volume of a cylinder.
% This function returns the volume in (units of the input variables)^3
% Input:
%   r - Radius [any length unit]
%   H - Height [same length unit as radius]
% Output:
%   V - Volume [units of input^3]
function [V] = volumeCylinder(r,H)
% Calculation of the volume of a cylinder
V = pi*H*r^2;
```

Typical Usage (can be typed in Command Window or can be a separate program):

```
>> Vol1=volumeCylinder(1,2)
Vol1 =
   6.2832
>> Vol2=volumeCylinder(2,1)
Vol2 =
   12.5664
```

Note that the order of arguments passed into the function matters. The order of variables expected by the function is based on the order they're listed in the function header. Therefore, the first input argument is the radius, followed by the height.

Comprehension Check 17-10

What is the output when you "pass in" the values 10 and 12 to the function in Exercise 17-10 above?

Assume you want to store the result of the previous function call in a variable called Cats. What is the command you would type in MATLAB??

EXAMPLE 17-11

Assume we are required to create a function, circleCalculations, to calculate the area and perimeter of a circle given the radius. The function should accept one input (radius) and return two outputs (area, perimeter). Any calculations in the functions must be suppressed.

Source Code

```
% Problem Statement: This function calculates the area and perimeter
of a circle.
% Input:
%   r - Radius [any length unit]
% Output:
%   A - Area [units of input^2]
%   P - Perimeter [units of input]
function [A,P] = circleCalculations(r)
% Calculation of the area of a circle
A = pi*r^2;
% Calculation of the perimeter of a circle
P = 2*pi*r;
```

Typical Usage (can be typed in Command Window or can be a separate program):

```
>> C=circleCalculations(5)
C =
   78.5398
>> [Area,Perimeter]=circleCalculations(5)
Area =
   78.5398
Perimeter =
   31.4159
```

Comprehension Check 17-11

What is the output when you "pass in" the value 10 to the function in Exercise 17-11 above?

EXAMPLE 17-12

Assume we are required to create a function, cylinderCalculations, to calculate the volume and lateral surface area of a cylinder given the radius and height. The function should accept two inputs (radius, height) and return two outputs (volume, surface area). For this example, you may assume that the user will input the radius and height in the same units. Any calculations in the functions must be suppressed.

Source Code

```
% Problem Statement: This function calculates the volume and surface
area of a cylinder.
% Input:
%    r - Radius [any length unit]
%    H - Height [same length unit as radius]
% Output:
%    V - Volume [units of input^3]
%    A - Surface Area [units of input^2]
function [V,A] = cylinderCalculations(r,H)
% Calculation of the volume of cylinder
V = pi*H*r^2;
% Calculation of the surface area of a cylinder
A = 2*pi*r*H;
```

Typical Usage (can be typed in Command Window or can be a separate program):

```
>> [Volume,Surface]=cylinderCalculations(3,2)
Volume =
   56.5487
Surface =
   37.6991
```

Comprehension Check 17-12

What is the output when you "pass in" the values 10 and 12 to the function in Exercise 17-12 above?

In-Class Activities

ICA 17-1

Create an algorithm (written and/or flowchart as specified by your instructor) that describes the steps necessary to create a peanut butter sandwich. You may assume that you are starting with a loaf of bread, jar of peanut butter, a knife, and a plate on the table in front of you. When you are finished, hand your algorithm to the instructor and wait for further instruction.

ICA 17-2

Create an algorithm (written and/or flowchart as specified by your instructor) that describes the steps necessary to create a paper airplane. You may assume that you are starting with a single sheet of $8\frac{1}{2} \times 11$ inch paper. When you are finished, hand your algorithm to the instructor and wait for further instruction.

ICA 17-3

Your instructor will provide you with a picture of a structure created using K'Nex™ pieces. Create an algorithm (written and/or flowchart as specified by your instructor) to recreate the picture. When you are finished, hand your algorithm to the instructor and wait for further instruction.

ICA 17-4

Create an algorithm (written and/or flowchart as specified by your instructor) to determine the factorial value of an input integer between 1 and 10. Add a statement to check for input range.

ICA 17-5

Create an algorithm (written and/or flowchart as specified by your instructor) to cook your favorite meal, including a starting materials list.

ICA 17-6

Create an algorithm (written and/or flowchart as specified by your instructor) to walk from your home or dorm to your engineering classroom without using any street names.

ICA 17-7

Create an algorithm (written and/or flowchart as specified by your instructor) that will repeatedly ask the user to draw a card from a deck of cards until the user pulls out the queen of hearts.

ICA 17-8

Create an algorithm (written and/or flowchart as specified by your instructor) to model the decision of a driver after looking at the color of a traffic light. You may assume that the traffic light contains three bulbs (red, yellow, green) to represent stop, slow down, and go.

ICA 17-9

Create an algorithm (written and/or flowchart as specified by your instructor) to determine whether a given altitude [meters] is in the troposphere, lower stratosphere, or the upper stratosphere. In addition, your algorithm should calculate and report the resulting temperature [°C] and pressure [kPa]. Refer to atmosphere model provided by NASA: http://www.grc.nasa.gov/WWW/K-12/airplane/atmosmet.html.

ICA 17-10

Create an algorithm (written and/or flowchart as specified by your instructor) to determine whether a given Mach number is subsonic, transonic, supersonic, or hyper sonic. Refer to the NASA page on Mach number: http://www.grc.nasa.gov/WWW/K-12/airplane/mach.html.

ICA 17-11

Do the following scripted exercise. The $\gg$ prompt indicates what you should type into the MATLAB Command Window.

>> clc	Press enter (this clears the screen ONLY).
>> 30/10	Press enter (this serves just as a calculator).
>> r = 30/10	Press enter and note the output.
>> r	Press enter (the variable "r" has now been assigned a value).
>> s = 10*r	Press enter (now "s" is defined as well).
>> v = sin(s)	Press enter (s is defined, it can be used to compute v). Why isn't sin(30)=0.5?
>> r	Press enter (r still has its original value).
>> clear r	Press enter.
>> r	Press enter (r is undefined; "clear r" deleted the variable r).
>> s	Press enter (we still have a value for s).
>> r = s;	Press enter (semicolon suppresses the display).
>> r	Press enter (the value of r was assigned even though display was suppressed).
>> clear	Press enter (this clears *all* variables).
>> help elfun	Gives elementary function list.
>> help sin	Provides info about sine function.
>> lookfor sine	The lookfor command will help you find things when you don't know what they might be called. Wait until MATLAB finds a few examples and then hold down the CTRL key and type the letter C to stop this. The program continue to search the documentation for the word "sine" unless we stop it.

ICA 17-12

Write a function that, given any two 2×2 arrays (**A** and **B**) as arguments, returns the product matrix $\mathbf{C} = \mathbf{AB}$. The function is not expected to work properly for any other matrix dimensions. You must calculate each term in **C** individually from terms in **A** and **B**—you may not use MATLAB's ability to multiply matrices directly. This function should feature:

- Standard header
- Calculations suppressed
- Solution matrix C printed in its entirety at the end.

Use the following values to test the program: A = [1 1; 2 2] and B = [3 3; 4 4]

ICA 17-13

Write a function that, given any two 2×2 arrays (**A** and **B**) as arguments, returns the sum of the two matrices, or **C = A + B**. The function is not expected to work properly for any other matrix dimensions. You must calculate each term in **C** individually from terms in **A** and **B**—you may not use MATLAB's ability to add matrices directly. This function should feature:

- Standard header
- Calculations suppressed
- Solution matrix C printed in its entirety at the end.

Use the following values to test the program: A = [1 1; 2 2] and B = [3 3; 4 4]

ICA 17-14

Write a MATLAB function that will implement the following mathematical expression:

$$f(x) = \frac{x^2 \cos(2x + 1)}{(6x) \log(x)}$$

All calculations in the function should be suppressed.

ICA 17-15

Write a MATLAB function that will implement the following mathematical expression:

$$P(x) = \frac{1}{\sigma \sqrt{2\pi}} e^{-(x-\mu)^2/(2\sigma^2)}$$

Note that this function requires three arguments in the function input: x, σ, and μ. All calculations in the function should be suppressed. This function calculates a *Gaussian normal distribution*.

ICA 17-16

Write a MATLAB function that will implement the following mathematical expression:

$$P(x) = \frac{e^{-\lambda} \lambda^x}{x!}$$

Note that this function requires two arguments in the function input: x and λ. All calculations in the function should be suppressed. This function calculates a *Poisson distribution*.

ICA 17-17

Write a function that receives an angle in degrees as an argument and returns the sine of that angle. Your function should do exactly what the built-in function sind does, *but you may not use the function sind as part of your new function.*

CHAPTER 18
INPUT/OUTPUT IN MATLAB

When writing a program, it's wise to make the program understandable and interactive. This chapter describes the mechanisms in MATLAB that allow the executor of a program to input values during runtime. In addition, this chapter describes a few different mechanisms for displaying clean, formatted output to the executor that can be read and interpret as if it were prepared by hand.

18.1 INPUT

If all programs required the programmer to enter the input quantities at the beginning of the program, everyone would need to be a computer programmer to use the program. Instead, MATLAB and other programming languages have ways of collecting input from the program's user at runtime. This section focuses on program-directed user input, using three different approaches.

Numerical Input Typed by the User

The `input` function allows the user to input numerical data into MATLAB's Command Window. The user is prompted to input values at the point the input statement occurs in the code. It is important to note the difference between user input and functional input. User input prompts the user in runtime to type in a value. Functional input are values passed in as arguments to a function before a function is executed. Traditionally, all user input occurs in programs instead of functions.

```
Variable = input ('String')
```
- `Variable`: The variable where the input value will be stored
- `'String'`: The text that will prompt the user to type a value

EXAMPLE 18-1

Imagine you are writing a program to calculate the speed of a traveling rocket. Prompt the user to input the distance the rocket has traveled.

Before writing an input statement, it's critical to remember that MATLAB cannot handle units, so it's up to the programmer to perform any unit conversions necessary in the algorithm. Imagine that the user wants to report the speed of the rocket in miles per hour. It would be wise to have the user input the distance traveled in miles and the travel time in hours.

```
Distance = input ('How far has the rocket traveled [miles]? ')
Time = input('How long has the rocket been in the air? [hours] ')
Speed = Distance / Time
```

In the preceding code segment, the functions were executed without a semicolon being at the end of the input call. If you insert a semicolon at the end of the input call, output of the stored variables to the screen is suppressed, which allows for cleaner display on the Command Window.

```
% Variable output suppressed
Distance = input('How far has the rocket traveled [miles]?');
Time = input('How long has the rocket been in the air [hours]?');
```

In addition to numerical input, the input function will allow the user input arrays or matrices directly into the MATLAB Workspace. This is *not recommended* because it requires the user of the program to understand how to represent an array or matrix in the correct syntax (e.g., "[1,2;3,4]").

> **Variable = input ('String','s')**
> - **Variable**: The variable where the input value will be stored
> - **'String'**: The text that will prompt the user to type a value
> - **'s'**: Input type is text, not numeric

Comprehension Check 18-1

- Write an input statement to ask for the user's height in inches.
- Write an input statement to ask the user to enter the temperature in degrees Fahrenheit.

Text Input Typed by the User

The `input` function also allows the user to input text data into MATLAB's Command Window. The programmer must include an additional argument (`'s'`) to the `input` function to tell MATLAB to interpret the user input as text. The term 's' is used to indicate a string.

EXAMPLE 18-2

Imagine you are required to ask for the user's name and for the name of the month they were born. Suppress all extraneous output.

```
Name= input('Type your full name and press enter: ','s');
DataMonth = input('What month were you born? ','s');
```

> **Variable = menu ('String','opt1', 'opt2', . . .)**
> - **Variable**: The variable where the ordinal value will be stored
> - **'String'**: The text that will prompt the user to select an item in the menu
> - **'optN'**: The options that appear as buttons in menu

Comprehension Check 18-2

- Write an input statement to ask the user for the color of their eyes.
- Write an input statement to ask the user to type the current month.

Menu-Driven Input

In the preceding example, the user was required to type the name of the month they were born, but there is no good way to predict the way the user will actually type the name of the month into MATLAB. For example, the user may abbreviate, represent the number numerically, or may misspell the name of the month completely. MATLAB has a built-in input function (`menu`) that allows the programmer to specify a list of options for user selection. The menu function will display a graphical prompt of a question and responses. The first argument of the menu function is always the question to display and the following arguments are the possible responses. The `menu` function does not store the actual text response, but rather the ordinal number of the response in the list. An example will help clarify this idea.

EXAMPLE 18-3

Suppose we want to ask the user to input a favorite color. However, we want to force the user to choose from either red, green, or blue. Suppress all extraneous output.

```
Color = menu('Favorite Color', 'Red', 'Green', 'Blue');
```

The value returned upon selection is an integer. For example, if the user were to select "Red" in the menu, the value stored in Color *would be 1 since "Red" is the first option available in the list.*

Comprehension Check 18-3

- Write a menu statement to ask the user to select the current month from a list.
- Given the menu statement below, what is stored in variable C when the user clicks "Red"?

```
C=menu('My favorite color:','Green','Blue','Yellow',
'Black','Red')
```

18.2 OUTPUT

As critical as input is to enhancing the versatility of a program, output is equally important—this is how the user gets feedback on the computer's solution. Furthermore, as discussed earlier, program output of intermediate calculations is an important diagnostic tool. Once a program is working correctly, you should eliminate superfluous program output by using the semicolon, as demonstrated earlier. The program user will also find it helpful if the remaining output is neatly formatted.

Using disp for User Output

```
disp (Variable)
```
- Variable: The variable to be displayed on the screen

One of the basic methods for displaying the contents of any variable is through the use of the disp function. Given some variable x defined in MATLAB's Workspace, disp(x) displays the value of x to the screen. However, the disp command cannot display a value and text on the same line. Furthermore, disp relies on using MATLAB's default numerical display format for displaying any numbers, so it does not provide a capability for changing the form in which the number is displayed.

Formatting Output with fprintf

```
fprintf ('String',
Var1, Var2, . . .)
```
- 'String': The formatted output string
- VarN: The variables referenced in the formatted output string

The formatted print command, fprintf, gives extensive control over the output format, including spacing. The fprintf command can intersperse precisely formatted numbers within text, guarantee table alignment, align the decimal points of a column of numbers, etc. The fprintf command uses control and format codes to achieve this level of precision, as shown in the tables below.

Table 18-1 Output control with fprintf

Control Character	Use	Example
\n	Inserts a new line	fprintf('Hello\n');
\t	Inserts a tab	fprintf('A\tB\tC\n');
\\	Inserts a backslash	fprintf('Hello\\World\n');
'' (two single quotes)	Inserts a single quote mark	fprintf('Bob''s car\n');
%%	Inserts a percent symbol	fprintf('25%%');

The biggest benefit of using `fprintf` is the ability to plug values contained in variables into formatted sentences. In addition to the formatting controls shown above, MATLAB contains a number of special controls to represent different types of variables that might be displayed within a sentence. Inserting a variable control character within a string tells MATLAB to plug in a variable at that location in the string.

Table 18-2 Variable control with `fprintf`

Assume two variables defined: `Variable` *(number) and* `TextVariable` *(text).*

Control Character	Use	Example
%0.Nf	Inserts a fixed point value with N decimal places	`fprintf('0.1%f',Variable);`
%e or %E	Inserts a number in exponential notation	`fprintf('%e,Variable);` `fprintf('%E,Variable);`
%s	Inserts text	`fprintf('%s',TextVariable);`

EXAMPLE 18-4

Assume that the following variables are defined in MATLAB's Workspace:

```
Age=20;
Average=79.939;
Food='pizza';
```

Display the variables in formatted `fprintf` statements. At the end of each formatted output statement, insert a new line.

Since age is a decimal value, it is best to use the `%d` control character:

```
fprintf('My age is %d\n',Age);
```

To display a numerical average with no decimal places, the user can use `%d`. Alternatively, the `%f` control character allows the programmer to hard-code the number of decimal places. For example, if the user wanted to display no decimal places:

```
fprintf('The average is %.0f\n',Average);
```

To display the average with two decimal places:

```
fprintf('The average is %.2f\n',Average);
```

Note that MATLAB rounds values when using `%f`.

To display a string within a sentence:

```
fprintf('My favorite food is %s',Food);
```

EXAMPLE 18-5

Assume the variables in Example 18-4 are still defined in the workspace. Create a single formatted output statement that displays each sentence with a line break between each sentence:

```
fprintf('My age is %d\nThe average is %.2f\nMy favorite food is
%s\n',Age,Average,Food);
```

In MATLAB, the previous line would be typed on a single line. Note that the order of the arguments to the function corresponds to the order of the variables in the string.

Comprehension Check 18-4

Assume that the variable M is stored in the workspace with the value 0.3539.

- What would be the control code used to display M with two decimal places?
- What would be the control code used to display M with three decimal places?
- What would be the output of M if a control code is used to display it to two decimal places?
- What would be the output of M if a control code is used to display it to three decimal places?

18.3 PLOTTING

MATLAB has many plotting options, not all of which concern us in this book. The simplest way to plot in MATLAB is the `plot` command.

Plotting Variables

If the vectors `time` and `distance` contain paired data, `plot(time,distance)` will plot the time data on the abscissa and the distance data on the ordinate.

In addition to `time` and `distance`, assume a second distance vector, `distance2` contains additional distance recordings at the same time values in `time`. To plot both `distance` and `distance2` on the same graph, the `plot` command will contain both the `distance` and `distance2` vectors: `plot(time,distance,time, distance2)`.

Plotting Theoretical Expressions

A handy feature is that MATLAB can plot a function by using `fplot`. For `x` as the independent variable: `fplot('sin(x)',[0 2*pi])` will plot sin(x) from 0 to 2π.

If it is necessary to plot multiple functions on the same graph, the `fplot` command accepts multiple functions in the string: `fplot('[sin(x),sin(2*x)]',[0 2*pi])` will plot sin(x) and sin(2x) from 0 to 2π.

Creating Proper Plots

A number of other functions allow the programmer to automatically insert information onto a graph generated in MATLAB. Many plot options are defined by separate commands, discussed below. These can be applied when the plot is created or afterwards as long as the plot is unchanged.

- `xlabel` – Insert a label on the abscissa.

  `xlabel('Time (t) [s]')`
- `ylabel` – Insert a label on the ordinate.

  `ylabel('Height (H) [m]')`

---

plot
(A,B,C, . . .)

- **A:** The vertical axis values
- **B:** The horizontal axis values
- **C:** Specification of the line/symbol type/color
- **A,B,C:** Can be repeated for multiple data series (plot (A1, B1, C1, A2, B2, C2))

---

fplot ('String', [Axis])

- **'String':** The theoretical expression(s) to be plotted
- **Axis:** The minimum and maximum values of the x (and optionally y) axis

- **title** – Insert a title.

  ```
  title('Flight Across The United States')
  ```

- **legend** – Insert a legend on the graph. The order of elements provided in the legend is based on the order they are passed into `plot` or `fplot`.

  ```
  legend('Airplane 1','Airplane 2')
  ```

- **grid** – Allows the grid to be turned on or off.

  ```
  grid on
  ```

- **axis** – Allows the setting of the minimum and maximum values on the abscissa and ordinate (in that order).

  ```
  axis([0 50000 0 90000])
  ```

Creating Figures in MATLAB

When MATLAB creates a new plot, it automatically draws the plot to a "Figure" window. If a Figure window is already open, MATLAB will replace the current plot with the new plot within this window. To create multiple figure windows, you type the command `figure` to create a new figure before drawing your next plot. Add all of your titles, x- and y-axis labels, etc., before you initiate the `figure` command because those functions will draw to the newest figure. If you need to reference a previous figure, you can type `figure(N)` where N is the number of the figure. By default, MATLAB names the first figure created Figure 1 and increases figure number by 1 as new figures are created.

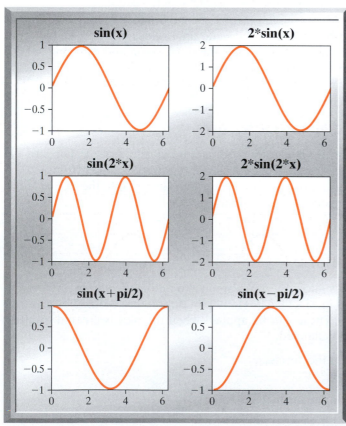

Using Subplots in MATLAB

In addition to generating single plots, MATLAB can generate multiple plots, called subplots, in a single window. This can make it easier to compare the results of different cases or to study a process from a variety of perspectives simultaneously. In this case, the `fplot` command has been used to study the effect of changing the input parameters to the general sine function A sin (Bx + C). Notice that even plots in MATLAB are arranged in matrices. The command to generate the grouping of plots is shown below. Note that parameter. A affects amplitude. Notice that B is a frequency. The remaining parameter, C, represents a phase shift.

The `subplot` command requires three numbers in parentheses. The first is the number of rows of graphs to appear in the window. The second is the number of columns of graphs to appear in the window. The third is the position of the graph about to be inserted (following a comma or on a new line). The graph positions are numbered starting in the top-left corner and going across each row.

```
subplot(3,2,1),fplot('sin(x)',[0 2*pi])        % plots sin(x) in top left
subplot(3,2,2),fplot('2*sin(x)',[0 2*pi])      % plots 2 sin(x) in top right
subplot(3,2,3),fplot('sin(2*x)',[0 2*pi])      % plots sin(2x) in middle left
subplot(3,2,4),fplot('2*sin(2*x)',[0 2*pi])    % plots 2 sin(2x) in middle right
subplot(3,2,5),fplot('sin(x+pi/2)',[0 2*pi])   % plots sin(x + π/2) in bottom
                                                % left
subplot(3,2,6),fplot('sin(x-pi/2)',[0 2*pi])   % plots sin(x − π/2) in bottom
                                                % right
```

Defining Terms Used in Sinusoidal Functions

The following terms are defined for *sinusoidal functions*: sine and cosine. Some of the terms also apply to other types of periodic functions. You may learn more about generalized periodic functions in future studies.

Amplitude: For a harmonic (repeated) motion, the amplitude (symbol A) is the maximum displacement of the curve from the equilibrium position.

Period: For a harmonic (repeated) motion, the period (symbol T) is the time elapsed between one condition and the time it takes for the system to return to that same condition (one cycle). As an example, for a pendulum, the period is the time required for the weight to move from the extreme position on one side to the other and back again.

Frequency: For harmonic (repeated) motion, the frequency is defined as the number of cycles the oscillating system goes through in a unit period of time. This is calculated as 1/T (and is reported as "cycles/second"); generally we convert it to angular frequency ("radians per second") by multiplying by 2π.

The period and the frequency of vibration can be determined by how often the curve crosses the "zero displacement" point. When the curve is above the zero displacement point, we call the displacement "positive" and when below, "negative." The frequency is often reported in units of hertz, where $1 \text{ Hz} = 1 \text{ s}^{-1}$.

In the figure, the amplitude (A) of Series 1 is 2 meters. The period (T) of Series 2 is 120 seconds, and the frequency is 0.0083 Hz.

Finding Zeros, Maxima, and Minima

To find a zero of a function requires seeking the value of independent variable for which the value of the dependent variable changes mathematical sign. A similar algorithm for finding a local maximum (peak) is to seek values of the independent variable for which the dependent variable goes from increasing to decreasing. Finding a local minimum would be just the reverse. These algorithms work fine with smooth data sets, but data sets with even a small amount of variability can interfere with this simple algorithm.

In the figure, the graph shows the function sin(x) where a random error of no more than ±0.02 has been added to the value of the function. If we apply the simple algorithm for finding a peak, the variability would cause us to think that there are many more maxima and minima than is truly the case. This difficulty can be managed by creating a more sophisticated algorithm or by "conditioning" the data.

Conditioning the data can remove variability due to random fluctuations ("noise"). This process is also called *smoothing*. To smooth the preceding case, a new data set was created that averages five consecutive data points from the original data set. After smoothing, the graph shown in the next figure results. This graph appears to be smooth enough for the simple algorithm to be used, yet the basic relationship between the independent and dependent variable remains unchanged by the smoothing.

Finding the Equilibrium Value

It is important to be able to identify the value of the dependent variable that corresponds to the center of oscillation, or the equilibrium value. In cases where the oscillation is "settling in" (the oscillation is damped), the value of the dependent variable at long times approaches the equilibrium value. Another approximation of the equilibrium value is the average of the first minimum and the first maximum. Both of these methods have failings: the first requires an algorithm that determines if the approximation is "close enough," based on how much oscillation remains; the second becomes less accurate the more the oscillation is damped. A simpler method is to take the average over all values of the independent variable. As long as the oscillation has gone through a number of cycles, this estimate is generally acceptable.

Finding the Period of Oscillation

The period is the time it takes to complete a full cycle. Based on the previous discussion, it would be possible to construct an algorithm that analyzes values of the dependent variable to find two successive maxima (peaks). Note that the time it takes for the dependent variable to go from the first maximum (which is the absolute maximum in the case of a damped oscillator) to the first minimum (again, the absolute minimum in a damped case) is one-half period (T/2). In this way, we can compute the period with simple operations. MATLAB's min function can return both the minimum term in a vector and the index of that value within the vector. The same options are available with the max function.

```
[ymax,maxposition]=max(y)       % returns maximum y value and index
[ymin,minposition]=min(y)       % returns minimum y value and index
Period = 2 * (x(minposition) - x(maxposition))
```

18.4   STATISTICS

Literally hundreds of functions are built into MATLAB. A few were introduced earlier. Here, we expand our use of the functions shown in Table 18-3.

Table 18-3 Common MATLAB statistical functions

MATLAB Function	Definition
ceil(X)	Rounds each element of X up to the next largest integer.
fix(X)	Rounds each element of X to the neighboring integer closest to 0.
floor(X)	Rounds each element of X down to the next smallest integer.
length(X)	If X is a vector, returns the number of elements in X. If X is a matrix, returns either the number of rows in X or the number of columns in X, whichever is larger.
max(X)	Finds the maximum value of X. If X is a matrix, returns the maximum value of the elements of each column in X.
mean(X)	Finds the mean or average of the elements of X. If X is a matrix, returns the mean of the elements in each column of X.
median(X)	Finds the median value of X. If X is a matrix, returns the median value of the elements of each column in X.
min(X)	Finds the minimum value of X. If X is a matrix, min will return the minimum value of the elements of each column in X.
polyfit(X,Y,N)	Calculates the coefficients of the N-order polynomial represented by the dependent values Y and the independent values X. For N=1, the first element of the array is the slope of a linear line and the 2nd element of the array is the y-intercept.
round(X)	Rounds each element of X to the nearest integer.
size(X)	Returns a vector of the number of rows and columns of X.
std(X)	Finds the standard deviation value of the elements of X. If X is a matrix, returns the standard deviation of the elements of each column of X.
var(X)	Finds the variance value of the elements of X. If X is a matrix, returns the variance of the elements of each column of X.

EXAMPLE 18-6

You are studying the number of fatal accidents that occur during different times of the day. Using MATLAB and the data shown, determine the following parameters. The data represent the number of accidents between midnight and 6 A.M. for nine consecutive weeks.

Week	Number of Fatal Accidents
A	190
B	202
C	179
D	211
E	160
F	185
G	172
H	205
I	177

Given the accident data:

```
>> accidents=[190 202 179 211 160 185 172 205 177];
```

Mean:

```
>> mean_accidents=ceil(mean(accidents))
mean_accidents =
    187
```

Median:

```
>> median_accidents=median(accidents)
median_accidents =
    185
```

Variance:

```
>> variance_accidents=var(accidents)
variance_accidents =
    281.9444
```

Standard deviation:

```
>> stdev_accidents=std(accidents)
stdev_accidents =
    16.7912
```

Using MATLAB for Statistical Analysis

NOTE

Basic statistical terms are discussed in detail in Chapter 9.

The outline below gives the steps necessary to use MATLAB for basic statistical analysis of a data set. This is presented as an example of the high and low temperature during the month of October 2006. The data are given in the table on the following page. This discussion focuses on the high temperatures, but would hold for the low temperatures as well.

1. Input the data. The first column is simply an identifier (in this case, the date). The second and third columns contain the actual raw data of high and low temperature each day, respectively.
2. Decide on the bin range. A rule of thumb is that the number of bins needed is approximately equal to the square root of the number of samples. While it is obvious in this example how many total samples are needed, the `length` function is often very useful.

```
>> number_bins=round(sqrt(length(high)))
number_bins =
    6
```

Date	High Temp [°F]	Low Temp [°F]
10/1/2006	82	58
10/2/2006	80	57
10/3/2006	81	60
10/4/2006	84	60
10/5/2006	86	61
10/6/2006	82	55
10/7/2006	71	49
10/8/2006	68	58
10/9/2006	80	60
10/10/2006	80	58
10/11/2006	79	61
10/12/2006	74	52
10/13/2006	67	44
10/14/2006	71	38
10/15/2006	68	37
10/16/2006	62	50
10/17/2006	65	53
10/18/2006	78	62
10/19/2006	73	66
10/20/2006	72	50
10/21/2006	66	45
10/22/2006	69	50
10/23/2006	56	38
10/24/2006	57	32
10/25/2006	56	35
10/26/2006	64	45
10/27/2006	60	50
10/28/2006	68	48
10/29/2006	74	43
10/30/2006	75	46
10/31/2006	64	49

3. Examine your data to determine the range of values. Using the `max` and `min` functions, you can determine that the highest high temperature and the lowest high temperature. Use these values to determine the range and number of points in each bin.

```
>> max_high=max(high)
max_high =
    86
>> min_high=min(high)
min_high =
    56

>> range = max_high - min_high
range =
   30
 >> bin_size=range/number_bins
 bin size =
    5
```

4. Determine the range of values that will appear in each bin. For example, the first bin will contain temperatures 55, 56, 57, 58 and 59; the second bin will contain temperatures 60–64. Create a vector with all of the center values of the ranges for each bin and a cell array with all of the bin range labels.

```
>> center_value=[57 62 67 72 77 82 87];
>> bin_range=['55-59'; '60-64'; '65-69';
'70-74'; '75-79'; '80-84'; '85-89'];
```

Using the `hist` Function

In this section, we illustrate three histograms created with the `hist` function.

- **One argument specified (data set):** By default, `hist` separates the data into 10 equally spaced bins and displays the histogram in a figure. Note that the following is not a proper plot.

```
>> hist(high)
```

- **Two arguments specified (data set, number of bins):** hist separates the data into the specified number of bins and displays the histogram in a figure. MATLAB divides the range between the minimum and maximum values of the data set into the number of bins specified by the user. Note that MATLAB does not line up the bins on powers of 5, 10, 100, etc., in order to create a reasonable graphing axis. Note, too, that the following is not a proper plot.

```
>> hist(high, 5)
```

- **Two arguments specified (data set, center value of bins):** hist separates the data into bins specified by the center value of each bin provided in the vector passed to the function and displays the histogram in a figure. Note that the following is not a proper plot.

```
>> hist(high,center_value)
```

In our discussion of the hist function, we have not addressed what data the hist function will return if we assign it to a variable. For example, if we save the result of hist using the centers of each bin and saving the result of the function call to the variable N, we see the following result.

```
>> N=hist(high,center_value)
     N =
        3    4    7    6    3    7    1
```

It is clear that the data stored in N is the same information conveyed by the histogram, but instead of a graphical representation, we now have integer values of the number of temperatures contained in each bin.

Creating a Cumulative Distribution Function

The Cumulative Distribution Function (CDF) is created in MATLAB by the following procedure.

1. **Create the cumulative sum of the histogram** data using the cumsum built-in function.

```
>> HTSum = cumsum(N)
     HTSum =
        3    7   14   20   23   30   31
```

2. **Normalize the cumulative sum.**
   To normalize our data, we divide by the total number of elements represented in the histogram (in this example, we are dealing with 31 temperatures). In general, we solve this by dividing by the maximum value of the cumulative sum and multiplying the result by 100 to create a CDF within the range of 0%–100%.

```
>> CDF=HTSum/max(HTSum)*100
   CDF =
   9.6774 22.5806 45.1613 64.5161 74.1935 96.7742
   100.0000
```

Creating CDF Graphs

> The command set (gca, ...) will draw things onto the figure. The first entry, gca, is an axis handle, used to indicate where to put the drawings. The second entry, 'XTickLabel', indicates which axis to put the entries. The final entry, bin_range, indicates where to get the values for the labels.

To create a bar graph of our CDF, use the `bar` function and the functions available for creating proper titles and axis labels.

```
>> bar(CDF)
>> set(gca,'XTickLabel',bin_range);
>> xlabel('Temperature (T) [deg F]')
>> ylabel('CDF Value [%]')
>> title('CDF of Daily High Temperature in Clemson, SC in October,
   2006')
```

To create a line graph of the CDF, use the `plot` function and the functions available for creating proper titles and axis labels.

```
>> plot(CDF)
>> set(gca,'XTickLabel',bin_range);
>> xlabel('Temperature (T) [deg F]')
>> ylabel('CDF Value [%]')
>> title('CDF of Daily High Temperature in Clemson, SC in October, 2006')
```

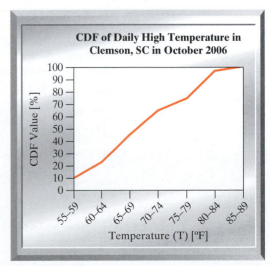

Comprehension Check 18-5

Repeat this analysis, using the daily low temperatures during October 2006.

In-Class Activities

ICA 18-1

Write a MATLAB statement that results in the input request shown in bold. The >> shows where your statement is typed, and the | shows where the cursor waits for input. The display must be correctly positioned. Each has a space before the cursor (shown as |). The input variable name and the variable type are shown at the right.

(a) >>
Enter the length of the bolt in inches: | (`bolt`, a number)
(b) >>
Enter the vibration frequency in kHz: | (`freq`, a number)
(c) >>
Enter the company's name: | (`Company`, text)
(d) >>
(In this statement, the window displayed will appear for the user to choose a color, and a number representing the color selected will be stored in the variable `LineColor`.)

ICA 18-2

For the following questions, z = 100/810. Write the MATLAB output that would result from each statement.

(a) >> `disp(z)`
(b) >> `disp('The value of z is '), disp(z)`
(c) >> `fprintf('%f',z)`
(d) >> `fprintf('%g',z)`
(e) >> `fprintf('%e',z)`
(f) >> `fprintf('%E',z)`

ICA 18-3

For the following questions, z = 100/810. Write the MATLAB input statement that displays z in the format shown in bold. In each case, the >> shows where your statement is typed. The display must be correctly positioned.

(a) >>
0.123457   (with no blank line after)
(b) >>
0.1   (with one blank line after)
(c) >>
0.123
(d) >>
1.235e-001
(e) >>
The value of z is 0.123.   (this must all appear on one line)
(f) >>
z is 0.123, so 10 z is 1.235.   (this must all appear on one line)
(g) Check the output of the statement below, and indicate how you would modify it so that the decimal points of the two numbers line up vertically with each other. You may only modify the two occurrences of `%.3f` in these statements—you may not use spaces to get these to line up.
`fprintf('\nThe value of z is %.3f\n',z), fprintf('1000z is equal to %.3f\n\n',1000*z)`

(h) Compare the output of the statements below. Why does adding 1 to this number change the last two digits from 78 to 90?

```
fprintf('\n%.17f\n\n',z)
fprintf('\n%.17f\n\n',z+1)
```

ICA 18-4

Joule's first law, also known as the **Joule effect**, relates the heat generated to current flowing in a conductor. It is named for James Prescott Joule, the same person for whom the unit of joule is named. The Joule effect states that electric power can be calculated as $P = I^2R$, where R is the resistance in [ohms, Ω] and I is the electrical current in [amperes, A]. Create a proper plot of the data. Create a proper plot of the experimental data.

Current (I) [A]	0.50	1.25	1.50	2.25	3.00	3.20	3.50
Power (P) [W]	1.20	7.50	11.25	25.00	45.00	50.00	65.00

ICA 18-5

The following experimental data was collected on the current (symbolized by I) in the positive direction and voltage (symbolized by V) across the terminals of two different thermionic rectifiers. Create a proper plot of the data.

Voltage (V) [volts]	Current (I) [milliamperes]	
	Rectifier A	Rectifier B
18	5	15
30	18	26
40	24	34
45	30	50

ICA 18-6

Your instructor will select several of the following functions. Plot them with subplot. The independent variable (angle) should vary from 0 to 360 degrees.

(a) $\sin(\theta)$
(b) $2\sin(\theta)$
(c) $-2\sin(\theta)$
(d) $\sin(2\theta)$
(e) $\sin(3\theta)$
(f) $\sin(\theta) + 2$
(g) $\sin(\theta) - 3$
(h) $\sin(\theta + 90)$
(i) $\sin(\theta - 45)$
(j) $2\sin(\theta) + 2$
(k) $\sin(2\theta) + 1$
(l) $3\sin(2\theta) - 2$

ICA 18-7

Your instructor will select several of the following functions. Plot them with `subplot`. The independent variable (angle) should vary from 0 to 360 degrees.

(a) $\cos(\theta)$
(b) $2\cos(\theta)$
(c) $-2\cos(\theta)$
(d) $\cos(2\theta)$
(e) $\cos(3\theta)$
(f) $\cos(\theta) + 2$
(g) $\cos(\theta) - 3$
(h) $\cos(\theta + 90)$
(i) $\cos(\theta - 90)$
(j) $2\cos(\theta) + 2$
(k) $\cos(2\theta) + 1$
(l) $3\cos(2\theta) - 2$

ICA 18-8

You want to create a graph showing the relationship of an ideal gas between pressure (P) and temperature (T). The ideal gas law relates the pressure, volume, amount and temperature of an ideal gas using the relationship: P V = n R T. The ideal gas constant, R, is 0.08207 atmosphere liter per mole kelvin. Assume the tank has a volume of 12 liters and is filled with nitrogen (formula, N_2; molecular weight, 28 grams per mole). Allow the initial temperature to be 270 kelvin at a pressure of 2.5 atmospheres. Create a graph, showing the temperature on the abscissa from 270 kelvin to 350 kelvin.

ICA 18-9

The decay of a radioactive isotope can be modeled with the following equation, where C_0 is the initial amount of the element at time zero and k is the decay rate of the isotope. Create a graph of the decay of Isotope A [k = 1.48 hours]. Allow time to vary on the abscissa from 0 to 5 hours with an initial concentration of 10 grams of Isotope A.

$$C = C_0 e^{-\frac{t}{k}}$$

ICA 18-10

To compare the popularity of different social networks on the Internet, it's often best to look at what percentage of the global Internet users visit a website. Below are usage data collected from Alexa, a company that collects and reports Internet usage statistics. Create a graph that shows the popularity of three social networks (Facebook, MySpace, Twitter) from July 2008 to May 2009. Experiment with different types of graphs to "best" illustrate the data.

	Jul-08	Aug-08	Sep-08	Oct-08	Nov-08	Dec-08	Jan-09	Feb-09	Mar-09	Apr-09	May-09
Facebook	9.0%	9.5%	10.0%	11.0%	11.5%	12.5%	14.0%	15.0%	15.5%	16.0%	17.0%
MySpace	7.0%	7.0%	6.5%	6.5%	6.0%	6.0%	4.5%	4.5%	5.0%	5.0%	5.0%
Twitter	0.0%	0.0%	0.0%	0.0%	0.0%	0.0%	0.0%	0.5%	1.0%	1.5%	2.0%

ICA 18-11

Create a graph to visualize the top five major league baseball players for 2008. Below is a chart the top five players according to their 2008 batting average (not shown), as well as the number of singles, doubles, triples, and home runs each player hit in 2008 according to ESPN (http://sports.espn.go.com/mlb/stats/batting?league=mlb&season=2008). Experiment with different types of graphs to "best" illustrate the data.

Player	Singles	Doubles	Triples	Home Runs
Chipper Jones	160	24	1	22
Albert Pujols	187	44	0	37
Manny Ramirez	183	36	1	37
Joe Mauer	176	31	4	9
Dustin Pedroia	213	54	2	17

ICA 18-12

Today, most traffic lights have a delayed green, meaning there is a short time delay between one light turning red and the light on the cross-street turning green. An industrial engineer has noticed that more people seem to run red lights that use delayed green. She conducts a study to determine the effect of delayed green on driver behavior. The following data were collected at several test intersections with different green delay times. These data represent only those drivers who continue through the intersection when the light turns red *before* they reach the limit line, defined as the line behind which a driver is supposed to stop. The data show the "violation time," defined as the average time between the light turning red and the vehicle crossing the limit line, as a function of how long the delayed green has been installed at that intersection. Graph the violation time (V, on the ordinate) and the time after installation (t, on the abscissa) for all three intersections on a single graph.

Time After Installation (t) [month]	Violation Time (V) [s]		
	Intersection #1	Intersection #2	Intersection #3
	1-second delay	2-second delay	4-second delay
2	0.05	0.1	0.5
5	0.1	0.5	1.5
8	0.3	1	2.5
11	0.4	1.3	3.1

ICA 18-13

The resistance of a typical carbon film resistor will decrease by about 0.05% of its stated value for each degree Celsius increase in temperature. Silicon is very sensitive to temperature, decreasing its resistance by about 7% for each degree Celsius increase in temperature. This can be a serious problem in modern electronics and computers since silicon is the primary material from which many electronic devices are fabricated. Most metals, on the other hand, increase their resistance as temperature increases, and seldom more than 0.5% per degree Celsius. The temperature coefficient of resistance (α) $[°C^{-1}]$ is negative if the resistance decreases with increasing temperature, and vice versa. Silicon, for example, has $\alpha = -0.07$ $[°C^{-1}]$ (a negative 7% decrease per degree Celsius as mentioned earlier). Gold, on the other hand, has $\alpha = 0.0037$ $[°C^{-1}]$ so its resistance increases less than 0.4% per degree Celsius. Most values of resistance and the temperature coefficient α are based on a reference resistance at 20 °C, more or less room temperature. For relatively small temperature differences from the reference temperature, this process is essentially linear. Create a proper plot to compare a carbon film resistor with a resistor fabricated from specially doped silicon ("doped" means impurities such as phosphorus or boron have been added to the silicon).

Temperature (T) [°C]	Carbon Film	Doped Silicon
15	10.025	9.9
20	10.000	10.0
25	9.975	10.1

ICA 18-14

You have been assigned to do preliminary calculations for a new 6-speed pickup truck transmission being designed to help maximize fuel efficiency. The same engine and differential will be used in three models, each of which has a different tire size. The differential is a specialized gearbox that connects the driveshaft from the transmission to the drive wheels. Your job is to produce graphs of vehicle speed as a function of gear ratio for the three different models. The gear ratio, expressed as A:B, gives the relationship between the speed at which the input (A) and output (B) shafts turn, typically given in RPM (revolutions per minute). Thus, a gear ratio of 3:1 means the input shaft is turning three times as fast as the output shaft. A gear ratio of 1:1 means the input and output shafts spin at the same rate. The engine selected is most efficient when running at 2400 rpm. The diameters of the three tires to be used are 24 inches, 28 inches, and 30 inches. Plot the speed of each vehicle with the engine running at 2400 RPM for transmission ratios of 1:1, 2:1, 3:1, and 4:1.

Vehicle Speed [mph]	Tire Diameter [in]		
Ratio (A:1)	24	28	30
1	55	64	69
2	28	32	35
3	18	21	23
4	14	16	17

ICA 18-15

Cadmium sulfide (CdS) is a semiconducting material with a pronounced sensitivity to light—as more light strikes it, its resistance goes down. In real devices, the resistance of a given device may vary over four orders of magnitude or more. An experiment was set up with a single light source in an otherwise dark room. The resistance of three different CdS photoresistors was measured when they were various distances from the light source. The farther from the source, the dimmer the illumination on the photoresistor. Plot the data.

Distance from Light (d) [m]	Resistance (R) [Ω]		
	A	B	C
1	79	150	460
3	400	840	2500
6	1100	2500	6900
10	2500	4900	15000

ICA 18-16

If the voltage across a semiconductor diode is held constant, the current through it will vary with changes in temperature. This effect can be (and has been) used as the basis for temperature sensors. Three different diodes were tested: a constant voltage (0.65 volts) was held across each diode while the current through each was measured at various temperatures. The following data were obtained. Plot the data.

Temperature (T) [K]	Current (I_D) [mA]		
	Diode A	Diode B	Diode C
275	852	2086	264
281	523	1506	179
294	194	779	81
309	69	390	35
315	47	301	26

ICA 18-17

The population of many living organisms increases exponentially as long as there are sufficient resources to support the increasing population. Here, we consider the species Homo sapiens (i.e., humans) to see if the population increase follows an exponential pattern. The following table lists population figures since 1750 for the planet Earth as well as for major geographical regions on Earth. Plot the four data sets with the largest 1998 values on one graph (World, Asia, Africa, and Europe).

Year	Population per Region			
	Africa	Asia	Europe	World
1750	1.06E+08	5.02E+08	1.63E+08	7.91E+08
1800	1.07E+08	6.35E+08	2.03E+08	9.78E+08
1850	1.11E+08	8.09E+08	2.76E+08	1.26E+09
1900	1.33E+08	9.47E+08	4.08E+08	1.65E+09
1950	2.21E+08	1.40E+09	5.47E+08	2.52E+09
1998	7.49E+08	3.59E+09	7.29E+08	5.90E+09

ICA 18-18

We ask 10 students to participate in a golf-putting contest. Each student is given five balls, placed a distance of 60 feet from the hole. They are to try to get each ball in the hole with a single putt. We then measure the number of feet that each ball is from the hole when it comes to rest. For simplicity, assume that none of the students make their putt.

Create a histogram of this situation (number of students on the ordinate and distance the putt stops from the hole on the abscissa), using the data given.

Distance from hole [ft]	0	1	2	3	4	5	6	7	8	9	10	11	12	13	14
Number of Balls	0	1	3	2	3	5	4	6	9	7	5	3	2	0	0

ICA 18-19

The table below lists the number of defective computer chips rejected during random testing over the course of a week on a manufacturing line. Use the following data to generate a histogram and CDF.

1	1	8	0	2	0
0	2	10	1	3	2
0	1	12	0	2	1
1	6	15	0	0	
3	8	1	2	5	

ICA 18-20

This exercise includes the measurement of a distributed quantity and the graphical presentation of the results. We are interested in determining how many "flexes" it takes to cause a paper clip to fail.

   Test the bending performance of 20 paper clips by doing the following:

1. Unfold the paper clip at the center point so the resulting wire forms an "S" shape.
2. Bend the clip back and forth at the center point until it breaks.
3. Record the number of "flexes" required to break the clip.

Use the table below to record the raw data for the paper clips you break. Summarize the data for the team by adding up how many clips broke at each number of flexes. Each team member should contribute 20 data points. Use the data to plot a histogram and cumulative distribution function.

Paper Clip Flexing Data

Paper Clip #	Flexes to Break
1	
2	
3	
4	
5	
6	
7	
8	
9	
10	
11	
12	
13	
14	
15	
16	
17	
18	
19	
20	

Summary of Data

No. of Flexes	No. of Clips

CHAPTER 19
LOGIC AND CONDITIONALS

Outside the realm of computing, logic exists as a driving force for decision-making. Logic transforms a list of arguments into outcomes based on a decision. Some examples of everyday decision-making:

- If the traffic light is red, stop. If the traffic light is yellow, slow down. If the traffic light is green, go.

    Argument: 3 traffic bulbs   Decision: is bulb lit?   Outcomes: stop, go, slow

- If the milk has passed the expiration date, throw it out; otherwise, keep the milk.

    Argument: expiration date   Decision: before or after?   Outcomes: garbage, keep

To bring decision-making into our perspective on problem solving, we need to first understand how computers make decisions. **Boolean logic** exists to assist in the decision-making process, where each argument has a binary result and the overall outcome exhibits binary behavior. **Binary behavior**, depending on the application, is any sort of behavior that results in two possible outcomes. That is,

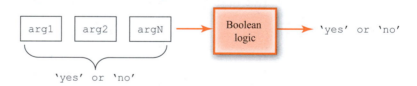

19.1   TRUTH TABLES

A tool called a **truth table** will help us determine the final outcome of our logical expression. To construct a truth table, we need to know in advance the logical expressions to be evaluated, as well as a list of all arguments used within the evaluations.

For example, assume that we are given the following statement:

If I don't follow the speed limit OR *I don't use proper turn signals, I am a danger on the highway.*

This statement can be broken down into the following logic sequence:

D=NOT(SL) OR *NOT(TS)* where D represents 'danger,' SL represents 'following the speed limit,' and TS represents 'properly using turn signals.'

In MATLAB, we would write this logical statement as

```
~SL || ~TS
```

To determine all possible outcomes of the logical sequence, we can construct a truth table:

Input Variables		Output Function
SL	**TS**	**D= ~SL ǁ ~TS**
0	0	1
0	1	1
1	0	1
1	1	0

NOTE

Generally, "true" is assigned the value of 1, and "false" is assigned the value of 0. If you later learn more about Boolean logic, you will discover that sometimes the opposite choice is made: "true" = 0 and "false" = 1. This is sometimes referred to as "negative logic." We assume "positive logic" throughout this text.

That is, when I don't follow the speed limit and don't use turn signals, I am considered a danger on the highway. Likewise, when I either follow the speed limit or use turn signals, but never both at the same time, I am considered a danger on the highway. The only case where I am not considered a danger on the highway is the single case where I both follow the speed limit and use proper turn signals.

The anatomy of a truth table is shown below.

Input

Input Variables		Output Function
SL	TS	D=~SL ǁ ~TS
0	0	1
0	1	1
1	0	1
1	1	0

Output

EXAMPLE 19-1

Imagine we have two variables that represent Cartesian coordinates, x and y. We want to determine whether these Cartesian coordinates are located inside, outside, or on the perimeter of the unit circle. To determine the location of coordinate, we calculate the distance from the center of the unit circle (0,0) to our coordinate with a distance equation:

$$d = \sqrt{x^2 + y^2}$$

To determine whether a coordinate is in, out, or on the circle:

- If $d > 1$, the coordinate is outside of the unit circle.
- If $d < 1$, the coordinate is outside of the unit circle.
- Otherwise, the coordinate is on the perimeter of the unit circle.

To make the decisions above, we could have formally prepared a truth table, as follows:

d > 1	d < 1	Outcome
0	0	On
0	1	In
1	0	Out
1	1	–

From this truth table, we recognize that we will need more than one decision to determine the location of the x-y coordinate. Note the use of a dash to indicate a condition that cannot occur.

THINK ABOUT IT!

Imagine you're making a drink selection by picking up two bottles with your hands (for example, tea in your left hand and a soda in your right hand) and selecting one of the drinks in your hand. If you want to decide between tea, soda, and coffee, you're required to bring in another person and let them hold the tea and soda while you hold the coffee in one hand and touch their shoulder with your free hand. Now you're able to choose between 3 drinks. If you want to choose between tea, soda, coffee, and milk, you need to bring in a third person, where both people are holding the 4 drinks and you're touching their shoulders. As we add more drink options, we're required to connect more people together to build our decision tree.

EXAMPLE 19-2

Imagine we have two variables that represent Cartesian coordinates, x and y. We want to determine whether these Cartesian coordinates are located in quadrant 1, 2, 3, or 4 of an x-y scatter plot.

$$\begin{array}{c|c} 4 & 1 \\ \hline 3 & 2 \end{array}$$

To make the decisions above, we could have formally prepared a truth table, as follows:

x > 0	y > 0	Outcome
0	0	3
0	1	2
1	0	4
1	1	1

From this truth table, we recognize that we will need more than one decision to determine the location of the x-y coordinate.

Comprehension Check 19-1

- Draw a truth table that represents the decisions made at a stop light (RED, GREEN, YELLOW).
- Draw a truth table that represents the decisions made to determine if water is frozen (< 0 degrees Celsius), steam (> 100 degrees Celsius), or liquid (otherwise).
- Draw a truth table that represents the decisions made to determine if a grade is an A, B, C, D, or F on a 10-point scale.

19.2　BINARY NUMBERS

To understand how an input sequence for variables is created, we must first understand how binary numbers work. In computing, integers are often represented as binary numbers. Binary numbers are composed strictly of 1's and 0's. The number system most

Table 19-1 Symbols used in different number systems

Binary (2)	Decimal (10)
0, 1	0, 1, 2, 3, 4, 5, 6, 7, 8, 9

familiar to people is the decimal system, which is composed of 10 different symbols representing different integers.

Using the different number systems shown in Table 19-1, we can represent the same numerical value by using a combination of the symbols available in each number system. For binary numbers, we are working with base 2 numbers, in contrast to working with base 10 numbers in the decimal system. For example, let's assume we want to communicate the value 13 in binary and the decimal system.

To do this, we use combinations of different powers of the number system base. Using a combination of 10^X values, we can communicate the number 13 in the decimal system by saying

$$1 * 10^1 + 3 * 10^0 = 13$$

Table 19-2 Decimal System (Base 10)

0^4	10^3	10^2	10^1	10^0
0	0	0	1	3

Note that the coefficients used in front of the 10^X terms (see Table 19-2) are the exact symbols necessary for communicating the required value.

In the binary system, we can communicate the value of 13 by using a combination of values 2^Y

$$1 * 2^3 + 1 * 2^2 + 0 * 2^1 + 1 * 2^0 = 1101 \text{ (which is the value of 13 in binary)}$$

Also note that in the binary example, the $0 * 2^1$ term was retained to properly extract the sequence of binary digits for representation of the number in binary, as shown in Table 19-3. (This is similar to the fact that 409 has 0 tens, being expressed as $4 * 10^2 + 0 * 10^1 + 9 * 10^0 = 409$.)

Table 19-3 Binary System (Base 2)

2^4	2^3	2^2	2^1	2^0
0	1	1	0	1

When entering input values in the truth table, the correct input sequence starts from the binary value of 0 and runs up to the binary value of 2^N, where N is the number of input variables. For example, if N = 3, the input sequence runs from 000 to 111. (The complete sequence is 000, 001, 010, 011, 100, 101, 110, 111.)

EXAMPLE 19-3

Convert the following numbers to binary numbers:

- 19
- 165
- 1060

To begin converting any number to its binary value, it's easiest to begin by determining the number of characters necessary to represent the binary value. To do this, we need to look at the calculated values of the first 11 positions in any binary value:

- $2^0 = 1$
- $2^1 = 2$
- $2^2 = 4$
- $2^3 = 8$
- $2^4 = 16$
- $2^5 = 32$
- $2^6 = 64$
- $2^7 = 128$
- $2^8 = 256$
- $2^9 = 512$
- $2^{10} = 1024$
- $2^{11} = 2048$

To convert 19 to a binary value, we will create a 5-digit binary value because 19 is between 16 (2^4) and 32 (2^5).

$$19 = 16 + 2 + 1$$

2^4	2^3	2^2	2^1	2^0
1	0	0	1	1

To convert 165 to a binary value, we will create an 8-digit binary value because 165 is between 128 (2^7) and 256 (2^8).

$$165 = 128 + 32 + 4 + 1$$

2^7	2^6	2^5	2^4	2^3	2^2	2^1	2^0
1	0	1	0	0	1	0	1

To convert 1060 to a binary value, we will create an 11-digit binary value because 1060 is between 1024 (2^{10}) and 2048 (2^{11}).

$$1060 = 1024 + 32 + 4$$

2^{10}	2^9	2^8	2^7	2^6	2^5	2^4	2^3	2^2	2^1	2^0
1	0	0	0	0	1	0	0	1	0	0

Therefore,

Base-10	Base-2
19	10011
165	10100101
1060	10000100100

19.3 LOGIC AND RELATIONAL OPERATORS IN MATLAB

And = `&&`
Or = `||`
Not = `~`

AND is true if and only if all arguments are true.

OR is true if at least one of the arguments is true.

To connect all of the Boolean arguments to make a logical decision, we turn to a few operators with which we can relate our arguments to determine a final outcome.

- **And:** The AND logical operator connects two Boolean arguments and returns the result as true if and only if *both* Boolean arguments have the value of TRUE. In MATLAB, we use the ampersand (`&&`) symbol (SHIFT+7 on a keyboard) to represent the AND logical operator.
- **Or:** The OR logical operator connects two Boolean arguments and returns the result as true if *either one* (or both) of the Boolean arguments has the value of TRUE. In MATLAB, we use two pipes (`||`) symbol (SHIFT+\ on a keyboard) to represent the OR logical operator.
- **Not:** The NOT logical operator inverts the value of a single Boolean argument or the result of another Boolean operation. In MATLAB, we use the tilde (`~`) symbol (SHIFT+' on a keyboard) to represent the NOT logical operator.

Operator	Meaning
>	Greater than
<	Less than
>=	Greater than or equal to
<=	Less than or equal to
==	Equal to
~=	Not equal to

In order to determine the relationship between two values (numbers, variables, etc.), we have a few operators with which we can compare two variables to determine whether or not the comparison is true or false.

These relational operators are usually placed between two different variables or mathematical expressions to determine the relationship between the two values. This expression of variable-operator-variable is typically called a **relational expression**. Relational expressions can be combined by logical operators to create a **logical expression**. If no logical operator is required in a particular decision, then the single relational expression can be the logical expression.

EXAMPLE 19-4

Often, it is necessary to express an inequality to describe a certain condition in a program or function. The following examples show the relationship between mathematical inequalities and how these are typically related in MATLAB code.

Mathematical inequality	**Relational expression**
$4 \leq X < 5$	4 <= X && X < 5
$10 < X \leq 20$	10 < X && X <= 20
$30 \leq X \leq 100$	30 <= X && X <= 100

In this example and others, MATLAB can't operate with multiple relational operators in a single expression—a common situation in mathematical inequalities. Instead, we must separate out the two relations and combine the expressions with the "and" symbol.

Comprehension Check 19-2

What are the relational expressions for the following mathematical inequalities?

- $5 \leq t < 10$
- $-30 < M \leq 20$
- $Y \leq 100$

19.4   CONDITIONAL STATEMENTS IN MATLAB

Conditional statements are commands by which users give decision-making ability to the computer. Specifically, the user asks the computer a question framed in conditional statements, and the computer selects a path forward (it provides an answer—either numerical or as a text comment) based on the answer to the question. Sample questions are given below:

- If the water velocity is fast enough, switch to an equation for turbulent flow!
- If the temperature is high enough, reduce the allowable stress on this steel beam!
- If the RPM level is above red line, issue a warning!
- If your grade is high enough on the test, state: You Passed!

In these examples, the comma indicates the separation of the condition and the action that is to be taken if the condition is true. The exclamation point marks the end of the statement. Just as in language, more complex conditional statements can be crafted with the use of "else" and similar words. In these statements, the use of a semicolon introduces a new conditional clause, known as a nested conditional statement. For example:

- If the collected data indicate the process is in control, continue taking data; otherwise, alert the operator.
- If the water temperature is at or less than 10 degrees Celsius, turn on the heater; or else if the water temperature is at or greater than 80 degrees Celsius, turn on the chiller; otherwise, take no action.

Single Conditional Statements

In MATLAB, a single conditional statement involves the `if` and `end` commands. If two distinct actions must occur as a result of a condition, the programmer can use the `else` command to separate the two actions. The basic structure of a single `if` statement is as follows:

```
if Logical_Expression
    % Actions if true
else
    % Actions if false
end
```

Note that the *Logical_Expression*, as shown in the structure of a single `if` statement, is a logical expression as described previously in the Logic section.

EXAMPLE 19-5

Write a MATLAB conditional statement to represent the following conditional statement.
If the water velocity is less than 10 meters per second, switch to an equation for laminar flow to determine the friction factor of the piping system.

English (Pseudocode)	MATLAB Code
If the water velocity is slow enough,	`if v < 10`
Switch to an equation for laminar flow	`f = 64/Re`
(period)	`end`

In this example, the comma indicates the separation of the condition and the action that is to be taken if the condition is true. The period marks the end of the statement.

EXAMPLE 19-6

Write a MATLAB conditional statement to represent the following conditional statement.
If the speed of the vehicle is greater than 65 mph, display "Speeding" to the Command Window, otherwise; display "OK".

English (Pseudocode)	MATLAB Code
If the speed is greater than 65,	`if Speed > 65`
display "Speeding"	`fprintf('Speeding')`
Otherwise;	`else`
Display "OK"	`fprintf('OK')`
(period)	`end`

In this example, the comma indicates the separation of the condition and the action that is to be taken if the condition is true. The semicolon after the word "otherwise" marks the beginning of the false action. The period marks the end of the statement.

Comprehension Check 19-3

Why isn't there a logical expression after the word "else" in Example 19-6?

EXAMPLE 19-7

Write a MATLAB conditional statement to represent the following conditional statement. If the water temperature is at or less than zero degrees Celsius, it is solid; else if the water temperature is at or greater than 100 degrees Celsius, it is vapor; else, it is liquid.

English (Pseudocode)	MATLAB Code
If the temperature is 0 degrees Celsius or less,	`if Tw <= 0`
Form is solid	`    form='solid'`
else if the temperature is 100 degrees Celsius or greater,	`elseif Tw >= 100`
Form is vapor	`    form='vapor'`
else	`else`
Form is liquid	`    form='liquid'`
(period)	`end`

Compound Conditional Statements

Even more complicated questions are possible with the use of logical operators AND, OR, and NOT.

- If the hot plate is on AND the beaker has no water in it, sound an alarm.
- If the stock price falls below $3.00 per share OR if the company's price/earnings ratio goes above $25, sell all shares of the stock.
- If the temperature in the oven falls below 350 degrees Fahrenheit and the internal temperature of the roast is not 250 degrees Fahrenheit, turn the heating element on.

Comprehension Check 19-4

Convert the compound conditional statements above into MATLAB code.

Format of Conditional Statements

Since conditional statements ask questions about data relationships, the first step in programming a conditional statement is defining all the possible relationships. In addition to less than ($<$), greater than ($>$), and equal to ($==$), there are natural extensions of these—less than or equal to ($<=$), greater than or equal to ($>=$), and not equal to ($\sim=$). Note that the conditional test for equality ($==$) is different from the assignment operator ($=$). The logical operator notation for AND ($\&\&$), OR ($||$), and NOT ($\sim$) allows compact logical statements that might look like this:

```
if ChargeLevel==discharged && BatteryStatus~=defective
      DeviceCheck(alternator)
end
```

if statements can be nested, implying the AND logic. Sometimes this makes the programming less complex:

```
if ChargeLevel==discharged
        DeviceCheck(battery)
        if BatteryStatus==defective
                DeviceReplace=battery
        else
                DeviceCheck(alternator)
                if AlternatorStatus==defective
                        DeviceReplace=alternator
                end
        end
end
```

Comprehension Check 19-5

Rewrite the above nested-if statement with the elseif control.
Rewrite the above nested-if statement with logical operators (AND, OR, NOT)

Equivalent Forms of Logic

Conditional statements are similar to traditional language in another way—there are many ways to express the same concept. The order, hierarchy (nesting), and choice of operators is flexible enough to allow the logic statement to be expressed in the way that makes the most sense. All three of the following logic statements are equivalent for integer values of NumTeamMembers, but all three read differently in English:

Conditional Statement	English Translation

`if NumTeamMembers>=4`
`  if NumTeamMembers<=5`
`    disp ('Team size OK.')`
`  end`
`end`
	If the number of team members is 4 or more and if the number of team members is 5 or fewer, the team size is okay.

`if NumTeamMembers>=4&&NumTeamMembers<=5`
`    disp('Team size OK.')`
`end`
	If the number of team members is 4 or more and 5 or fewer, the team size is okay.

`if NumTeamMembers==4||NumTeamMembers==5`
`    disp('Team size OK.')`
`end`
	If the number of team members is 4 or 5 the team size is okay.

`if NumTeamMembers < 4||NumTeamMembers> 5`
`    disp('Adjust team size')`
`end`
	If the number of team members is fewer than 4 or more than 5, adjust the team size.

`if NumTeamMembers~=4&&NumTeamMembers~=5`
`    disp('Adjust team size')`
`end`
	If the number of team members is not 4 and not 5, adjust the team size.

In-Class Activities

ICA 19-1

Choose from the list below to answer the questions that follow.

(a) `++`   **(b)** `&&`   **(c)** `%%`   **(d)** `||`   **(e)** `!!`   **(f)** `~`   **(g)** `#`

i.   Which of the above operators represents AND?
ii. Which of the above operators represents OR?
iii. Which of the above operators represents NOT?

ICA 19-2

Choose from the list below to answer the questions that follow.

(a) `=`   **(b)** `!`   **(c)** `~`   **(d)** `>`   **(e)** `>>`   **(f)** `==`   **(g)** `!!`   **(h)** `~~`
(i) `>=`   **(j)** `=>`   **(k)** `<>`   **(l)** `><`   **(m)** `!=`   **(n)** `~=`   **(o)** `!>`   **(p)** `~>`

i.   Which of the above operators represents EQUAL TO?
ii. Which of the above operators represents NOT EQUAL TO?
iii. Which of the above operators represents GREATER THAN OR EQUAL TO?

ICA 19-3

Write a program that prompts the user to input a temperature of water in degrees Celsius and that returns the phrase "Ice," "Liquid," or "Steam," depending on the input temperature.

ICA 19-4

Write a program that displays a menu of traffic light colors. Depending on which button is pressed, the user should then be prompted to start, slow, or stop.

ICA 19-5

Write a program that prompts the user for both the distance and the time at which a vehicle starts and finishes. The input values should be stored in variables `StartDist` (m), `StartTime` (s), `FinishDist` (m), and `FinishTime` (s). If the finish time is not greater than the starting time, an appropriate error message should be generated and the program should terminate. Otherwise, the program should return the phrase "Backing up at ___ m/s," "Not moving," or "Moving forward at ____ m/s" as appropriate (including numbers where there are blanks).

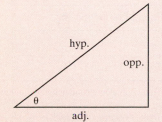

ICA 19-6

Write a program that asks the user to enter the length of the hypotenuse, opposite, and adjacent sides of a right triangle with respect to an angle (θ). After the user has entered all three values, a menu should pop up asking if the user wants to calculate $\sin(\theta)$, $\cos(\theta)$, $\tan(\theta)$, $\cot(\theta)$, $\csc(\theta)$, or $\sec(\theta)$. If the user inputs lengths that don't work with a right triangle, an error message should appear and the program should terminate.

ICA 19-7

This graph shows the change due if an item is purchased with $1. Region I and Region II represent distinctly different situations.

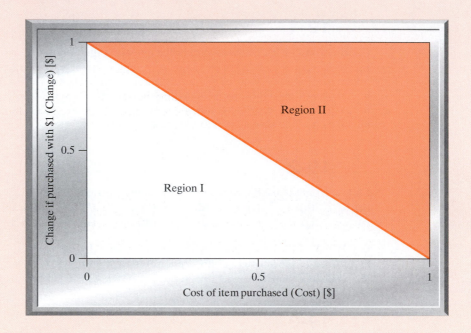

Write a function that takes as input the item cost and the change received and gives as output an appropriate message depending on whether the change given is on the line, in Region I, or in Region II. In Region I or II, the message should include the extra amount due the seller, or the extra amount of change due the buyer. Assume the purchase price is one dollar or less.

ICA 19-8

For the protection of both the operator of a zero-turn radius mower as well as the mower itself, several safety interlocks must be implemented. These interlocks and the variables that will represent them are listed below.

Interlock Description	Variable Used	State
Brake Switch	Brake	True if brake on
Operator Seat Switch	Seat	True if operator seated
Blade Power Switch	Blades	True if blades turning
Left Guide Bar Neutral Switch	LeftNeutral	True if in neutral
Right Guide Bar Neutral Switch	RightNeutral	True if in neutral
Ignition Switch	Ignition	True if in run position
Motor Power Interlock	Motor	True if motor enabled

For the motor to be enabled (thus capable of running), the following conditions must all be true.

- The ignition switch must be set to "Run".
- If the operator is not properly seated, both guide levers must be in the locked neutral position.
- If either guide lever is not in the locked neutral position, the brake must be off.
- If the blades are powered, the operator must be properly seated.

(a) Write a function that will accept all of the above variables except `Motor`, decide whether the motor should be enabled or not, and place true or false in `Motor` as the returned variable.

(b) Assume the operator has been mowing and desires to dismount from the mower with the engine running. What sequence of actions to the controls (brake, blade power, guide bars, ignition switch) must be taken before the operator dismounts and in what order if the order matters. Note that the operator does not necessarily need to do something with all of the controls.

ICA 19-9

We go to a state-of-the-art amusement park. All the rides in this amusement park contain biometric sensors that measure data about potential riders while they're standing in line. Assume the sensors can detect a rider's age, height, weight, heart problems, and possible pregnancy. For each ride at the park, help the engineers write the conditional statement based on their safety specifications.

Variable Definitions:

```
Age               % the age of the potential rider (as an integer)
Height            % the height of the potential rider (as an integer)
Weight            % the weight of the potential rider (as an integer)
HeartCondition    % HC status ('yes' if the person has a heart condition,
                  %            'no' otherwise)
Pregnant          % pregnancy status (1 if pregnant, 0 otherwise)
```

(a) The Spinning Beast: All riders must be 17 years or older, more than 62 inches tall, must not be pregnant or have a heart condition.

```
if _____
            fprintf('Sorry, you cannot ride this ride');
end
```

(b) The Lame Train: All riders must be 8 years or older and must not be pregnant.

```
if _____
            fprintf('Sorry, you cannot ride this ride');
end
```

(c) The MATLAB House of Horror: All riders must be 17 years or older and must not have a heart condition.

```
if _____
            fprintf('Welcome to the MATLAB House of Horror!');
end
```

(d) The Neck Snapper: All riders must be 16 years or older, more than 65 inches tall, must not be pregnant or have a heart condition.

```
if _____
            fprintf('Welcome to The Neck Snapper!');
end
```

ICA 19-10

A phase diagram for cobalt and nickel is shown below. If it is assumed the lines shown are linear, the mixture has the following characteristics:

- Above 1500 degree Celsius, it is a liquid (L).
- Below 1500 degree Celsius, there are two possible phases: face-centered cubic (FCC) phase and hexagonal close-packed (HCP) phase.

Write a conditional statement to determine the phase given the weight % nickel in a variable w and the temperature in a variable T. Call the phases "HCP", "FCC", "L", "FCC+LL" (on the line), and "HCP+FCC" (on the line) for simplicity. Store the phase as text in a variable called `Phase`.

ICA 19-11

A phase diagram for carbon and platinum is shown. If it is assumed the lines shown are linear, the mixture has the following characteristics:

- Below 1700 degree Celsius, it is a mixture of solid platinum (Pt) and graphite.
- Above 1700 degree Celsius, there are two possible phases: a Liquid (L) phase and a Liquid (L) + Graphite phase. The endspoints of the division line between these two phases are labeled on the diagram.

Write a conditional statement to determine the phase given the weight % carbon in a variable w and the temperature in a variable T. Call the phases "Pt + G", "L", "L+G", "Pt+L+LG" (on the line), and "L+L+G" (on the line) for simplicity. Store the phase as text in a variable called `Phase`.

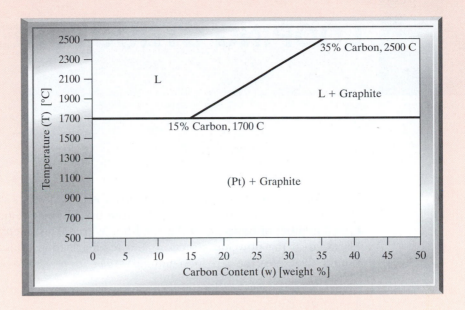

ICA 19-12

Most resistors are so small the actual value would be difficult to read if printed on the resistor. Instead, colored bands denote the value of resistance in ohms. Anyone involved in constructing electronic circuits must become familiar with the color code, and with practice, one can tell at a glance what value a specific set of colors means. For the novice, however, trying to read color codes can be a bit challenging (see Table 19-6).

You are to design a program that, when a user enters a resistance value, the program will display (as text) the color bands in the order they'll appear on the resistor.

The resistance will be entered in two parts: the first two digits, and a power of 10 by which those digits will be multiplied. You must include an appropriate input message and an appropriate warning if the user enters an invalid number. The user should be able to either type in values or select them from a menu. Note that you must use the two input values to calculate the total resistance, the first digit, the second digit, and the number of zeros.

As examples, a resistance of 4700 ohms has first digit 4 (yellow), second digit 7 (violet), and 2 zeros following (red). A resistance of 56 ohms would be 5 (green), 6 (blue), and 0 zeros (black); 1,000,000 ohms is 1 (brown), 0 (black), and 5 zeros (green). Particularly note that if the second digit is zero, it does not count in the multiplier value (see Tables 19-4 and 19-5). There are numerous explanations of the color code on the web if you want further information or examples.

Table 19-4 Standard Resistor Values

10	12	15	18	22	27	33	39	47	56	68	82

Table 19-5 Standard Multipliers

1	10	100	1000	10000	100000	1000000

Table 19-6 Color Codes

0	1	2	3	4	5	6	7	8	9
Black	Brown	Red	Orange	Yellow	Green	Blue	Violet	Grey	White

CHAPTER 20
LOOPING STRUCTURES

The key to making an algorithm compact is to exploit its variables by means of a loop. A loop is a programming structure that allows a segment of code to execute a fixed number of times or until a condition is true. This phenomenon was previously described in the flowcharts section as a feedback loop. This chapter describes two looping structures: the `for` loop and the `while` loop.

20.1   for LOOPS

```
for counter =
  start : step :
    finish
    % executable
    statements
end
```

You should have noticed that a number of the algorithms we have already studied have parts that are repetitive. Sometimes, we want to execute an algorithm a certain number of times:

- For every pressure sensor, record the pressure reading.
- For each employee, check the date of the employee's last safety training.

It may be useful to sample less than every possible point, as in:

- For every fifth data point, enter its value into an array for plotting.

The ability to count backwards is also useful:

- For every second counting down from 10 to 0 seconds, announce the time remaining until launch.

The syntax of a `for` loop is illustrated below:

```
for counter = start : step : finish
        % executable statements
    end
```

This programming structure functions as follows: execute the part of the program that is between the `for` statement and the `end` statement for every value of `counter` from `start` to `finish` counting by `step`. The `counter` variable can be used in the loop, and most often is: to calculate other variables, to index an array, to set function parameters, etc.

A sample `for` loop is shown below. This loop will display on the screen the numbers from 1 to 5, each on a separate line.

```
for k=1:1:5
        disp(k)
    end
```

UNDERSTANDING THE for LOOP

Most program statements do *exactly* the same type of thing every time, although they may be manipulating different information. The for loop is slightly different. A for statement can be arrived at during program execution in one of two ways:

1. It can be executed immediately following a statement that was outside the loop, typically a statement immediately above it.
2. It can be executed upon returning from the end statement that marks the end of the loop after the loop has executed one or more times.

What the for statement does in these two cases is different, and crucial to understanding its operation.

1. When the for statement is arrived at from another statement *outside* the loop, the for does the following:
   - Places the start value into the counter variable.
   - Decides whether or not to do the instructions inside the for loop.

2. When the for statement is arrived at from its corresponding end statement below, the for does the following:
   - Adds the step value to the counter variable.
   - Decides whether or not to do the instructions inside the for loop.

The second step is the same in both cases and requires a bit more explanation. To determine whether or not to execute the instructions inside the loop (those between the for and its corresponding end), the for statement asks the question "Has the counter gone beyond the finish value?" (See comment about what "beyond" means.)

- If the answer is no, execute the statements inside the loop.
- If the answer is yes, skip over the loop and continue with the first statement (if any) immediately following the corresponding end statement.

If the loop is counting up (step > 0), then we could ask "Is the counter greater than the finish value?"

If the loop is counting down (step < 0), we could use "less than" instead. Several simple examples are given below.

NOTE

Since the loop can count either up or down, we use the word "beyond."

EXAMPLE 20-1

Simple for loops:

(a) The statements in the loop will be executed four times, once each with i = 1, 2, 3, and 4:

```
for i=1:1:4
      % loop statements
end
```

(b) The statements in the loop will be executed two times, once each with time = 1 and 3. When time is incremented to 5, it is "beyond" the finish value:

```
for time=1:2:4
      % loop statements
end
```

(c) The statements in the loop will be executed three times, once each with gleep = 4, 2, and 0:

```
for gleep=4:-2:0
      % loop statements
end
```

(d) The statements in the loop will be executed three times, once each with `raft` = 13, 7, and 1. When time is decremented to −5, it is "beyond" the finish value:

```
for raft=13:-6:-4
        % loop statements
end
```

(e) The statements in the loop will not be executed. `k` is "beyond" the finish value when the loop is first entered:

```
for k=6:1:4
        % loop statements
end
```

(f) The statements in the loop will be executed five times, once each with `wolf` = 5, 12, 19, 26, and 33:

```
S1=5;
S2=7;
for wolf=S1:S2:S1^2+S2+1
        % loop statements
end
```

Arithmetic Sequences

Changing the `for` loop control parameters allows for more intuitive design of loop function. Some of the sequences above can be achieved by changing the `start`, `step`, or `finish` values in the above sample loop. Specifically, any **arithmetic sequences**—those that require only changing `k` by adding or subtracting a number from it—can be achieved by changing the loop controls.

Referring back to the first sample for loop, let us examine how the same sequences can be generated using either the loop controls or calculations inside the loop itself. The first example is repeated here for convenience.

```
for k=1:1:5
        disp(k)
end
```

This generates the sequence k = 1, 2, 3, 4, 5

In each or the following examples, a sequence of five numbers is given, with two solutions shown to generate that sequence. The first code segment in each example uses the `for` loop to count from 1 to 5, and the values desired are calculated inside the loop. The second segment of code accomplishes the same goal by manipulating the `start`, `step`, and `finish` values instead.

EXAMPLE 20-2

We desire the sequence to be 3, 6, 9, 12, 15.

```
for k=1:1:5
        disp(3*k)
end
```

or

Given that k *goes through the sequence k =1, 2, 3, 4, 5, the desired sequence could be calculated as 3k.*

```
for k=3:3:15
        disp(k)
end
```

This illustrates the idea that multiplication and division of a sequence by a constant to generate a new sequence affects all three loop control parameters: start, step, *and* finish.

EXAMPLE 20-3

We desire the sequence to be 3, 4, 5, 6, 7.

```
for k=1:1:5
        disp(k+2)
end
```

or

This can be calculated as k + 2

```
for k=3:1:7
        disp(k)
end
```

Adding and subtracting a constant to create a new sequence affects only the start *and* finish *values.*

EXAMPLE 20-4

We desire the sequence to be 4, 6, 8, 10, 12.

Note that standard order of operations applies, and the multiplication must be done first.

```
for k=1:1:5
        disp(2*k+2)
end
```

or

This can be calculated as 2k + 2.

```
for k=4:2:12
        disp(k)
end
```

Note that all three loop parameters are affected: start *and* finish *are subject to both multiplication and addition, but* step *is only affected by multiplication.*

Comprehension Check 20-1

(a) Write a for loop to display every even number from 2 to 20 on the screen.
(b) Write a for loop to display every multiple of 5 from 5 to 50 on the screen.
(c) Write a for loop to display every odd number from 13 to −11 on the screen.

Using Variable Names to Clarify Loop Function

Using variables with meaningful names as loop controls makes it easier for the programmer (and those trying to interpret the program) to keep track of what the loop is doing. The following loop keeps track of the position of an object moving at a constant speed. By defining the step distance as `Speed*TimeStep` (thus an increment in position), the number of steps in the loop is flexible as well. Note that these variables would need to be defined earlier in the program.

```
for Position=StartPosition:Speed*TimeStep:FinalPosition
        disp(Position)
end
```

Comprehension Check 20-2

Rewrite the `for` loop above using shorter variable names. Be sure that the shorter variable names are clear enough to characterize the purpose of the `for` loop.

Using a `for` Loop in Variable Recursion

An important use of loops is for **variable recursion**—passing information from one loop execution to the next. A simple form of recursion is to keep a running total. Some of the sequences discussed in Comprehension Check 20-1 require recursion. The loop below totals the sales for all the vendors at a football game. The loop assumes that the array `Sales` contains the total sales for each vendor, as might be uploaded from Excel. The number of vendors is computed when the loop begins, using the `length` command, and the total sales are accumulated in `TotalSales`. Note that `TotalSales` must be initialized before the loop is entered, since it appears on the right-hand side of the equation:

```
TotalSales = 0;
for Vendor=1:1:length(Sales)
        TotalSales = TotalSales + Sales(Vendor);
end
```

Manipulating a `for` Loop Counter

Many kinds of computations can be performed in a loop. The loop above will output the value of `k` to the screen each time through the loop. Other sequences can be achieved by having an expression other than `k` in the executable part of the loop.

Comprehension Check 20-3

Consider the following table of values. Figure out the formulae that represent columns A through J below. In some cases, the formulas in rows 1-10 will be similar. In other cases, the entry in the first row will be a number rather than a formula.

Create a `for` loop to display the values; use tabs in a formatted `fprintf` statement to create each column.

k	A	B	C	D	E	F	G	H	I	J
1	2	1	1	1	7.25	10	1	2	1	1.00
2	4	3	4	3	7.50	9	−1	4	2	.50
3	6	5	9	6	7.75	8	1	8	6	.33
4	8	7	16	10	8.00	7	−1	16	24	.25
5	10	9	25	15	8.25	6	1	32	120	.20
6	12	11	36	21	8.50	5	−1	64	720	.17
7	14	13	49	28	8.75	4	1	128	5020	.14
8	16	15	64	36	9.00	3	−1	256	40320	.13
9	18	17	81	45	9.25	2	1	512	362880	.11
10	20	19	100	55	9.50	1	−1	1024	3628800	.10
formula										

Using the Counter Variable as an Array Index

The example above used the counter variable Vendor to access each element of the Sales array in sequence to add them together. If we wanted to add only the odd-numbered elements, the only change necessary would be to change the step variable to 2. If we wanted every fourth element of the array beginning with the fourth element, the for statement would become

```
for Vendor=4:4:length(Sales)
```

and so forth.

To access elements in a two-dimensional array, we could use a single index since MATLAB will handle this, but we would have to be very careful to make certain which row and column is being accessed. For example, if A is a 3 by 2 matrix, A(4) = A(1,2). Since this can be quite confusing, it is usually better to use the double index notation when accessing values in a two-dimensional array. One method to accomplish this would be to use nested for loops, that is, a for loop inside another for loop.

Let's modify the football sales example above as an example.

Assume we have a two-dimensional array named Sales where each row represents a specific vendor and each column represents a different category of item, such as drinks, hot dogs, hats, etc. Now if we want the total sales, we need to step through every element in every column (or every element in every row) and add them all up:

```
% Initialize the sum of all sales to 0
TotalSales = 0;
% determine the number of vendors (rows) and items (columns)
[NumbVendors, NumbItems]=size(Sales)
for Vendor=1:1:NumbVendors    % this will index the rows
    for Item=1:1:NumbItems    % this will index the columns
        TotalSales = TotalSales + Sales(Vendor, Item);
    end
end
```

Note that the inner for loop (Item loop) goes through all items (columns) before exiting, at which point the outer for loop increments to the next vendor (row) and the inner loop resets to the first column and steps through all items again, but for a different vendor.

NOTE

Matrices are described dimensionally as the number of rows "by" the number of columns. Given a matrix as input, the size function saves the number of rows into the first output variable and the number of columns into the second output variable.

If there were four vendors (rows) and three items (columns), the order in which the entries in the `Sales` array would be added to `TotalSales` would be

(1,1); (1,2); (1,3); (2,1); (2,2); (2,3); (3,1); (3,2); (3,3); (4,1); (4,2); (4,3);

Comprehension Check 20-4

Modify the example program above for football sales to create a column vector called `VendorSales`, each element of which contains the total sales for the corresponding vendor.

Modify the example program again to create a row vector called `ItemSales`, each element of which contains the total sales of each corresponding item.

20.2   `while` LOOPS

Once a `for` loop has started, it executes a specified number of times, regardless of what happens in the loop. At other times, we want to continue executing a loop until a particular condition is satisfied. This requires the use of a `while` loop, which executes until the specified condition is false. For example,

- While the calculation error is unacceptable, refine the calculations to improve accuracy.
- While the cell phone call is in progress, keep track of the duration of the call.

```
while error >= 0.01
        % require less than 1% error
        % perform another iteration of the calculation
        % recalculate error
end
```

The conditional part of the `while` statement has the same syntax as that of the `if` statement.

```
while Logical_
  Expression
  % executable
  statements
end
```
- `Logical_Expression`: A valid MATLAB logical expression

Converting "Until" Logic for Use in a `while` Statement

In some cases, it may make more sense to use the word *until* in phrasing a conditional loop, but MATLAB does not have such a structure. As a result, it is sometimes necessary to rephrase our conditions to fit the `while` structure.

```
while Logical_Expression
        % executable statements
end
```

- `Logical_Expression`: A valid MATLAB logical expression

"Until" logic condition	`while` logic translation

until cows == home
    party
end
	while cows ~= home
	    party
	end

until homework == done
    TelevisionPower = off
end
	while homework ~= done
	    TelevisionPower = off
	end

Initializing `while` Loop Conditions

In `for` loop constructions, the loop initializes and keeps track of the loop counter. In a `while` loop, it is necessary to make sure that variables are properly initialized.

```
while error >= 0.01
        % require less than 1% error
        % perform another iteration of the calculation
        % recalculate error
end
```

If `error` is not initialized, a syntax error results. The `error` term does not exist when MATLAB tries to compare it to 0.01. If `error` is incorrectly initialized to a value less than 0.01, the loop never executes.

EXAMPLE 20-5

We want to input a number from the user that is between (and including) 5 and 10, but reject all other numbers. All extraneous output should be suppressed.

```
X=0;
while X < 5 || X > 10
        X=input('Enter a number between 5 and 10');
end
```

Comprehension Check 20-5

Write a `while` loop that requires the user to input a number until a non-negative number is entered.

EXAMPLE 20-6

We have a number stored in a variable `T` that we want to repeatedly divide by 10 until `T` is smaller than 3×10^{-6}. After determining the first value of `T` that meets the condition, it should be displayed in a formatted `fprintf` statement where the value of `T` is displayed in exponential notation. All extraneous output should be suppressed.

```
T=3993;
while T > 3E-6
        T=T/10;
end
fprintf('%E\n',T);
```

EXAMPLE 20-7

Given a vector of positive integers, `V`, we want to create 2 new vectors: vector `E` will contain all of the even values of `V`; vector `O` will contain all of the odd values of `V`. To begin solving this problem, we need to seek out a MATLAB function that will help us to determine whether or not a given value is even or odd. The built in function `rem` accepts two input arguments ($Z = rem(X,Y)$) such that `Z` is the remainder after dividing X by Y. For example, if we type `rem(3,2)`, the function would return 1 because $3/2 = 1$ with a remainder of 1.

```
clear,clc;
V = [ 2, 3, 91, 87, 5, 8];
E = [];
O = [];  % initially, E and O are both empty vectors
fprintf('V ='),disp(V);

while length(V) > 0
    if rem(V(1),2) == 0 % is the first element of V even?
        E=[E V(1)];      % add it to the end of E
    else                 % else first element of V is odd
        O=[O V(1)];      % add it to the end of O
    end
    V(1)=[];             % delete the first element of V
end

fprintf('E ='),disp(E);
fprintf('O ='),disp(O);
```

The output of this code segment would be:

```
V =     2     3    91    87     5     8
E =     2     8
O =     3    91    87     5
```

In-Class Activities

ICA 20-1

Write a program containing a `for` loop to compute the factorial of any positive integer. Include an `if...else...` statement that displays the phrase, "Factorial of #A is #B" (where #A is the number for which the factorial is being computed, and #B is the computed factorial) if the computed factorial is equal to the one computed by MATLAB's "factorial" function. If the computed factorial is not correct, the program should display, "Computational error. Check algorithm".

ICA 20-2

Write a function for computing the sum of a vector of data. It should receive an array as an argument and return the sum of the values.

ICA 20-3

Write a function for computing the average of a vector of data. It should receive an array as an argument and return the average of the values.

ICA 20-4

Write a function that receives as an argument an amount of change to be returned to a customer and returns the number of dollars, quarters, dimes, nickels, and pennies to be given to the customer. Be sure to include the change returned reported in a neatly formatted table.

ICA 20-5

A symmetric matrix is a square matrix, A, that is equivalent to its transpose, $A = A^T$. The entities of a symmetric matrix are symmetric with respect to the main diagonal (top left to bottom right). An example of a symmetric matrix is shown below. Write a MATLAB function, `IsSymmetric`, that determines if the input matrix A is a symmetric matrix.

$$\begin{bmatrix} 1 & 2 & 3 \\ 2 & 4 & -8 \\ 3 & -8 & 9 \end{bmatrix}$$

ICA 20-6

An array named `bolts` has two rows and contains measured lengths and diameters of large bolts being manufactured in a factory. The lengths are stored in row 1, the diameters in row 2. The bolts have a nominal diameter of one inch, and a nominal length of 15 inches. As part of quality control, it is desired to know how many bolts are more than 0.5% out of the nominal specifications, and what type of difference it is. Write a MATLAB program that will print the following information:

- Total bolts out of specification is XXXX.
- Total bolts too long is XXXX.
- Total bolts too short is XXXX.
- Total bolts with diameter too small is XXXX.
- Total bolts with diameter too large is XXXX.
- Total bolts with both length and diameter out of specifications is XXXX.

ICA 20-7

A materials engineer is testing the uniformity of a large sheet of a new high-strength transparent material by measuring the refractive index, η, (a measure of how fast light travels in the material) in every square centimeter of the sheet. The measured values are placed in a two-dimensional array named Refract, each element of which corresponds to the locations on the sheet, for analysis.

Write a program that will generate two outputs.

- A variable named TotalHigh that equals the total number of measurements with a refractive index greater than that of common window glass ($\eta = 1.5$).
- An array named GridLoc containing two columns, and TotalHigh rows. Each row in GridLoc will contain the row and column indices of the locations with $\eta > 1.5$.

ICA 20-8

Assume you are given a square array, Dragon. Write a program using only one for loop that will create a new two column matrix, Diagons. The first column contains the elements of the main diagonal of Dragon; the second column contains the elements of the secondary diagonal.

Example:
$$\text{Dragon} = \begin{bmatrix} 5 & 6 & 7 \\ 0 & 1 & 2 \\ -3 & -4 & -5 \end{bmatrix} \quad \text{Diagons} = \begin{bmatrix} 5 & -3 \\ 1 & 1 \\ -5 & 7 \end{bmatrix}$$

ICA 20-9

> **NOTE**
>
> To create a new matrix of zeros, MATLAB has a built in function zeros, where M is the number of rows and N is the number of columns in the desired matrix of zeros.

Given an integer N, write a program to create an N × N matrix, PascalTri, which contains Pascal's triangle in the upper-right diagonal half (including the main diagonal). The other elements should equal 0.

Example with N = 5
$$\text{PascalTri} = \begin{bmatrix} 1 & 1 & 1 & 1 & 1 \\ 0 & 4 & 3 & 2 & 1 \\ 0 & 0 & 6 & 3 & 1 \\ 0 & 0 & 0 & 4 & 1 \\ 0 & 0 & 0 & 0 & 1 \end{bmatrix}$$

ICA 20-10

In a desperate attempt to raise his dismal student evaluations, Professor Vapid has instituted a new grade-calculation policy and has asked you to write the code to implement his scheme.

The name of each student is in a column vector, StudentNames. The grades for these students are in a two-dimensional array, StudentGrades. Each row represents the student from the corresponding element of StudentNames, and each column represents a specific assignment.

Write a program to create a column vector named StudentAvg, with each element corresponding to the students in StudentNames. The average for each student is to be computed by two methods and the larger of the two calculated values, rounded to the nearest integer, placed in the proper element of StudentAvg.

The first method calculates the average grade for that student after both the highest and the lowest grades for that student are deleted. For example, if the student in row 1 had grades 78, 91, 58, 95, 49, and 62, the calculated average for that student would be (78 + 91 + 58 + 62)/4 =72.25.

The second method calculates the average grade for that student after the two lowest grades for that student are deleted. For example, if the student in row 1 had grades 78, 91, 58, 95, 49, and 62, the calculated average for that student would be (78 + 91 + 95 + 62)/4 = 81.5.

A report should be printed, one student per line, giving the student's name and average. For example, the first line of the report might read:

Ainsley Aardvark: final average = 82

ICA 20-11

Write a program for a simple guessing game. The program should first prompt the user for a number between 1 and 100. The program should then generate a single random number `TargetNum` between 0 and N. This can be done with `TargetNum = N*rand(1)`.

Next, the program should prompt the user to guess the number. After each guess, the program should inform the user whether the guess is too high or too low. When the user guesses the value of `TargetNum` to within 1% (`TargetNum ±1*N`), the program should print the following, where XXXX represents the appropriate values. You will need to consider how many significant digits you should print.

```
TargetNum = XXXX. Your guess of XXXX is within XXXX%. This
required XXXX guesses.
Would you like to play again?
```

If the user responds with "N", the program should terminate.

If the user responds with "Y", the program should prompt the user for a new value of N, generate a new random number, etc.

ICA 20-12

You are programming a controller to automatically turn off the lights in a room when a motion sensor detects no movement for 10 minutes and turn them back on when motion is detected. In addition, the program should sound an audible alarm 15 seconds before turning off the lights. You may assume that the alarm sound is stored in a vector `Warning`. To play the alarm sound, use `sound(Warning)`. You may also assume you have the following functions available to use:

- `Delay(n)`   Waits n milliseconds [ms] before proceeding to the next program line.
- `Motion`   If motion detected in the last 500 ms returns 1; otherwise returns 0.
- `LightsOn`   Turns lights on.
- `LightsOff`   Turns lights off.

Assume that the lights are on when the program begins.

ICA 20-13

You are designing a simple home thermostat to turn a furnace on and off according to the thermostat setting and the room temperature.

Assume you have the following functions available:

- `SetTemp`   Returns the thermostat setting in degrees Fahrenheit.
- `RoomTemp`   Returns the temperature of the room in degrees Fahrenheit.
- `ToggleHeat`   If furnace is on, turns it off. If furnace is off, turns it on.
- `Status`   Returns 1 if thermostat is set to "Heat". Returns 0 if set to "Off".

The thermostat must be set to "Heat" for the furnace to turn on.

If the room temperature falls more than 2 degrees Fahrenheit below the thermostat setting, the furnace should turn on. If the room temperature rises more than 2 degrees Fahrenheit above the thermostat setting, the furnace should turn off.

Note that the thermostat temperature setting and status may be changed at any time; thus, you will need to continually recheck these.

Also note that when the system is first activated, the system must be able to deal with all possible cases.

1. In the game of chess, each piece can move a different way across the board. A diagram of a chessboard at the beginning of a game is shown for reference. Write an algorithm to show how the following pieces can move. The pieces can begin at any location on the board (not necessarily in their starting position) and will proceed forward as far as possible (within the limitations of their piece or until the end of the board is reached or until it encounters another piece).

   (a) Pawn
   (b) Bishop
   (c) Knight

2. You want to calculate your grade point ratio (GPR). Create an algorithm to input your courses for this semester, the semester hours for each course, the letter grade you expect to earn, and number of grade points earned for each course (such as A = 4). Calculate the average GPR for the semester using the following formula: *summation of the product of the fall semester hours and number of points earned divided by the summation of fall semester hours*. After you have written out the algorithm, convert your narrative algorithm to a flowchart.

3. Construct a flowchart to determine the class rank (freshman, sophomore, junior or senior) of students enrolled in this course. Output the total number of students in each rank.

4. Construct a flowchart to make a hamburger, offering the user a choice of 4 toppings (lettuce, tomato, onions, cheese), 3 condiments (mustard, mayonnaise, ketchup) and 3 ways to cook the burger (medium, medium well, well done).

5. Construct a flowchart of how to interpolate, given two ordered pairs of data.

6. Construct a flowchart to draw a histogram in Excel.

7. Write the MATLAB code to evaluate the following mathematical expression:

$$x = \frac{b + \sqrt{b - 4a}}{c^7}$$

8. What is the output as seen on the Command Window after the following MATLAB code segment is executed?

```
R = 4
S = 3;
T = R/S;
fprintf('%4.2f',T);
fprintf('%4.3e',S)
```

9. Rewrite the following MATLAB statements, fixing all syntax errors:

```
r=('Enter the radius');
V = (4/3)(pi)(R^3);
fprintf('\nV='%f,V)
if(V = 4/3 pi r)
disp('The value of V is %V and r is %r')
```

10. What is the value of `W` after this MATLAB program is finished?

```
X = 2;
Y = 4;
Z = 2;
W = (Y+Z/X+Z^1/2)
```

11. Consider the following: C = A + B, where A and B are 2×2 matrices. Write a *function* to add two 2×2 matrices, A and B, to create the resulting matrix, C. The function should take two inputs (A and B) and should return, as function output, C. All calculations must be suppressed.

12. Write a function that implements the quadratic equation. Given three inputs (**a**, **b**, and **c**), calculate the roots (**r1** and **r2**) of the quadratic formula. Recall the quadratic equation:

$$r = \frac{-b \pm \sqrt{b^2 - 4ac}}{2a}$$

13. Write a function that implements the Pythagorean theorem. Recall that the theorem states that the length of the hypotenuse (**z**) can be calculated by the sum of the squares of the adjacent sides (**x** and **y**), or

$$z = \sqrt{x^2 + y^2}$$

14. Write a program that asks a user to input (one at a time) the four numbers to fill a 2×2 matrix. For each number entered, make sure the user knows which number is being entered. Display the completed matrix so the user can check the entries and verify the submission. The function should feature:

   ▪ Standard header. Since a description of what the function does will be explained to the user, a function description does not need to be part of the standard header.
   ▪ Output statements describing the program's function and giving instructions to the user.
   ▪ Input statements that gather matrix elements (output suppressed).
   ▪ Verify the matrix assembled correctly.
   ▪ Matrix displayed to user with explanation.

As a reminder, interpolation is covered in Chapter 8.

15. Write a function that gathers two (x , y) data pairs in a 2×2 matrix and then asks the user to input another value of x for which a value of y will be interpolated or extrapolated. The function should feature:

   ▪ Standard header. Since a description of what the function does will be explained to the user, a function description does not need to be part of the standard header.
   ▪ Output statements describing the program's function and giving instructions to the user.
   ▪ Input statement gathering value of x and determining if the method used is interpolation or extrapolation.
   ▪ Calculation of the value of y from matrix elements.
   ▪ Report of the value of y in the correct sentence below (with numbers in the blanks).

```
Given x = _____, interpolation finds that y = _____ or Given
x = _____, extrapolation finds that y = _____.
```

16. If an object is heated, the temperature of the body will increase. The energy (Q) associated with a change in temperature (ΔT) is a function of the mass of the object (m) and the specific heat (C_p), which is the rate at which the object will lose or gain energy. Specific heat is a material property, and values are available in literature. In an experiment, heat is applied to the end of an object, and the temperature change at the other end of the object is

recorded. This leads to the theoretical relationship shown. An unknown material is tested in the lab, yielding the following results:

$$\Delta T = \frac{Q}{m \, C_p}$$

Heat Applied (Q) [J]	12	17	25	40	50	58
Temp Change (ΔT) [K]	1.50	2.00	3.25	5.00	6.25	7.00

(a) Graph the experimental temperature change (ΔT, ordinate) versus the heat applied (Q, abscissa) using `plot`. This plot should be a proper plot.

(b) Graph the theoretical model that represents the temperature change (ΔT, ordinate) versus the heat applied (Q, abscissa) using `fplot`. This plot should be a proper plot. Consider the mass (m) to be 5 kilograms and the specific heat (Cp) to be copper at 0.39 joules per gram kelvin.

17. When one attempts to stop a car, both the reaction time of the driver and the braking time must be considered. The following data were taken from www.highwaycode.gov.uk.

Vehicle Speed (v) [mph]	Distance (d) [m]	
	Reaction (d_r)	Braking (d_b)
20	6	6
30	9	14
40	12	24
50	15	38
60	18	55
70	21	75

Plot the data using `plot`. This plot should be a proper plot.

18. A *pitot tube* is a device that measures the velocity of a fluid, typically the airspeed of an aircraft. The failure of a pitot tube is credited as the cause of Austral Lineas Aéreas flight 2553 crash in October 1997. The pitot tube had frozen, causing the instrument to give a false reading of slowing speed. As a result, the pilots thought the plane was slowing down, so they increased the speed and attempted to maintain their altitude by lowering the wing slats. Actually, they were flying at such a high speed that one of the slats ripped off, causing the plane to nosedive; the plane crashed at a speed of 745 miles per hour [mph].

In the pitot tube, as the fluid moves, the velocity creates a pressure difference between the ends of a small tube. The tubes are calibrated to relate the pressure measured to a specific velocity, using the speed as function of the pressure difference (P [=] Pa) and the density of the fluid (ρ [=] kg/m³).

$$v = \left(\frac{2}{\rho}\right)^{0.5} P^m$$

Plot the data using `plot`. This plot should be a proper plot.

Pressure (P) [Pa]	50,000	101,325	202,650	250,000	304,000	350,000	405,000	505,000
Velocity Fluid A (v) [m/s]	11.25	16.00	23.00	25.00	28.00	30.00	32.00	35.75
Velocity Fluid B (v) [m/s]	9.00	12.50	18.00	20.00	22.00	24.00	25.00	28.00
Velocity Fluid C (v) [m/s]	7.50	11.00	15.50	17.00	19.00	20.00	22.00	24.50

19. A growing field of inquiry that poses both great promise and great risk for humans is nanotechnology, the construction of extremely small machines. Over the past couple of decades, the size that a working gear can be made has consistently gotten smaller. The table shows milestones along this path.

Years from 1967	0	5	7	16	25	31	37
Minimum Gear Size [mm]	0.8	0.4	0.2	0.09	0.007	2E-04	8E-06

Plot the data using `plot`. This plot should be a proper plot.

20. Capillary action draws liquid up a narrow tube against the force of gravity as a result of surface tension. The height the liquid moves up the tube depends on the radius of the tube. The following data were collected for water in a glass tube in air at sea level.

Radius (r) [cm]	0.01	0.05	0.10	0.20	0.40	0.50
Height (H) [cm]	14.0	3.0	1.5	0.8	0.4	0.2

Plot the data using `plot`. This plot should be a proper plot.

21. In a turbine, a device used for mixing, the power required by the turbine depends on the size and shape of impeller. In the lab, we have collected the following data:

Diameter (D) [ft]	0.5	0.75	1	1.5	2	2.25	2.5	2.75
Power (P) [hp]	0	0.04	0.13	0.65	3	8	18	22

Plot the data using `plot`. This plot should be a proper plot.

22. You will be given an Excel spreadsheet that contains salary data from major league baseball in 2005, titled "Baseball Data." Use the data provided to determine:
  (a) Salary mean, median, and standard deviation for all players.
  (b) Salary mean, median, and standard deviation for the Arizona Diamondbacks.
  (c) Salary mean, median, and standard deviation for the Atlanta Braves.
  (d) Salary mean, median, standard deviation for all pitchers.
  (e) Salary mean, median, standard deviation for all outfielders.
  (f) Which team had the highest average salary?
  (g) Which team had the lowest average salary?
  (h) Which position had the highest average salary?
  (i) Which position had the lowest average salary?
  (j) Draw a histogram and CDF in Excel for all players.
  (k) What percentage of players earned more than $1 million?
  (l) What percentage of players earned more than $5 million?

23. The following set of data was collected by Ed Fuller of the NIST Ceramics Division in December 1993. The data represent the polished window strength, measured in units of ksi [kilopounds per square inch] and were used to predict the lifetime and confidence of airplane window design. Use the data set to generate a histogram and CDF. (http://www.itl.nist.gov/div898/handbook/eda/section4/eda4291.htm).

24. The following information was taken from the report of the EPA on the U.S. Greenhouse Gas Inventory (http://yosemite.epa.gov/oar/globalwarming.nsf).

Greenhouse gas emission inventories are developed for a variety of reasons. Scientists use inventories of natural and anthropogenic emissions as tools when developing atmospheric models. Policy makers use inventories to develop strategies and policies for emissions reductions and to track the progress of those policies. Regulatory agencies and corporations rely on inventories to establish compliance records with allowable emission

> Importing data from Excel is a process that keeps changing, due to updates in both Excel and MATLAB operating systems.
>
> For the most current information on how to import data from and Excel (XLS) file or a comma-separated value file (CSV), search MATLAB for `xlsread` or `csvread`.

rates. Businesses, the public, and other interest groups use inventories to better understand the sources and trends in emissions.

The Inventory of U.S. Greenhouse Gas Emissions and Sinks provides important information about greenhouse gases, quantifies how much of each gas was emitted into the atmosphere, and describes some of the effects of these emissions on the environment.

In nature, carbon is cycled between various atmospheric, oceanic, biotic, and mineral reservoirs. In the atmosphere, carbon mainly exists in its oxidized form as carbon dioxide (CO_2). Carbon dioxide is released into the atmosphere primarily as a result of the burning of fossil fuels (oil, natural gas, and coal) for power generation and in transportation. It is also emitted through various industrial processes, forest clearing, natural gas flaring, and biomass burning.

The EPA website provides data on emissions. The data found in the file "Carbon Dioxide Emissions Data" were taken from this website for the year 2001 for all 50 states and the District of Columbia.

(a) Use the data on all 51 states (including the District of Columbia) provided in the file "Carbon Dioxide Emissions Data" to create a histogram with an appropriate bin size.

(b) Determine the mean and median of the data. Indicate each of these values on your graph for part a.

(c) Which value more accurately describes the data? Indicate your choice (mean or median) and the value of your choice. Justify your answer in words.

25. Choose *one* of the following options and collect the data required.

(A) On a campus sidewalk, mark two locations 50 feet apart. As people walk along, count how many steps they take to go the 50 feet. Do this for 125 individuals.

(B) Select 250 words at random from a book (fiction). Record the number of letters in each word. Alternatively, you can count the words in 250 sentences and record these.

(C) Go to one section of the library, and record the number of pages in 125 books, all of which are in the same section.

(D) Interview 125 people and determine how far away (in miles) their home is from the university.

For the data source you select, you are to do the following:

(a) Construct a histogram, including justification of bin size.

(b) Determine the mean, median, variance and standard deviation values.

(c) Construct a cumulative distribution function.

26. During the month of November, the HVAC system in your apartment appeared to be broken. To prove this, you record the following daily high temperatures in degrees Fahrenheit, taken every other day.

58	54	60	58	55	55	60	60
86	80	85	85	82	85	91	93

(a) Calculate the mean and median of the data.

(b) Draw the associated histogram and CDF for the data, using temperature on the abscissa and days on the ordinate.

(c) When you take your complaint to the apartment manager, he fails to see the problem; according to his heating bill, your apartment had an average temperature of 72°F. Show both the mean and median on the graph drawn for part b. Which is a better presentation of the data, part a or part b? Justify.

27. Write a program that takes as input StartDist (m), FinishDist (m), and TotalTime (s) and returns the phrase "Backing up at _____ m/s," "Not moving," or "Moving forward at _____ m/s" as appropriate (including numbers where there are blanks).

28. A phase diagram (from http://www.eng.ox.ac.uk/~ftgamk/ engall_tu.pdf) for copper-nickel, shows that three phases are possible: a solid alpha phase (a solid solution), a liquid solution, and a solid-liquid combination.

Write a program that takes as input a mass % of nickel and a temperature and gives as output a message indicating what phases are present. Assume the region borders are linear, and determine a method for dealing with any points that lie directly on the lines.

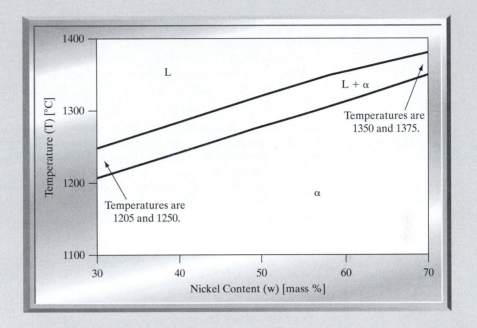

29. This generic phase diagram, based on the temperature and composition of elements A and B, is taken from <http://www.soton.ac.uk/~pasr1/build.htm>, where a description of how phase diagrams are constructed can also be found. The alpha and beta phases represent solid solutions of B in A and of A in B, respectively. The eutectic line represents the temperature below which the alloy will become completely solid if it is not in either the alpha or beta region. Below that line, the alloy is a solid mixture of alpha and beta. Above the eutectic line, the mixture is at least partially liquid, with partially solidified lumps of alpha or beta in the labeled regions.

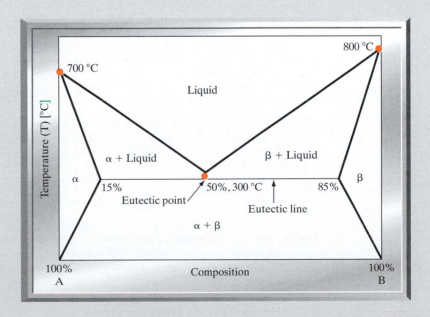

Using the simplified phase diagram, which substitutes straight lines for the curved lines typical of phase diagrams, write a MATLAB program that takes as input the mass % of B and the temperature and returns the phases that may exist under those conditions. Assume the following:

- The melting point of pure A is 700 degrees Celsius.
- The melting point of pure B is 800 degrees Celsius.
- The eutectic line is at 300 degrees Celsius. and the eutectic point occurs when the composition is 50% B.
- The ends of the eutectic line are at 15% B and 85% B.

Your MATLAB program should include special notes if the provided conditions are on the eutectic line or at the eutectic point.

30. Soil texture can be classified with a moist soil sample and the questions in the table below from the Soil Texture Key found at <http://www.denislindsell.demon.co.uk/pasture/soils/>. Each answer results in either a soil classification or directions to continue. Write a program that implements this decision tree. You may find it useful to download the key itself from the above address. You may also want to compare your results to an online implementation of the key found at the same website. Your program should include use of menu windows to gather user responses.

Question	Yes	No
1. Does the soil feel or sound noticeably sandy?	Go to Q2	Go to Q6
2. Does the soil lack all cohesion?	SAND	Go to Q3
3. Is it difficult to roll the soil into a ball?	LOAMY SAND	Go to Q4
4. Does the soil feel smooth and silky as well as sandy?	SANDY SILT LOAM	Go to Q5
5. Does the soil mould to form a strong ball that smears without taking a polish?	SANDY CLAY LOAM	SANDY LOAM
6. Does the soil mould to form an easily deformed ball and feel smooth and silky	SILT LOAM	Go to Q7
7. Does the soil mould to form a strong ball that smears without taking a polish?	Go to Q8	Go to Q10
8. Is the soil also sandy?	SANDY CLAY LOAM	Go to Q9
9. Is the soil also smooth and silky?	SILTY CLAY LOAM	CLAY LOAM
10. Does the soil mould like plasticine, polish and feel very sticky when wetter?	Go to Q11	Restart unless organic
11. Is the soil also sandy?	SANDY CLAY	Go to Q12
12. Is the soil also smooth and buttery?	SILTY CLAY	CLAY

31. You are part of a team developing the guidance programming for a robot vacuum. One of several modes the robot can be in is "wall following"—proceeding forward next to a wall to clean the interface where wall and floor meet. However, it is possible for the robot to enter wall-following mode when it bumps into a column, table leg, or other item that is out in the area being vacuumed. Your task is to implement code to prevent the robot from becoming trapped endlessly circling a column or table leg.

The robot can also be in a mode called "pentacleaning," where it deliberately circles in one spot five times to clean particularly heavy dirt. This mode must be allowed without letting the robot get trapped going around a column.

After the robot has been going straight for at least three feet, a variable called `TurnTotal` is set to zero. If the robot then begins a turn, `TurnTotal` is updated to contain the number of degrees turned since it last went straight for three feet or more. Turns to the right add to the total, turns to the left subtract from the total. For this simple example, assume all turns are 90°. If `TurnTotal` reaches ±360°, the robot has gone in a "circle" (although its path may look more like a square or such) less than three feet across.

Write a function that accepts two variables, `TurnTotal` and `CleanMode`. If the robot is in pentaclean mode, `CleanMode` equals 5. If the robot is in wall-following mode, `CleanMode` equals 2. The function should return a variable called `ChangeMode`. `ChangeMode` should be set to TRUE in the following circumstances:

(a) The robot is in wall-following mode and has gone in more than two complete "circles."
(b) The robot is in pentaclean mode and has gone in more than five complete "circles."
(c) The robot is in any other mode and has gone more than half a complete "circle."

32. One of your coworkers at the robot vacuum design lab has been given the task of writing the wall-following algorithm (see Question 31). In testing, the left wall-following algorithm does not work correctly—the robot keeps getting stuck in corners. Your colleague seeks you out to ask if you can find the error in the code.

The sensors mounted on the robot that are used in the wall-following algorithm are:

- Two side sensors detect the presence of a wall. These sensors are mounted near the rear of the robot, one on each side. If the robot is within three inches of a wall to the side, a variable is set to 1; otherwise, it is set to 0. A setting of zero indicates that the robot should turn in that direction if it is following the wall. There are two such variables, one for each side, named `LeftWall` and `RightWall`.
- There is a front collision sensor. If the robot bumps into something in front of it, a variable named `FrontWall` is set to 1; otherwise, it is 0. To follow the wall on the left, it should turn right when it encounters a wall in the front.

There are also several functions for moving the robot around:

- `WallForward2` moves the robot forward 2 inches in a straight line, maintaining a maximum distance of 2 inches from the wall.
- `WallForward4` moves the robot forward 4 inches in a straight line, maintaining a maximum distance of 2 inches from the wall.
- `FwdTurnRight` turns the robot 90° to the right (pivots on the rear-right corner).
- `FwdTurnLeft` turns the robot 90° to the left (pivots on the rear-left corner).
- `BackTurnRight` turns the robot 90° to the right (pivots on the rear-left corner).
- `BackTurnLeft` turns the robot 90° to the left (pivots on the rear-right corner).
- `CenterTurnRight` turns the robot 90° to the right (pivots on the rear center).
- `CenterTurnLeft` turns the robot 90° to the left (pivots on the rear center).

The robot itself is shaped like a semicircle, with the flat side being the rear. The radius of the semicircle is 8 inches.

The problem has been isolated to the following code fragment, which is executed after every use of one of the WallForward functions above. Explain why this code malfunctions, and write a corrected version.

```
if LeftWall==1
WallForward2
elseif FrontWall==1
BackTurnRight
WallForward2
else
FwdTurnLeft
WallForward4
end
```

Vac-U-Bot

33. The variable grade can have any real number value from 0 to 100. Write an if-elseif-else statement that displays the letter grade (any format) corresponding to a numerical grade. A typical range of grades:

A: $90 \leq$ grade
B: $80 \leq$ grade < 90
C: $70 \leq$ grade < 80
D: $60 \leq$ grade < 70
F: grade < 60

34. Write a function for finding the maximum value in a vector of data. It should receive an array as an argument and return the maximum of the values without using MAX or other MATLAB built-in functions.

35. Write a function for sorting numerical values in a vector of data. It should receive an array as an argument and return an array with the same values sorted without using any MATLAB built-in functions. It should ask the user if the data is to be sorted in ascending or descending order.

36. Below is the same table presented in the for loop chapter. Implement a for loop that computes a new value for A–J and displays the next row of the table as output each time through the loop. Before loop execution, write a separate output statement to create the table header. Since some of the columns use recursion, reuse the previous value of the variables in those columns. Columns that use recursion will also require the initialization of their values to ensure proper computation of the first entry in the column.

k	A	B	C	D	E	F	G	H	I	J
1	2	1	1	1	7.25	10	1	2	1	1.00
2	4	3	4	3	7.50	9	−1	4	2	.50
3	6	5	9	6	7.75	8	1	8	6	.33
4	8	7	16	10	8.00	7	−1	16	24	.25
5	10	9	25	15	8.25	6	1	32	120	.20
6	12	11	36	21	8.50	5	−1	64	720	.17
7	14	13	49	28	8.75	4	1	128	5020	.14
8	16	15	64	36	9.00	3	−1	256	40320	.13
9	18	17	81	45	9.25	2	1	512	362880	.11
10	20	19	100	55	9.50	1	−1	1024	3628800	.10

37. Write a function that receives as arguments two matrices A and B (in that order) and returns the product C = AB. The program must use for loops and matrix indices to compute each term in C, and cannot use MATLAB's ability to multiply either matrices or vectors. If the matrices A and B cannot be multiplied because their inner matrix dimensions do not agree, you must give an error message. The function should include:

- Determination of argument matrix dimensions using the length command; this determines if there is an error and establishes the loop limits.
- Error reported if inner matrix dimensions do not agree.
- for loops addressing each term in C.
- Correct vector product calculations of each term in C.

38. It is sometimes necessary to "smooth" a data set in order to simplify certain analyses. Write a function that receives as an argument a vector of any length and returns a smoothed vector. The function should include:

- Each array term computed as the average of the five terms surrounding the original term.
- First and last terms computed as special cases, that is, compute the smoothed value by counting the original term three times in the average (and the neighboring points once each).
- Second and next-to-last terms computed as special cases, that is, compute the smoothed value by counting the original term and the last term 1.5 times each in the average (and neighboring points once each).
- Smoothed vector returned.

39. Write a function that receives a number as its argument and returns the square root of that number. You may not use the built-in function `sqrt`. We use two important observations in this algorithm. The first is that the square root of a number is always between the number and 1. The second is that if we make a guess at the square root, the real square root lies between the guess and the number itself divided by the guess.

(a) Make a first guess as the average of the number and 1. (Related to the first observation above.)

(b) Divide the original number by the guess.

(c) Compute $|(\text{quotient}^2 - \text{guess}^2)|/\text{number}$. (This is a measure of the error.)

(d) If a smaller error is desired, our new guess is the average of the guess and the quotient. (Related to the second observation above.)

(e) Repeat from step (b) until step (c) tells us we are close enough; report average of guess and quotient as the root.

Testing the Algorithm

Given the number 3, for which we want the square root,

(a) Make a first guess as the average of the number and 1.

First guess is average of 3 and 1, which is 2.

(b) Divide the original number by the guess.

Quotient is $\frac{3}{2}$, which is 1.5.

(c) Compute (quotient − guess)/number. (This is a measure of the error.)

$|(\text{Quotient}^2 - \text{Guess}^2)|/\text{Number} = |(2.25 - 4)|/3 = 58\%$ error.

(d) If a smaller error is desired, our new guess is the average of the guess and the quotient.

A smaller error is desired. New guess = average of 1.5 and 2 = 1.75.

(e) Repeat from step (b) until step (c) tells us we are close enough.

  (b) Quotient is $\frac{3}{1.75}$, which is 1.71.

  (c) $|(\text{Quotient}^2 - \text{Guess}^2)|/\text{Number} = |(3.06 - 2.94)|/3 = 4.1\%$ error.

  (d) A smaller error is desired. New guess = average of 1.75 and 1.71 = 1.73.

    (b) Quotient is $\frac{3}{1.730}$, which is 1.734.

    (c) $|(\text{Quotient}^2 - \text{Guess}^2)|/\text{Number} = |(3.01 - 2.99)|/3 = 0.7\%$

    (d) Report root as average of 1.730 and 1.734, which is 1.732.

You must include the following features:

- `if` statement that checks if the user inputs a number less than zero; `if` this occurs, the program will issue the following statements to the user (and then do the specified action):

This program cannot compute the square root of negative numbers. The square root of the absolute value of that number will be computed.

- ▪ `while` statement that loops until an accuracy of 1% has been achieved; use the error estimate discussed in the example covering the square root calculation algorithm;
- ▪ Print out, for each iteration, the current guess and the accuracy of that guess, with program output as follows:

  A guess of ___ gives an accuracy of ___ %.
  A guess of ___ gives an accuracy of ___ %.

- ▪ At program completion, the solution declared in a simple sentence, for example,

  The square root of 5 is 2.2361.

COMPREHENSION CHECK ANSWERS

Chapter 3

CC 3-1

Criterion	Computer	Automobile	Bookshelf
Inexpensive	Less than $300	Less than ⅕ the median annual U.S. family income	Less than $30
Small	Folds to the size of a DVD case	Two can fit in a standard parking space	Collapses to size of briefcase
Easy to assemble	No assembly: just turn it on	All parts easily replaceable	Requires only a screwdriver
Aesthetically pleasing	Body color options available	Looks like the Batmobile©	Blends well with any decor
Lightweight	Less than one pound	Less than one ton	Less than 5% of the weight of the books it can hold
Safe	Immune to malware	Receives 5 star rating in NCAP crash tests	Stable even if top-loaded
Durable	Survives the "Frisbee® test"	200,000 mile warranty	Immune to cat claws
Environmentally friendly	Contains no heavy metals	Has an estimated MPG of at least 50	Made from recycled materials Low VOC finish

CC 3-2

(This list is not prioritized)

- Materials
  - Durable
  - Made recycled material or renewable resources
- Landscaping
  - Drought tolerant plants
  - No invasive species
  - Provides cover and food for wildlife
  - Minimizes runoff and erosion
- Location
  - Previously developed property (doesn't destroy pristine area)
  - Close to basic necessities (grocery, medical care, etc.)
  - Does not interfere with migration paths, endangered species, etc.
- Water Usage
  - Greywater system
  - Rainfall collection system
  - High efficiency plumbing fixtures (low flush volume, auto-cutoff faucets)

- Energy usage
  - On-demand water heaters
  - Alternative energy collection (solar, wind, etc.)
  - Energy efficient appliances
- Indoor air quality
  - No outgassing materials
  - Radon control
  - HEPA filters in HVAC system
  - Sufficient outside air exchange
- Construction
  - Minimize construction waste
  - Minimize soil disruption and plant destruction during construction

CC 3-3

- Decrease aerodynamic drag
  - Remove unnecessary protrusions
  - Redesign body based on wind-tunnel testing
- Decrease weight
  - Use carbon-composite materials
  - Remove unnecessary material
- Increase engine efficiency
  - Use ceramic parts (e.g. valves and pistons) so engine can run hotter, thus more efficiently
  - Change to hybrid design with regenerative braking
  - Limit maximum acceleration if consistent with safety
- Manage friction
  - Improved bearings in wheels, engine, etc.
  - Redesign tires for less slippage on road surface
- Encourage fuel-efficient driving
  - Include displays for real-time and cumulative fuel usage (MPG)

CC 3-4

PAT

Hindrance: Tended to slow the team down due to difficulty understanding things

Helpful: Always available and willing to do assigned tasks. The individuals who helped explain things to Pat probably developed a deeper understanding of the material in the process, thus this was helpful to them personally, though not to the team directly. Probably the second most useful team member.

CHRIS

Hindrance: Often absent, seldom prepared, seldom contributes, offers excuses for poor performance. Definitely the worst team member.

Helpful: Teaches the other team members about the real world, and how to deal with slackards.

TERRY

Hindrance: Not a team player, impatient, not encouraging to others, wants to dominate the team. Despite cleverness, probably the second worst team member.

Helpful: Solves problems quickly

ROBIN

Hindrance: No obvious hindrance

Helpful: The real leader of the group, kept things going, encouraging to weaker members. The most useful team member.

Chapter 5

CC 5-1

(a)	Significant figures: 2	Decimal places: 4
(b)	Significant figures: 4	Decimal Places: 0
(c)	Significant figures: 3	Decimal Places: 2
(d)	Significant figures: 3	Decimal Places: 0
(e)	Significant figures: 2	Decimal Places: 0 or 3
(f)	Significant figures: 5	Decimal Places: 0
(g)	Significant figures: 7	Decimal Places: 0
(h)	Significant figures: 5	Decimal Places: 3

CC 5-2

(a) -58.9
(b) 247
(c) 2.47
(d) 0.497

CC 5-3

(a) Elevator capacity may require a conservative estimate of 180 pounds.
(b) Serving sizes are usually measured in integer cups, so estimate at 1 cup, but to make sure the bowl is full, round up to 1.3 cups.
(c) If recording, conservative might be to round up to 33 seconds. If it doesn't matter, 30 seconds is fine.

CC 5-4

(a) Scientific notation: 5.809×10^7 Engineering notation: 58.093×10^6
(b) Scientific notation: 4.581×10^{-3} Engineering notation: 4.581×10^{-3}
(c) Scientific notation: 4.268×10^7 Engineering notation: 42.677×10^6

Chapter 6

CC 6-1

H = 15 ft
W = 18.5 ft
L = 25 ft

CC 6-2

Objective:

Determine the mass of the gravel in units of kilograms

Observations:

- Assume the container is full, thus the height of the gravel is 15 feet
- Assume the gravel is flat on the top and not mounded up
- Neglect the space between the gravel pieces

CC 6-3

List Variables and constants:

- ρ = density $\quad [=] \text{lb}_m/\text{ft}^3 = 97 \text{ lb}_m/\text{ft}^3$ – given
- V = volume of $\quad [=] \text{ft}^3$
  container
- m = mass of $\quad [=] \text{kg}$
  gravel
- L = length $\quad [=] \text{ft} \quad = 25 \text{ ft}$ – given
- W = width $\quad [=] \text{ft} \quad = 18.5 \text{ ft}$ – given
- H = height $\quad [=] \text{ft} \quad = 15 \text{ ft}$ – assumed

CC 6-4

Equations:

1. Density in terms of mass & volume: $\quad \rho = m/V$
2. Volume of a rectangle: $\quad V = L \times W \times H$
3. Unit conversion: $\quad 1 \text{ kg} = 2.205 \text{ lb}_m$

CC 6-5

- First, find the volume of the container; use equation (2)
  $V = L \times W \times H$
- Substituting L, W, and H:
  $V = (25 \text{ ft})(18.5 \text{ ft})(15 \text{ ft}) = 6937.5 \text{ ft}^3$

- Second, find the mass of gravel; use equation (1)
  $\rho = m/V$
- Substituting the density and the volume:
  $(97 \text{ lb}_m/\text{ft}^3) = m \, (6937.5 \text{ ft}^3) \quad m = 672937.5 \text{ lb}_m$
- Convert from pound-mass to kilograms:
  $$\frac{672{,}937.5 \text{ lb}_m}{} \left| \frac{1 \text{ kg}}{2.205 \text{ lb}_m} \right. = 305{,}881 \text{ kg} \cong 306{,}000 \text{ kg}$$

Chapter 7

CC 7-1

- No title
- No axis labels
- Is year 1 1993 or 2000 or . . .?
- Need different line types
- This may be better shown as a column chart instead of a line graph

CC 7-2

- No title
- Poor abscissa scale
- No units or description of axis quantities (what is IC?)
- Need different symbols and line types
- Shouldn't use light colors
- No legend
- Need gridlines if wish to read data

CC 7-3

- Title could be easier to understand
- Percent of what? Unclear
- Symbols too small to distinguish
- Is abscissa value average for decade, or value in that yer?
- Colors hard to see on background; need different line types, darker line colors, and white background

CC 7-4

- Title rather cryptic
- No ordinate quantity description
- No vertical axis line
- No gridlines
- No abscissa labels
- Use legend instead of labels on lines
- Different line types needed

CC 7-5

(a) Equal to a constant positive value
(b) Equal to zero
(c) Equal to a constant negative value
(d) Increasing at a constant rate
(e) Increasing at an increasing rate
(f) Increasing at a decreasing rate

CC 7-6

Velocity:

Distance:

CC 7-7

(a) Machine 2
(b) Machine 2
(c) 1×10^6 feet
(d) Machine 2
(e) Machine 1

Chapter 8

CC 8-1

$-24.8\ °C$

CC 8-2

22,151 feet

CC 8-3

$-35.8\ °F$

CC 8-4

$-38.4\ °F$

Chapter 9

CC 9-1

Mean: 9.3
Median: 9
Variance: 12.9
Standard Deviation: 3.6

CC 9-2

Mean: 102.6
Median: 104
Variance: 295.4
Standard Deviation: 17.2

CC 9-3

(a) Population size has decreased
(b) Variance has increased
(c) Mean has increased

CC 9-4

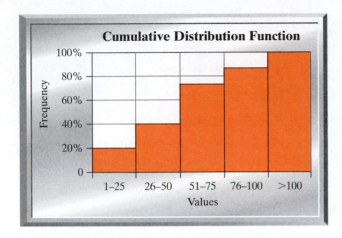

Cumulative Distribution Function

CC 9-5

Not in control
Rule 2—More than nine consecutive points below mean after 26 h

Rule 3—Downward trend between 25 h and 37 h

Rule 5—Two out of three in zone A – Several after 29 h

Rule 6—Four out of five on same side of mean in zone B—start at 33 h

Rule 8—Eight consecutive points outside of zone C—after 28 h

CC 9-6

Machine A not in control: Rule 7—Fifteen consecutive points in zone C

Machine B possibly not in control: Possibly Rule 3—six points with downward trend

Chapter 10

CC 10-1

(a) 5×10^{-9} picoliters
(b) 8×10^{-14} microliters
(c) 3×10^{-19} liters

CC 10-2

(a) Incorrect: 5 s
(b) Incorrect: 60 mm
(c) Incorrect: 3,800 mL OR 3.8 L

CC 10-3

(a) 5.5 mi
(b) 6.78 mi
(c) 328 ft

CC 10-4

2 km/min

CC 10-5

231 in^3

CC 10-6

(a) 84 ft^3
(b) 0.042 m^3
(c) 1058 L

CC 10-7

$L^{0.5}T^{-1}$

CC 10-8

1.73×10^5 mm^2

CC 10-9

2.72 gal

Chapter 11

CC 11-1

13.3 N

CC 11-2

11 lb$_f$

CC 11-3

3,214 lb$_m$/ft^3

CC 11-4

SG = 1.61

CC 11-5

194. 5 K

CC 11-6

$0.000357 \frac{\text{BTU}}{\text{g K}}$

CC 11-7

221 in Hg

CC 11-8

1.22 atm

CC 11-9

220 kPa

CC 11-10

2.4 L

CC 11-11

16.9 m

CC 11-12

20 K

CC 11-13

0.83 hr

CC 11-14

61.3 W

CC 11-15

Efficiency is dimensionless and unitless

Chapter 13

CC 13-1

(a) Relative addressing; cell F23 displays 13
(b) Absolute addressing; cell H26 displays 30
(c) Mixed addressing; cell D30 displays 30; cell F28 displays 5
(d) Mixed addressing; cell G30 displays 5; cell I28 displays 30

CC 13-2

A1.	12
A2.	11
A3.	10
A4.	32
A5.	3.14159265
A6.	110
A7.	48
A8.	1.57079633
A9.	1
A10.	0.89399666

A11.	0.78540775
A12.	45
A13.	#NAME? (Excel does not have a function named "cubrt")

CC 13-3

NO; possible answers given

(a) `=IF(B3>B4,B3 - B4,1)` OR
`=IF((B4 - B3)>=0,1,B3 2 B4)`
(b) `=IF(B3<=B4,'MAX',B4)` OR
`=IF((B3 - B4)>0,B4,'MAX')`
(c) `=IF((B2 + B3 1 B4)<100,`
`'Too Low','')`

CC 13-4

NO; two possible answer given

```
=IF(B4>0,If(B4,100<'liquid',
  'steam'),'ice')
=IF(B4>=100,'steam',IF
  (B4<=0,'ice','liquid')
```

CC 13-5

NO; two possible answers given

```
=IF(B2>=80,IF(B3>=80,IF
  (B4>=80,'OK','Recycle'),
  'Recycle'),'Recycle')
=IF(OR(B2<80,B3<80,B4<80),
  'Recycle','OK')
```

CC 13-6 ▶

In cell E7: =VLOOKUP(B7, G8:H17,2,FALSE)

This assumes the table of planetary gravities is in G7:H17 (header in row 7)

The name of the planet is typed into cell B7.

** See video in online materials

CC 13-7 ▶

** See video in online materials

CC 13-8

City ascending: Fountain Inn

City descending first, Site Name descending second: Rochester Property

Contaminant ascending first, Site Name ascending second: Sangamo Weston

Chapter 14

CC 14-1 ▶

** See video in online materials

CC 14-2 ▶

** See video in online materials

Chapter 15

CC 15-1

(a) atm/K
(b) 35.6 g
(c) 14.1 L

CC 15-2

(a) 63.1 lb_m/ft^3
(b) 3.36×10^{-4} $lb_m/(ft\ s)$
(c) 5×10^{-4} Pa s
(d) 0.455 St

CC 15-3

NOTE: Units on all values are N/m

Singles: 1, 2, 3

2 in parallel: 3, 4, 5

2 in series: 0.667, 0.75, 1.2

3 in parallel: 6

3 in series: 0.545

Parallel combination of 2 in series with third: 1.5, 1.33, 0.833

Series combination of 2 in parallel with third: 3.67, 2.75, 2.2

CC 15-4

NOTE: Units on all values are ohms $[\Omega]$

Singles: 2, 3

2 in series: 4, 5

2 in parallel: 1, 1.2

3 in series: 7

3 in parallel: 0.75

Series combination of 2 in parallel with third: 1.71, 1.43

Parallel combination of 2 in series with third: 4 (same as 2 + 2 in series), 3.2

CC 15-5

(a) L atm
(b) 11.4 g
(c) 387 °C

CC 15-6

(a) 35 g
(b) 1 h^{-1}

Chapter 17

CC 17-1

Known:
- The minimum value in the sum will be 2.
- The maximum value in the sum will be 20.
- Only even numbers will be included

Unknown:
- The sum of the sequence of even numbers

Assumptions:
- [None]

CC 17-2

Known:
- The minimum value in the product of powers of 5 will be 5
- The maximum value in the product of powers of 5 will be 50

Unknown:
- The product of the sequence of powers

Assumptions:
- Only include integer values

Concerning the actual operation to be performed, there are two alternate assumptions, since the wording is intentionally unclear.

1. Multiply all values between 5 and 50 that are an integer power of 5. In other words the product of 5 and 25.
2. Multiply all integer powers of 5 with a power between 5 and 50 inclusive. In other words $5^5 * 5^6 * 5^7 * \cdots * 5^{49} * 5^{50}$

CC 17-3

Known:

- Lowest power of 5 in product is 5
- Highest power of 5 in product is 50

Unknown:

- Product of all integer powers of 5 between the lowest and highest powers

Assumptions:

- The result will not exceed the maximum representable number in the computer. Note that by the time the sequence gets to x = 17, the product is approximately one googol (10^{100})

Algorithm:

1. Set product = 1
2. Set x = 5
3. If x is less than or equal to 50
    - **(a)** Multiply product by 5^x
    - **(b)** Add 1 to x
    - **(c)** Return to Step 3 and ask the question again
3. End the process

CC 17-4

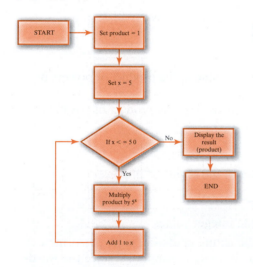

CC 17-5

c, d, e, g

CC 17-6

A = 1
B = 1.2000
C = 1.0000 e^{-10}
D = 2.0000 + 7.0000i
E = 2
The semicolon suppresses the output to the screen.

CC 17-7

Hello to Command Window	`'Hello'` in workspace
Hello World! to Command Window	`'Hello World!'` in workspace
It's 3:00 PM to Command Window	`'It's 3:00 PM'` in workspace

Two consecutive single quotes stores a single quote.

CC 17-8

A = [1,2,3]
B = [1,2,3]
C = [2;4;6]
D = [2,3,4,5,6]
E = [1,2,3;4,5,6]
F = [1,2,3;4,5,6]
G = [1,2,3;4,5,6]
H = [1,2,3;4,5,6]

- The colon produces a sequence of consecutive integers from the value on the left of the colon to the value on the right of the colon, inclusive.
- A comma inside square brackets marks the end of one value in a row and the beginning of the next value in that row.
- A blank space does the same thing as the comma. Spaces after a bracket or after a semicolon are ignored.

CC 17-9

314.1953
`Dogs=areaCircle(10);`

CC 17-10

3.7699e3
`Cats=volumeCylinder(10,12);`

CC 17-11

314.1953
62.8319

CC 17-12

3.7699e3
753.9822

Chapter 18

CC 18-1

```
UserHeight=input('Please enter your height
in inches. ');
CurrentTemp=input('Please enter the current
temperature in degrees F. ');
```

CC 18-2

```
EyeColor=input('Please enter the color of
your eyes. ','s');
ThisMonth=input('Please enter the name of
the current month. ','s');
```

CC 18-3

```
Month=menu('The current month is:', 'Jan',
'Feb', 'Mar', 'Apr', 'May', 'Jun', 'Jul',
'Aug', 'Sep', 'Oct', 'Nov', 'Dec');
```
(all on same line) C = 5

CC 18-4

%.2f
%.3f
0.35
0.354

CC 18-5

```
% assumes the low temperatures in array LowT
max_low=max(LowT);
min_low=min(LowT);
range=max_low-min_low;
number_bins=round(sqrt(length(LowT)));
bin_size=range/number_bins;
% assumes center_value and bin_range defined
N=hist(LowT,center_value)
LTSum=cumsum(N);
CDF=LTSum/max(LTSum)*100;
bar(CDF)
set(gca,'XTickLabel',bin_range);
xlabel('Temperature [T] [deg F]');
ylabel('CDF Value [%]');
title('CDF of Daily Low Temperature in
Clemson, SC in October 2006')
```

Chapter 19

CC 19-1

Red	Yellow	Green	Action
0	0	0	Power out – Caution
0	0	1	Continue through intersection
0	1	0	Caution – Decide to stop or go
0	1	1	–
1	0	0	Stop
1	0	1	–
1	1	0	–
1	1	1	–

T < 0°C	T > 100°C	State
0	0	Liquid
0	1	Steam
1	0	Ice
1	1	–

Score < 60	Score < 70	Score < 80	Score < 90	Grade
0	0	0	0	A
0	0	0	1	B
0	0	1	0	C
0	0	1	1	C
0	1	0	0	D
0	1	0	1	D
0	1	1	0	D
0	1	1	1	D
1	0	0	0	F
1	0	0	1	F
1	0	1	0	F
1	0	1	1	F
1	1	0	0	F
1	1	0	1	F
1	1	1	0	F
1	1	1	1	F

CC 19-2

$5 <= t \text{ && } t < 10$
$-30 < M \text{ && } M <= 20$
$Y <= 100$

CC **19-3**

There is not a logical expression after the "else" because you know (or the program "knows") that the statement following the else is only executed if the conditional statement in the if statement was false. Since Speed > 65 is false, this implies that your speed is 65 or less—you do not need to ask the question—you already know the answer.

CC **19-4**

```
% Assume 'on', 'empty', and 'sound' have been
% appropriately defined.
if HotPlate == on && Beaker = empty
      AlarmStatus=sound
end

% Assume 'MinPrice', and 'PtoEMax' have been
% appropriately defined.
if Price < MinPrice || P_E_Ratio >  PtoEMax
      disp('Sell All Stock.')
end

% Assume 'done' and 'HeaterStatus' are
% appropriately defined
if  Temp < 350 && FoodStatus ~= done
      HeaterStatus(on)
end
```

CC **19-5**

```
if ChargeLevel==discharged
      DeviceCheck(battery)
DeviceCheck(alternator)
      if BatteryStatus==defective
      DeviceReplace=battery
      elseif AlternatorStatus==defective
            DeviceReplace=alternator
      end
 end

DeviceCheck(battery)
DeviceCheck(alternator)
if ChargeLevel==discharged &&
  BatteryStatus==defective
      DeviceReplace=battery
end
      if ChargeLevel==discharged &&
AlternatorStatus==defective
      DeviceReplace=alternator
end
```

Chapter 20

CC **20-1**

```
for evens=2:2:20
      disp(evens)
end

for fives=5:5:50
      disp(fives)
end

for odds=13:-2:-11
      disp(odds)
end
```

CC **20-2**

```
for Pos=StartPos:Spd*TStep:EndPos
      disp(Pos)
end
```

CC **20-3**

Formulae:
$A(k) = 2 * k$
$B(k) = 2 * k - 1$
$C(k) = k \wedge 2$
$D(1) = 1 \quad$ for k>1, $D(k) = D(k-1) + k$
$E(k) = 7 + k/4$
$F(k) = 11 - k$
$G(k) = -1 \wedge (k + 1)$
$H(k) = 2 \wedge k$
$I(1) = 1 \quad$ for k>1, $I(k) = I(k-1) * k$
$J(k) = 1/k$

COMPLETE PROGRAM

```
%  Print Header row
fprintf('\tk\tA\tB\tC\tD\tE\tF\tG\tH\tI\tJ\
   n')
% Calculate one row at a time and print that
% row
for k = 1 : 10
    A(k) = 2 * k;
    B(k) = 2 * k - 1;
    C(k) = k ^ 2;
    E(k) = 7 + k/4;
    F(k) = 11 - k;
    G(k) = -1 ^ (k+1);
    H(k) = 2 ^ k;
    J(k) = 1/k;
    %  Handle initialization of first
    %  elements of D and I
```

```
    if k == 1
          D(k) = 1;
          I(k) = 1;
    else
          D(k) = D(k - 1) + k;
          I(k) = I(k - 1) * k;
    end
    %  Print the row
    fprintf('\t%d\t%d\t%d\t%d\t%d\t%.2f\t%d\
       t%d\t%d\t%d\t%.2f\n',
k,A(k),B(k),C(k),D(k),E(k),F(k),G(k),H(k),
    I(k),J(k))
end
```

CC 20-4

```
% determine the number of vendors (rows) and
% items (columns)
[NumbVendors,NumbItems]=size(Sales)
for Vendor=1:1:NumbVendors
% this will index the rows
      % Initialize the sum of all sales for
      % current vendor to 0
      VendorSales(Vendor,1) = 0;
      for Item=1:1:NumbItems
          % this will index the columns
          VendorSales(Vendor,1) = Vendor-
          Sales(Vendor,1) + Sales
      (Vendor,Item);
      end
end
% determine the number of vendors (rows) and
% items (columns)
[NumbVendors,NumbItems]=size(Sales)
for Item=1:1:NumbItems
    % this will index the columns
      % Initialize the sum of all sales for
      % current item to 0
      ItemSales(Item) = 0;
      for Vendor=1:1:NumbVendors
          % this will index the rows
                ItemSales(Item) = ItemSales
(Item) + Sales(Vendor,Item);
      end
end
```

CC 20-5

```
UserNumb=-1;
while UserNumb < 0
      UserNumb=input('Please enter a
        non-negative number. ')
end
```

INDEX

Excel Functions

Operator	Meaning	Operator	Meaning
>	Greater than	<=	Less than or equal to
<	Less than	=	Equal to
>=	Greater than or equal to	<>	Not equal to

Function as Written in Excel	Definition
ACOS (cell)	Calculate the inverse cosine of a number
ASIN (cell)	Calculate the inverse sine of a number
AVERAGE (cells)	Finds the mean or average value of a list of cells
COS (angle in radians)	Calculate the cosine of an angle
COUNT (cells)	Counts number of cells that are non-blank and do not contain an error
COUNTIF (cells, criteria)	Counts number of cells which meet the stated criteria, such as a numerical value, text, or a cell reference
DEGREES (angle in radians)	Convert radians to degrees
EXP (cell)	Raise e (base of the natural log) to the power "cell"
IF (logical test, value if true, value if false)	Conditional statement format of a single logical expression
INTERCEPT (y values, x values)	Calculates linear line for range of (x, y) pairs and returns the intercept value of y (where x = 0)
MAX (cells)	Finds the maximum value in a list of cells
MEDIAN (cells)	Finds the median value of a list of cells
MIN (cells)	Finds the minimum value in a list of cells
PI()	Calculate pi (π) to about 15 significant figures
POWER (cell, power)	Raise the cell to the "power"
PRODUCT (cells)	Find the product of a list of cells
RADIANS (angle in degrees)	Convert degrees to radians
ROUND (cell, number of decimal places)	Rounds a number to a specific number of decimal places
SIN (angle in radians)	Calculate the sine of an angle
SLOPE (y values, x values)	Calculates linear line for range of (x, y) pairs and returns the slope value
SQRT (cell)	Find the square root of "cell"
STDEV (cells)	Finds the standard deviation value of a list of cells
SUM (cells)	Find the sum of a list of cells
TRUNC (cell, number of decimal places)	Truncates a number to a specific number of decimal places
VAR (cells)	Finds the variance value of a list of cells

MATLAB Functions

Operator	Meaning	Operator	Meaning
>	Greater than	<=	Less than or equal to
<	Less than	==	Equal to
>=	Greater than or equal to	~=	Not equal to

Function as Written in MATLAB	Definition
acos (variable)	Calculate the inverse cosine of a variable, result in radians
asin (variable)	Calculate the inverse sine of a variable, result in radians
ceil (variable)	Rounds the value in the variable up to an integer
cos (variable)	Calculate the cosine of a variable [radians]
exp (variable)	Raise e (base of the natural log) to the power "variable"
floor (variable)	Rounds the value in the variable down to an integer
length (variable)	Counts number of elements in a vector saved in a variable
max (variable)	Determines the maximum value of a variable
mean (variable)	Finds the mean or average value of a variable
median (variable)	Finds the median value in a variable
min (variable)	Determines the minimum value of a variable
polyfit (variable$_1$, variable$_2$, N)	Calculates the coefficients of the N-order polynomial represented by the dependent values in variable$_2$ and the independent values in variable$_1$. For N = 1, the first element of the array is the slope of a linear line and the 2nd element of the array is the y-intercept.
power (variable$_1$, variable$_2$)	Raise each element of variable$_1$ (value, vector, matrix) to the power of variable$_2$
round (variable)	Rounds the value in the variable to the nearest integer
sin (variable)	Calculate the sine of a variable [radians]
size (variable)	Determines the dimension of a matrix saved in a variable
sqrt (variable)	Find the square root of the values in a variable
sum (variable)	Find the sum of values in a variable
std (variable)	Finds the standard deviation of the values in a variable
var (variable)	Finds the variance of the values in a variable

fprintf Control Characters	Use	Example V = numerical variable (=0.125) T = text variable (= 'Hello')	Output
\n	Inserts a new line	fprintf ('Hello\n');	Hello
\t	Inserts a tab	fprintf('A\tB\tC\n');	A B C
\\	Inserts a backslash	fprintf('Hello\\World\n');	Hello\World
'' (two single quotes)	Inserts a single quote mark	fprintf('Bob''s car\n');	Bob's car
%%	Inserts a percent symbol	fprintf('25%%');	25%
%0.Nf	Insert a fixed point value with N decimal places	fprintf('%0.1f',V);	0.1
%e or %E	Insert a number in exponential notation	fprintf('%e',V); fprintf('%E',V);	1.25e-001 1.25E-001
%s	Insert text	fprintf('%s',T);	Hello

Greek Letters [Name, Greek Form, English Form, Common Uses]

alpha	A, α	A, a	α angle, angular acceleration, Neper frequency, thermal diffusivity
beta	B, β	B, b	β bulk expansion coefficient, coefficient of thermal expansion, transistor current, gain
gamma	Γ, γ	G, g	γ specific weight; Euler's constant
delta	Δ, δ	D, d	Δ change; tip deflection of a beam δ small change, impulse function
epsilon	E, ε	E, e	ε dielectric constant, pipe roughness, porosity, strain
zeta	Z, ζ	Z, z	ζ dampening ratio of spring system
eta	H, η	H, h	η efficiency
theta	Θ, θ	Q, q	Θ dimensionless time θ thermal resistance, Euler angle
iota	I, ι	I, i	ι imaginary unit, consecutive integers
kappa	K, κ	K, k	κ permeability, thermal conductivity
lambda	Λ, λ	L, l	λ wavelength, arrival rate, eigenvalue
mu	M, μ	M, m	μ SI prefix micro (10^{-6}), statistical mean, dynamic viscosity, coefficient of friction
nu	N, ν	N, n	ν frequency, kinematic viscosity, Poisson's ratio, specific volume
xi	Ξ, ξ	X, x	ξ permeability, damping coefficient in circuit
omicron	O, o	O, o	
pi	Π, π	P, p	Π product π 3.141592... (circumference/diameter of circle), osmotic pressure
rho	P, ρ	R, r	ρ density, resistivity
sigma	Σ, σ	S, s	Σ summation σ conductivity, standard deviation, surface tension, stress, Stefan-Boltzmann constant
tau	T, τ	T, t	τ time constant, shear stress, torque
upsilon	Y, υ	U, u	
phi	Φ, ϕ or φ	F, f	Φ potential ϕ, φ Euler angle, Golden ratio
chi	X, χ	C, c	χ electronegativity
psi	Ψ, ψ	Y, y	ψ Euler angle
omega	Ω, ω	W, w	Ω unit of ohm ω angular frequency, angular velocity

Nomenclature [common units]

a	acceleration [m/s^2]	L	inductance [H]	w	weight [N]
A	altitude [m]		length [m]	W	width [m]
	area [m^2]	m	mass [kg]		work [J]
C	capacitance [F]	$\dot{m}$	mass flowrate [kg/s]	x	displacement [m]
	circumference [m]	Ma	Mach Number [-]		
	concentration [g/L]	n	number of moles [mol]	α	neper frequency [Hz]
C$_d$	coefficient of drag [-]		shaft speed [Hz]		thermal diffusivity [m^2/s]
C$_p$	specific heat [J/(g °C)]	P	perimeter [m]	β	coefficient of thermal
d	distance [m]		power [W]		expansion [°C^{-1}]
D	diameter [m]		pressure [Pa]	γ	specific weight [N/m^3]
E	energy [J]		vertical load of beam [N]	Δ	delta; change in quantity
	Young's modulus [Pa]	PE	potential energy [J]		tip deflection of beam [m]
F	force [N]	Q	electric charge [C]	ε	pipe roughness [mm]
f	frequency [Hz]		thermal energy [cal]		strain [-]
	friction factor [-]		volumetric flowrate [m^3/s]	η	efficiency [-]
G	conductance [S]	r	radius [m]	θ	angle [rad]
h	heat transfer coefficient [W/(m^2 °C)]	R	resistance [Ω]	μ	coefficient of friction [-]
h$_L$	energy loss per unit mass [J/kg]	Re	Reynolds Number [-]		dynamic viscosity [cP]
H	height [m]	S	arc length [m]	ν	kinematic viscosity [St]
I	current [A]	SG	specific gravity [-]	ρ	mass density [kg/m^3]
	moment of inertia [kg m^2]	T	period [s]		resistivity [Ω m]
k	spring constant [N/m]		temperature [K]	σ	stress [Pa]
	thermal conductivity [W/(m °C)]	t	time [s]		surface tension [J/m^2]
K	beam stiffness [N/m]	v	velocity [m/s]	τ	shear stress [Pa]
KE	kinetic energy [J]	V	volume [m^3]	ω	angular frequency [rad/s]
			voltage [V]		

Unit Abbreviations

°C	degree Celsius	ft	foot	mm Hg	millimeters mercury
°F	degree Fahrenheit	g	gram	mol	mole
°R	degree Rankine	gal	gallon	mph	miles per hour
A	ampere	gpm	gallons per minute	N	newton
amu	atomic mass unit	H	henry	Pa	pascal
atm	atmosphere	h	hour	psi	pound-force per square inch
AU	astronomical unit	hp	horsepower	rad	radian
BTU	British thermal unit	Hz	hertz	s	second
C	coulomb	in Hg	inches of mercury	S	seimens
cal	calorie	in	inch	St	stoke
cc	cubic centimeter	J	joule	t	ton
cd	candela	K	kelvin	tbsp	tablespoon
cP	centipoise	kg	kilogram	tsp	teaspoon
d	day	kWh	kilowatt-hour	V	volt
Da	dalton	L	liter	W	watt
deg, °	degree	lb$_f$	pounds-force	yd	yard
eV	electron volt	lb$_m$	pounds-mass	yr	year
F	farad	m	meter	Ω	ohm
fl oz	fluid ounce	mi	mile		
ft H$_2$O	feet of water	min	minute		

Miscellaneous Equations

Antoine equation

$$P = 10^{\left(A - \frac{B}{T+C}\right)}$$

Cantilever Beam Deflection

$$P = K\,\Delta$$

Discharging Capacitor

$$V = V_0\, e^{-t/RC}$$

Energy in an Inductor

$$E = \tfrac{1}{2} L\, I^2$$

Moore's Law

$$T = T_0\, 2^{\frac{t}{2}}$$

Oscillating spring

$$f = \frac{1}{2\pi}\sqrt{\frac{k}{m}}$$

Poiseuille's Equation

$$R = \frac{8\mu L}{\pi r^4}$$

Pythagoras' Theorem

$$d = \sqrt{(x_2 - x_1)^2 + (y_2 - y_1)^2}$$

Resonant frequency for one inductor and one capacitor

$$f_R = \frac{1}{2\pi\sqrt{L\,C}}$$

Shockley Equation

$$I = I_0\left(\frac{V_D}{e^{nv_t} - 1}\right)$$

Torricelli's Law

$$v = \sqrt{2\,g\,H}$$

van der Waals Equation

$$\left(P + \frac{an^2}{V^2}\right)(V - b\,n) = n\,R\,T$$

Bingham plastic

$$\tau = \mu\frac{\Delta v}{\Delta y} + \tau_0$$

Dimensionless Number Equations

Coefficient of Drag

$$C_d = \frac{F}{\frac{1}{2}\rho v^2 A_p}$$

Friction factor: Laminar Flow

$$f = \frac{64}{Re}$$

Friction factor: Blasius Approximation

$$f = 0.316\,(Re)^{-1/4}$$

Friction factor: Colebrook

$$\frac{1}{\sqrt{f}} = -2\,\log\left(\frac{\varepsilon/D}{3.7} + \frac{2.51}{Re\sqrt{f}}\right)$$

Mach Number

$$Ma = \frac{\text{speed}_{object}}{\text{speed}_{sound\ in\ air}}$$

Power Number

$$N_P = \frac{P}{\rho n^3 D^5}$$

Reynolds Number: Use with Coefficient of Drag

$$Re = \frac{D_p \rho v}{\mu}$$

Reynolds Number: Pipe Flow

$$Re = \frac{\rho D v}{\mu}$$

$Re < 2000$ = laminar flow
$2000 < Re < 10{,}000$ = transition flow
$Re > 10{,}000$ = turbulent flow

Reynolds Number: Use with Power Number

$$Re = \frac{D^2 n \rho}{\mu}$$

Statistics Equations

Mean

$$\overline{X} = \frac{1}{N}(X_1 + X_2 + \cdots + X_N) = \frac{1}{N}\sum_{j=1}^{N} X_j$$

Median
Place data in ascending order.
If N = odd, median = central value.
If N = even, median = mean of two central values.

Variance

$$V_x^2 = \frac{1}{N-1}\left((\overline{X} - X_1)^2 + \cdots + (\overline{X} - X_N)^2\right) = \frac{1}{N-1}\sum_{j=1}^{N}(\overline{X} - X_j)^2$$

Standard Deviation

$$SD_X = \sqrt{V_x^2}$$

Equations

Circuits: Ohm's Law

$V = I R$

Circuits: Parallel Resistors

$$R_{eff} = \left(\frac{1}{R_1} + \frac{1}{R_2} + \cdots + \frac{1}{R_{N-1}} + \frac{1}{R_N} \right)^{-1}$$

Circuits: Series Resistors

$R_{eff} = R_1 + R_2 + \cdots + R_{N-1} + R_N$

Density

$\rho = m/V$

Efficiency

$\eta = $ output/input

Work

$W = F \, \Delta x$

Potential Energy

$PE = m \, g \, \Delta H$

Kinetic Energy, translational

$KE_T = \frac{1}{2} m \, (v_f^2 - v_i^2)$

Kinetic Energy, rotational

$KE_R = \frac{1}{2} I \, (\omega_f^2 - \omega_i^2)$

Kinetic Energy, total

$KE = KE_T + KE_R$

Thermal Energy

$Q = m \, C_p \, \Delta T$

Force: Newton's 2nd Law

$F = m \, a$

Interpolation

$$\frac{(A - A_1)}{(A_2 - A_1)} = \frac{(B - B_1)}{(B_2 - B_1)}$$

Power

$P = E/t$

Power: Joule's First Law

$P = I^2 R$

Pressure

$P = F/A$

Pressure, Hydrostatic: Pascal's Law

$P_{hydro} = \rho \, g \, H$

Pressure, Total

$P_{total} = P_{surface} + \rho \, g \, H$

Pressure: Ideal Gas Law

$P V = n R T$

Specific Gravity

$SG = \rho_{object}/\rho_{water}$

Specific Weight

$\gamma = w/V$

Springs: Hooke's Law

$F = k \, x$

Springs: Parallel

$k_{eff} = k_1 + k_2 + \cdots + k_{N-1} + k_N$

Springs: Series

$$k_{eff} = \left(\frac{1}{k_1} + \frac{1}{k_2} + \cdots + \frac{1}{k_{N-1}} + \frac{1}{k_N} \right)^{-1}$$

Temperature

$$\frac{T\,[°F] - 32}{180} = \frac{T\,[°C] - 0}{100}$$

$T\,[K] = T\,[°C] + 273$

$T\,[°R] = T\,[°F] + 460$

Temperature: Newton's Law of Cooling

$T = T_0 \, e^{-kt}$

Viscosity, Dynamic: Newton's Law of Viscosity

$\tau = \mu \, (\Delta v/\Delta y)$

Viscosity, Kinematic

$\nu = \mu/\rho$

Weight

$w = m \, g$

Young's Modulus

$\sigma = E \, \varepsilon$

Geometric Formulas

Rectangle

Area = a b

Perimeter = 2 a + 2 b

Rectangular Parallelepiped

Volume = a b c

Surface area = 2 (a b + a c + b c)

Circle

Area = $\pi \, r^2$

Perimeter = $2 \pi \, r$

Sphere

Volume = $4/3 \, \pi \, r^3$

Surface area = $4 \, \pi \, r^2$

Triangle

Area = $\frac{1}{2}$ b H

Right Circular Cone

Volume = $1/3 \, \pi \, r^2$ H

Right Circular Cylinder

Volume = $\pi \, r^2$ h

Lateral Surface Area = $2\pi \, r$ H

Fundamental Dimensions [Base SI Units]

I	electric current [A]	M	mass [kg]	T	time [s]	
J	light intensity [cd]	N	amount of substance [mol]	Θ	temperature [K]	
L	length [m]					

Common Derived Units in the SI System

Dimension	SI Unit	Base SI Units	Derived from
Force (F)	newton [N]	$1\,N = 1\dfrac{kg\,m}{s^2}$	$F = m\,a$ Force = mass times acceleration
Energy (E)	joule [J]	$1\,J = 1\,N\,m = 1\dfrac{kg\,m^2}{s^2}$	$E = F\,d$ Energy = force times distance
Power (P)	watt [W]	$1\,W = 1\dfrac{J}{s} = 1\dfrac{kg\,m^2}{s^3}$	$P = E/t$ Power = energy per time
Pressure (P)	pascal [Pa]	$1\,Pa = 1\dfrac{N}{m^2} = 1\dfrac{kg}{m\,s^2}$	$P = F/A$ Pressure = force per area
Voltage (V)	volt [V]	$1\,V = 1\dfrac{W}{A} = 1\dfrac{kg\,m^2}{s^3\,A}$	$V = P/I$ Voltage = power per current

Physical Constants [value and units]

c	speed of light in a vacuum	$= 3 \times 10^8$ m/s
e	Euler's number (base of natural logarithm)	$= 2.71828\ldots$
	elementary charge of an electron	$= 1.602 \times 10^{-19}$ C
φ	Golden ratio	$= 1.61803\ldots$
F	Faraday's constant	$= 9.65 \times 10^4$ C/mol
g	acceleration due to gravity	$= 9.8$ m/s^2 on Earth
G	gravitational constant	$= 6.67 \times 10^{-11}$ (N m^2)/kg^2
h	Planck constant	$= 6.62 \times 10^{-44}$ J
k	Boltzmann constant	$= 1.48 \times 10^{-24}$ J/K
N_A	Avogadro's number	$= 6.022 \times 10^{23}$ mol^{-1}
π	ratio of circle circumference to diameter	$= 3.14159\ldots$
R	gas constant	$= 8314$ (Pa L)/(mol K) $= 0.08206$ (atm L)/(mol K)
σ	Stefan-Boltzmann constant	$= 5.67 \times 10^{-8}$ W/(m^2 K^4)